Polymer Surface Modification: Relevance to Adhesion

Editor:
K.L. Mittal

Utrecht, The Netherlands, 1996

VSP BV
P.O. Box 346
3700 AH Zeist
The Netherlands

First published in 1996

ISBN 90-6764-201-0

CIP-DATA KONINKLIJKE BIBLIOTHEEK, DEN HAAG

Polymer

Polymer surface modification: relevance to adhesion / ed.:
K.L. Mittal. - Utrecht : VSP
Publ. embodies the proceedings of the International
Symposium on Polymer Surface Modification: Relevance to
Adhesion, held in Las Vegas. - Previously publ. in four
special issues of the Journal of Adhesion Science and
Technology.
ISBN 90-6764-201-0 bound
NUGI 812
Subject headings: surface modification / polymers /
adhesion.

Printed in The Netherlands by Koninklijke Wöhrmann BV, Zutphen.

Contents

Polymer Surface Modification: Relevance to Adhesion, pp. ix–x
K. L. Mittal (Ed.)

Preface

This book embodies the proceedings of the International Symposium on Polymer Surface Modification: Relevance to Adhesion held under the auspices of Skill Dynamics, an IBM Company in Las Vegas, Nevada, November 3–5, 1993. Apropos, these papers were earlier published as special issues of the *Journal of Adhesion Science and Technology* as: Volume 8, number 10 (1994), Volume 9, number 3 (1995), Volume 9, number 5 (1995), and Volume 9, number 9 (1995). In addition to these special issues, four papers were also published in Volume 9, issue 12. As many researchers and technologists evinced considerable interest in acquiring these proceedings separately, so we decided to make available this hard-bound volume.

Polymeric materials are used for a legion of applications in a host of technological areas. However, polymers are innately hydrophobic, low surface energy materials and thus do not adhere well to other materials brought in contact. This necessitates their surface modification/treatment to render them adhesionable. By the way, surface modification of polymers is carried out not only to improve their adhesion characteristics, but for a variety of other reasons too, for example, to increase their hydrophobicity, to modify their tribological behavior, to render them flame resistant, etc. However, in this symposium the emphasis was on surface modification of polymers with relevance to adhesion. In light of the high tempo of R&D activity in surface modification of polymers coupled with the tremendous technological interest in this topic, we decided that it was very opportune to hold this symposium.

This symposium was organized with the following objectives in mind: (i) to bring together the scientists and technologists with active interest in this topic; (ii) to provide a forum for discussion of latest developments; (iii) to provide an opportunity for cross-pollination of ideas; and (iv) to identify the unsolved problems as well as the surface modification techniques which offered good promise and vindicated vigorous pursuit. The very first announcement of this symposium elicited a tremendous response which was a testimonial to the brisk R&D activity and much interest in this topic; concomitantly, the technical program comprised 69 papers reflecting overviews and original research results. A wide array of surface modification techniques, ranging from simple to sophisticated, wet to dry, vacuum to nonvacuum for a host of polymeric materials were covered.

Now turning to this volume, here the papers are arranged in a more logical manner vis-a-vis the order in which they appeared in special issues. The book is divided into four parts as follows: Part 1: Plasma Surface Modification Techniques; Part 2: Laser Surface Modification Techniques; Part 3: Other/Miscellaneous Surface Modification Techniques; and Part 4: General Papers. The topics covered include: plasma surface treatment of a number of polymers: laser surface treatment of various polymers; corona, flame, UV, ozone, UV/ozone, photochemical, photografting, chemical grafting, and chemical methods of polymer surface modification; modification of polyamide surfaces by microorganisms; effect of polymer surface modification on metal/polymer adhesion; barrier properties of surface treated polymers; ageing study of surface treated polymers; physico-chemical properties of surface-modified polymers; application of inverse gas chromatography in the characterization of polymers; and surface acoustic wave sensor to study polymer surface treatments.

I certainly hope anyone interested in polymer surfaces and adhesion will find this book, representing a repository of latest information on this topic, of great interest and use.

K. L. Mittal

Part 1

Plasma Surface Modification Techniques

Polymer Surface Modification: Relevance to Adhesion, pp. 3–15
K. L. Mittal (Ed.)

Plasma treatment of polydimethylsiloxane

MICHAEL J. OWEN* and PATRICK J. SMITH
Dow Corning Corporation, Midland, MI 48686-0994, USA

Revised version received 29 October 1993

Abstract—Plasma treatment of silicone surfaces is a useful way of increasing wettability to improve adhesion and a first step in producing various organosilicon thin-film composites. Despite numerous earlier studies, there is no consensus on the effect of plasma treatment nor on the mechanism of the subsequent hydrophobic recovery. X-ray photoelectron spectroscopy (XPS) and scanning electron microscopy (SEM) were used to study the effect of plasma treatments of polydimethylsiloxane elastomer using four different plasma gases: argon, helium, oxygen, and nitrogen. In each case, the surface was oxidized to produce a thin, wettable, brittle silica-like layer. These surfaces progressively recover their hydrophobicity by diffusion of untreated polymer chains through cracks in the treated layer. Angle-resolved XPS detected the untreated, diffused layer and SEM revealed the common occurrence of cracks in the treated layer, although conditions could be found for each gas where the surface becomes completely wettable by water but is free from cracks.

Keywords: Polydimethylsiloxane; plasma treatment; X-ray photoelectron spectroscopy; scanning electron microscopy; wettability.

1. INTRODUCTION

The use of plasma treatment to increase the wettability of polymers has been known for a considerable time [1]. Biomedical applications of this phenomenon include tissue culture surfaces with improved cell attachment and growth characteristics [2], contact lens materials with enhanced wettability by tears and altered protein and lipid deposition [3], and double catheter systems with reduced polymer/polymer friction [4]. The most familiar application of polymer plasma treatment is to improve the bondability of polymers to dissimilar materials while producing no change in their bulk properties. Silicones, particularly polydimethylsiloxane (PDMS), feature in these various applications. Plasma treatment of PDMS is also an important first step in producing a variety of organosilicon-based, thin-film composites. For example, it enables self-assembled monolayers of hydrolyzed alkyltrichlorosilanes to be chemisorbed [5],

---

*To whom correspondence should be addressed.

producing model systems to study adhesion between polymer surfaces. Plasma treatment is also useful in producing organosilicon thin films on other substrates of interest in fundamental adhesion and friction studies, notably mica [6].

Clark *et al.* [7] have reviewed the effects of plasma treatment on the surface of polymers. The earliest studies were those of Hall *et al.* [8], who treated PDMS elastomers in an ammonia plasma. Bonding results with a urethane adhesive were erratic, attributed to a weak boundary layer of low-molecular-weight (MW) PDMS fractions. ATR-IR (attenuated total reflection infrared) spectroscopy of a treated PDMS elastomer gum revealed no NH_2 bonds. This work was followed by a major mechanistic contribution by Hollahan and Carlson [9]. Their IR studies of the effect of oxygen plasma suggested that the polar groups produced in the surface region were predominantly $SiCH_2OH$ groups. More recent workers have contested this and there is still no consensus on the nature of the chemical changes produced by the plasma treatment nor on the mechanism of the subsequent hydrophobic recovery process once the plasma treatment ceases. Until a more complete understanding of these processes is available, informed new product development in this area will not be possible.

There are at least five recent studies of plasma-treated PDMS surfaces. Triolo and Andrade [4] used XPS (X-ray photoelectron spectroscopy), contact angle, and SEM (scanning electron microscopy) on helium plasma-modified PDMS; Morra *et al.* [10] used XPS, ATR-FTIR, and SSIMS (static secondary ion mass spectrometry) on ^{18}O-modified PDMS; Stewart and Urban [11] used ATR-FTIR and DMA (dynamic mechanical analysis) on argon, carbon dioxide, and ammonia plasma-treated samples; Ikada *et al.* [12] used contact angle on argon plasma-treated PDMS; and Gaboury and Urban [13] used ATR-FTIR to study nitrogen and argon plasma-modified surfaces. Triolo and Andrade [4] concluded that a silica-like surface was produced. They offered no evidence for the presence of silanol groups but suggested oxidized carbon species such as carboxyl and aldehyde groups. A silica surface from an oxygen plasma treatment was suggested earlier by Feneberg and Krekeler [14]. Morra *et al.* [10] specifically refute the presence of aldehyde groups. They found surface silanol groups and a considerable number of unaltered $SiCH_3$ groups. Gaboury and Urban [13] (nitrogen and argon plasmas) found carbonyl groups, low MW uncrosslinked PDMS chains, and evidence for reduced silicon in the form of SiH groups. Although most workers [4, 10, 14] claim increased crosslinking in the surface, Stewart and Urban [11] found depolymerization using DMA.

The use of plasmas with widely different excitation frequencies must account for some of these major differences. Urban and co-workers [11, 13] used a microwave plasma, whereas the other studies were carried out with radio frequency (RF) plasmas. It may also be that the different gases used in these RF plasma studies and the different conditions employed produce different effects, although there is no consensus even when the same gas plasma is used. For this reason, we resolved to examine a variety of plasma gases (helium, argon, oxygen, and nitrogen) to learn more about the changes produced and the subsequent hydrophobic recovery effect. The possible mechanisms for hydrophobic recovery appear to be:

(A) reorientation of surface hydrophilic groups away from the surface (also called 'overturn' of polar groups in the polymer surface [12]);
(B) migration of treated polymer chains from the surface to the bulk;
(C) migration of untreated polymer chains from the bulk to the surface;
(D) loss of volatile oxygen-rich or other polar entities to the atmosphere;
(E) surface silanol condensation preventing chain reorientation [15];
(F) changes in surface roughness; and
(G) external contamination of the polymer surface.

Note that the Lee and Homan [15] silanol condensation suggestion is taken from the corona discharge treatment literature. The close parallel between plasma and corona treatment effects was first pointed out by Hollahan and Carlson [9] and has been developed more recently by others, including Morra *et al.* [10]. Information gained in one area can be directly used to help in the other. We have recently published some studies on corona-treated PDMS samples [16].

2. EXPERIMENTAL

The PDMS used was 0.5 mm thick elastomer sheeting. This is a platinum-catalyzed SiH/Si vinyl crosslinked material containing 20% by weight of fumed silica filler. These sheets are post-cured for 8 h at 175°C and cleaned by Soxhlet extraction in methanol for 6 h. The quasistatic advancing and receding contact angles of distilled water are 104° and 85°, respectively (19° hysteresis), with a smooth surface topography judged by SEM. Water contact angle hysteresis for filled silicone elastomers is typically in the 20°–40° range [3, 17]; unfilled PDMS is much lower; for example, a peroxide-cured PDMS gum has a water contact angle hysteresis of 3°.

Plasma treatment was carried out in a Branson IPC Series 2000 chamber, which generates a low pressure, RF, or cold plasma. The conditions affecting the degree of treatment that are under experimental control are the treatment time, RF power, chamber pressure, and gas composition. Typical times are 3 s to several minutes; typical powers are 10–400 W, and chamber pressure is variable from 40 to 90 Pa. The lower limit for power is a function of the gas used. For example, a He plasma can be run at 10 W but an O_2 plasma will not ignite until 20 W.

XPS spectra were obtained with a Perkin-Elmer model 550 XPS/AES instrument using a Mg anode operated at 15 kV and 300 W. All spectra were recorded at a 25 eV pass energy. Angular resolution was used for some spectra. The grazing take-off angle enhances the signal from the surface and the normal take-off angle enhances the signal from the bulk. A low-energy electron flood gun was used to neutralize sample charging. Atomic composition data were calculated using empirical sensitivity factors [18]. XPS atomic composition, binding chemical shifts, and angle-resolved experiments were all useful in analyzing these plasma-treated PDMS surfaces.

Surface structure was observed by optical microscopy and SEM. A Zeiss transmitted light microscope and a Nikon microscope that operates with either transmitted or reflected light were used. The Nikon microscope was equipped with a camera attachment. SEM was carried out with a JEOL T300 instrument. Samples were sputter-coated with Au/Pd to minimize sample charging. It showed details of cracking that could not be observed optically and its higher magnification gave better verification that plasma treatment did not always result in a cracked surface. Contact angles were measured using a Ramé-Hart, Inc. NRL Model A-100 contact angle goniometer. All the angles reported are the average of at least three measurements taken on both sides of the drop.

3. RESULTS AND DISCUSSION

Table 1 shows XPS atomic composition data for oxygen plasma treatment of the PDMS elastomer surface as a function of the treatment time. The different plasma conditions used for Tables 1, 2, and 3, and Figs 1, 2, and 3 were chosen to produce uncracked surfaces (see later discussion). Longer treatment of these substrates produced cracked surfaces but no significant change in XPS composition. However, it is important to note that all surfaces were found to be cracked upon removal from the XPS instrument, even those originally uncracked. The theoretical atomic composition of a PDMS surface is 25% O, 50% C and 25% Si. We routinely obtained oxygen values *ca.* 2% higher and silicon values *ca.* 2% lower than expected for the untreated

Table 1.
XPS surface analysis of oxygen plasma (70 W, 70 Pa)-treated PDMS

Treatment time (s)	Atomic composition (%)		
	O	C	Si
0	27.1	50.3	22.6
1	43.2	34.4	22.4
5	43.5	34.7	21.8
10	45.1	32.0	22.9
30	46.5	29.7	23.8

Table 2.
Angle-resolved XPS surface analysis of oxygen plasma (70 W, 70 Pa, 5 s)-treated PDMS

Sample	Angle of resolution	Atomic composition (%)		
		O	C	Si
Untreated	Grazing	25.9	50.2	23.1
Untreated	Normal	26.1	49.8	24.1
Treated	Grazing	32.0	44.4	23.6
Treated	Normal	46.7	31.1	22.1

Table 3.
Angle-resolved XPS surface analysis of PDMS treated with various plasmas

Plasma	Angle of resolution	Atomic composition (%)			
		O	N	C	Si
Oxygen	Grazing	32.4	ND	44.1	23.5
(70 W, 70 Pa, 30 s)	Normal	46.7	ND	30.7	22.6
Argon	Grazing	30.1	ND	42.0	27.9
(10 W, 70 Pa, 30 s)	Normal	44.7	ND	34.1	21.2
Helium	Grazing	29.8	ND	43.4	26.8
(10 W, 70 Pa, 30 s)	Normal	43.8	ND	32.6	23.6
Nitrogen	Grazing	28.6	1.2	40.8	29.4
(70 W, 50 Pa, 30 s)	Normal	48.7	1.9	28.9	20.5

ND = Not detected (< 0.5%).

samples. Hydrocarbon contamination would not leave the carbon content unchanged, neither can there be silica filler present as this would increase both the silicon and the oxygen values and is not supported by binding energy values. It probably reflects errors in the empirical sensitivity factors. The treated surfaces show increased oxygen, decreased carbon, and constant silicon by comparison with the control. The amount of oxygen increases and the amount of carbon decreases with increased plasma treatment. Similar atomic composition values have been reported earlier for plasma-treated PDMS [4].

The data in Table 1 were taken without angular resolution. In this usual mode, all take-off angles are analyzed. Table 2 contains data taken from signals recorded at the normal 90° angle and at a grazing angle of 15°. We estimate that at the normal angle the signal from as deep as 8–10 nm is maximized relative to that from the surface layer, whereas at the grazing angle the signal from the top 2 nm surface region is greatly enhanced. The grazing angle data show higher carbon and lower oxygen values than the normal angle data. Since the plasma oxidation occurs from the gas phase, these results are not as expected. Table 3 shows that a similar result is obtained whichever of the four plasma gases is used except that a small amount of nitrogen is incorporated in the nitrogen plasma case.

Since different chemical species of the same element bind electrons at slightly different energies, such states can be detected from chemical shift data, i.e. shifts in the binding energy of the photoelectron peaks. Figures 1, 2, and 3 show high-resolution spectra for oxygen plasma-treated and untreated PDMS for the Si 2p, C 1s, and O 1s regions, respectively. Similar spectra were obtained for all the gases used. These spectra were obtained at 25 eV analyzer pass energy, giving a binding energy resolution of $\Delta E = 0.5$ eV. The Si 2p peak at 101.5 eV for the untreated sample is consistent with other PDMS studies having a typical peak width at half-maximum of 1.8 eV. The Si 2p peak for the treated PDMS is significantly broader (peak width

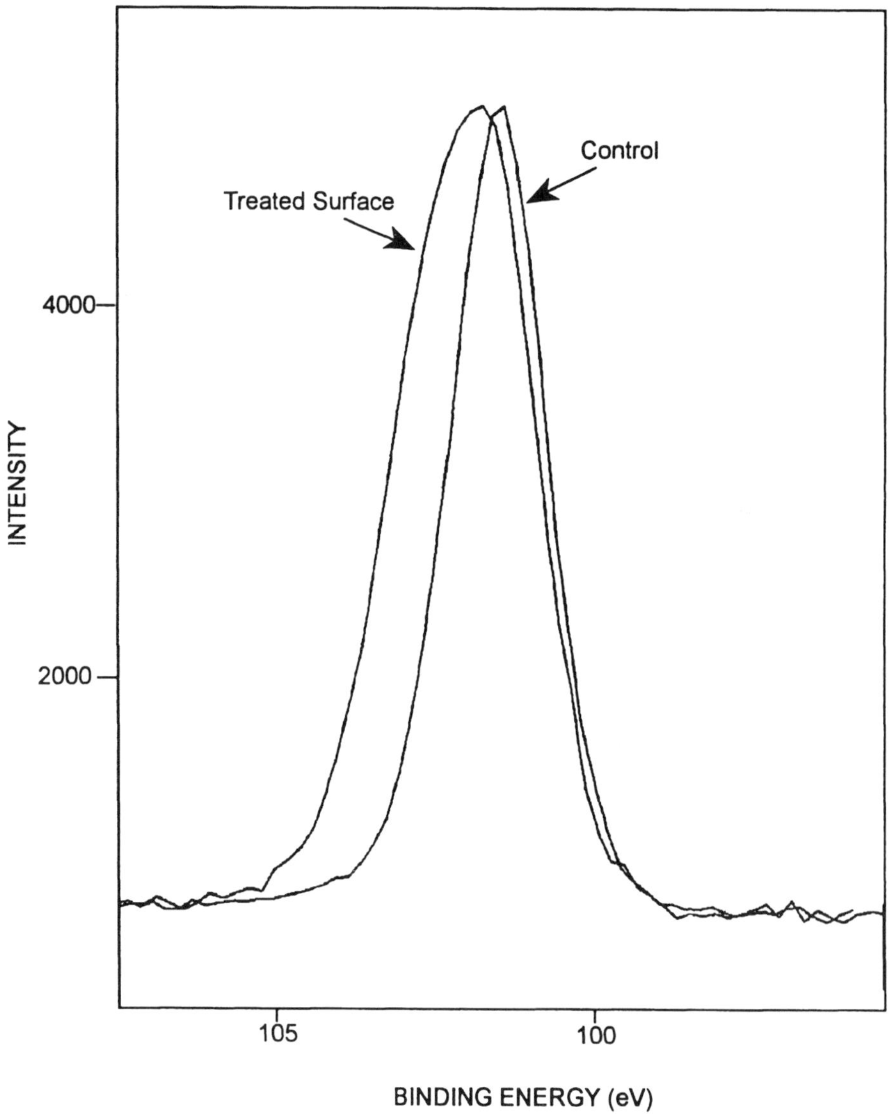

Figure 1. XPS Si 2p region of untreated PDMS and oxygen plasma (70 W, 70 Pa, 15 s)-treated PDMS.

at half-maximum 2.6 eV) and can be resolved into two peaks, one at 101.5 eV as before and one at 103.5 eV consistent with silica. The methyl carbon peak at 285 eV dominates the C 1s spectra of both treated and untreated samples but minor higher binding energy peaks and shoulders of oxidized carbon can be seen in the treated case. At 0.5 eV resolution, alcohol carbon and methyl carbon cannot be distinguished, but there is a clear peak at 290 eV corresponding to carboxylate carbon. The 0.4 eV broadening of the treated O 1s peak is also suggestive of an increase in the number of chemical states of oxygen present but no assignment is possible in this unresolved case.

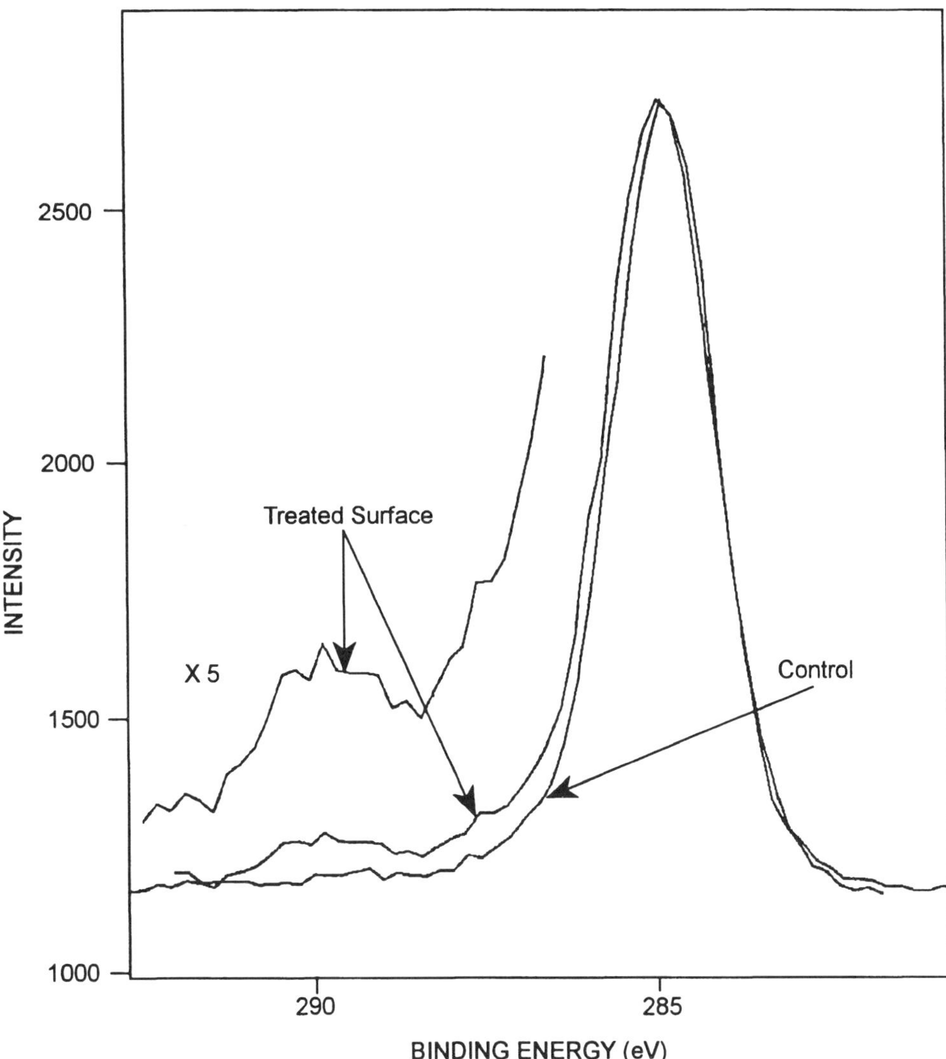

Figure 2. XPS C 1s region of untreated PDMS and oxygen plasma (70 W, 70 Pa, 15 s)-treated PDMS.

Figure 4 is a micrograph at ×500 magnification of untreated PDMS. The surface is smooth with the occasional fragment deposited as shown in the figure plus a few dimples. The lack of features in the surface made the detection of cracks, our primary concern, very simple. Figures 5 and 6 are pictures at two magnifications (×1500 and ×4000) of oxygen plasma treatment under conditions (400 W, 70 Pa, 10 min) that cause considerable cracking. Some of the cracks can be seen to offset another crack line. For example, this can be clearly seen in Fig. 6, where the crack running from side to side of the micrograph is offset where it passes between the two cracks running from top to bottom of the micrograph. This side-to-side crack must have been made before the top-to-bottom cracks. In all cases, displaced cracks appear to be not as

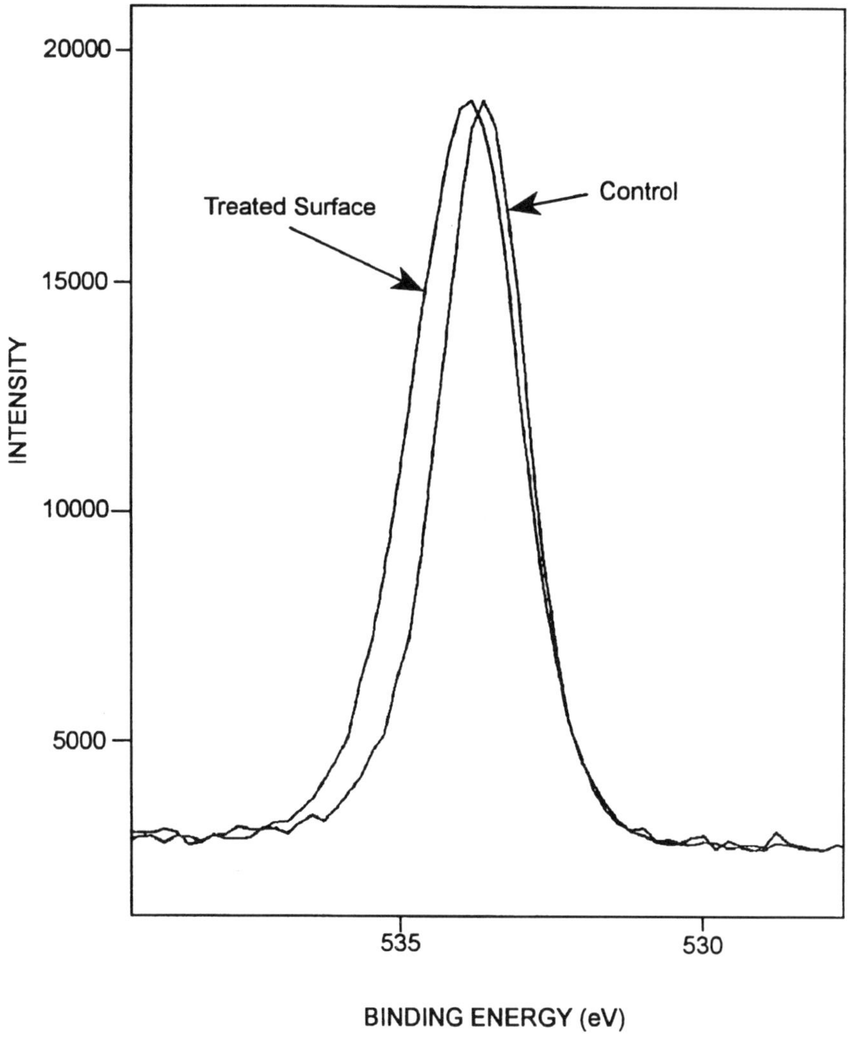

Figure 3. XPS O 1s region of untreated PDMS and oxygen plasma (70 W, 70 Pa, 15 s)-treated PDMS.

deep as the cracks that displaced them. We take this to mean that the treated layer becomes thicker with increasing treatment time. This is a qualitative observation as we have not measured crack depths.

Although surface cracking seems a widespread consequence of plasma treatment, this is not always so, as shown in Fig. 7. This is a ×500 magnification micrograph of a helium plasma-treated sample (10 W, 70 Pa, 3 min). The treatment is sufficient to permit complete water wettability with no trace of cracking, although deposited fragments and surface dimpling or pitting are more extensive than in Fig. 4. In general, less harsh, lower RF power, and shorter treatment times produced uncracked surfaces. All four plasmas could produce completely wettable surfaces without cracking, although the conditions to achieve this state varied from one plasma gas to another.

Figure 4. SEM micrograph of untreated PDMS.

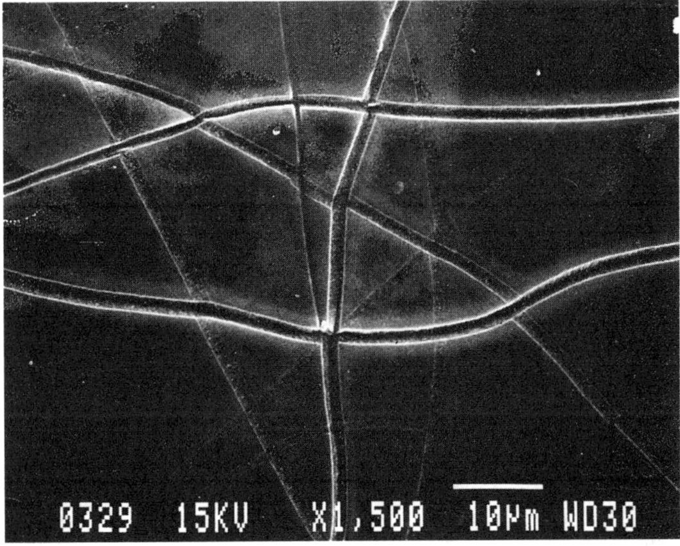

Figure 5. SEM micrograph (×1500) of oxygen plasma (40 W, 70 Pa, 10 min)-treated PDMS.

Another type of cracking, due to cracking of the Au/Pd coating used in the SEM sample preparation, was also occasionally seen. The cracks are very different from the silica layer cracks; they appear only in isolated areas of the surface and can be eliminated by careful sample handling. Note again that however carefully the samples were handled, cracking was always detected after XPS characterization. This may be

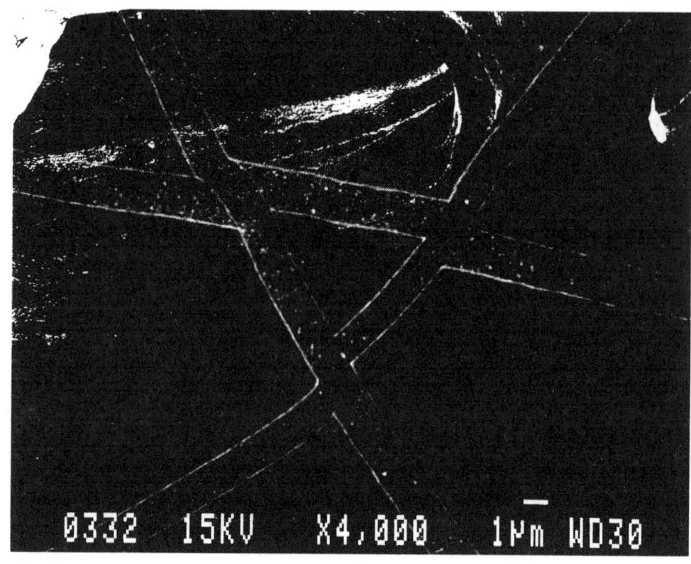

Figure 6. SEM micrograph (×4000) of oxygen plasma (40 W, 70 Pa, 10 min)-treated PDMS.

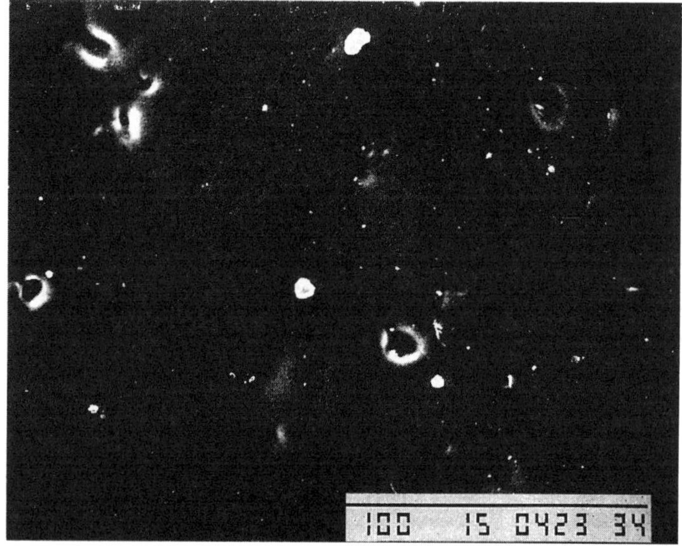

Figure 7. SEM micrograph of helium plasma (10 W, 70 Pa, 3 min)-treated PDMS.

the result of X-ray or UHV exposure and must be recognized as an analysis artifact in these and presumably, other studies. As untreated PDMS does not crack during XPS analysis, we assume that these cracks originate in the brittle silica-like surface layer suggested by the XPS data.

Of the mechanisms for hydrophobic recovery listed in the Introduction, the last two can be eliminated from consideration. The consistency of our data suggests that random contamination has been adequately controlled and, apart from the cracking phenomenon discussed previously, no changes in overall surface roughness were evident. Mechanism (A), reorientation or overturn, requires a polymer chain that can freely rotate. Similarly, the migration mechanisms [(B) and (C)] are possible only for mobile polymer chains. Untreated PDMS chains certainly meet this requirement. Loss of volatile oxygen-rich or other polar species, mechanism (D), is most likely for carbon-containing functional groups or for water, the by-product of mechanism (E). This surface silanol condensation will only occur when two silanol groups are sufficiently close for reaction, which may also require a certain degree of polymer mobility.

Plasma treatment is evidently a progressive oxidation of the surface. The source of oxygen in the nitrogen, argon, and helium plasmas may be residual air in the chamber; from water vapor, which is also very difficult to eliminate by pumping; or oxygen retained in the polymer itself. At mild oxidation, where only a few of the polymer side groups have been oxidized and the polymer backbone is still intact with no crosslinking taken place, the matrix chains would be mobile and any of the mechanisms (A) through (E) would apply. The more usually encountered situation is a more complete oxidation of PDMS to silica. This is evident in the present studies and in much of the earlier reported studies. This layer would be wettable and brittle but it would not be mobile. Therefore, only mechanism (C) and to a certain extent mechanism (E), silanol condensation and the reverse hydration reaction of this equilibrium, are possible. Neither of these reactions would make the silica surface hydrophobic, whereas it is commonly observed that virtually total recovery of hydrophobicity is possible if sufficient time is allowed or the sample is heated or boiled in water.

Our hypothesis is that the topmost surface, which receives the highest plasma exposure, oxidizes to a silica-like state. If this cracks, as it frequently does, these cracks will provide a locus for the surface-tension-driven migration of free PDMS chains from the bulk to the surface, mechanism (C), to cover the polar silica-like surface. The angle-resolved XPS data are strongly supportive of this free PDMS layer concept. Note also that below this silica-like layer, which must be thinner than the XPS sampling depth, there may be a region of oxidized carbon such as the $SiCH_2OH$ groups originally proposed by Hollahan and Carlson [9]. We see such oxidized carbon in only angle-resolved studies normal to the surface where signals from as deep as possible in the sample are favored. Such entities as $SiCH_2OH$ may seem unlikely because of their instability, but the overlying silica-like layer may provide adequate protection. This puts a limit on the thickness of the treated layer of less than 8–10 nm.

The cracks must run to the untreated PDMS to provide a pathway for free PDMS migration. That these cracks are involved in the hydrophobic recovery process can be simply demonstrated by treating two samples of PDMS to complete wettability under the same conditions and then deliberately bending one sample to form cracks in the surface. The advancing water contact angle of the cracked sample immediately jumps to around 40–60° and continues to increase until complete recovery occurs after a few hours, while the other sample recovers more slowly. These difficulties mean that any quantitative study of the hydrophobic recovery process must include careful control and assessment of surface cracks and this has yet to be done. Clearly XPS cannot be used in such a study. One inference that we wish to test is that an uncracked treated layer would not recover its original hydrophobicity.

4. CONCLUSIONS

The effect of RF plasma treatment of PDMS elastomer is broadly similar whether the gas used is argon, helium, oxygen, or nitrogen. In each case, a thin, brittle, silica-like layer is produced on the surface. Hydrophobic recovery of this surface is due to migration of untreated polymer chains from the bulk to the surface through cracks in the silica-like layer. Cracking is a difficult parameter to control but needs to be taken into account in future quantitative recovery studies. Cracking always occurs during XPS analysis of treated PDMS samples, rendering this technique unsuitable for such studies.

Acknowledgements

We thank Jennifer Fritz and Rebecca Durall for their help in the SEM work. We are also grateful to Tom Brodhagen and Manoj Chaudhury for provision of materials and valuable discussions.

REFERENCES

1. M. Hudis, in: *Techniques and Applications of Plasma Chemistry*, J. R. Hollahan and A. T. Bell (Eds), pp. 113–147. John Wiley, New York (1974).
2. C. F. Amstein and P. A. Hartman, *J. Clin. Microbiol.* **2**, 46–54 (1975).
3. F. J. Holly and M. J. Owen, in: *Physicochemical Aspects of Polymer Surfaces*, K. L. Mittal (Ed.), Vol. 2, pp. 625–636. Plenum Press, New York (1983).
4. P. M. Triolo and J. D. Andrade, *J. Biomed. Mater. Res.* **17**, 129–147 (1983).
5. M. K. Chaudhury and M. J. Owen, *J. Phys. Chem.* **97**, 5722–5726 (1993).
6. M. K. Chaudhury and M. J. Owen, *Langmuir* **9**, 29–31 (1993).
7. D. T. Clark, A. Dilks and D. Shuttleworth, in: *Polymer Surfaces*, D. T. Clark and W. J. Feast (Eds), pp. 185–211. John Wiley, New York (1978).
8. J. R. Hall, C. A. L. Westerdahl, A. T. Devine and M. J. Bodnar, *J. Appl. Polym. Sci.* **13**, 2085–2096 (1969).
9. J. R. Hollahan and G. L. Carlson, *J. Appl. Polym. Sci.* **14**, 2499–2508 (1970).
10. M. Morra, E. Occhiello, R. Marola, F. Garbassi, P. Humphrey and D. Johnson, *J. Colloid Interface Sci.* **137**, 11–24 (1990).
11. M. T. Stewart and M. W. Urban, *Polym. Mater. Sci. Eng.* **59**, 334–338 (1988).
12. Y. Ikada, T. Matsunaga and M. Suzuki, *Nippon Kagaku Kaishi* **6**, 1079–1086 (1985).
13. S. R. Gaboury and M. W. Urban, *Polym. Commun.* **32**, 390–392 (1990).
14. P. Feneberg and U. Krekeler, US Patent 3 959 105 (1976).
15. C.-L. Lee and G. R. Homan, in: *Annual Report (81CH1668-3) Conference on Electrical Insulation and Dielectric Phenomena*, pp. 435–443. IEEE, Piscataway, NJ (1981).
16. P. J. Smith, M. J. Owen, P. H. Holm and G. A. Toskey, in: *Proc. IEEE CEIDP Conf.*, Victoria, BC, Canada, pp. 829–836. Piscataway, NJ (1992).

17. A. Baszkin, M. M. Boissonade, J.-E. Proust, S. Tchaliovska, L. Ter-Minassian-Saraga and G. Wajs, *J. Bioeng.* **2**, 527–537 (1978).
18. C. D. Wagner, L. E. Davis, M. V. Zeller, J. A. Taylor, R. H. Raymond and L. H. Gale, *Surface Interface Anal.* **3**, 211–225 (1981).

Polymer Surface Modification: Relevance to Adhesion, pp. 17–32
K. L. Mittal (Ed.)

Atmospheric silent discharge versus low-pressure plasma treatment of polyethylene, polypropylene, polyisobutylene, and polystyrene

O. D. GREENWOOD, R. D. BOYD, J. HOPKINS and J. P. S. BADYAL*
Department of Chemistry, Science Laboratories, Durham University, Durham DH1 3LE, UK

Revised version received 21 June 1994

Abstract—Polyethylene, polypropylene, polyisobutylene, and polystyrene films have been exposed to high- and low-pressure non-equilibrium electrical air discharges. The modified surfaces have been characterized by X-ray photoelectron spectroscopy (XPS) and atomic force microscopy (AFM). Atmospheric silent discharge treatment causes a greater level of topographical disruption, whereas surface oxygenation is dependent on the chemical nature of the polymer substrate and its reactivity towards the electrical discharge medium. Oxygen incorporation occurs much more readily for the unsaturated polystyrene surface than for the saturated polyethylene, polypropylene, and polyisobutylene substrates.

Keywords: Silent discharge; plasma; polyethylene; polypropylene; polyisobutylene; polystyrene; XPS; AFM.

1. INTRODUCTION

Both atmospheric and low-pressure electrical discharges are used commercially for enhancing printability and adhesion properties of polymeric substrates [1–3]. This is believed to arise from alteration of the chemical [4] and/or topographical [5, 6] nature of the substrate.

Two types of atmospheric electrical discharge are in common usage [7]. The *corona* discharge can be identified as bright filaments extending from a sharp high voltage electrode towards the substrate. Alternatively, a parallel plate dielectric barrier configuration can result in a more uniform electrical breakdown, and this is known as a *silent* discharge. Electron impact at the substrate surface is important during air corona treatment [8, 9], whereas a combination of electron/photon excitation and ozone chemistry governs the characteristics of an air dielectric barrier reactor (the ozone concentration is typically more than 100 times higher than that of electrons

*To whom correspondence should be addressed.

or ions, and ten times greater than that of any other excited molecular species, e.g. N_2, O_2, singlet oxygen, etc.) [10–12].

Vacuum-ultraviolet light and oxygen atoms are reported to be the chemically prominent species in low-pressure non-equilibrium air plasmas [13, 14]. In this case, highly energetic photons are capable of rupturing organic bonds [15]. The major limitation of this technique is that a vacuum system is required, which restricts it to mainly batch mode operation.

Numerous investigations have been devoted to the mechanistic aspects of electrical discharge modification of polymeric surfaces and their dependence on process parameters. However, most of these studies have tackled only one or two substrates at a time for a fixed set of experimental conditions. This article quantitatively compares high- and low-pressure non-isothermal electrical discharge treatments of a range of hydrocarbon polymers listed below:

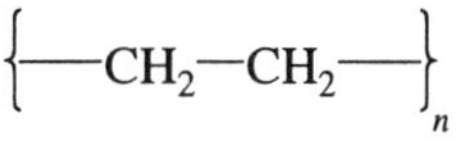

$\{-CH_2-CH_2-\}_n$

Polyethylene (PE)

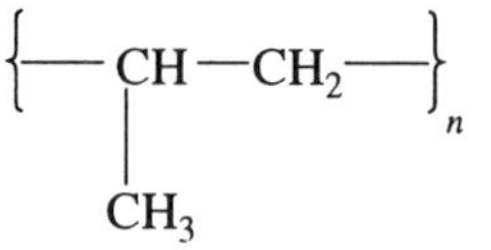

$\{-CH(CH_3)-CH_2-\}_n$

Polypropylene (PP)

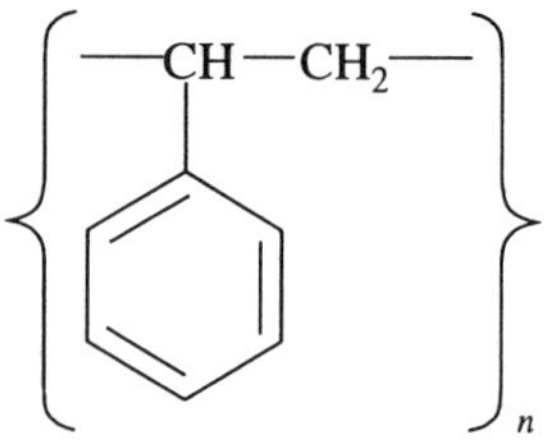

$\{-CH(C_6H_5)-CH_2-\}_n$

Polystyrene (PS)

$\{-CH_2-C(CH_3)_2-\}_n$

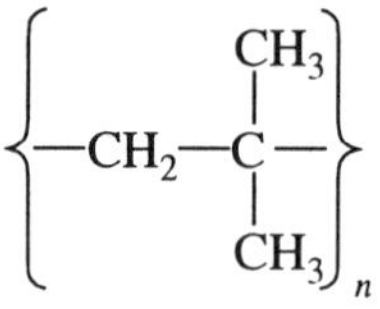

Polyisobutylene (PIB)

2. EXPERIMENTAL

Small (2 cm × 1 cm) strips of low-density polyethylene (ICI, 50 μm), polypropylene (ICI, 50 μm), and polystyrene (Goodfellows, 1 mm) were rinsed in a non-polar (heptane) and then a polar (isopropyl alcohol) solvent. Polyisobutylene (Exxon) was spin-coated onto clean glass slides from 2% weight/volume toluene solution.

Low-pressure plasma treatments were carried out in an electrodeless cylindrical reactor (4.5 cm diameter, 515 cm^3 volume, with a leak rate better than 4×10^{-3} cm^3 min^{-1}) enclosed in a Faraday cage [16]. This was fitted with a gas inlet, a Pirani pressure gauge, and a 27 $l\,min^{-1}$ two-stage rotary pump attached to a liquid nitrogen cold trap. A matching network was used to inductively couple a copper coil (4 mm diameter, 9 turns, spanning 8–15 cm from the gas inlet) wound around the reactor to a 13.56 MHz radio frequency (RF) generator. All joints were grease-free. Gas and leak mass flow rates were measured by assuming ideal gas behaviour [17]. A typical experimental run comprised initially scrubbing the reactor with detergent, rinsing with isopropyl alcohol, and oven drying, followed by a 60 min high-power (50 W) air plasma cleaning treatment. At this stage, a strip of polymer was inserted into the centre of the reactor (i.e. glow region), which was then evacuated down to a base pressure of 1.5×10^{-3} Torr. Subsequently air was introduced into the reaction chamber at 2×10^{-1} Torr pressure and a flow rate (F_V) of 1.0 $cm^3\,min^{-1}$. After allowing 5 min for purging, the glow discharge was ignited at 20 W for 30 s. Upon termination of the treatment, the RF generator was switched off and the system was let up to atmosphere.

Silent discharge treatment of each sample in air was carried out for 30 s using a home-built parallel-plate dielectric barrier discharge cell operating at 3 kHz, 11 kV, with an electrode gap of 3.00 ± 0.05 mm (Fig. 1). The silent discharge electrodes were made out of aluminium (top diameter = 4.5 cm, bottom diameter = 3.5 cm). Each electrode was chemically polished and then degreased using isopropyl alcohol before use. Polyethylene film was used as the dielectric material to cover the lower electrode. The high voltage supply unit comprised a capacitor which charged and then discharged through the primary of an induction coil; this produced ~3 kHz oscillation in the secondary coil. The current and voltage waveforms were measured using an oscilloscope to be dampened sinusoids of 2 ms half-life with very short current pulses (~1 μs duration) superimposed on top of the decaying waveform. The total power dissipated in the discharge was calculated to be 0.6 W.

A Kratos ES300 electron spectrometer equipped with a Mg K_α X-ray source (1253.6 eV), and a hemispherical analyser, was used for surface analysis by X-ray

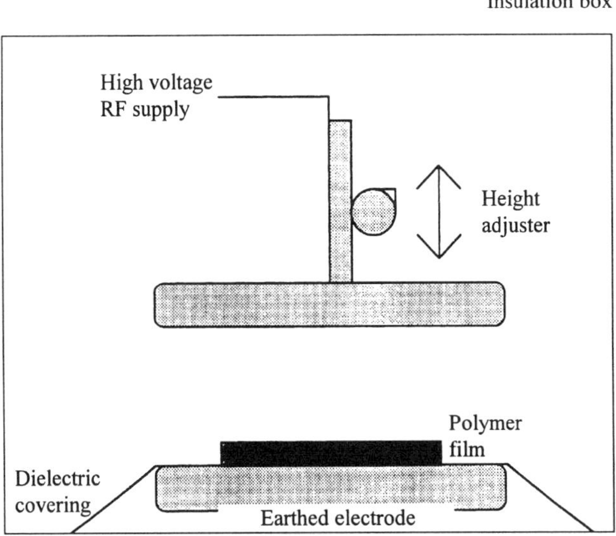

Figure 1. Apparatus used for silent discharge treatment of polymer film.

photoelectron spectroscopy (XPS). Photo-emitted electrons were collected at a take-off angle of 30° from the substrate normal, with electron detection in the fixed retarding ratio (FRR, 22:1) mode. XPS spectra were accumulated on an interfaced IBM PC computer. Instrumentally determined sensitivity factors were taken as $C(1s):O(1s)$ equals 1.00:0.62. Gross and experimental errors were calculated for each electrical discharge treatment. The cleanliness of each substrate and the absence of any surface-active inorganic additives were verified by XPS. Subsequently, each polymer was treated by the respective electrical discharge and inserted into the XPS spectrometer for analysis (usually within 60 s).

Atomic force microscopy (AFM) offers structural characterization of surfaces in the 10^{-4}–10^{-10} m range without the prerequisite of special sample preparation (e.g. metallization). A Digital Instruments Nanoscope III atomic force microscope was used for examining the topographical nature of each substrate surface before and after electrical discharge exposure. All of the AFM images were acquired in air using the Tapping mode [18] and are representative of each treated surface. This technique employs a stiff silicon cantilever oscillating at a large amplitude near its resonance frequency (several hundred kHz). The RMS amplitude is detected by an optical beam system. A large RMS amplitude is used to overcome the capillary attraction of the surface layer, whilst the high oscillation frequency allows the cantilever to strike the surface many times before being displaced laterally by one tip diameter. These features offer the advantage of low contact forces and no shear forces.

3. RESULTS

Exposure times of 30 s were used for both the silent discharge and the low-pressure plasma treatments. Wide-scan XPS spectra of the treated and untreated polymer substrates yielded only carbon and oxygen features for low-pressure plasma modification and dielectric barrier treatment. C(1s) XPS spectra were fitted with Gaussian peaks of equal full width at half-maximum (FWHM) [19], using a Marquart minimization computer program. Energies distinctive of different types of oxidized carbon moieties were referenced to the hydrocarbon peak ($-\underline{C}_xH_y-$) at 285.0 eV [20, 21]: carbon adjacent to a carboxylate group ($\gg\underline{C}-CO_2-$) at 285.7 eV, carbon singly bonded to one oxygen atom ($\gg\underline{C}-O-$) at 286.6 eV, carbon singly bonded to two oxygen atoms or carbon doubly bonded to one oxygen atom ($>\underline{C}=O/-O-\underline{C}-O-$) at 287.9 eV, carboxylate groups ($-O-\underline{C}=O$) at 289.0 eV, and carbonate carbons ($-O-\underline{C}O-O-$) at 290.4 eV. The $\pi-\pi^*$ shake up satellite around 291.7 eV for polystyrene was fitted with a Gaussian peak of different FWHM in order to assess the level of aromaticity present before and after electrical discharge exposure (see Fig. 2).

3.1. *Polyethylene*

For low-density polyethylene, both types of electrical discharge used in this study gave similar O:C ratios and comparable concentrations of the various types of oxidized carbon functionalities (Tables 1 and 2). The O:C values are approximately the same

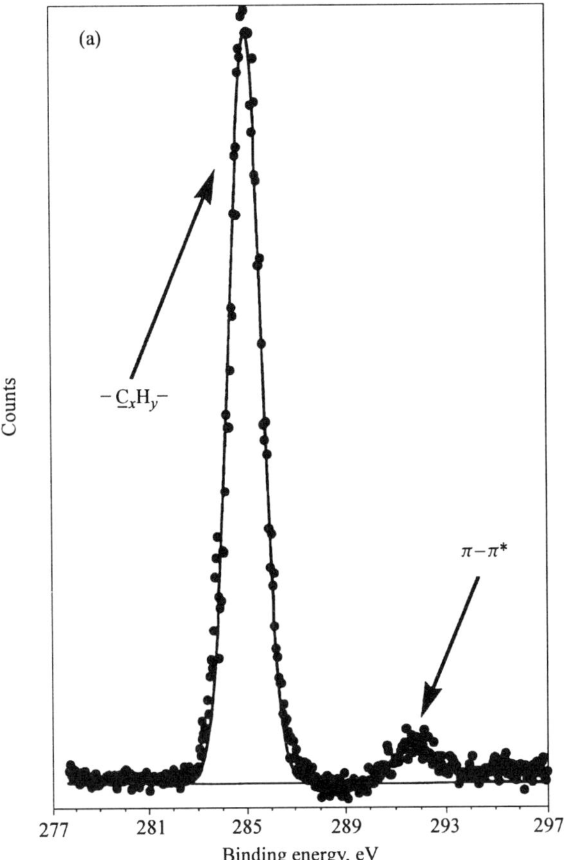

Figure 2(a). C(1*s*) XPS spectrum of clean polystyrene.

Table 1.
Compilation of O:C ratios following electrical discharge treatment (±0.01)

Treatment	PE	PP	PIB	PS
Clean	0.00	0.00	0.00	0.00
Low-pressure plasma	0.21	0.29	0.05	0.42
Silent discharge	0.21	0.29	0.12	0.34

as those reported previously for coronas [22, 23] and low-pressure oxygen plasma [24] treatments. However, in contrast to the lack of any topographical change reported for oxygen plasma modification [24], there is a definite variation in the surface roughness of polyethylene following the electrical discharge exposures employed in this study (Fig. 3). The untreated polymer substrate was found to be very rough (Table 3). Air plasma treatment attenuates this macro-roughness; this effect becomes even more pronounced for the silent discharge treatment, where a globular appearance is clearly evident.

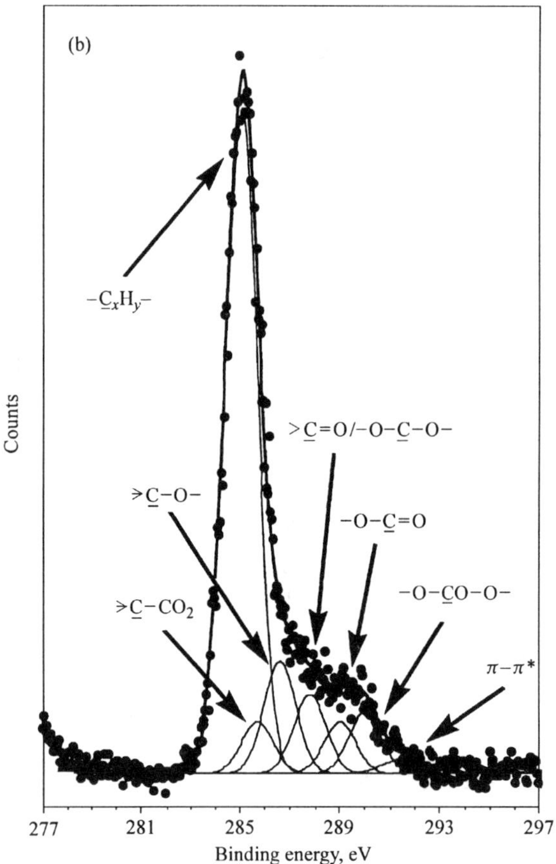

Figure 2(b). C(1*s*) XPS spectrum of silent discharge-treated polystyrene.

Table 2.
Relative % amounts of carbon functionalities following electrical discharge treatment (±0.8)

Treatment	$-\underline{C}-H-$	$-\underline{C}-CO_2-$	$\geqslant\underline{C}-O-$	$>\underline{C}=O/ -O-\underline{C}-O-$	$-O-\underline{C}=O$	$-O-\underline{C}O-O-$	$\pi-\pi^*$
Plasma PE	78.1	4.0	7.7	3.9	4.2	2.3	—
SD PE	76.2	4.5	8.5	4.5	4.5	1.8	—
Plasma PP	70.5	5.3	9.8	5.1	5.3	4.0	—
SD PP	67.1	5.9	11.4	7.7	6.0	1.9	—
Plasma PIB	90.8	1.1	5.3	1.5	1.1	0.2	—
SD PIB	83.2	2.5	8.6	2.9	2.5	0.3	—
Plasma PS	63.9	4.1	13.2	7.0	4.1	4.8	2.9
SD PS	67.3	3.9	9.3	6.1	4.0	6.6	2.8

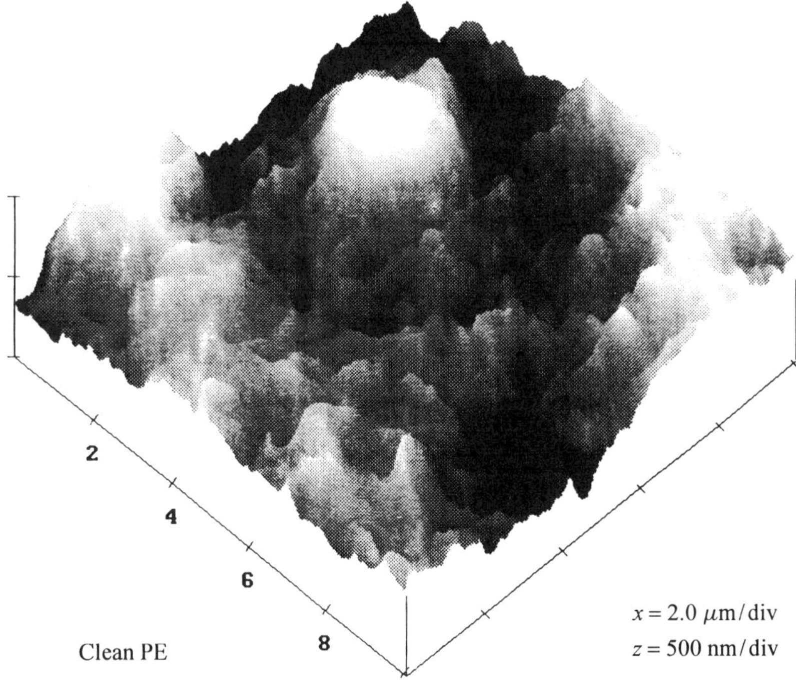

Figure 3(a). Atomic force micrograph of clean polyethylene. (NB. x refers to the horizontal plane and z refers to the vertical axis.)

Table 3.
Surface roughness following electrical discharge treatment

Substrate	Area analysed (μm)	RMS roughness (nm)
Clean PE	10 × 10	129
Plasma PE	10 × 10	117
Silent discharge PE	10 × 10	98
Clean PP	1 × 1	9
Plasma PP	1 × 1	24
Silent discharge PP	1 × 1	45
Clean PIB	5 × 5	6
Plasma PIB	5 × 5	19
Silent discharge PIB	5 × 5	23
Clean PS	1 × 1	7
Plasma PS	1 × 1	9
Silent discharge PS	1 × 1	13

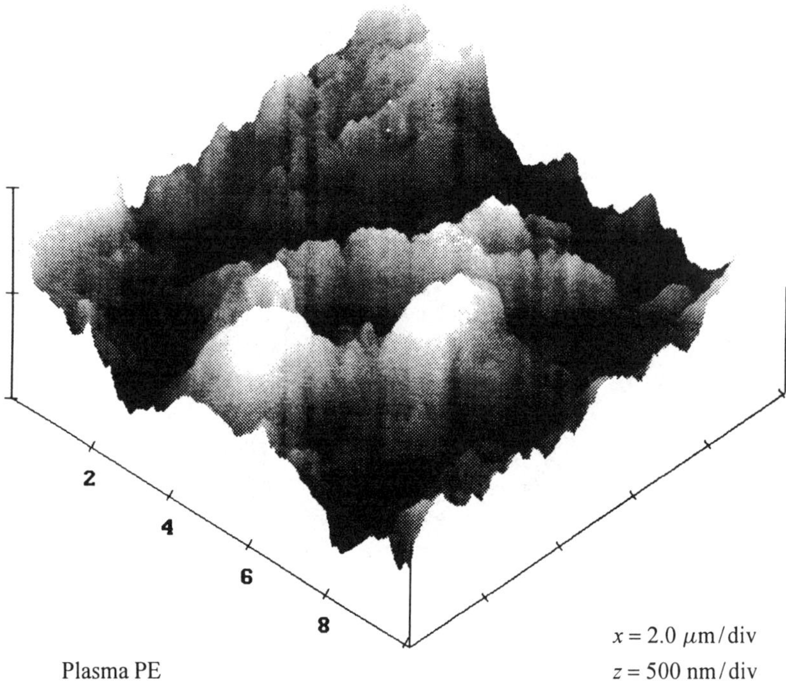

Figure 3(b). Atomic force micrograph of low-pressure plasma-treated polyethylene. (NB. x refers to the horizontal plane and z refers to the vertical axis.)

3.2. Polypropylene

Similar O:C ratios and comparable concentrations of the various types of oxidized carbon functionalities were measured for both types of plasma treatment. The O:C ratios are much higher in this study than those previously reported for corona treatment of polypropylene [6, 25–27]. Both silent discharge and low-pressure plasma exposure gave rise to carbon singly bonded to oxygen (—C̲—O—) as the major oxidized carbon functionality.

Surface roughness on the local scale is again eradicated to give way to roughness at the macroscopic level (Fig. 4). This effect is most prominent for the dielectric barrier treatment, where bubble formation is observed. Corona treatment can also lead to such globular features [6, 27] and has been attributed to either agglomeration of low-molecular-weight oxidized material [28] or local melting under the electrical discharge [29].

3.3. Polyisobutylene

Polyisobutylene was found to be more susceptible to chemical and topographical attack by the silent discharge treatment than by low-pressure plasma exposure. There

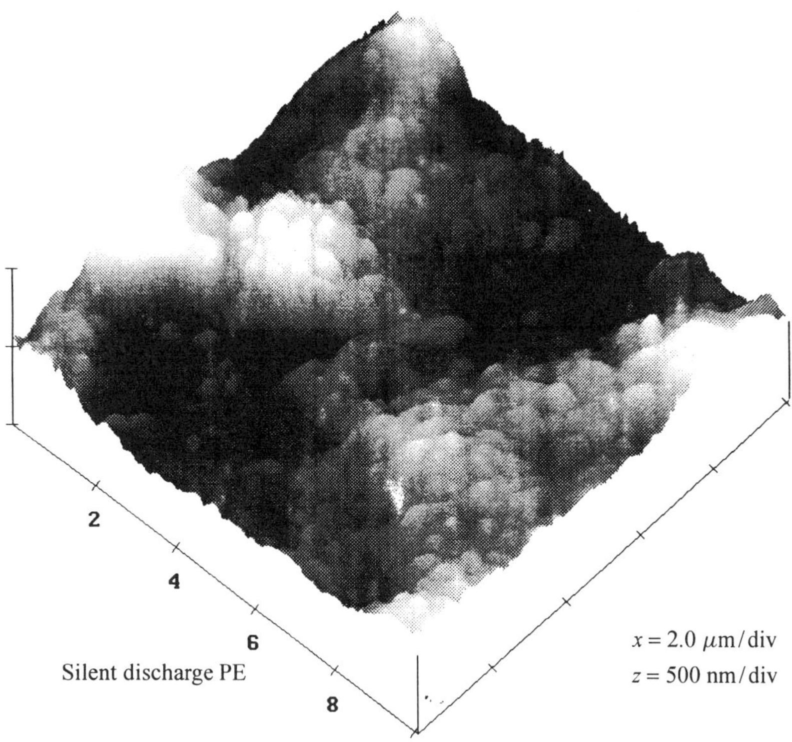

Figure 3(c). Atomic force micrograph of silent discharge-treated polyethylene. (NB. x refers to the horizontal plane and z refers to the vertical axis.)

appeared to be a greater level of oxygen incorporation into the polymer during dielectric barrier modification along with the emergence of a Swiss-cheese-like structure (Fig. 5).

3.4. Polystyrene

Clean polystyrene spectra showed a pure hydrocarbon peak with about 6% of the total intensity due to the aromatic π–π^* shake-up satellite (Fig. 2). Both silent discharge and low-pressure plasma treatments of polystyrene had a strongly oxidizing effect at the polymer surface, whilst the aromaticity was approximately halved in each case. Surprisingly, there was a greater level of oxygen incorporation during the low-pressure plasma treatment; this was mainly in the form of $-\underline{C}-O-$ linkages. As with the saturated polyolefins, $-\underline{C}-O-$ groups are the predominant oxidized carbon functionality. The carbonate group ($-O-\underline{C}O-O-$) concentration is greater for treated polystyrene than that measured for the other modified polymers. A greater degree of oxidation at the polystyrene surface was observed in this study in comparison with a previously reported corona treatment of polystyrene, where no carbonate species were detected [30].

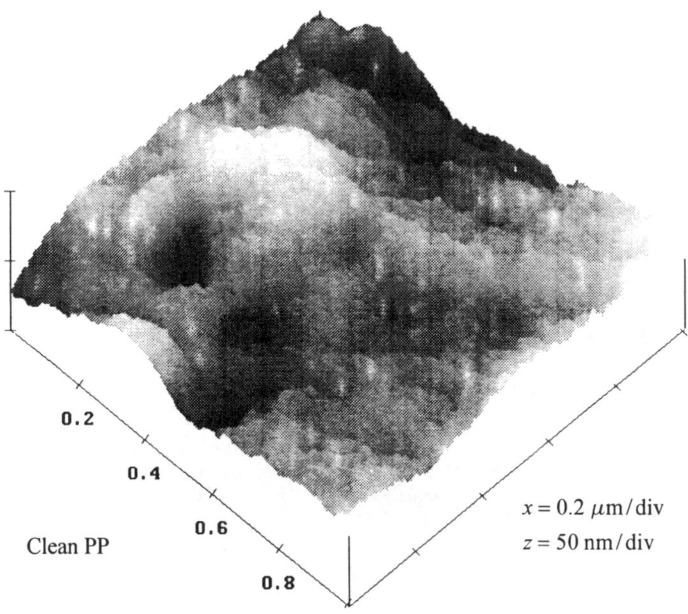

Figure 4(a). Atomic force micrograph of clean polypropylene. (NB. x refers to the horizontal plane and z refers to the vertical axis.)

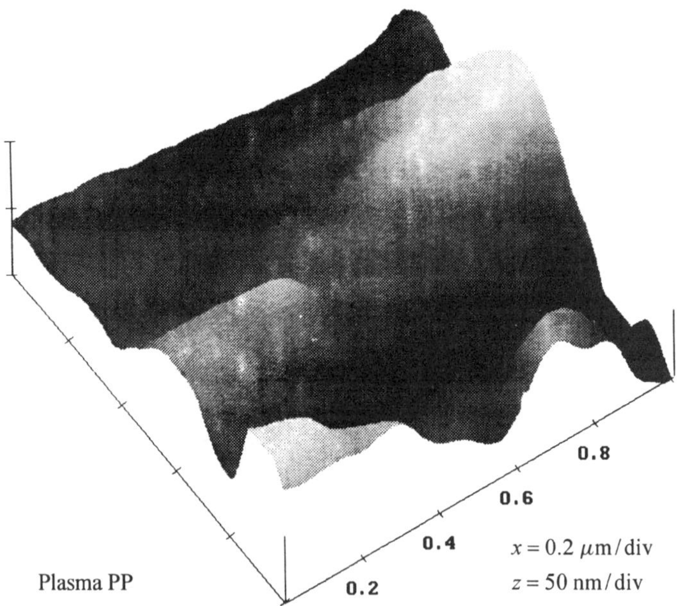

Figure 4(b). Atomic force micrograph of low-pressure plasma-treated polypropylene. (NB. x refers to the horizontal plane and z refers to the vertical axis.)

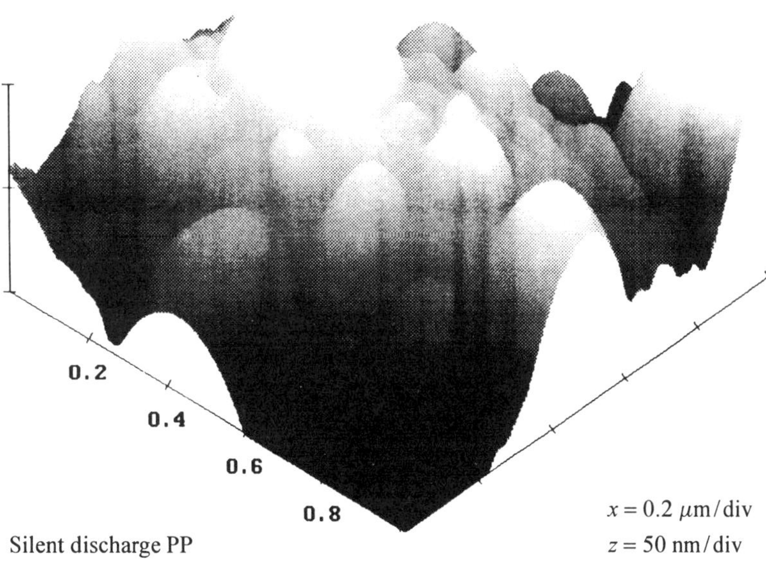

Figure 4(c). Atomic force micrograph of silent discharge-treated polypropylene. (NB. x refers to the horizontal plane and z refers to the vertical axis.)

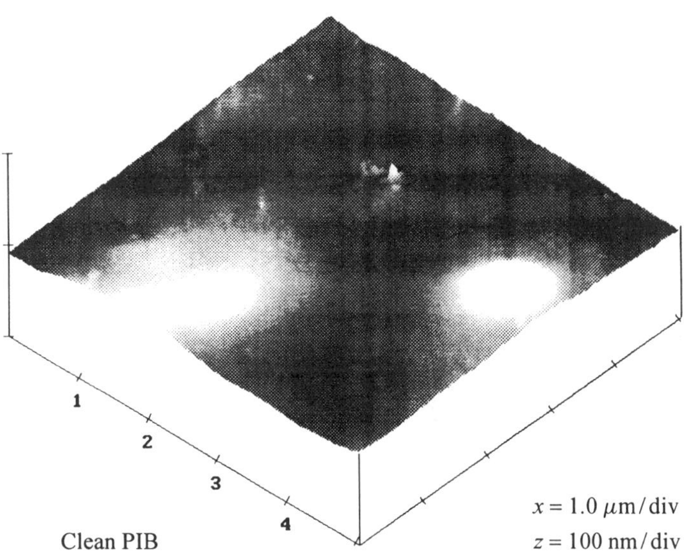

Figure 5(a). Atomic force micrograph of clean polyisobutylene. (NB. x refers to the horizontal plane and z refers to the vertical axis.)

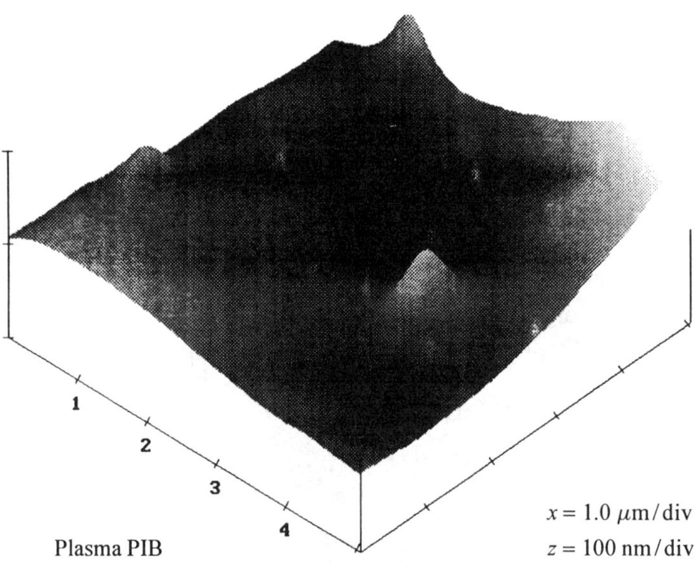

Figure 5(b). Atomic force micrograph of low-pressure plasma-treated polyisobutylene. (NB. x refers to the horizontal plane and z refers to the vertical axis.)

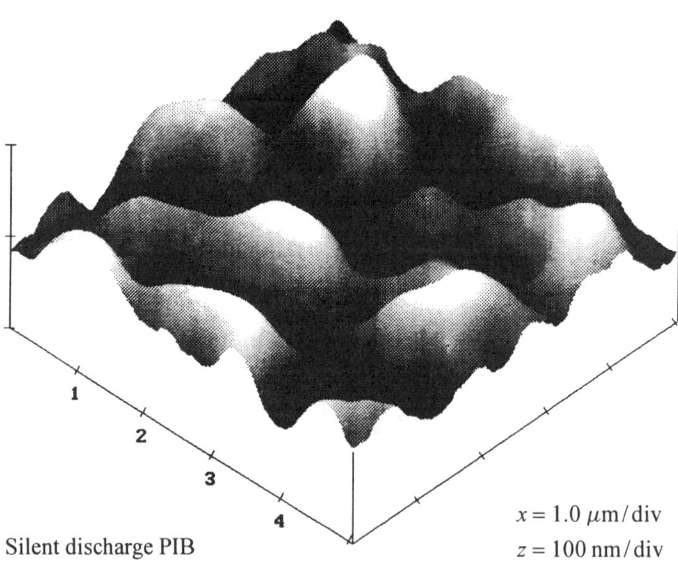

Figure 5(c). Atomic force micrograph of silent discharge-treated polyisobutylene. (NB. x refers to the horizontal plane and z refers to the vertical axis.)

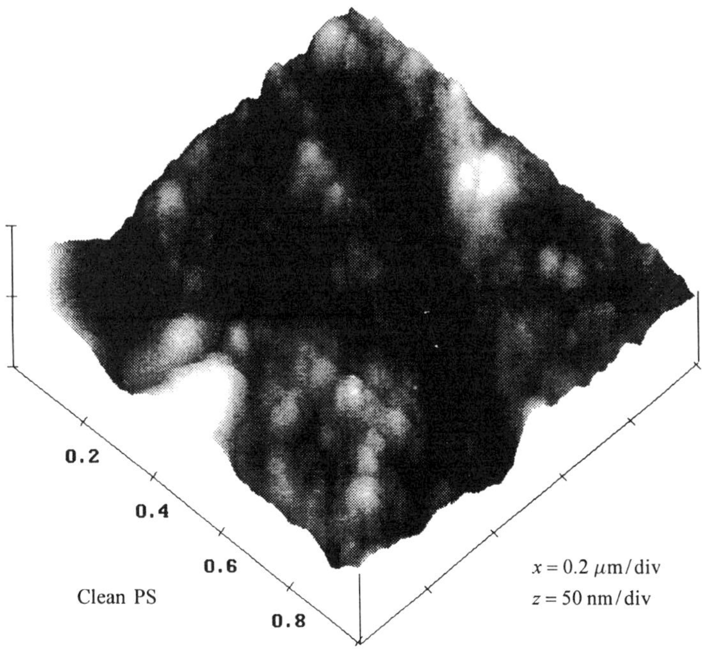

Figure 6(a). Atomic force micrograph of clean polystyrene. (NB. x refers to the horizontal plane and z refers to the vertical axis.)

Atomic force microscopy illustrates the more energetic nature of the silent discharge treatment compared with low-pressure plasma modification of polystyrene (Fig. 6). The globular features are much smaller in the latter case. These results are in contrast to previous SEM studies concerning the oxygen plasma treatment of polystyrene, where no change in surface topography was reported [31].

4. DISCUSSION

Carbon singly bonded to oxygen ($-\underline{C}-O-$) was found to be the predominant oxidized carbon functionality for all of the electrical discharge/polymer substrate combinations examined in this study. The general trend appears to be that silent discharge exposure gives rise to a greater degree of surface roughening compared with the low-pressure plasma treatment. However, careful attention needs to be paid to the fact that the untreated polyethylene substrate is very rough at the macroscopic level, and therefore this partially masks the topographical changes occurring on the small scale (Table 3).

Electrical discharges contain electrons, ions, excited neutrals, and photons. All of these species can react with a polymer surface; however, their relative concentrations and energies vary with the type of discharge employed and can have a strong influence on the mode of surface modification.

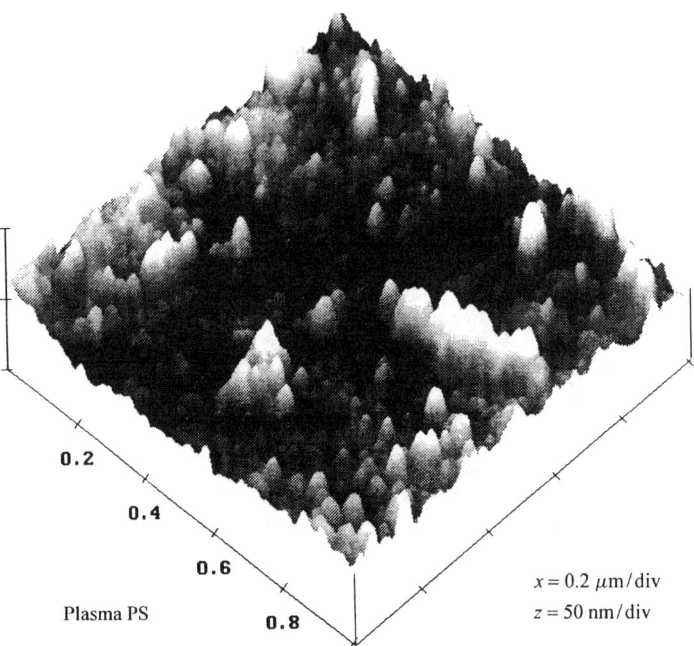

Figure 6(b). Atomic force micrograph of low-pressure plasma-treated polystyrene. (NB. x refers to the horizontal plane and z refers to the vertical axis.)

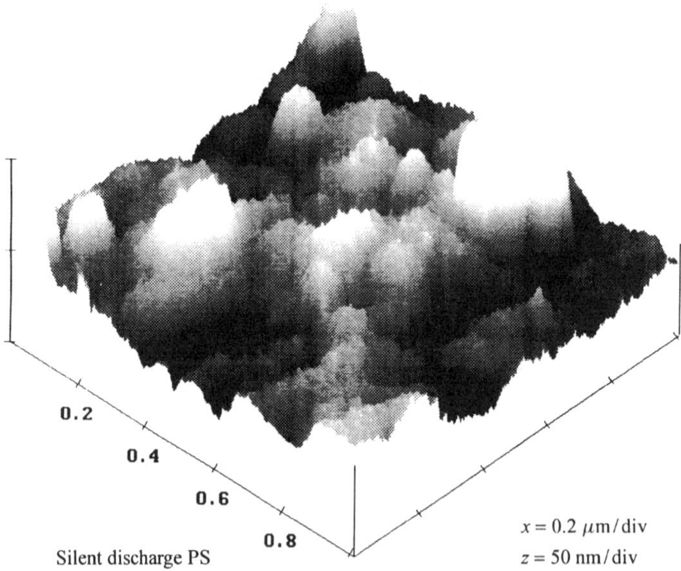

Figure 6(c). Atomic force micrograph of silent discharge-treated polystyrene. (NB. x refers to the horizontal plane and z refers to the vertical axis.)

Corona treatment of polymeric substrates proceeds via free-radical mechanisms [8, 9], whereas reaction of ozone with the excited substrate is an important feature of silent discharge treatments [10–12, 32]. Both of these processes are highly energetic; this is clearly evident in the case of dielectric barrier modification of polyethylene, polypropylene, polyisobutylene, and polystyrene, where AFM has detected a significant amount of topographical disruption at the polymer surface.

Vacuum-ultraviolet radiation and oxygen atoms are important components of a low-pressure air glow discharge [13]. In this case, highly energetic photons are able to cleave saturated bonds as well as excite unsaturated bonds at the polymer surface [14, 15]. The greater oxygen incorporation into polystyrene during low-pressure plasma treatment may be attributed to its chromophoric phenyl rings, and these can readily undergo photo-oxidation and crosslinking during vacuum-ultraviolet irradiation [33].

5. CONCLUSIONS

Electrical air discharge treatment of polymers results in both textural and chemical changes at the substrate surface. Silent discharge modification induces a much greater degree of morphological disruption than low-pressure plasma treatment for all of the polymers screened in this study. In chemical terms, saturated polymers (e.g. polyethylene, polypropylene, and polyisobutylene) yield a less oxidized surface, whereas a larger amount of oxygen is incorporated into polystyrene.

Acknowledgements

O.D.G. thanks PIRA International for financial support and British Petroleum Ltd. for provision of equipment.

REFERENCES

1. J. A. Lanauze and D. L. Myers, *J. Appl. Polym. Sci.* **40**, 595 (1990).
2. M. Morra, E. Occhiello and F. Garbassi, in: *Metallized Plastics 2, Fundamentals and Applied Plastics*, K. L. Mittal (Ed.), p. 363. Plenum Press, New York (1991).
3. W. L. Wade, R. J. Mammone and M. Binder, *J. Appl. Polym. Sci.* **43**, 1589 (1991).
4. D. K. Owens, *J. Appl. Polym. Sci.* **19**, 3315 (1975).
5. P. Blais, D. J. Carlssen and D. M. Wiles, *J. Appl. Polym. Sci.* **15**, 129 (1971).
6. C. Y. Kim and D. A. I. Goring, *J. Appl. Polym. Sci.* **15**, 1357 (1971).
7. J.-S. Chang, P. A. Lawless and T. Yamamoto, *IEEE Trans. Plasma Sci.* **19**, 1152 (1991).
8. J. Skalny, M. Luknarova and D. Dindosova, *Czech. J. Phys.* **B38**, 329 (1988).
9. H. Steinhauser and G. Ellinghorst, *Angew. Makromol. Chem.* **120**, 177 (1984).
10. B. Eliasson, M. Hirth and U. Kogelschatz, *J. Phys. D, Appl. Phys.* **20**, 1421 (1987).
11. K. Honda and Y. Naito, *J. Phys. Soc. Jpn* **10**, 1007 (1955).
12. E. Landers, *Proc. IEE* **125**, 1069 (1978).
13. D. T. Clark and A. Dilks, *J. Polym. Sci., Polym. Chem. Ed.* **15**, 2321 (1977).
14. A. G. Shard and J. P. S. Badyal, *Macromolecules* **25**, 2053 (1992).
15. E. M. Liston, *J. Adhesion* **30**, 199 (1989).
16. A. G. Shard, H. S. Munro and J. P. S. Badyal, *Polym. Commun.* **32**, 152 (1991).

17. C. D. Ehrlich and J. A. Basford, *J. Vac. Sci. Technol.* **A10**, 1 (1992).
18. Q. Zhong, D. Inniss, K. Kjoller and V. B. Elings, *Surface Sci.* **290**, L688 (1993).
19. J. F. Evans, J. H. Gibson, J. F. Moulder, J. S. Hammond and H. Goretzki, *Fresenius Z. Anal. Chem.* **319**, 841 (1984).
20. D. T. Clark and A. Dilks, *J. Polym. Sci., Polym. Chem. Ed.* **16**, 991 (1978).
21. G. Johansson, J. Hedman, A. Berndtsson, M. Klasson and R. Nilsson, *J. Electron Spectrosc. Relat. Phenom.* **2**, 295 (1973).
22. J. H. Lee, H. G. Kim, G. S. Khang, H. B. Lee and M. S. Jhon, *J. Colloid Interface Sci.* **152**, 563 (1992).
23. L. J. Gerenser, J. F. Elman, M. G. Mason and J. M. Pochan, *Polymer* **26**, 1162 (1985).
24. L. J. Gerenser, *J. Adhesion Sci. Technol.* **1**, 303 (1987).
25. M. Strobel, C. S. Lyons, J. M. Strobel and R. S. Kapaun, *J. Adhesion Sci. Technol.* **6**, 429 (1992).
26. M. Strobel, C. Dunatov, J. M. Strobel, C. S. Lyons, S. J. Perron and M. C. Morgen, *J. Adhesion Sci. Technol.* **3**, 321 (1989).
27. J. M. Strobel, M. Strobel, C. S. Lyons, C. Dunatov and S. J. Perron, *J. Adhesion Sci. Technol.* **5**, 119 (1991).
28. R. M. Overney, H.-J. Guntherodt and S. Hild, *J. Appl. Phys.* **75**, 1401 (1994).
29. R. M. Overney, R. Luthi, H. Haefke, J. Frommer, E. Meyer, H.-J. Guntherodt, S. Hild and J. Fuhrmann, *Appl. Surface Sci.* **64**, 197 (1993).
30. E. C. Onyiriuka, L. S. Hersh and W. Hertl, *J. Colloid Interface Sci.* **144**, 98 (1991).
31. E. Occhiello, M. Morra, P. Cinquina and F. Garbassi, *Polymer* **33**, 3007 (1992).
32. B. Eliasson, *IEEE Trans. Plasma Sci.* **19**, 309 (1991).
33. R. K. Wells, I. W. Drummond, K. S. Robinson, F. J. Street and J. P. S. Badyal, *Polymer* **34**, 3611 (1993).

Polymer Surface Modification: Relevance to Adhesion, pp. 33–48
K. L. Mittal (Ed.)

Hydrophobic recovery of repeatedly plasma-treated silicone rubber. Part 1. Storage in air

EMMANUEL P. EVERAERT,* HENNY C. VAN DER MEI, JOOP DE VRIES and HENK J. BUSSCHER

Laboratory for Materia Technica, University of Groningen, Bloemsingel 10, 9712 KZ Groningen, The Netherlands

Revised version received 21 March 1995

Abstract—Silicone rubber is used for a wide variety of biomedical and industrial applications due to its good mechanical properties, combined with a hydrophobic surface. Frequently, however, it is desirable to alter the surface hydrophobicity of silicone rubber. Often this is done by plasma treatments but the effects are usually transient. In this study, surfaces of medical grade silicone rubber have been repeatedly modified by means of oxygen, argon, carbon dioxide, and ammonia RF plasma treatments with a 24 h time interval in between treatments. Treated samples were stored in air prior to surface characterization by water contact angle measurements, X-ray photoelectron spectroscopy (XPS), streaming potential measurements, and profilometry for surface roughness. The carbon percentage of the surfaces decreased after plasma treatment, while the silicon and oxygen percentages increased irrespective of the plasma used. The formation of Si—O—Si bridges between siloxane chains after plasma treatment was demonstrated by the appearance of a new component in the Si_{2p} peak but the degree to which this occurred differed per gas. Streaming potential measurements in a 10 mM potassium phosphate buffer indicated a more negatively charged surface for treated samples compared to untreated samples (−23.3 mV at pH 7.0). Surface roughness increased slightly for repeatedly plasma-treated samples from $R_A = 0.35$ μm to $R_A = 0.46$ μm, while scanning electron microscopy showed the presence of several 'cracks' spanning the surface after repeated treatment. Argon, carbon dioxide, and ammonia plasmas significantly reduced the advancing water contact angle from 115° to 58°, 72°, and 85°, respectively, on a more permanent basis (especially when the treatments were repeated after recovery). Oxygen plasma effects on water contact angles generally disappeared within 5 h, also after repeated treatment.

Keywords: Hydrophobic recovery; silicone rubber; plasma treatment; water contact angle; X-ray photoelectron spectroscopy; surface roughness; streaming potential.

1. INTRODUCTION

Silicone polymers exhibit good mechanical properties for a variety of biomedical and industrial applications. For instance, silicone rubber has been used for voice

*To whom correspondence should be addressed.

prostheses [1, 2], urinary catheters [3], contact lens materials [4], and icing coating material [5]. However, their inherently high hydrophobicity limits certain applications of this material, despite its favorable mechanical properties [6]. Plasma treatment of silicone polymers may affect their hydrophobicity, and thereby their bondability to other materials, without affecting the bulk properties. Plasma treatment often involves progressive oxidation of the surface and crosslinking of surface molecular groups, which inhibits the migration of low molecular weight oligomers from the bulk to the surface.

Various gases have been used to modify silicone polymers by plasma treatment, such as oxygen [6, 7], helium [6, 8], carbon dioxide [9], ammonia [9], nitrogen [6, 10], and argon [6, 9, 10]. Frequently a thin crosslinked, sometimes water-washable, silica-like surface layer was produced by plasma treatment, but there is no consensus about the nature of the chemical groups produced at the outermost surface. It has been suggested that the polar entities created might be silanol groups or other oxidized carbon species, e.g. aldehyde or carboxylic groups [7].

The surface hydrophilicity created by plasma treatment is often lost over time [6, 7, 11]. This so-called hydrophobic recovery can be influenced by the storage conditions, whether in air or in liquid, temperature or subsequent adsorption of a surfactant [12]. A number of mechanisms for the hydrophobic recovery of plasma-treated silicone rubber have been proposed by Owen and Smith [6], including:

(1) reorientation of surface hydrophilic groups away from the surface;

(2) migration of treated polymer chains from the surface to the bulk;

(3) migration of untreated polymer chains from the bulk to the surface;

(4) loss of volatile oxygen-rich or other polar entities to the atmosphere;

(5) surface silanol condensation preventing chain reorientation;

(6) changes in surface roughness; and

(7) external contamination of the polymer surface.

Recently, Owen and Smith [6] reported that the effects of RF treatments of a polydimethylsiloxane (PDMS) elastomer were broadly similar for argon, helium, oxygen, and nitrogen. All treatments yielded a thin, brittle, silica-like layer on the surface, as concluded from XPS and electron microscopy. Although no contact angles were measured in this study, which is necessary to determine directly the hydrophobic recovery of plasma-treated surfaces, it was suggested that hydrophobic recovery originated from the migration of untreated polymer chains from the bulk to the surface through cracks in the silica-like layer.

Van der Mei *et al.* [11], hypothesizing that a thick treated layer might inhibit the migration of hydrophobic groups towards the surface, reported that repeated oxygen plasma treatment of polyethylene, with a 7-day interval in between, was more effective in creating a permanently hydrophilized surface than employing a higher RF power or a longer duration of the treatment.

In this study, we modified the surface of medical grade silicone rubber by repeated RF plasma treatments using various discharge gases including oxygen, argon, carbon dioxide, and ammonia in an attempt to create more permanent effects. The temporal behavior of the effects on the physico-chemical properties of the silicone rubber was investigated using different surface characterization techniques, including water contact angle measurements, X-ray photoelectron spectroscopy (XPS), streaming potential measurements, profilometry (for surface roughness), and scanning electron microscopy.

2. EXPERIMENTAL

2.1. Materials

A Silastic Medical Grade Silicone Rubber (Q7-4750, Dow Corning) kit was purchased and 1-mm-thick 50 × 70 mm and 2.8-mm-thick 25 × 76 mm plates were produced following the procedures suggested by the manufacturer. Briefly, equal proportions of part A and part B were thoroughly blended together and injected into a mold at room temperature through a 3 mm diameter opening with a force of 3000 kg. Subsequently, the silicone rubber was immediately cured at 200°C for 50 min. Finally, samples were cleaned in a 5% RBS 35 (Omnilabo International B.V., The Netherlands) detergent solution under simultaneous sonication (5 min, 150 W) and thoroughly rinsed in Millipore grade water and absolute ethanol (> 96%).

2.2. Plasma treatment

The silicone rubber samples were repeatedly glow-discharged in a PLASMOD instrument (Tegal Corporation, Richmond, CA, USA). The PLASMOD is a commercially available, inductively coupled (13.56 MHz RF) instrument equipped with a cylindrical quartz-made reaction chamber (inner diameter 8 cm, length 15 cm). Pumping down was done with a Balzers rotary pump (320 l/min) using a liquid nitrogen cold trap. All plasma treatments were done under 3.7 Torr gas pressure for 60 s and at a RF power of 50 W. Oxygen (99.5%), argon (99.996%), carbon dioxide (99.99%), and ammonia (99.98%) gases were obtained from Hoekloos Nederland B.V., The Netherlands.

The treated samples were stored in air in disposable Petri dishes and used immediately for water contact angle measurements, and after 1 month for XPS characterization. As the hydrophobic recovery of the treated silicone rubber occurred within a few hours, plasma treatments were repeated every 24 h. This cycle was repeated six times. Samples once used for surface characterization were not used again. All values given in this paper are the means of experiments on two separately prepared samples.

2.3. Contact angle measurements

Advancing and receding water contact angles were measured at room temperature with an image analyzing system, using the sessile drop technique. The advancing and receding angles were obtained by keeping the needle in the water droplet after positioning on the surface and by carefully moving the sample until the advancing angle appeared maximally. Each value was obtained by averaging the results of at least ten droplets on one sample.

2.4. Elemental surface composition

X-ray photoelectron spectroscopy (XPS) was performed on each treated sample using an S-Probe spectrometer (Surface Science Instruments, Mountain View, CA, USA) equipped with an aluminum anode (10 kV, 22 mA) and a quartz monochromator. The direction of the photoelectron collection angle was 55° with the normal to the sample and the electron flood gun was set at 10 eV. A survey scan was made with a 1000×250 μm spot and a pass energy of 150 eV. Detailed scans of the C_{1s}, O_{1s}, N_{1s}, and Si_{2p} lines were obtained using a pass energy of 50 eV. Binding energies were determined by setting the binding energy of the C_{1s} component due to carbon involved in siloxane (C—Si—O—Si) bonds at 284.5 eV [13]. The experimental peaks were integrated after nonlinear background subtraction and the peaks were decomposed assuming a Gaussian/Lorentzian ratio of 85/15 by using the SSI PC software package. All Si_{2p} bands were fitted by fixing the distance between $Si_{2p(3/2)}$ and $Si_{2p(1/2)}$ at 0.7 eV and by imposing an intensity ratio of 2.0. Elemental surface compositions were expressed in atomic %, setting % C + % O + % Si + % N to 100%.

2.5. Surface roughness

The surface roughness R_A was determined on a Perthometer C5D (Perthen, Germany) equipped with a 2 μm stylus (opening angle of 90°). The R_A value indicates the average distance of the roughness profile to the center line of the profile. The R_A values reported were obtained by averaging ten scans on one sample. Furthermore, the influence of the plasma treatments on the surface topography was occasionally studied by scanning electron microscopy. To this end, the specimens were mounted on stubs and sputter-coated with gold (15 nm).

2.6. Streaming potential

Streaming potentials (V_{str}) in a 10 mM potassium phosphate solution (pH 3–8) were measured employing rectangular platinum electrodes (5.0×25.0 mm) located at both ends of a parallel plate flow chamber [14]. Two samples of silicone rubber (length 76 mm, width 25 mm, and thickness 2.8 mm), separated by a 0.2 mm Teflon gasket, constituted the top and bottom plates of the chamber.

Zeta potentials (ζ) were derived from the pressure dependence of the streaming potentials (V_{str}) measured, according to

$$\zeta = \frac{\eta}{\varepsilon} K_b \frac{\Delta V_{str}}{\Delta P}, \tag{1}$$

where ε is the dielectric permittivity, η is the viscosity, and K_b is the solution conductivity. Note that the use of equation (1) implies that surface conduction is neglected. The conductivity of the electrolyte, K_b, was measured after each experiment, using a

Knick Konduktometer 702 (Berlin, Germany). Measurements were done at ten different pressures ranging between 37.5 and 150 Torr and each pressure was applied for 10 s in both directions. The streaming potentials measured were independent of the flow direction, and linearity between V_{str} and the applied pressure was always observed, at least for untreated silicone rubber.

3. RESULTS

3.1. Water contact angles

Figure 1 shows examples of advancing and receding water contact angles on repeatedly treated silicone rubber in an argon and oxygen plasma as a function of the storage

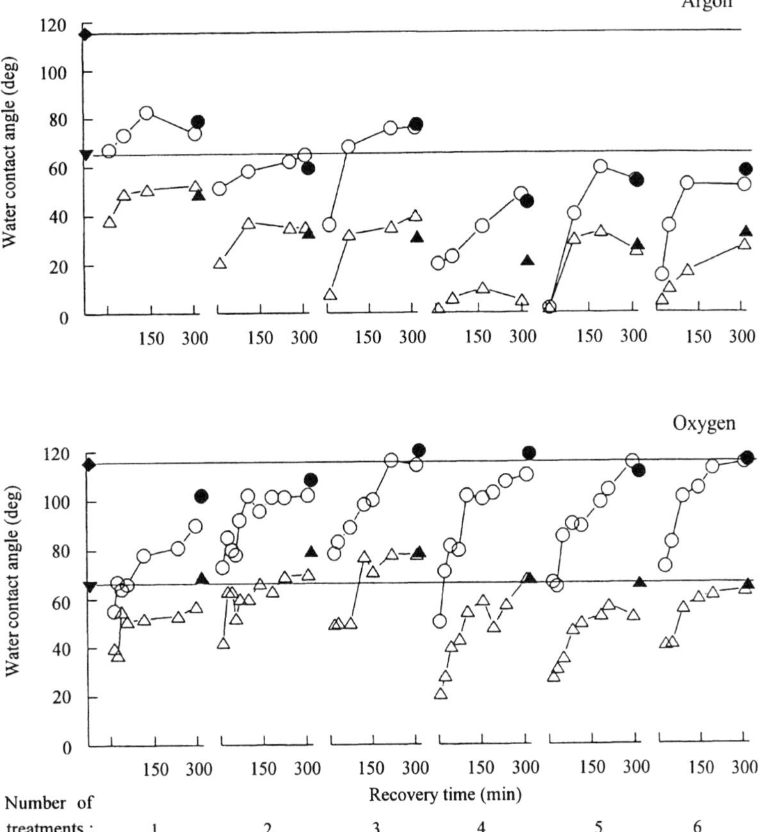

Figure 1. Advancing (○) and receding (△) water contact angles (degrees) on repeatedly argon plasma- and oxygen plasma-treated silicone rubber (50 W, 3.7 Torr, 60 s) stored in air. The open symbols represent contact angles measured up to 320 min after treatment, while the filled symbols represent contact angles measured after 24 h of aging. The ◆ and ▼ symbols represent the advancing and receding water contact angles for untreated silicone rubber. The standard deviation over ten measurements on one sample amounted to 3° on average, while the results of two separately prepared samples generally coincided within 6°.

time and the number of plasma treatments. Immediately after each plasma treatment, the water contact angles showed a large decrease in surface hydrophobicity. As can be seen, however, all surfaces treated regained some degree of hydrophobicity after storage in air. As complete hydrophobic recovery of oxygen-treated samples occurred within a few hours, water contact angles were measured at regular intervals during the storage period up to 320 min. An additional measurement was carried out after 24 h to observe the long-term recovery. In contrast to the hydrophobic recovery of oxygen plasma-treated silicone rubber, argon plasma-treated samples did not show complete recovery.

Figures 2 and 3 show the advancing and receding water contact angles, respectively, on silicone rubber repeatedly treated in oxygen, argon, carbon dioxide, and ammonia plasmas as measured after different storage times. Oxygen plasma reduced the advancing water contact angles to only 60° after 5 min, but within 5 h the surface had regained its original hydrophobicity. Argon and carbon dioxide plasmas yielded a fully wettable surface after 1–5 treatments, while the hydrophobic recovery was

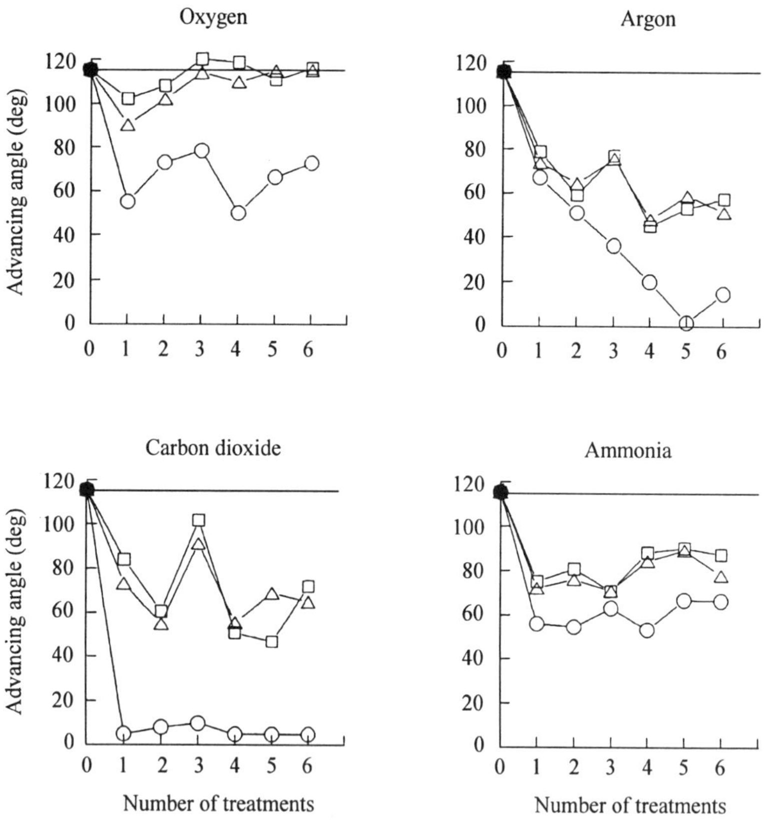

Figure 2. Advancing water contact angles (degrees) on silicone rubber repeatedly plasma-treated in oxygen, argon, carbon dioxide, and ammonia, after storage times of 5 min (○), 5 h (△), and 24 h (□). The ● symbol represents the advancing contact angle for untreated silicone rubber. The standard deviation over ten measurements on one sample amounted to 3° on average, while the results of two separately prepared samples generally coincided within 6°.

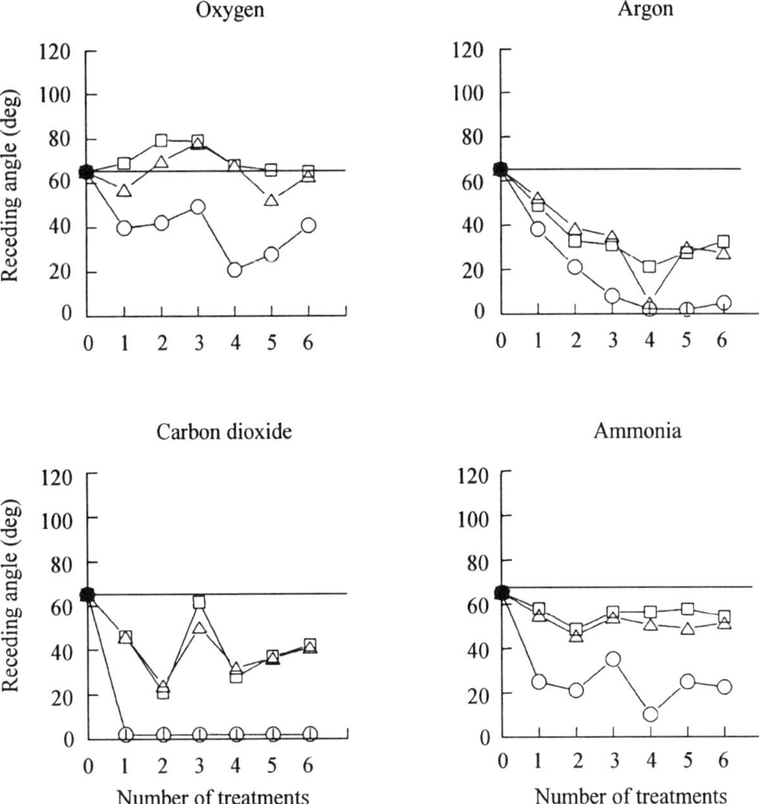

Figure 3. Receding water contact angles (degrees) on silicone rubber repeatedly plasma-treated in oxygen, argon, carbon dioxide, and ammonia, after storage times of 5 min (○), 5 h (△) and 24 h (□). The ● symbol represents the receding contact angle for untreated silicone rubber. The standard deviation over ten measurements on one sample amounted to 3° on average, while the results of two separately prepared samples generally coincided within 6°.

only partial, even after 24 h. Note that for argon plasma, the degree of hydrophobic recovery decreased with the number of treatments. Ammonia plasma made a slightly wettable surface, with, however, only a minor hydrophobic recovery. From Figs 2 and 3 it can be seen that the advancing and receding contact angles on RF plasma-treated samples behaved almost the same in relation to their hydrophobic recovery, with a contact angle hysteresis of around 30° on average, which is low compared with the hysteresis on untreated silicone rubber (50°).

3.2. Elemental surface composition

Figure 4 summarizes the elemental surface compositions by XPS for untreated and repeatedly plasma-treated silicone rubber. The carbon content decreased with increasing number of plasma treatments, while the silicon and oxygen amounts increased, irrespective of the plasma used. The decrease in carbon content differed among the various gases used, from an initial value of 51.5% to 42.6% for oxygen, 38.0% for

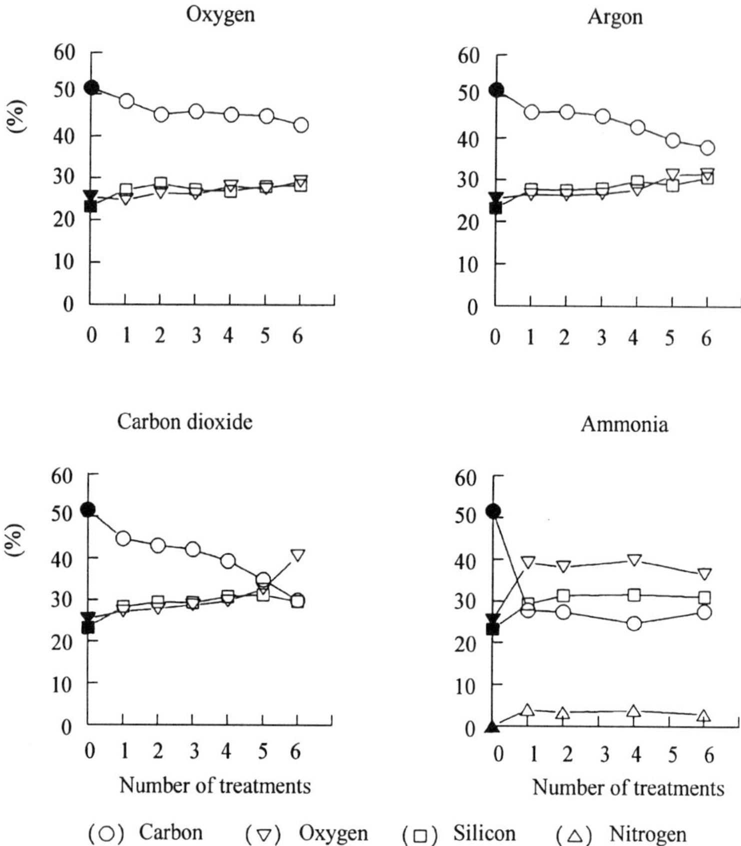

Figure 4. Elemental surface composition of silicone rubber repeatedly treated with oxygen, argon, carbon dioxide, and ammonia plasmas (50 W, 3.7 Torr, 60 s) as a function of the number of treatments. The closed symbols represent the values for untreated silicone rubber. The standard deviation over three measurements on one sample amounted to 1% on average for untreated silicone rubber, while the results for two separately treated samples coincided within 3%.

argon, 29.8% for carbon dioxide, and 27.5% for ammonia. Nitrogen was not detected by XPS except for the ammonia-treated samples, for which the nitrogen content measured ranged from 2 to 4%.

The C_{1s}, O_{1s}, and Si_{2p} peaks presented multiple components. Usually the C_{1s} peak had two components at 284.5 and 285.6 eV for carbon involved in methyl groups of the siloxane bond and in crosslinking between the siloxane chains, respectively. After 5–6 repeated plasma treatments, two additional C_{1s} components for carbon involved in oxygen bonds appeared at 287.6 and 289.8 eV. The O_{1s} peak had two components, one at 532.0 eV, due to the oxygen atoms in siloxane bonds, and the other at 533.2 eV, attributed to silanol groups and other Si–O bonds. The Si_{2p} peak showed a new component as well. Consequently, the $Si_{2p(3/2)}$ had a component at 101.9 eV indicative of the silicon in siloxane polymer chains and at 103.4 eV attributed to silicon in other multiple oxygen bonds, as in silica. In Fig. 5, it can

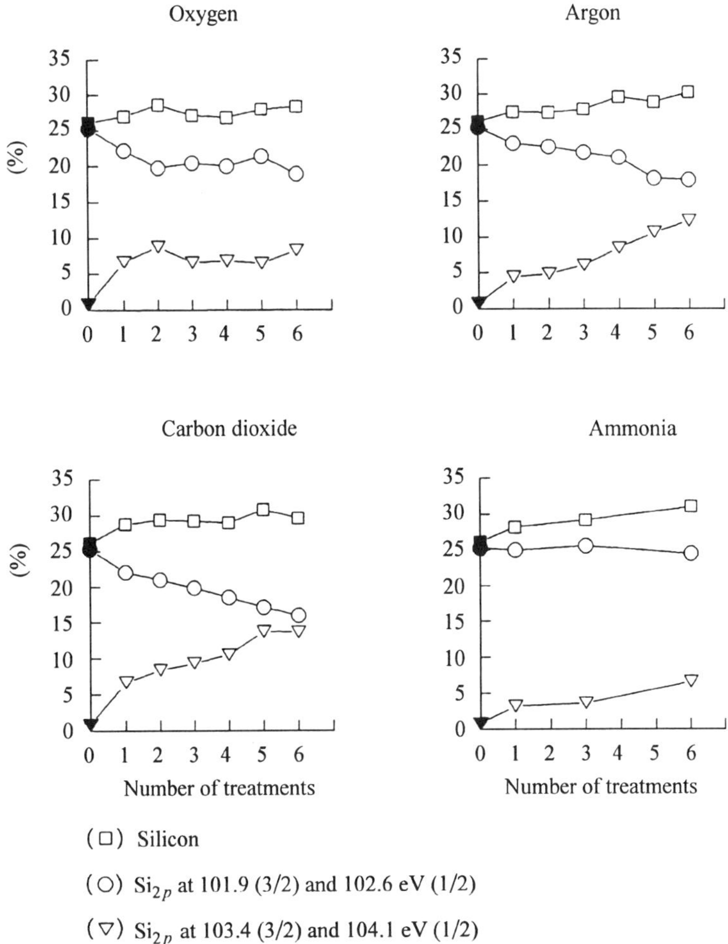

Figure 5. Si_{2p} elemental surface concentration and Si_{2p} components of untreated (filled symbols) silicone rubber and of oxygen, argon, carbon dioxide, and ammonia plasma-treated silicone rubber (open symbols) at 50 W, 3.7 Torr, 60 s and stored in air as a function of the number of plasma treatments. The standard deviation over three measurements on one sample amounted to 1% on average for untreated silicone rubber, while the results for two separately treated samples coincided within 3%.

be seen that the relative prevalence of the new Si_{2p} component increases with the number of plasma treatments, irrespective of the type of plasma used.

3.3. Surface roughness

The surface roughness increased slightly from its original smooth surface value $R_A = 0.35$ μm to approximately $R_A = 0.46$ μm after six repeated plasma treatments, independent of the type of plasma employed. However, scanning electron microscopy showed the presence of several 'cracks' spanning the surface (see Fig. 6) which remained nevertheless undetected by profilometry and which were equally present on all treated samples.

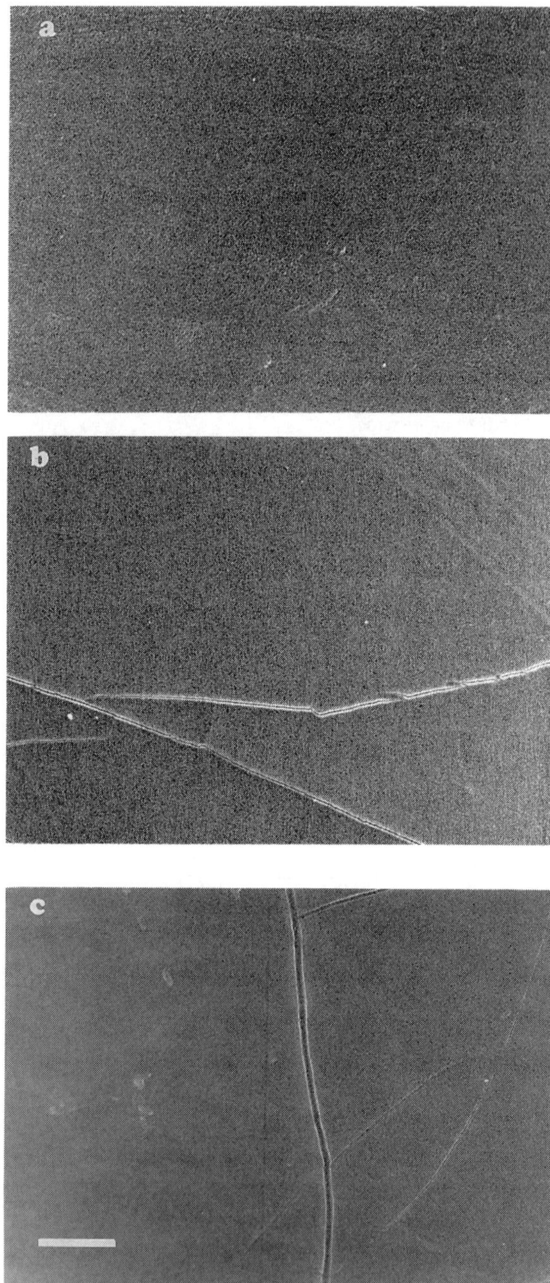

Figure 6. Scanning electron micrographs of (a) untreated silicone rubber, (b) silicone rubber treated six times in an argon plasma with a 24 h recovery period in between each treatment, and (c) silicone rubber treated six times in an oxygen plasma with a 24 h recovery period in between each treatment. The bar equals 10.0 μm.

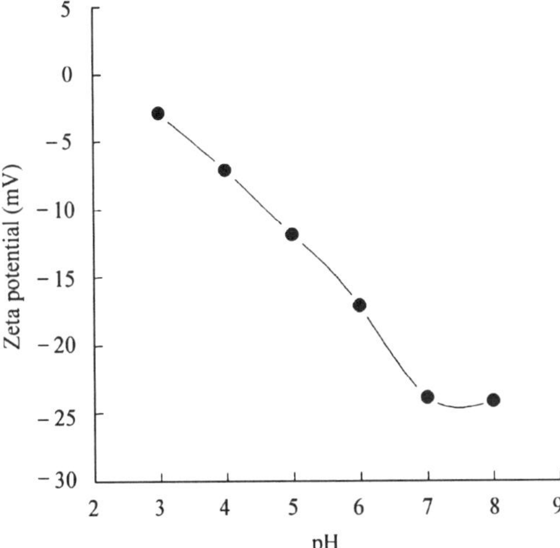

Figure 7. Zeta potentials in a 10 mM potassium phosphate solution of untreated silicone rubber as a function of the pH and measured by streaming potentials. The results of two separately prepared samples coincided within 3 mV.

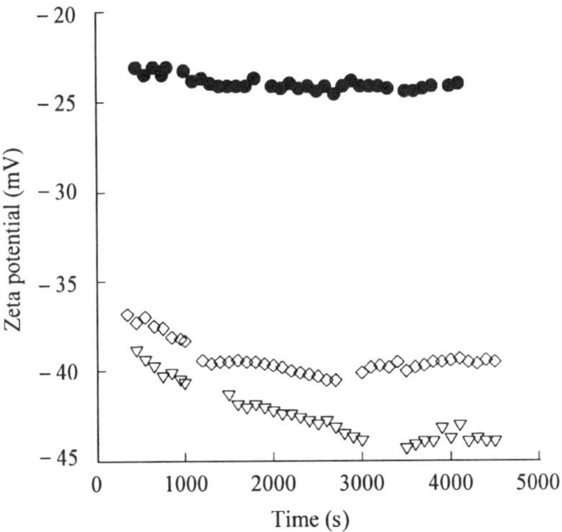

Figure 8. Zeta potentials in a 10 mM potassium phosphate buffer (pH 7.0) of untreated silicone rubber (●) and of argon (◇) and carbon dioxide (▽) plasma-treated silicone rubber as a function of the time of application of a pressure of 112.5 Torr alternating in both directions. The values of two separately prepared samples coincided within 4 mV.

3.4. Streaming potential

Figure 7 shows the zeta potentials of untreated silicone rubber as a function of the pH. The zeta potential of untreated silicone rubber ranged from −24.2 mV at pH 8.0 to −2.9 mV at pH 3.0. For RF plasma-treated samples, a constant zeta potential could not be observed during application of a certain pressure (see also Fig. 8), but the values measured indicated that plasma-treated silicone rubber was up to 20 mV more negatively charged over the entire pH range than untreated silicone rubber.

4. DISCUSSION

4.1. Water contact angle

There is a major water contact angle hysteresis on both RF plasma-treated and untreated silicone rubbers, despite the fact that the surface roughness is in the submicrometer range. Therefore [15], it is likely that this hysteresis is related to the high flexibility of the siloxane backbone of the elastomer. It has been reported [16] that the rotational freedom of the siloxane backbone allows the exposure of the methyl groups to their best effect, i.e. away from the surface in water and towards the surface in air. Consequently, this yields a relatively low receding water contact angle and a high advancing contact angle. The high mobility of polymer chains in silicone rubber may also explain why the hydrophobic recovery of plasma-treated samples occurs in several hours. By comparison, Van der Mei *et al.* [11] reported that hydrophobic recovery of repeated oxygen plasma-treated polyethylene was seen in more than 8 days. Obviously, the mobility of untreated polyethylene chains from the bulk to the surface is much slower than that of untreated siloxane polymer chains, in accordance with the difference in the glass transition temperatures of polyethylene (−35 °C) and siloxane polymer chains (−125 °C).

It is unclear why the kinetics of hydrophobic recovery differ for the various gases used. Although cracks have been suggested as a starting point for hydrophobic recovery by migration of untreated chains to the surface [6], no significantly different amounts of cracks were observed for silicone rubber samples treated with different gases. Possibly, the thickness, density, and composition of the argon and carbon dioxide plasma-treated silicone rubber samples allow less migration of untreated chains than when oxygen or ammonia plasma is employed.

Generally, the water contact angle hysteresis on RF plasma-treated silicone rubber was smaller than that of untreated silicone rubber. This may be due to a loss of the rotational freedom of the treated polymer chains. Such loss can be caused by the formation of new chemical groups and crosslinking, as detected by XPS.

4.2. Elemental surface composition

The elemental surface composition of our untreated silicone rubber is identical to the surface composition of PDMS found by Owen and Smith [6] and close to the values found by Ratner *et al.* [17] and Raimondi *et al.* [18]. The general trends of the elemental surface composition changes observed (Fig. 4) upon RF plasma treatment are similar to that reported by Owen and Smith [6]. Although Owen and Smith concluded that for a single plasma treatment the effects were similar for the four different gases used, we found that the degree of hydrophobic recovery, as revealed also by XPS, differs per gas (see Figs 4 and 5).

After a single plasma treatment, new components in the C_{1s}, O_{1s}, and Si_{2p} peaks appeared, as are also formed after repeated treatments [6]. Interestingly, the O_{1s} component at 533.2 eV (attributed to the oxygen included in new Si−O−Si or silica-like bonds) correlated linearly with the new Si_{2p} component with a slope that varies

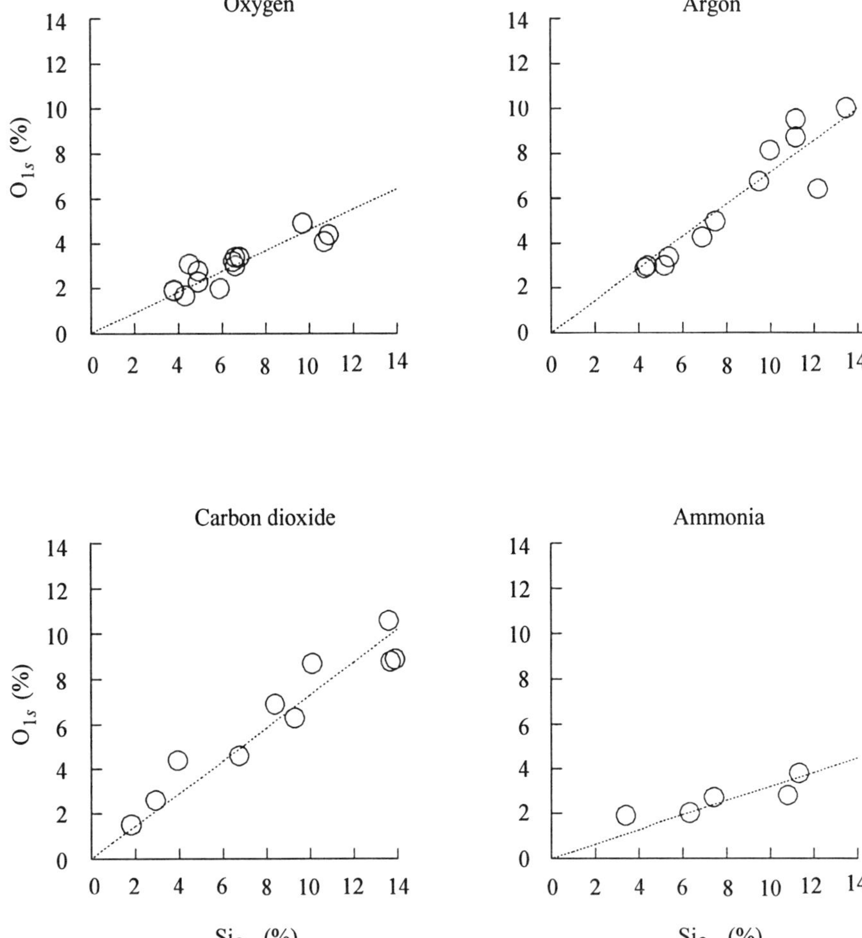

Figure 9. The component of the O_{1s} peak at 533.2 eV as a function of the new Si_{2p} component at 103.4 eV of oxygen, argon, carbon dioxide, and ammonia plasma-treated silicone rubber. Linear regression analyses of the data yielded the dotted lines, with slopes of 0.46, 0.72, 0.73, and 0.32 for oxygen, argon, carbon dioxide, and ammonia, respectively.

per gas (see Fig. 9). This suggests that new Si—O—Si bonds have been formed at the treated surface, by condensation of silanol groups [19] leading to bridging between siloxane chains, but that the extent to which this occurs differs per gas.

4.3. Surface roughness

The effects of plasma treatments on the R_A values of repeatedly RF plasma-treated silicone rubber are minor, irrespective of the gas used. On average, after six repeated treatments, R_A increased slightly by 0.1 μm, which is not enough to account for the changes in contact angle hysteresis encountered. Obviously occasional cracks are not reflected in the R_A values. Although cracks spanning RF plasma-treated silicone rubber surfaces have been suggested to be a starting point for migration mechanisms during hydrophobic recovery [6], this suggestion is likely inadequate for explaining why the time scale of hydrophobic recovery differs per gas used.

4.4. Streaming potentials

The zeta potentials of untreated silicone rubber as derived from streaming potentials under the conditions employed are negative. No stable zeta potentials could be measured for the plasma-treated surfaces, possibly because the surface layers on silicone rubber created by RF plasma treatment are partially water-washable. Although not pursued any further, the change in time of streaming potentials of RF plasma-treated silicone rubber during measurement might be a useful tool for studying the exact physico-chemical nature of the layer.

5. CONCLUSION

Single RF plasma treatment of silicone rubber usually yields transient effects due to, amongst others, the high mobility of the siloxane backbone and the migration of untreated polymer chains from the bulk to the surface during hydrophobic recovery. Also, the effects are described as being similar for different gases [6].

This study shows that the hydrophobic recovery of silicone rubber after RF plasma treatment can be slowed down considerably by repeating the treatment, whereas the kinetics of hydrophobic recovery after repeated plasma treatment differ per gas, indicating that there is an effect of the type of gas.

No evidence was found that the difference in the degree of cracking of surfaces treated by various gases could account for the difference in the degree of hydrophobic recovery observed. Thus, it is suggested that the various gases used for plasma treatment of silicone rubber surfaces yield top layers on the silicone rubber with different physico-chemical properties.

To analyze further the top layer created by plasma treatment, we examined the top layer of ammonia plasma-treated silicone rubber by angle-resolved XPS. Ammonia was chosen for this single-run experiment as it yielded the incorporation of nitrogen, an element normally not found in silicone rubber. As the presence of carbon contamination on surfaces often obscures the interpretation of angle-resolved XPS data, untreated silicone rubber was also examined as a control. Figure 10 shows that the composition of untreated silicone rubber was similar over the range of take-off angles employed, whereas a clear accumulation of elements due to the plasma treatment in the deeper surface layers can be seen. Interestingly, also the relative prevalence of the new component in the Si_{2p} peak indicative of the new Si–O–Si or silica-like bonds is most pronounced in the deeper surface layer (lowest take-off angle). These data can be taken as evidence to support the hypothesis that hydrophobic recovery is due to the migration of untreated chains from the bulk to the surface.

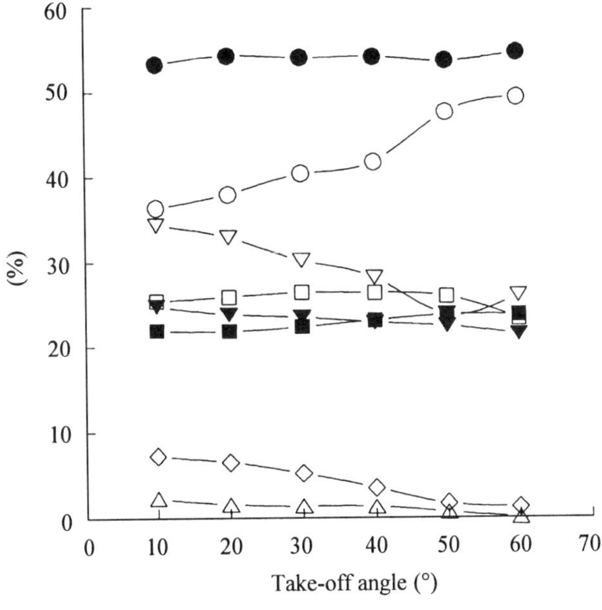

(○) Carbon (▽) Oxygen (□) Silicon (△) Nitrogen

(◇) Si_{2p} at 103.4 eV (3/2) and 104.1 eV (1/2)

Figure 10. Elemental surface composition of untreated silicone rubber and ammonia RF plasma-treated silicone rubber (50 W, 3.7 Torr, 60 s) as a function of the electron take-off angle. The closed symbols represent the values for untreated silicone rubber.

Acknowledgement

This project was funded, in part, by a grant from the European Community, Eureka project EU72310.

REFERENCES

1. J. D. Cooper, F. G. Pearson, G. A. Patterson, T. R. J. Todd, R. J. Ginsberg, M. Golberg and P. Waters, *Ann. Thorac. Surg.* **47**, 371–378 (1989).
2. T. R. Neu, H. C. van der Mei, H. J. Busscher, F. Dijk and G. J. Verkerke, *Biomaterials* **14**, 459–464 (1993).
3. B. F. Farber and A. G. Wolff, *J. Biomed. Mater. Res.* **27**, 599–602 (1993).
4. F. J. Holly and M. J. Owen, in: *Physicochemical Aspects of Polymer Surfaces*, K. L. Mittal (Ed.), Vol. 2, pp. 625–636. Plenum Press, New York (1983).
5. L.-O. Andersson, C.-L. Golander and S. Persson, *J. Adhesion Sci. Technol.* **8**, 117–132 (1994).
6. M. J. Owen and P. J. Smith, *J. Adhesion Sci. Technol.* **8**, 1063–1075 (1994).
7. M. Morra, E. Occhiello, R. Marola, F. Garbassi, P. Humphrey and D. Johnson, *J. Colloid Interface Sci.* **137**, 11–24 (1990).
8. P. M. Triolo and J. D. Andrade, *J. Biomed. Mater. Res.* **17**, 129–147 (1983).
9. M. T. Stewart and M. W. Urban, *Polym. Mater. Sci. Eng.* **59**, 334–338 (1988).
10. S. R. Gaboury and M. W. Urban, *Polym. Commun.* **32**, 390–392 (1990).

11. H. C. van der Mei, I. Stokroos, J. M. Schakenraad and H. J. Busscher, *J. Adhesion Sci. Technol.* **5**, 757–769 (1991).
12. M. I. van Dyke, H. Lee and J. T. Trevors, *Biotechnol. Adv.* **9**, 241–252 (1991).
13. G. Beamson and D. Briggs, *High Resolution XPS of Organic Polymers. The Scienta ESCA 300 Database*, pp. 268–269. John Wiley, Chichester (1992).
14. R. A. van Wagenen and J. D. Andrade, *J. Colloid Interface Sci.* **76**, 305–314 (1980).
15. H. J. Busscher, A. W. J. Van Pelt, P. De Boer, H. P. De Jong and J. Arends, *Colloids Surfaces* **9**, 319–331 (1984).
16. M. J. Owen, *J. Appl. Polym. Sci.* **35**, 895–901 (1988).
17. B. D. Ratner, D. Leach-Scampavia and D. G. Castner, *Biomaterials* **14**, 148–152 (1993).
18. M. L. Raimondi, C. Sassara, I. R. Bellobono and L. Matturri, *J. Biomed. Mater. Res.* **29**, 59–63 (1995).
19. J. R. Hollahan and G. L. Carlson, *J. Appl. Polym. Sci.* **14**, 2499–2508 (1970).

Polymer Surface Modification: Relevance to Adhesion, pp. 49–72
K. L. Mittal (Ed.)

The improvement in adhesion of polyurethane–polypropylene composites by short-time exposure of polypropylene to low and atmospheric pressure plasmas

JÖRG FLORIAN FRIEDRICH,[1,*] WOLFGANG UNGER,[2] ANDREAS LIPPITZ,[2] THOMAS GROSS,[2] PAUL ROHRER,[3] WOLFGANG SAUR,[3] JÜRGEN ERDMANN[4] and HANS-VIKTOR GORSLER[4]

[1]*Institut für Angewandte Chemie Berlin-Adlershof, Rudower Chaussee 5 (Haus 4.1), D-12484 Berlin-Adlershof, Germany*

[2]*Bundesanstalt für Materialforschung und -prüfung (BAM), Labor 5.33, Unter den Eichen 80–85, D-12205 Berlin, Germany*

[3]*Gurit-Essex, AG CH-8807 Freienbach, Switzerland*

[4]*Ahlbrandt System GmbH, Vogelsbergstraße 45–47, D-36339 Lauterbach, Germany*

Revised version received 1 August 1994

Abstract—The surfaces of polymers, namely polypropylene, copolymers and blends, were exposed to low pressure oxygen and atmospheric pressure air plasmas to improve their adhesion to polyurethane adhesives. A correlation is attempted between lap shear strengths of polypropylene–polyurethane composites and the relevant XPS, AFM and NEXAFS data. It was found that plasma functionalization improved the adhesion to maximum values even when the time of exposure was low: 1 to 10 seconds for low pressure plasmas, and 0.1 to 1 seconds in case of atmospheric plasma jet treatments. Thus, high lap shear strengths were obtained at relatively small oxygen contents. The improvement in shear strength at short time plasma exposures seems to be correlated to the complete smoothening of the supermolecular structure of stretched polypropylene foils as shown by AFM. Valence band XPS and derivatization techniques revealed more details of the oxygen functionalization on polypropylene. NEXAFS experiments confirmed a re-orientation of bonds and segments of the macromolecules by plasma exposure which are assumed to be responsible for adhesion improvement.

Keywords: Polyurethane–polypropylene adhesion; low pressure plasma pretreatment; atmospheric pressure plasma treatment; short-time exposure; relevance of functional groups to adhesion; molecular orientation at polymer surfaces.

*To whom correspondence should be addressed.

1. INTRODUCTION

Low pressure glow discharge techniques are well established in improving the wettability and adhesion of low surface energy polymers. In the past, significant efforts were made to use this type of plasma modification to improve the adhesion strength in polymer composites. We have utilized this very efficient method to increase the adhesion of polyurethane adhesives to polypropylene, polyethylene and copolymers.

However, because of the batch-wise vacuum process associated with its high cost and incompatibility to automated processes, the use of the low pressure glow discharge technology has found only limited interest within industry (e.g. automotive). Therefore, we evaluated different kinds of hyposonic jets (arc, corona, or spark discharges) at ambient conditions. The spark jet provides maximum adhesion strength results. It works at rather mild conditions, e.g. at gas temperatures between 60 and 80 °C. Short exposure times together with the rather low working temperature provide a favourable opportunity to apply the spark jet plasma to polyolefin pretreatment.

The results of adhesion strength improvement by low pressure plasma modification are briefly reported as well as the dependence of its efficiency on some important plasma or process parameters and the correlation between lap shear strength measurements and selected surface analytical data. Relying on the experience gained in the investigation of low pressure plasma application, an attempt is made to interpret the results of the atmospheric pressure plasma treatments.

It is often reported that plasma treatment of a polymer surface results in the formation of a broad range of different surface functional groups and radicals. These groups are viewed to be very important within the framework of several models for adhesion phenomena [1]. For a more detailed study we selected polypropylene foils and thicker sheets as the substrate and polyurethane as the adhesive.

XPS (X-ray photoelectron spectroscopy) is known to be a very powerful tool when quantitative data on the overall functionalization of a polymer surface are of interest [2]. The presence of different functionalities can be detected by peak fitting the respective C 1s photoelectron spectra. Owing to the lack of specificity in corelevel binding energies of carbon (many functional groups give rise to similar C 1s binding energy shifts) it is necessary to overcome this problem by using derivatizing agents which react selectively with a functional group or radical and label it with a distinctive element. This method for enhancing the analytical capabilities of XPS has been reviewed by Chilkoti and Ratner [3], and Batich [4]. Phenyl isocyanate and *n*-butyl isocyanate are used as derivatizing reagents to measure the concentration of OH functionalities on a plasma treated polypropylene surface. From another point of view, this labelling reaction with short-chain isocyanates is a model for the reaction of isocyanate end groups of the polyurethane prepolymer adhesive with a plasma treated polypropylene surface. The principal pathways for reaction between *n*-butyl isocyanate and functional groups were investigated by IR in [5]. Another approach to identify functionalities on a polymer surface is to analyze the photoemission valence band region spectra [6, 7]. A typical valence band region spectrum of a polymer represents the photoelectrons emitted from delocalized or bonding molecular orbitals. NEXAFS (near edge X-ray absorption fine structure) spectroscopy is a surface analytical method which provides specific molecular information (qualitative analysis) together with data on the orientation of macromolecules (or parts of these) [8]. Application of this method is very attractive for the characterization of plasma-treated polymer surfaces because orientation effects are important in adhesion phenomena. An indication of the important role of molecular orientation effects in the resulting surface energy of polypropylene was reported by Andre *et al.* [9]. It was found from contact angle measurements that the surface free energy was increased even only at millisecond exposure to an oxygen plasma by increasing the dispersive part of the surface energy [9]. We interpret this unusual phenomenon to be associated with molecular and supermolecular rearrangements of the macromolecules. At a millisecond treatment the number of polar groups was found to be rather small [9]. We have used AFM (atomic force microscopy) to provide information on changes in the supermolecular structures on polypropylene surfaces after short-time exposure to low and atmospheric oxygen plasmas.

2. EXPERIMENTAL

2.1. Plasma treatments

The low pressure glow discharge technique is well established and is described elsewhere [1]. The plasma equipment is schematically shown in Fig. 1(a). The d.c. treatment was carried out within a reactor based on a Leybold Univex 300 system (cf. [10, 11]).

The plasma was generated as d.c. glow discharge where low energy inputs and low pressure (6 Pa) at medium gas consumption (3.5 $dm^3/h \sim 60$ sccm) were used. The polymer substrates rotate on a cylindrical sample holder through the plasma zone and the dark space in the plasma chamber. Thus the exposure time was determined from the period of residence in the plasma zone.

The atmospheric plasma jets were generated using different types of plasmas (arc, corona, and spark, see Fig. 1(b)). The principle of a plasma jet is that the plasma gas (air, oxygen $\rightarrow$ 'spark, corona' and Ar, N_2, H_2 $\rightarrow$ 'arc') is streaming with a

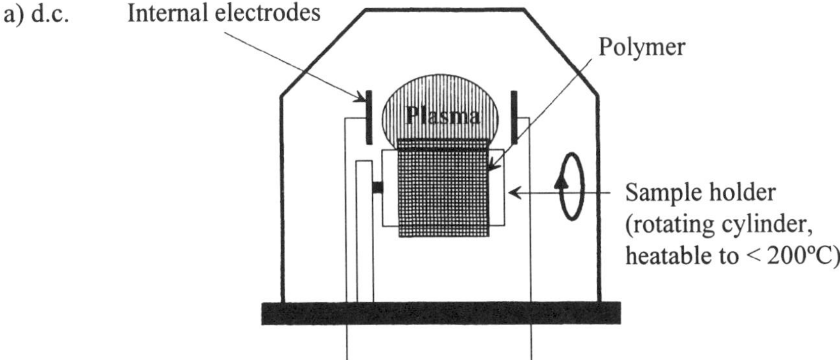

Figure 1(a). Low-pressure plasma reactor configuration (the heat source is placed within the rotating sample holder).

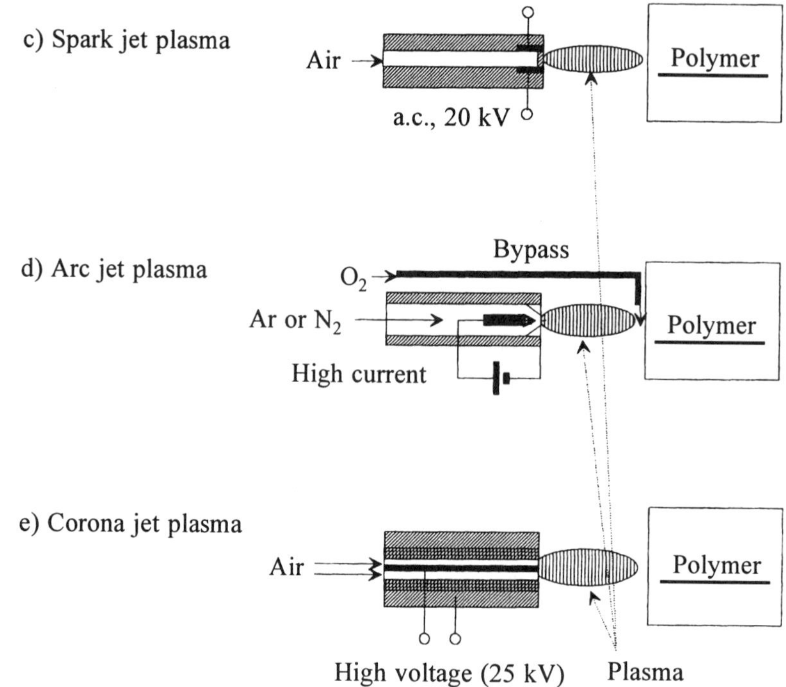

Figure 1(b). Schematics of different atmospheric plasma jets devices evaluated in our investigations.

very high velocity through the electrode zone generating a 'plasma flame'. The gas temperature, an important jet parameter, is *ca.* 80 to 150°C for corona and spark discharges and 7000 to 14 000°C for the arc plasma [12]. Spark discharge offers the most suitable parameters for polymer surface modification. At exposure times leading to high adhesion strength, i.e. in the range 0.01 to 2 seconds, we measured temperatures between 60 and 80°C. Different plasma jet configurations are shown in Fig. 1(b). The sample surface was scanned by the plasma jet at a linear speed of 320 mm/s. The activated zone was 15 × 50 mm. The known disadvantage of the plasma jet approach, i.e. the formation of the so-called streamer craters, was minimized in our experimental approach using the '3d Treater' by Ahlbrandt System GmbH (Germany).

Relevant plasma and process parameters of the atmospheric and the low pressure plasma setups together with information on sample handling are listed in Table 1.

2.2. Adhesion specimens, and samples for analytical purposes

Two plasma modified polymer specimens were bonded with an overlapping bond line using a polyurethane adhesive in laboratory air. The overlapped area was 2.5 cm^2, the bond line thickness was 1 mm. The isocyanate-terminated polyurethane adhesive was cured under ambient humidity. The specimens were aged 7 days at 70°C and 100% humidity and subjected to an additional cold shock at −20°C for 16 h (*Cataplasma* artificial ageing). The lap shear tests were performed at 180°C. Many commercial

Table 1.
Plasma conditions and sample preparation

Parameters	Glow discharge	Spark jet
Generation	d.c. (1 kV) r.f. (13.6 MHz)	a.c. (20 kV, 25 kHz)
Electrodes	internal or external (capacitively coupled)	steel or tungsten (gap 10 mm)
Power	5–60 W	500 W
Pressure	6 Pa	10^5 Pa
Flow rate	4 litre/h	*ca.* 5000 litre/h
Gas	group 1 (surface functionalizing plasmas): oxygen, ammonia, argon etc. (cf. Fig. 2) group 2 (plasma polymer depositing monomers): silanes, ethanol	air, oxygen
Sample position	within the 'positive column' — floating potential	10 mm from the spark in the zone of 'diffuse plasma'
Sample holder	rotating cylinder (0.5 cycles/s)	rotating cylinder (2 cycles/s)
Sample cleaning	rinsing with ether (analytical samples) dipping in *n*-hexane (lap shear strength specimen)	rinsing with ether (analytical samples) dipping in *n*-hexane (lap shear strength specimen)
Cleaning after plasma treatment	with and without rinsing with acetone	with and without rinsing with acetone
Plasma exposure time	*ca.* 1–3 s	*ca.* 0.1–1 s

polyolefins, other polymers, copolymers, blends, fibre or mineral filled composites were subjected to this test.

20 μm (Hercules) and 30 μm thick (Hoechst) polypropylene foils were used as model substrates for XPS, AFM and NEXAFS investigations. By using XPS we found the Hercules foil surface to be free of surface contamination. The Hoechst foils were cleaned before plasma treatment by rinsing in ether because O and Si signals were detected in the XPS spectra of 'as-received' samples. The effect of a post-plasma treatment rinsing (in acetone) of short-time treated polypropylene on oxygen content was checked but no effect was found by XPS. Long-time plasma exposed samples, investigated for comparison, were washed in acetone to remove weakly bonded polar fragments of molecular chains at the surface.

2.3. XPS analysis

XPS measurements were carried out with a VG Scientific ESCALAB 200X electron spectrometer. XPS narrow scan spectra were recorded using Mg Kα excitation (300 W) in the constant retarding ratio mode. The spectrometer energy scale was calibrated by employing the procedure and binding energy (BE) reference data recommended by Anthony and Seah [13].

The polypropylene foils were prepared on standard VG sample stubs with the help of double-sided adhesive tapes. Before recording the spectra the samples were degassed in a separate chamber (extended VG preplock) of the ESCALAB (oil free vacuum) at 10^{-5} Pa. Analysis of the XP spectra was done by peak fitting using the data processing unit of the VGS 5250 data system. C 1s spectra were fitted with symmetric Gauss–Lorentz product functions after subtraction of non-linear backgrounds by applying the Shirley method [14]. For the oxygen plasma treated samples a maximum number of four peaks (CH_x, C–O, C=O and O–C=O) was used to fit the spectra.

Derivatization of a plasma-treated polypropylene sample was carried out immediately after treatment within the reactor chamber (3 h, 10 to 100 Pa vapour pressure of derivatizing agent, room temperature). Phenyl isocyanate and *n*-butyl isocyanate react with surface hydroxyl groups forming urethane species thus giving a N label for the surface OH groups. The derivatizing reactions are as follows:

$$\text{phenyl}{-}\text{N}{=}\text{C}{=}\text{O} + \text{OH}{-}\text{polymer} \rightarrow \text{phenyl}{-}\text{NH}{-}\text{CO}{-}\text{O}{-}\text{polymer}$$

$$\text{CH}_3{-}\text{CH}_2{-}\text{CH}_2{-}\text{CH}_2{-}\text{N}{=}\text{C}{=}\text{O} + \text{OH}{-}\text{polymer} \rightarrow$$

$$\text{CH}_3{-}\text{CH}_2{-}\text{CH}_2{-}\text{CH}_2{-}\text{NH}{-}\text{CO}{-}\text{O}{-}\text{polymer}$$

A concurrent reaction of *n*-butyl isocyanate with carboxylic groups (*ca.* 25% of the total consumption of the isocyanate at 140°C) was reported in an IR spectroscopy study [5]. Therefore, we checked our results by applying the trifluoroacetic anhydride (TFAA) derivatization method which provides a F labelling for the surface OH groups [2].

2.4. Valence band spectra

Valence band spectra were obtained using a Surface Science Instruments SSX 100 photoelectron spectrometer. We used monochromatised Al Kα radiation, a fixed energy pass at 50 eV and a take-off angle of 35°. At these conditions the Au $4f_{7/2}$ full width at half maximum was 1.08 eV and the binding energy was found to be 83.98 eV. The spectrometer energy scale of this instrument was calibrated following [12]. To reduce the charging, a flood gun releasing 2 eV electrons was used. Additionally the analysis chamber of the spectrometer was filled with high purity (99.9999%) Ar to 3×10^{-5} Pa. Under these conditions the charging can be minimised (cf. [15]). A carbon 1s binding energy of 284.8 eV was used for static charge referencing. The valence band spectra were fitted by Gauss–Lorentz sum functions where the Gaussian fraction was fixed at 80%. A linear background was subtracted before peak fitting. The number of peaks to be fitted was decided based on the respective theoretical and experimental results published in [4, 14].

2.5. NEXAFS analysis

The NEXAFS experiments were carried out on the HE-PGM2 beamline of the synchrotron storage ring device BESSY I (Berlin, Germany) at a resolution better than 0.8 eV at the C 1s edge. The angle of incidence of the X-ray photons was varied between 90° (E-vector in the surface plane, 'normal incidence') and 20° (E-vector nearly parallel to the surface normal, 'grazing incidence'). Calibration of the energy scale was performed by recording the Au 4d and Cu 2p absorption edges in first and second order, respectively. The C edge spectra were taken in the Total Electron Yield mode (TEY) and were normalised to the incident photon flux by dividing the raw data by the monochromator transmission function, which was determined for a freshly sputtered, clean gold surface.

2.6. AFM investigations

The surface topography of plasma-treated polypropylene films monitored by AFM. The AF microscope was an air-AFM (Omicron-Vakuumphysik GmbH). The cantilever (ultralever, feather constant 0.03...0.06 N/m, top radius 10 nm) was used in the constant force mode of some mN.

3. RESULTS

3.1. Lap shear strength and process parameter

Only a short time (0.1 to 10 s) of *low pressure plasma* exposure was required to achieve optimum results regarding the initial adhesion as well as the performance after heat and humidity ageing. This holds true for different plasma gases. Important results are summarised in Fig. 2.

Similar high lap shear strengths were measured when the *plasma jet* pretreatment was used (cf. Fig. 3). The exposure to oxygen containing atmospheric pressure plasmas results in cohesive failure in the polyurethane adhesive of the sheared polypropylene–polyurethane composites. The addition of chemically inert gases such as nitrogen or argon to the air jet plasma was found to lower the resulting lap shear strengths.

The oxygen content increases quickly with exposure time (Fig. 4). Considering the influence of the plasma induced oxygen content on the adhesion of polyurethane to polypropylene it was found that rather low oxygen contents (1.5 to 5 at%) in the surface layer gave high lap shear strengths. Therefore, the optimum treatment times are in the range of 0.05 to 2 s for the *atmospheric spark jet* and 1 to 10 s for the *low-pressure glow discharge* in oxygen (cf. Fig. 4).

To establish a process parameter field the influence of selected parameters on the resulting lap shear strengths was analyzed. In Figs 5(a) and 5(b), two-parameter maps, correlating lap shear strengths, plasma exposure times and XPS oxygen surface concentration, are presented which were helpful for further process parameter optimisation. As shown in Fig. 5(a) the oxygen low pressure plasma possesses a broad

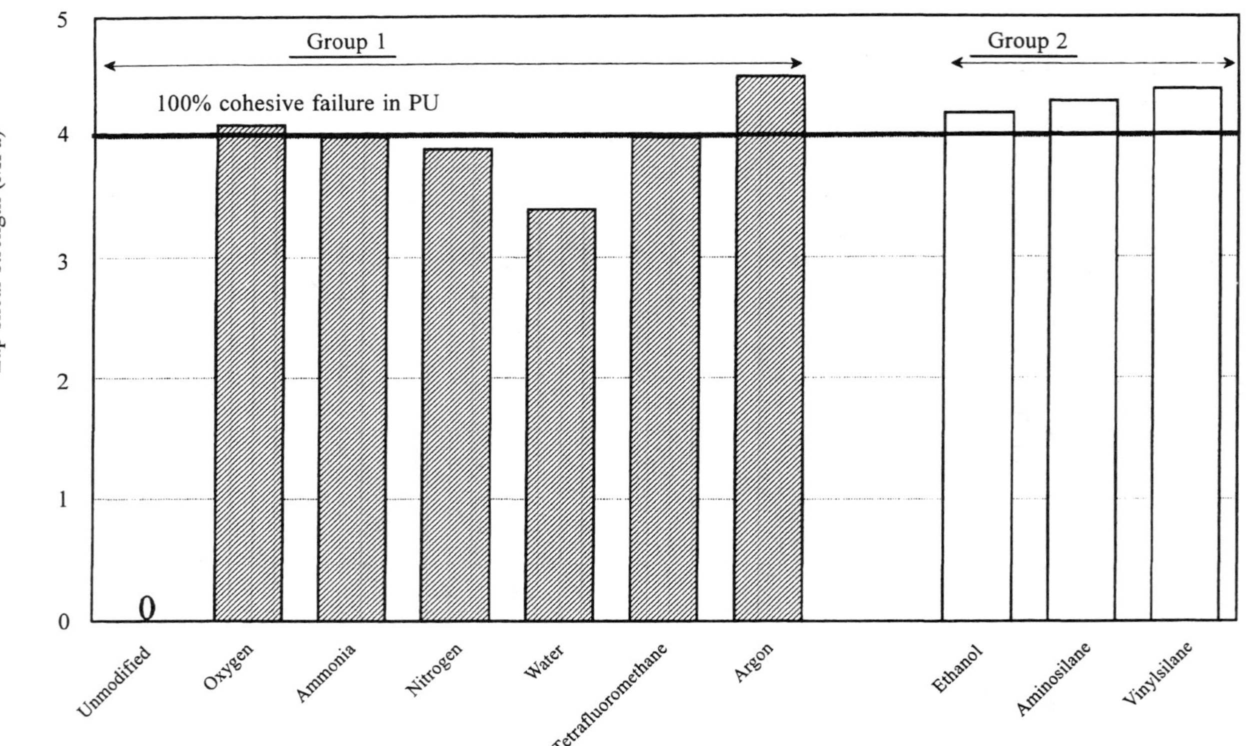

Figure 2. Lap shear strengths of low pressure plasma pretreated polypropylene (8 s) coated with polyurethane (first group: modification by functionalization (O_2, NH_3, N_2, H_2O, CF_4), etching (O_2) and crosslinking (Ar); second group: modification by deposition of plasma polymers (ethanol, γ-aminopropyltriethoxy silane, vinyltriethoxy silane; 18 s)).

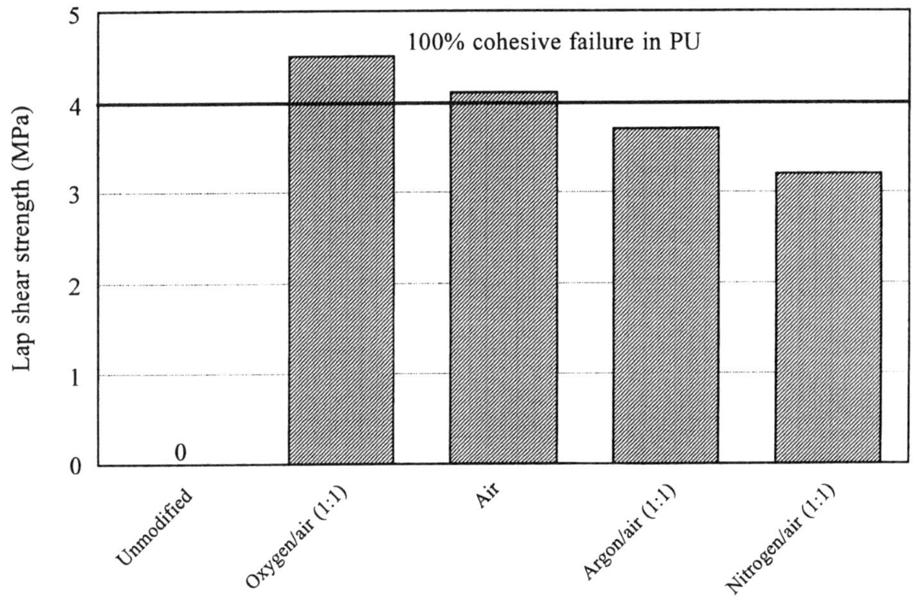

Figure 3. Lap shear strengths of atmospheric pressure plasma pretreated polypropylene (1 s) coated with polyurethane.

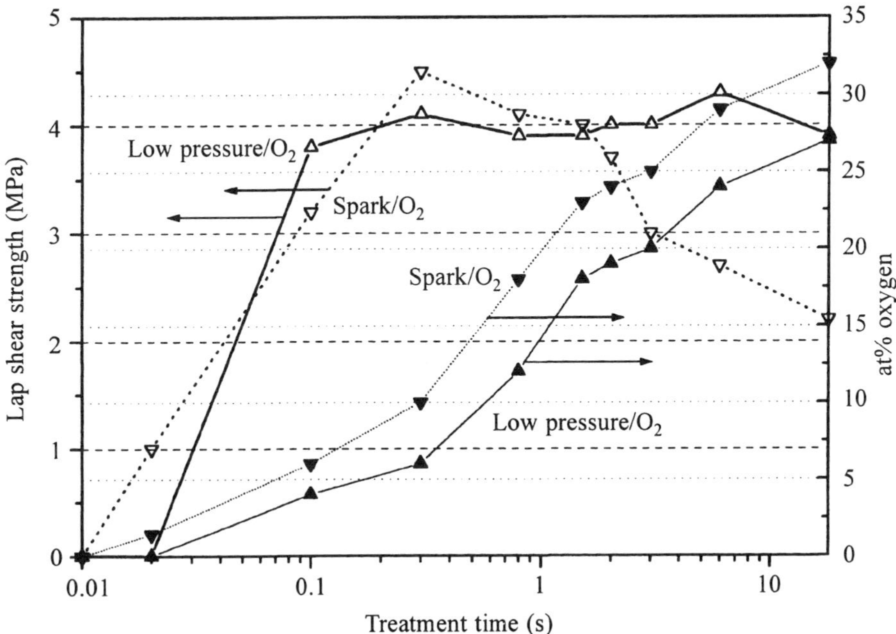

Figure 4. Dependence of lap shear strengths for polypropylene–polyurethane specimens and oxygen surface concentration (XPS data) on the treatment time.

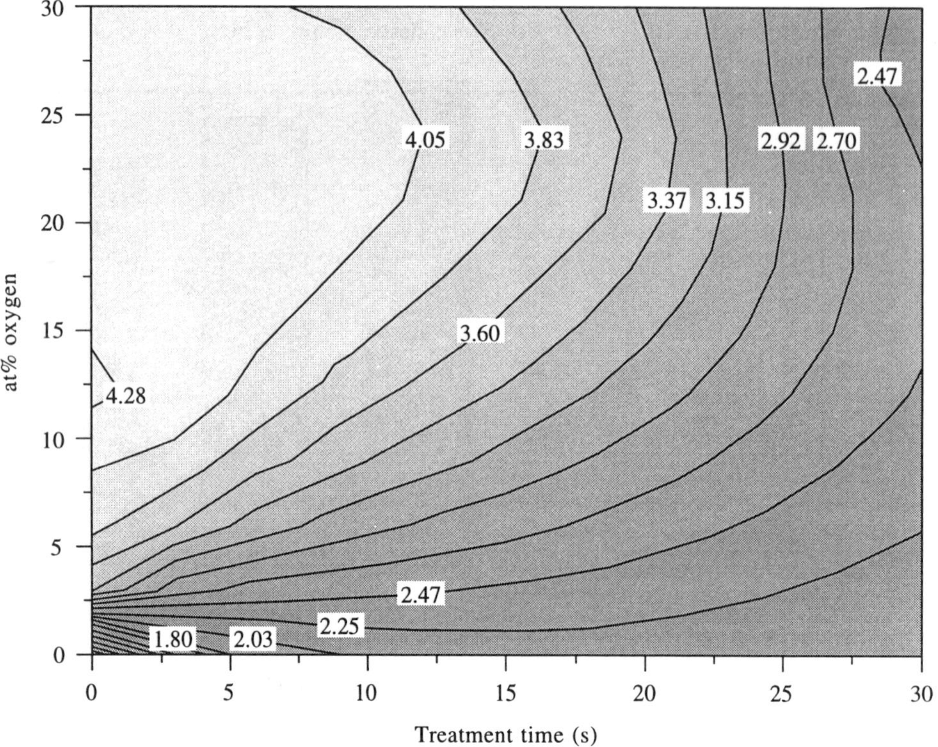

Figure 5(a). Correlation between a process parameter (treatment time), the XPS elemental analysis of the polypropylene surface (at% oxygen) and the results of mechanical testing (lap shear strength, numbers in MPa) for the case of oxygen low pressure plasma modification.

field of maximum lap shear strength (numbers shown in the diagram represent lap shear strength data in MPa) in dependence on treatment time and oxygen content. The 1:1 oxygen/air mixture in the *atmospheric plasma* treatment process has also a broad range of treatment times and oxygen contents with high lap shear strengths (Fig. 5(b)).

Technologically important polymer substrates were pretreated by low pressure plasma: polypropylene (PP), polyethylene (PE), polycarbonate (PC) and polybutyleneterephthalate (PBT); different blends: e.g. PP–PE, PE–PBT, PBT–RTS (RTS = synthetic rubber), PC–ABS (ABS = acrylnitrile–butadiene–styrene copolymer); elastomers: polypropylene elastomer, EPDM (terpolymer of ethylene and propylene); and fibre (GF = glass fibre) or mineral filled materials. As shown in Fig. 6 the *low pressure plasma* pretreatment, carried out with oxygen, results in rather high lap shear strengths for many different polymer–polyurethane combinations. For nearly all composites mentioned in Fig. 6 the adhesion strength is enhanced from zero to a maximum lap shear strength of about 4 MPa. This value is identical to the cohesive strength of the polyurethane adhesive itself. The *spark jet* approach was also applied but only to pure polypropylene, copolymers and some mineral filled polypropylene (Fig. 6).

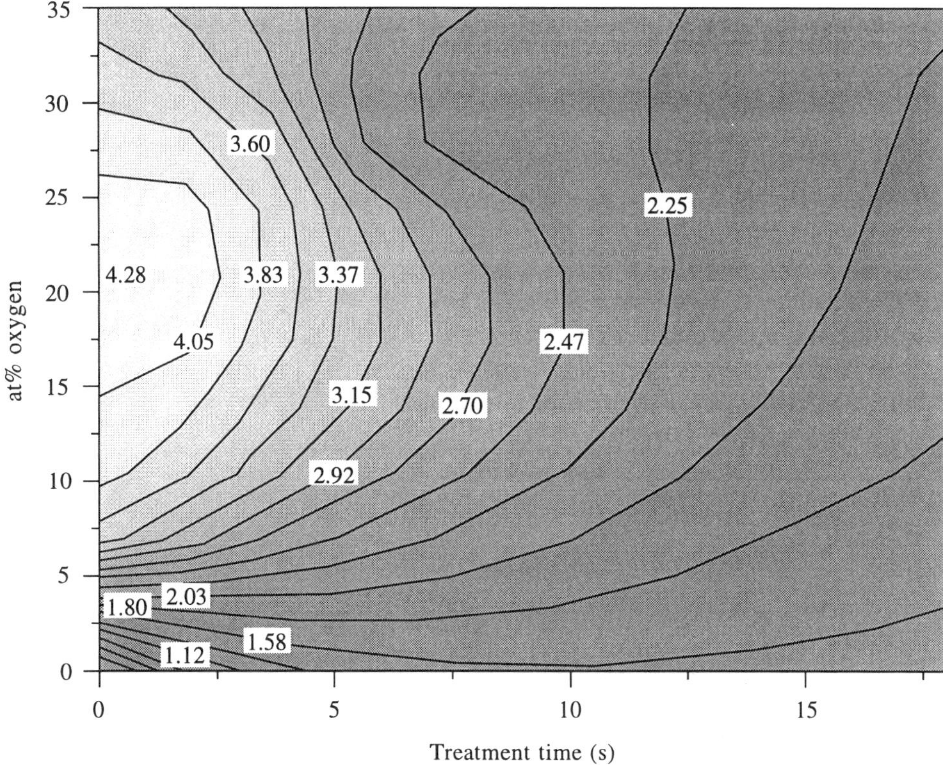

Figure 5(b). Correlation between a process parameter (treatment time), the elemental analysis of the polypropylene surface (XPS data, at% oxygen) and the results of mechanical testing (lap shear strength, numbers in MPa) for the case of spark jet modification.

3.2. Surface analysis

A qualitative comparison of the surfaces of low oxygen pressure and atmospheric pressure air plasma treated polymers revealed only slightly different oxygen functionalization effects, varying only a little with the type of polymer.

As known, an analysis of the XPS C 1s and O 1s photoemission signals cannot provide a detailed identification of species at the plasma modified polypropylene surface [2]. This approach is limited to an identification of some types of carbon–oxygen bonds (C–O, C=O, O–C=O). Figures 7(a) and 7(b) represent the C 1s peaks of oxygen low pressure plasma and atmospheric spark jet plasma (O_2/air mixture) modified polypropylene surfaces, respectively, with the difference spectra with untreated polypropylene. The principal positions of the most important functional groups related to C 1s subpeaks are given in Figs 7(a) and 7(b). At equal times of exposure, the effect of an atmospheric plasma on the polypropylene surface is stronger in comparison to the effect of a glow discharge plasma. The same conclusion can also be drawn from the data presented in Fig. 4.

The results of the OH group derivatization (cf. [3]) at plasma treated polypropylene surfaces by employing low molecular isocyanates and TFAA are presented in Table 2.

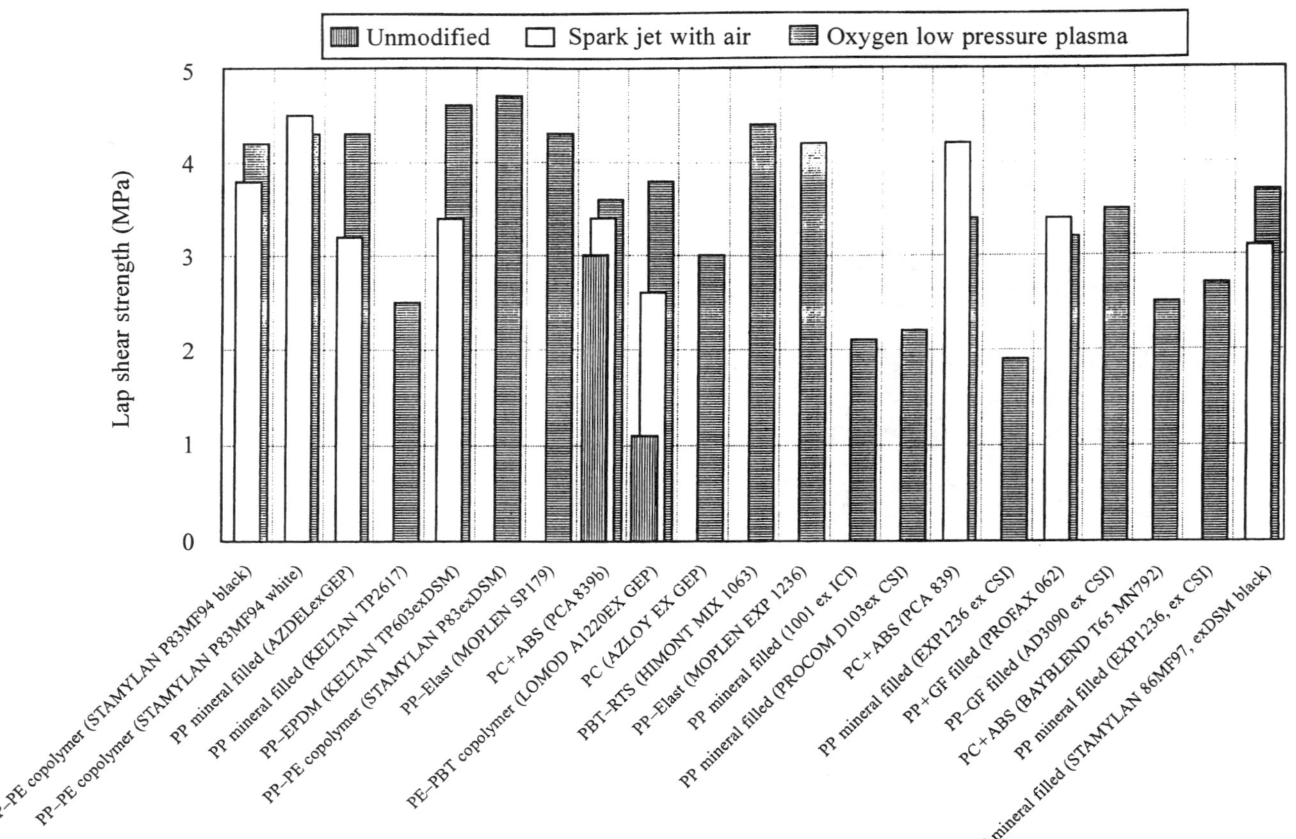

Figure 6. Plasma pretreatment of polypropylene, other polymers, blends and mineral and fibre filled composites (oxygen low pressure plasma: 8 s, 6 Pa, d.c. type; spark jet in air, 1 s).

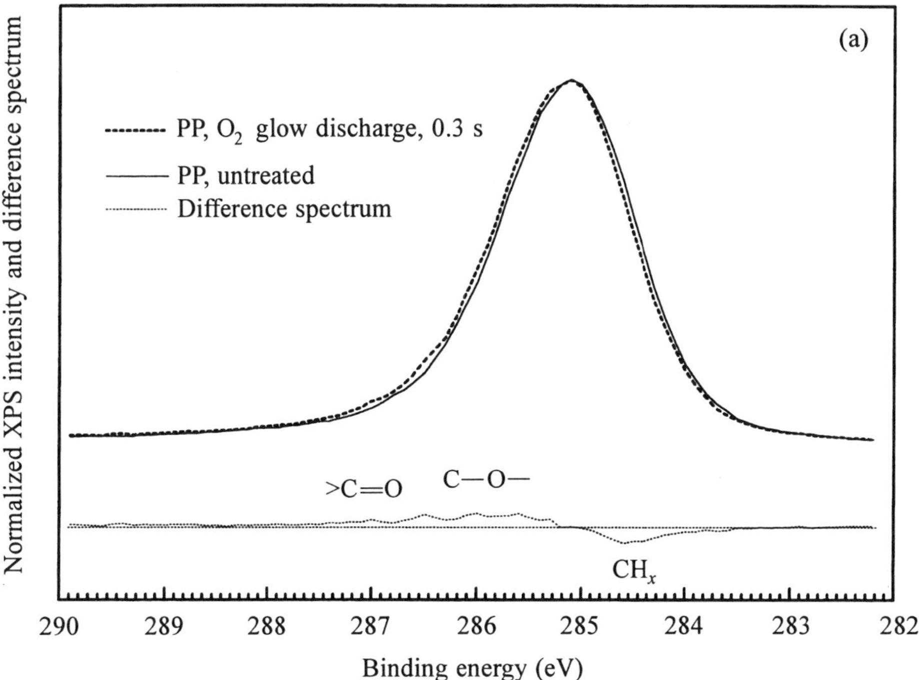

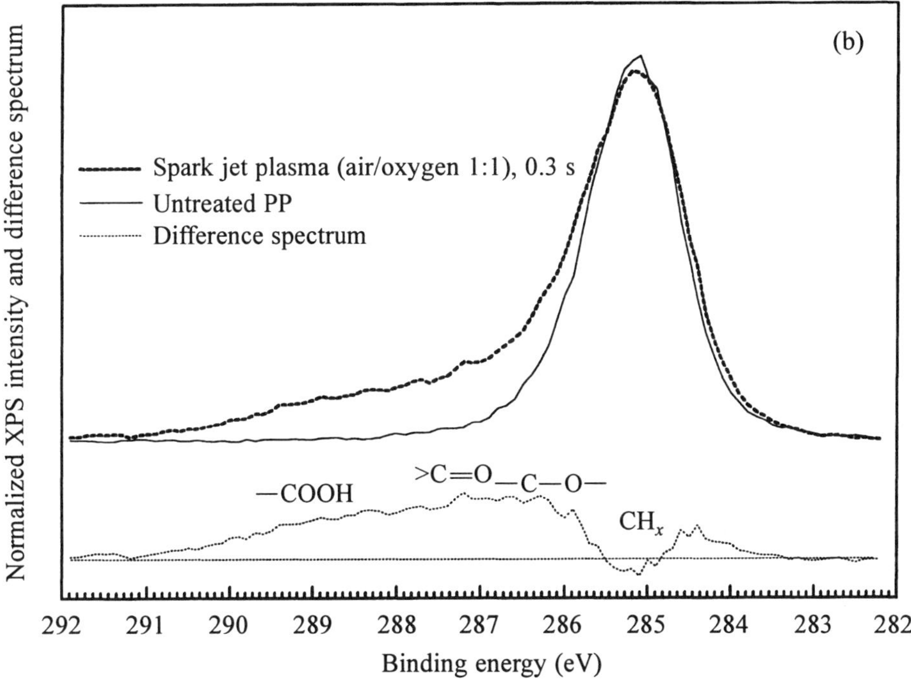

Figure 7. C 1s XPS peaks and difference spectra for plasma treated and untreated polypropylene surfaces (low pressure glow discharge (a) and spark jet (b) treatment times: 0.3 s).

Table 2.
Species at an oxygen d.c. glow discharge treated polypropylene surface (18 s) as obtained by peak fitting the C 1s spectrum and labelling by derivatization with TFAA (trifluoroacetic anhydride), BIC (*n*-butyl isocyanate) and PIC (phenyl isocyanate). The surface concentration ratios obtained after derivatization were not corrected for the effect of derivatizing reagents (this means that the carbon atoms of the derivatizing molecule are additionally involved in the XPS surface analysis)

Sample	O/C surface concentration ratio	O functionalized C atoms (%)				F/C surface concentration ratio	N/C surface concentration ratio
		Total	C−O	C=O	−O−C=O		
PP	0	0	0	0	0	0	0
PP + O_2-plasma	0.28	0.24	0.12	0.08	0.04		
PP + O_2-plasma + TFAA	0.31	0.28				0.05	
PP + O_2-plasma + BIC	0.31	0.28					0.05
PP + O_2-plasma + PIC	0.31	0.27					0.05

Table 3.
Valence band peaks (see Fig. 8) of unmodified and 1 s oxygen low pressure d.c. plasma modified polypropylene

Peak	Binding energy (eV)		Percentage of total intensity	
	unmodified	modified	unmodified	modified
1	18.8	19.1	42.6	21.0
2	13.3	13.1	21.2	18.7
3	16.1	16.1	22.8	27.7
4	9.1	8.9	5.4	5.0
5	6.2	6.1	8.0	4.4
6	—	7.5	—	4.3
7	—	4.9	—	1.8
8	—	17.9	—	5.4
9	—	10.2	—	1.3
10	—	24.9		10.3

3.3. Valence band spectra

More insight into plasma chemical functionalization of the surface of polypropylene was also gained from X-ray excited valence band region spectra. Spectra of untreated polypropylene and short-time low pressure oxygen glow discharge treated polypropylene are presented in Fig. 8. With the clean polypropylene sample we obtained a valence region spectrum very similar to that published by Foerch *et al.* in [6]. In detail (see Table 3) one can see signals related to backbone (polyethylene) C 2s–C 2s bonding (1) and antibonding (2) molecular orbitals, a methyl group

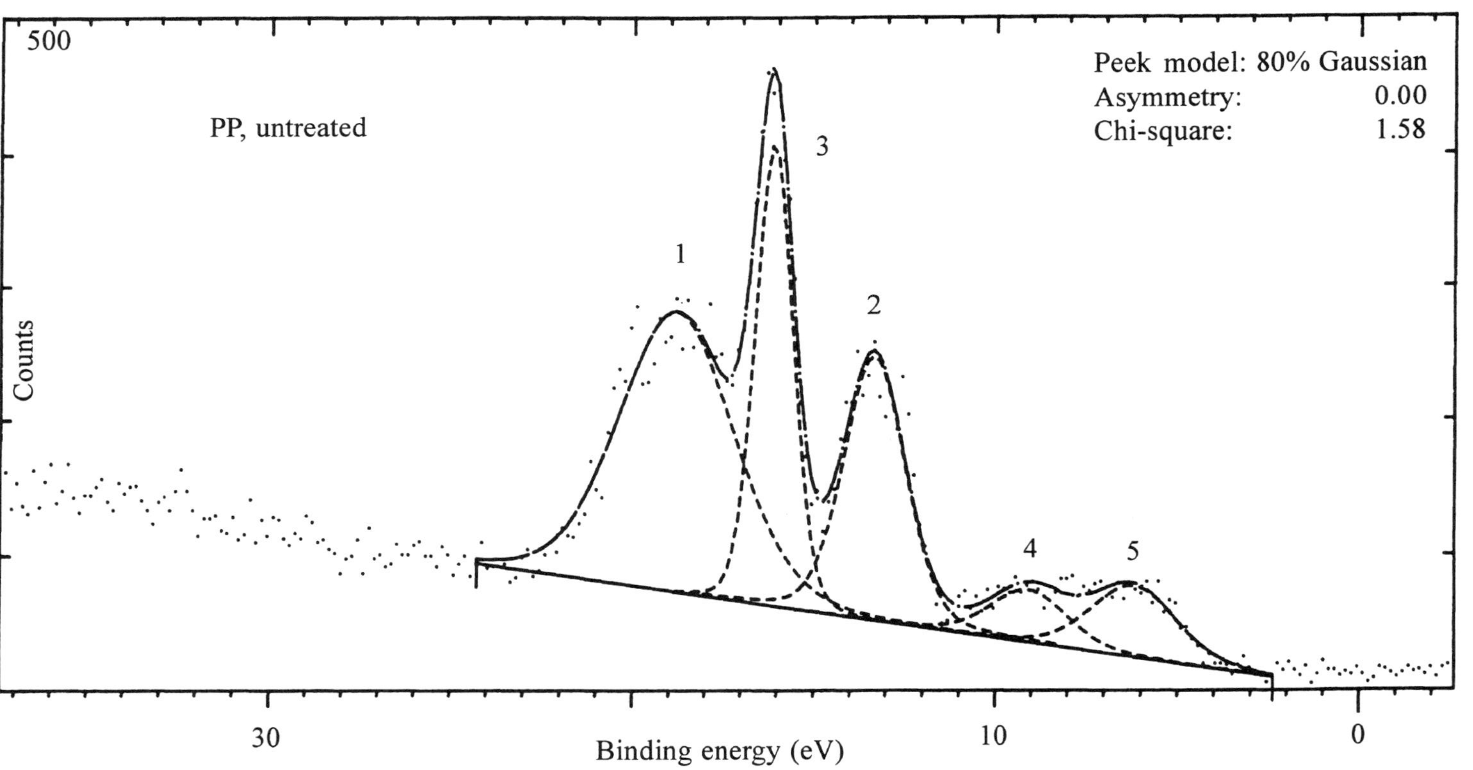

Figure 8(a). Polypropylene valence band spectra (untreated: 1 s low pressure O_2 d.c. plasma).

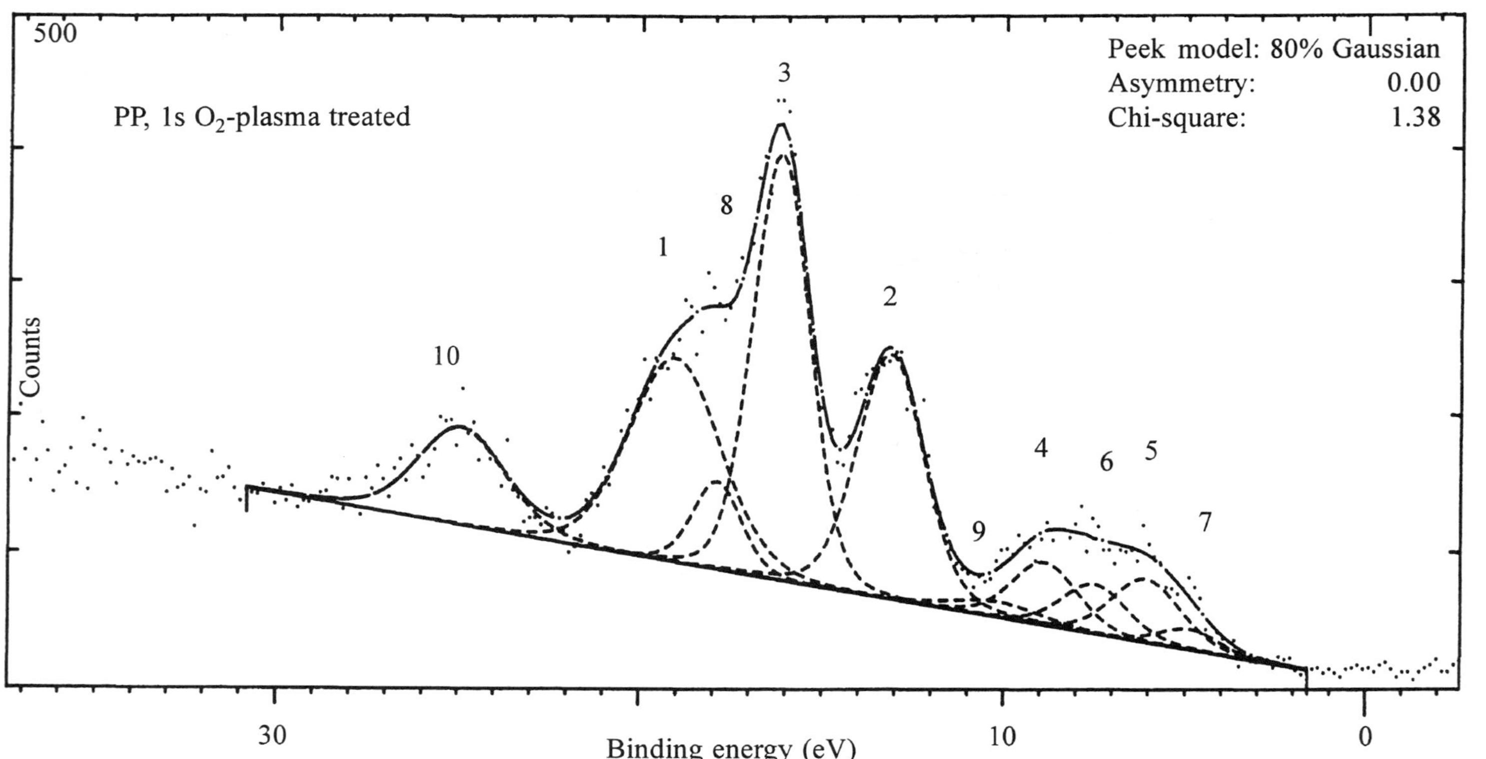

Figure 8(b). Polypropylene valence band spectra (treated: 1 s low pressure O_2 d.c. plasma).

C 2s level (3) and methyl group (4) or backbone (6) C 2p–H 1s bonding interactions. Analysis of the spectrum of the 1s oxygen plasma treated polypropylene surface reveals new peaks together with the peaks observed with untreated polypropylene. The new features were interpreted by employing the ideas of Orti *et al.* outlined in [16]. The most prominent, new, oxygen related peak occurs due to the O 2s electronic level (10). Considering the result of the short time (1 second) low pressure d.c. plasma treatment we find methoxy group (O—C) related features, (5), (7) and (8), and only a very small carbonyl related signal (9). After treatment the probed polypropylene surface layer provides fingerprints of C—O functionalities and a pure polypropylene. This means that the original polypropylene sequences still exist within the information depth of this method ($3\lambda \sim 30$ nm [17]).

3.4. NEXAFS results obtained with plasma treated polypropylene samples

As known, stretched PP foils are partially crystalline in nature. Orientation phenomena at a molecular level can be identified with the help of NEXAFS. The spectra of the untreated PP foil (cf. Fig. 9) reveal a C—H* resonance at 287.5 eV, labelled with '2', also observed with paraffins and used as an energy reference, broad C—C σ^* resonances at photon energies >290 eV, and a C=C related π^* resonance labelled with '1'. The reason for the surprising occurrence of this C=C π^* resonance in the unmodified polypropylene should be the olefinic double bonded end groups terminating the chains. Comparison of the spectra obtained at different electric field vector angle (90° or 20°) relative to the surface normal reveals differences in the C—H and

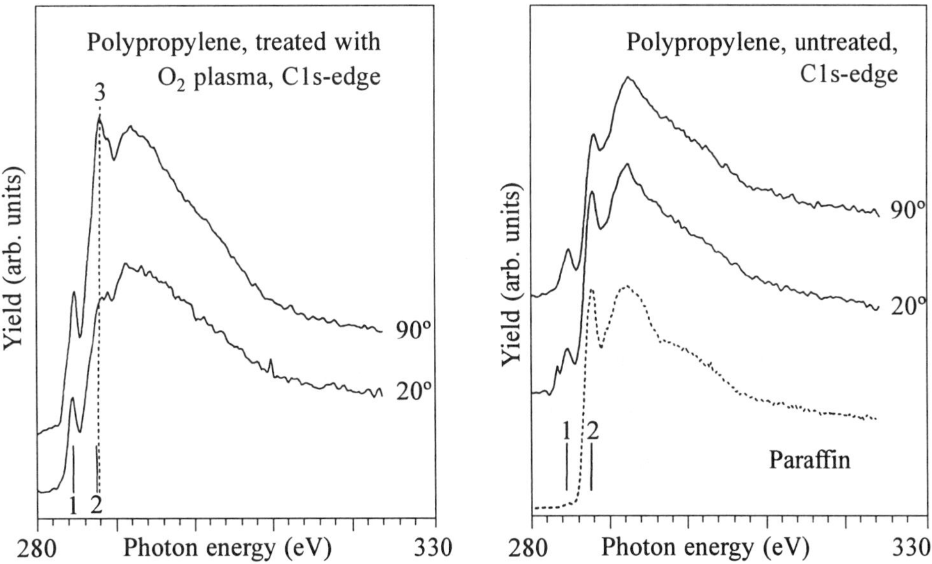

Figure 9. NEXAFS spectra of untreated polypropylene and paraffin (right, incident synchrotron radiation electric field vector angle 90° or 20° relative to the surface normal, paraffin is used as reference for saturated aliphatic molecules); O_2 low pressure plasma modified polypropylene (left, 3 s).

C—C features. The resulting difference spectrum is similar but less pronounced to that presented in [8] for oriented PE. Our data suggest a slightly non-statistical orientation (anisotropy) of macromolecular segments in the surface region (see [18]). NEXAFS on short time treated PP samples reveals a decrease of the C—H* resonance intensity, an increase of the C=C* resonance intensity, and a disappearance of the slight orientation effects. That means the surface region becomes isotropic. Oxygen related features cannot be resolved from the C K-edge spectra because of the co-existence of great number of O functionalized species. C K-edge NEXAFS with a longer treated PP sample (O/C atomic ratio = 0.22, cf. Fig. 9) reveals a significant decrease of the C—H* resonance intensity, a strong increase of the C=C π^* resonance intensity, and the appearance of a new resonance at $\approx$287.8 eV (labelled '3') attributed to a carbonyl bond species [18]. Its intensity maximum at an 'in surface plane' **E** polarisation clearly suggests a preferential orientation of these C=O bonds perpendicularly to the surface because a C 1s $\rightarrow$ C=O π^* transition has maximum intensity when **E** points along the direction of maximal orbital amplitude.

3.5. Surface topography

3-D surface mapping of the polypropylene foil by AFM shows a structure which is typical of biaxially stretched polypropylene films (Fig. 10). The surface roughness amplitude is of the order of 100 nm. Obviously, by the *glow discharge* treatment, the surface topography is efficiently smoothened within a short time (here, 3 seconds).

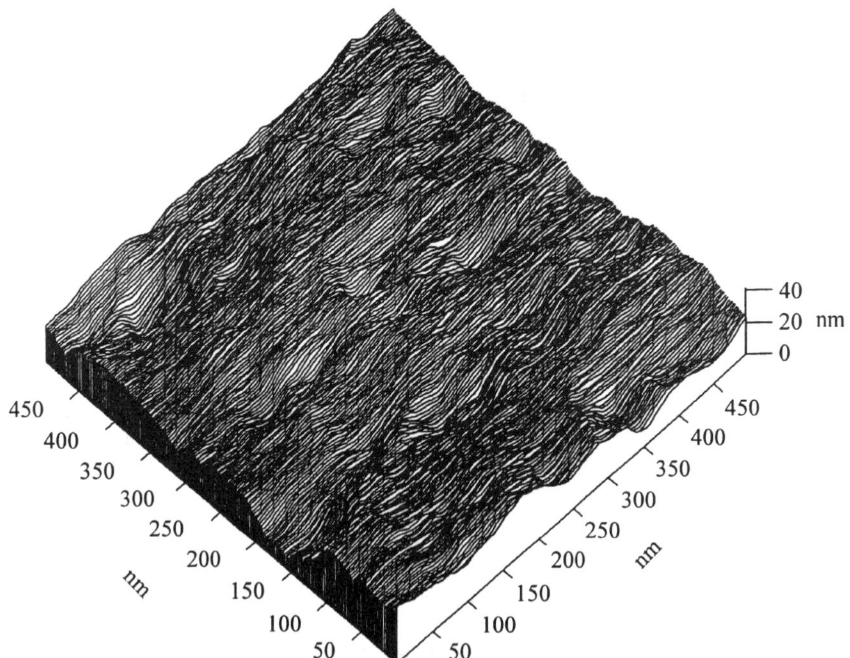

Figure 10(a). AFM pseudo 3-D plots of polypropylene foils (in nm); untreated.

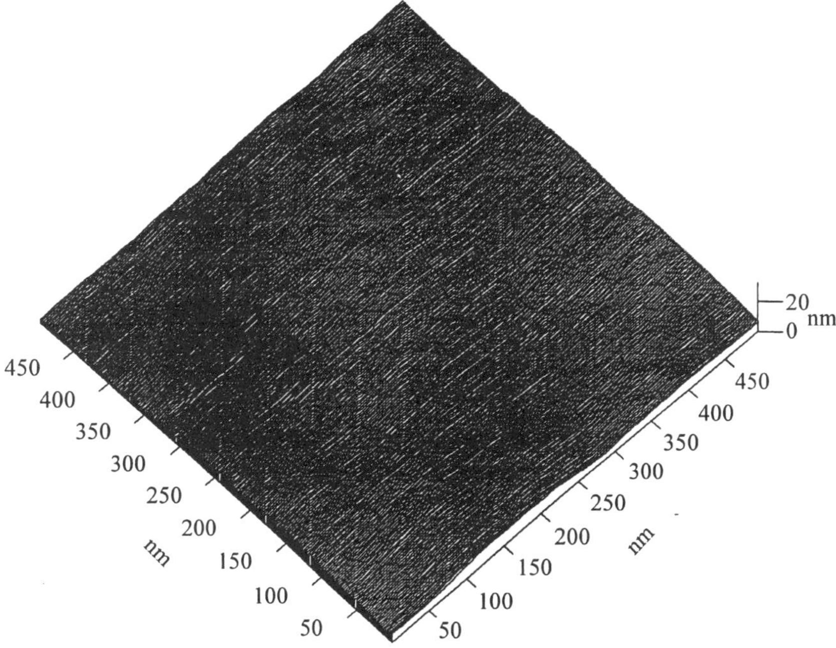

Figure 10(b). AFM pseudo 3-D plots of polypropylene foils (in nm); low-pressure O_2 plasma, 3 s, 6 Pa.

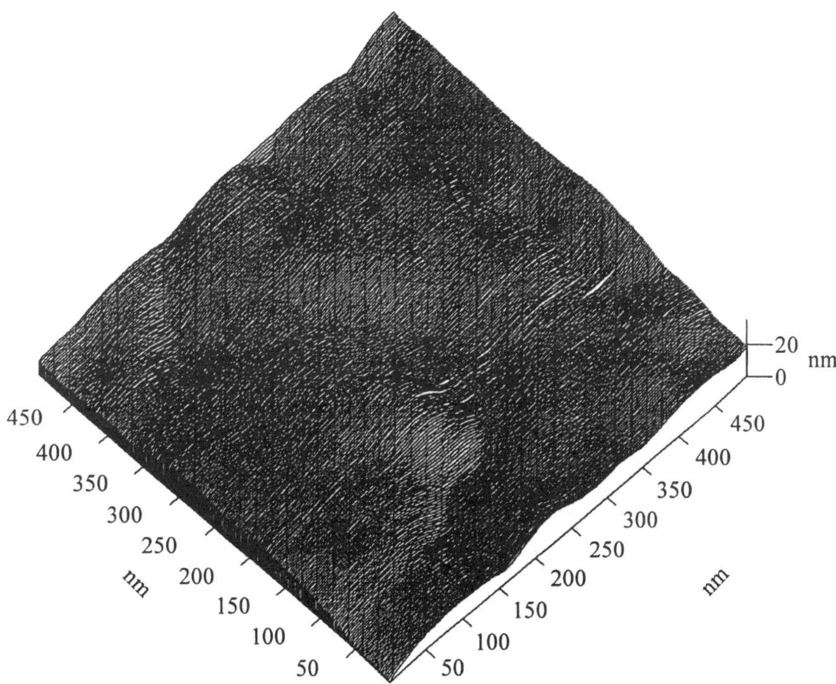

Figure 10(c). AFM pseudo 3-D plots of polypropylene foils (in nm); spark jet, air, 0.7 s, 10^5 Pa.

The result is a rather plain polymer surface. When the *spark jet* is applied to the same polymer surface there is also a distinct, but reduced, smoothening of the surface after a short-time treatment.

3.6. Locus of failure in the composites

High values of lap shear strength correspond to an exclusively cohesive failure in the polyurethane adhesive. This was determined by visual inspection of both surfaces of the fractured composites. XPS analysis of the sheared polypropylene surface of three composites characterised by different lap shear strengths are presented in Fig. 11. The polypropylene substrates were pretreated in the spark jet plasma at different times. The locus of failure can be derived from the existence of additional subpeaks in the C 1s signal which cannot occur with polypropylene alone. They can be assigned to the polyurethane adhesive because there are C—N, C—O, and C=O groups. Additionally the $\pi - \pi^*$ shake-up satellite characteristic of aromatic groups in the polyurethane molecule was identified. Composites with lap shear strength values greater than 4 MPa show C 1s peaks exclusively related to polyurethane, clearly indicating cohesive failure (Fig. 11).

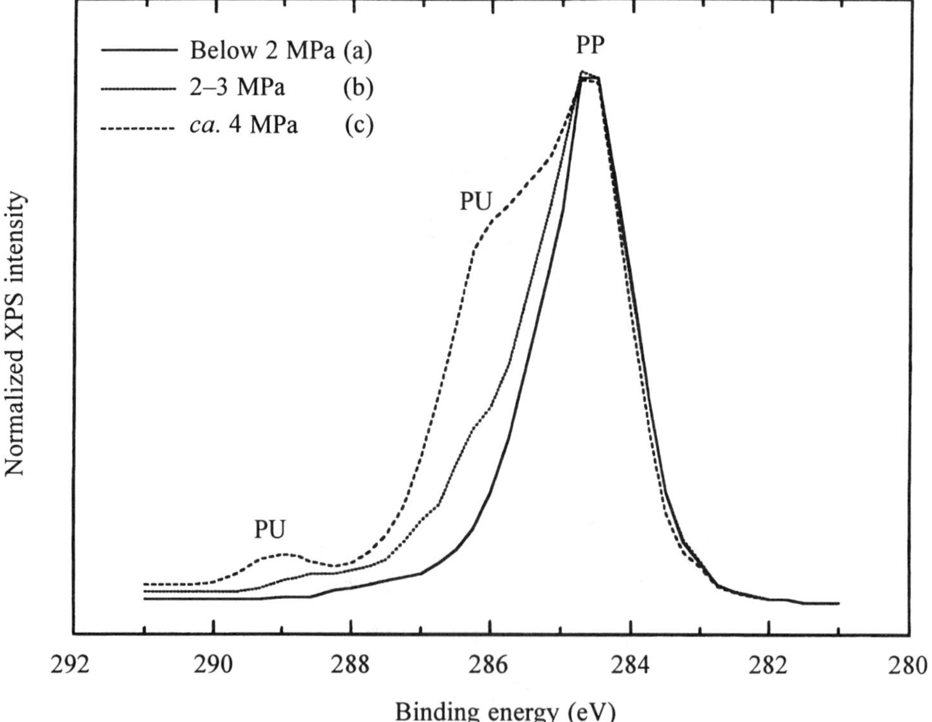

Figure 11. XPS analysis of the polypropylene side of three types of sheared polypropylene–polyurethane composites: (a) low adhesion (spark jet treated, 0.05 s); (b) medium adhesion (spark jet treated, 18 s); (c) high adhesion (spark jet treated, 0.3 s).

4. DISCUSSION

The fact reported in this communication that short exposure to the O_2 plasma only results in low oxygen surface concentrations and little functionalization on the polymer surface on the one hand and high adhesion strengths on the other hand is not easy to understand in terms of the common theoretical approaches to adhesion. Obviously, rather low percentages of polar groups on the surface of polypropylene ($<5\%$ of all C-atoms are involved) are sufficient to promote optimum adhesion strengths for the polypropylene–polyurethane composite. One explanation could be that the low amount of polar groups is concentrated exclusively in the outermost layer. Another explanation could be that an orientation of the macromolecules or parts of them within the outermost surface layer is a key factor for the adhesion promotion effect observed in our study. Furthermore, the plasma-induced changes in supermolecular structure at the surface may also contribute to the increased adhesion measured as lap shear strengths.

As a more general conclusion we propose the following picture of the first steps of the oxygen plasma modification of polymers under consideration. This scheme is supported by the results of Andre *et al.* [9].

first processes ($\rightarrow \approx 0.1$ s)	• plasma radiation and particle impact induced desorption of low molecular species • re-arrangements (and re-orientation) at the molecular scale
second process ($\rightarrow \approx 30$ s)	• re-arrangements at the supermolecular scale
third process ($\rightarrow \approx 30$ s)	• formation of O-functionalized groups
extended plasma treatment	• scissions of side groups and chains followed by etching, ($\rightarrow >30$ s) • crosslinking and melting/pyrolysis (in the case of atmospheric plasma treatment)

It was striking that the same oxygen concentration level in the surface layer of polypropylene generated by low or atmospheric plasmas did not result in similar lap shear strengths (cf. Fig. 12). The small amount of oxygen on the surface obtained after short-time low pressure plasma treatment gives rather high lap shear strengths. To reach the same lap shear strength the spark jet treatment requires a significantly higher surface concentration of oxygen. On the other hand, the model reactions of low molecular isocyanates with OH groups at the plasma treated polypropylene surface confirm the formation of chemical bonds in the same manner as with the polyurethane prepolymer:

$$\sim\sim\sim\text{NCO} + \text{HO}-\{\{\{ \longrightarrow \sim\sim\sim\text{NHCOO}-\{\{\{$$

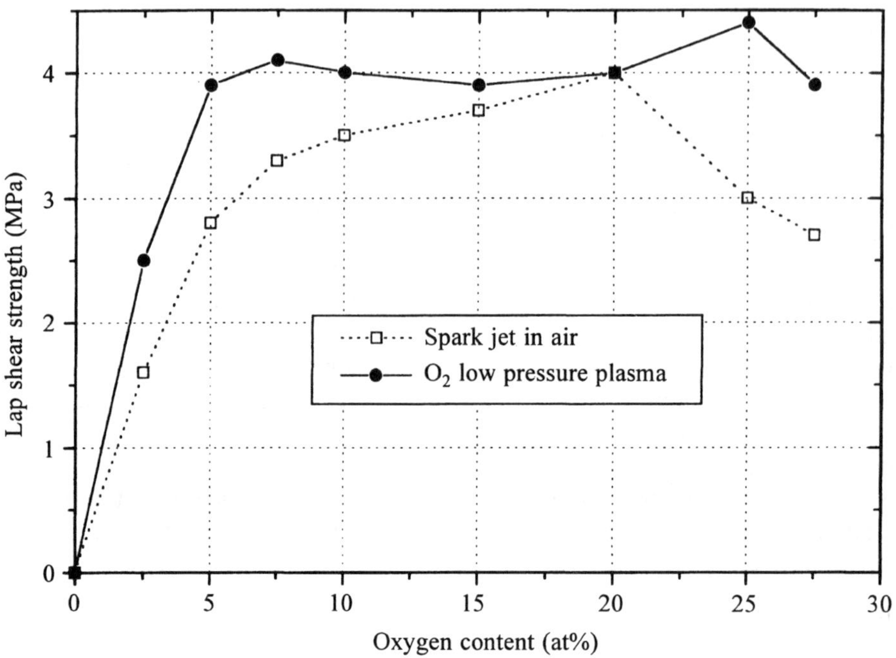

Figure 12. Lap shear strength of polypropylene–polyurethane composites vs. oxygen surface concentration (XPS data) for polypropylene after plasma pretreatment.

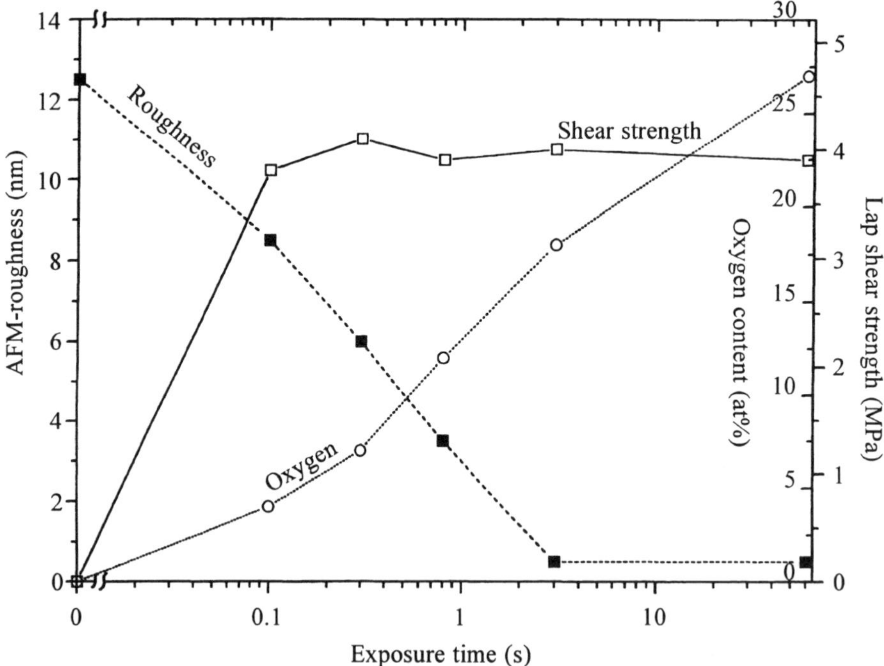

Figure 13. Lap shear strength, oxygen content and surface roughness (obtained by AFM) vs. time of exposure for the low pressure oxygen d.c. plasma treated polypropylene.

However, the high viscosity of the prepolymer and the small number of isocyanate endgroups lower the chance of forming chemical bonds. Another mechanism, which could have importance for the adhesion strengths, is the possibility to form dipole–dipole or H–bridge interactions between the plasma-induced polar groups at the polypropylene surface and the urethane-rich hard segments of the polyurethane adhesive.

Atomic force microscopy of low pressure oxygen plasma treated polypropylene reveals a rapid smoothening of the original supermolecular structure to a rather smooth surface. This smoothened polypropylene surface provides maximum adhesion strengths. Figurc 13 illustrates that for low oxygen pressure plasma treated polypropylene the lap shear strength rapidly increases with time of exposure whereas the surface roughness correspodingly decreases. In contrast to that, the increase of the oxygen surface concentration does not strictly follow the shear strength enhancement; it is rather moderate.

One avenue for gaining more insight into the phenomena occurring due to plasma treatments can be the use of the NEXAFS spectroscopy. With the help of this method, information on the orientation phenomena of functional groups or bonds of the macromolecules (or parts of these) at the polymer surface was obtained. Our first results show that some olefinic double bond species exist in layers near the surface when the untreated sample is considered. Additional C=C groups were formed by the plasma treatment. Another relevant result is the observed weak preferential orientation of the C—O and C=O bond axes of species formed by the plasma treatment [18].

5. CONCLUSIONS

Low pressure plasma modification of polypropylene, other polymers, copolymers, blends, and mineral and fibre-filled composite materials results in high adhesion to polyurethane adhesives. Only a short exposure to an oxygen plasma is required to achieve optimum results regarding initial adhesion as well as performance after exposure to heat and humidity. This holds true also for other plasma gases and vapours.

As an economic alternative to the low pressure plasma process different kinds of hyposonic plasma jets (arc, corona, spark discharges) at ambient conditions were tested. The spark jet was found to be superior regarding adhesion strength results. It is a low cost approach which could be integrated into automated high speed assembly lines within the industry. So it overcomes some of the main disadvantages of the low pressure plasma processes.

XPS analysis revealed that a rather small oxygen uptake is sufficient (1.5 to 5%) to improve the adhesion strenght to optimum values. Derivatization with phenyl isocyanate and TFAA was successfully used to estimate the amount of OH functionality on a plasma treated polypropylene surface. The same reaction (with isocyanates) was used to model the interaction of an isocyanate terminated polyurethane prepolymer wiht plasma treated polypropylene surfaces.

Atomic force microscopy of low pressure oxygen plasma treated polypropylene revealed a rapid smoothening of the original texture to a rather smooth surface which is characterized by optimal adhesion data. The application of a spark jet results in a less pronounced surface smoothening.

Short time exposure to oxygen plasma gives only little functionalization but high adhesion strengths. This cannot be interpreted in terms of the common adhesion models. Therefore, we assume that orientation phenomena at the molecular and supermolecular level in the treated surface layer could be relevant to the adhesion phenomena.

Acknowledgements

We express our gratitude to Th. Gesang and R. Höper (IfAM Bremen) for the AFM results and Priv.-Doz. Dr Ch. Wöll (Universität Heidelberg) for collaboration during the NEXAFS investigations.

REFERENCES

1. K. L. Mittal (Ed.), *Physicochemical Aspects of Polymer Surfaces*. Vols 1 and 2. Plenum Press, New York (1983).
2. D. Briggs and M. P. Seah, *Practical Surface Analysis by Auger and X-ray Photoelectron Spectroscopy*. Wiley, Chichester (1990).
3. A. Chilkoti and B. D. Ratner, in: *Surface Characterization of Advanced Polymers*, L. Sabbattini and P. G. Zambonin (Eds). Verlag Chemie, Weinheim (1993).
4. C. D. Batich, *Applied Surface Science* **32**, 57 (1988).
5. I. Munteanu, J. M. Herdan and G. Valeanu, *Revue Romaine de Chimie* **33**, 467 (1988).
6. R. Foerch, G. Beamson and D. Briggs, *Surf. Interface Anal.* **17**, 842 (1992).
7. J. J. Pireaux, J. Riga and J. J. Verbist, in: *Photon, Electron and Ion Probes for Polymer Structure and Properties*, D. W. Dwight, T. J. Fabish and H. R. Thomas (Eds), p. 169. ACS Symposium Series No. 162, Amer. Chem. Soc., Washington, DC (1981).
8. J. Stöhr, *NEXAFS Spectroscopy*. Springer Verlag, Heidelberg (1992).
9. V. Andre, F. Tchoubineh, F. Arefi and J. Amouroux, in: *Plasma-Surface Interactions and Processing of Materials*, O. Auciello (Ed.), pp. 89–99. Kluwer, Dordrecht (1990).
10. J. Friedrich, J. Elger, W. Unger, A. Lippitz, Th. Gesang, R. Höper and H. V. Gorsler, in: *Plasmatec '93*, H. Nowack and E. Schindel-Bidinelli (Eds), pp. 107–131. Print Service Mühlheim (1993).
11. J. Friedrich, W. Unger, A. Lippitz, W. Saur and P. Rohrer, in: *Kleben–Swissbonding '93*, E. Schindel-Bidinelli (Ed.), pp. 92–105. Print Service Mühlheim (1993).
12. G. Hertz and R. Rompe, *Einführung in die Plasmaphysik und ihre technische Anwendung*. Akademie-Verlag, Berlin (1968).
13. M. T. Anthony and M. P. Seah, *Surf. Interface Anal.* **6**, 95 (1984).
14. A. D. Shirley, *Phys. Rev.* **B5**, 4709 (1972).
15. X.-R. Hu and H. Hantsche, *J. Electr. Spectr. Relat. Phenomena* **50**, 19 (1990).
16. E. Orti, R. Virruela, J. L. Bredas and J. J. Pireaux, in: *Polymer-Solid Interfaces*, J. J. Pireaux, P. Bertrandt and J. L. Bredas (Eds), pp. 325–336. Institute of Physics Publishing, Bristol (1992).
17. M. P. Seah and W. A. Dench, *Surf. Interface Anal.* **1**, 1 (1979).
18. Th. Gross, A. Lippitz, J. F. Friedrich, Ch. Wöll and W. Unger, *Polymer Communications* (in press).

Polymer Surface Modification: Relevance to Adhesion, pp. 73–85
K. L. Mittal (Ed.)

Fluoropolymer surface modification for enhanced evaporated metal adhesion

M. K. SHI,[1] A. SELMANI,[1] L. MARTINU,[2,*] E. SACHER,[2] M. R. WERTHEIMER[2] and A. YELON[2]

[1]*Department of Chemical Engineering, and* [2]*Groupe des Couches Minces and Department of Engineering Physics, Ecole Polytechnique, Box 6079, Station Centre-Ville, Montreal, Quebec H3C 3A7, Canada*

Revised version received 18 April 1994

Abstract—Adhesion of evaporated Cu to Teflon PFA (polytetrafluoroethylene-co-perfluoroalkoxy vinyl ether) was greatly enhanced by plasma pretreatment. The efficiency of the treatment decreased in the following order: $N_2 > O_2 > (N_2 + H_2) > (O_2 + H_2) > H_2$. X-ray photoelectron spectroscopy (XPS) showed the loss of fluorine and the incorporation of oxygen and nitrogen at the polymer surface. Among the gases, H_2 was found to be the most efficient for fluorine elimination, and $(N_2 + H_2)$ for surface functionalization. Based on this investigation, it is proposed that Cu reacts with both oxygen and nitrogen to form, respectively, Cu—O and Cu—N bonds at the interface but no reaction occurs with carbon and fluorine. While greater enhancement in polymer surface wettability and stronger interfacial reactions can account for the higher performance of N_2 over O_2 in improving adhesion, these effects cannot explain the lower efficiency of H_2. Several possibilities are discussed, including surface cleaning, oxygen incorporation and the formation of weak boundary layers.

Keywords: Teflon PFA; plasma; surface modification; Cu evaporation; adhesion; XPS.

1. INTRODUCTION

Fluoropolymers have many desirable properties, such as high thermal stability, low dielectric constants and low dissipation factors, and surface inertness, which make them ideal for microelectronic and optical applications [1, 2]. Previous studies of the metallization of fluoropolymers demonstrated that both surface defluorination and the formation of metal carbide species contribute to metal adhesion [3–5]. However, these effects do not provide sufficient adhesion strength for practical applications.

Activation of polymer surfaces prior to metal deposition has been shown to enhance metal adhesion [6–12]. Along with wet chemical treatment, ion bombardment

*To whom correspondence should be addressed.

and X-ray irradiation, cold plasma treatment is considered one of the most efficient activation techniques (for a recent critical review, see [13]).

The interaction of a plasma with a polymer surface includes physical bombardment by energetic ions, U.V. irradiation and chemical reactions at the surface; these result in cleaning (etching), crosslinking and the incorporation of new functional groups. Plasma-introduced groups were found to serve both as metal nucleation and chemical reaction sites, leading to relatively uniform metal overlayers [11, 14–16] and to the formation of chemical linkages at the interface [6, 11]. Burkstrand [6] reported that the adhesion of Cu to polystyrene was increased by plasma treatment in water vapour, due to the formation of Cu–O bonds through the reaction of Cu with plasma-introduced hydroxyl groups. Adhesion improvement was also observed for the Ag/polyethylene system following O_2, Ar and N_2 plasma treatments [11]. Moreover, Ag reacted more readily with oxygenated groups introduced by plasma than with the original ester groups at the poly(ethylene terephthalate) surface [17].

Fluoropolymers are resistant to oxygen attack. Both surface degradation and oxygen incorporation were found to be small for O_2-plasma-treated Teflon PTFE (polytetrafluoroethylene) [18–20]. In contrast, H_2 plasma was reported to be very efficient in eliminating fluorine atoms from the surface, due, probably, to the formation of volatile HF compound [21, 22]. In this paper, we compare the effects of different plasmas, such as N_2, O_2, H_2 and their mixtures, on the adhesion of Cu to Teflon PFA (polytetrafluoroethylene-co-perfluoroalkoxy vinyl ether).

2. EXPERIMENTAL

DuPont Teflon PFA $[(CF_2-CF_2)_n-(CF_2-CFOC_3F_7)_m]$ ($n/m = 39$) films (50.8 μm thick) were used as received. X-ray photoelectron spectroscopy, obtained at a take-off angle of 30° from the sample plane, showed an F/C ratio of 1.86, lower than the theoretical ratio of 2.00 [23]. The oxygen content in the PFA is very low, making the oxygen peak too small to be measured with precision (the expected O/C atomic ratio is close to 0.01).

Surface treatment was performed in a microwave (MW, 2.45 GHz) plasma reactor, described in greater detail elsewhere [12]. In the present study, the sample (15×15 cm^2) was mounted on a grounded electrode. After evacuation to 10^{-3} mTorr, gas was admitted to the reactor. The gas flow rate, measured using a Dynamass F-4 mass flow controller, and the pressure, measured with an MKS Baratron gauge, were then adjusted. Typical treatment conditions include a microwave power of 100 W, a flow rate of 50 sccm, a pressure of 200 mTorr and a treatment time of 60 s.

N_2, O_2, H_2 and their mixtures were studied. It was expected that (O_2+H_2) (volume flow ratio = 1:2) and (N_2+H_2) (volume flow ratio = 1:3) mixtures would introduce OH and NH_x ($x = 1, 2$) groups at the surface. These groups have been found to be more reactive with different metals than C–O, C=O, C–N or C=N [6, 17, 24, 25]. The (N_2+3H_2) mixture, though numerically equivalent to ammonia (NH_3) in its H/N atomic ratio, is more attractive than ammonia for practical application because it is non-toxic. The use of ($O_2 + 2H_2$), on the other hand, permits a better control of the gas pressure than the use of water vapour and avoids the heating process required for water vaporization.

After treatment, the sample was exposed to the laboratory atmosphere for several minutes while being transferred for surface analysis and for metal depositions. It is well known that, upon exposure to air, the chemical composition of plasma-treated surfaces can be modified due to reaction with oxygen and water vapour and to the reorientation of different functionalities in order to minimize the surface energy [13, 26, 27]. Post-oxidation is the main process of oxygen incorporation at PFA surfaces treated with N_2, H_2 and (N_2+H_2) plasmas. Several samples (about 1×1 cm^2), treated under the same conditions, were introduced into the XPS chamber for surface analysis and for metallization. Submonolayers of Cu was deposited by electron-beam evaporation in the XPS preparation chamber at a pressure of 2×10^{-8} Torr. The deposition rate was typically 0.07 Å/s. The thickness of the deposited Cu was estimated using a quartz crystal microbalance (QCM). Due to the different sticking probabilities of Cu on PFA and on the QCM [5], the thicknesses given throughout the paper are only nominal.

In situ XPS analysis was performed in a VG ESCALAB MKII system, using a Mg anode (1253.6 eV), operated at 12 kV and 20 mA. Survey spectra (0–1000 eV) were obtained using a pass energy of 50 eV. High-resolution spectra were obtained using a pass energy of 20 eV. The pressure during analysis was maintained at about 10^{-9} Torr. Charging effects were compensated for by setting the CF_2 C1s peak to 291.8 eV [5, 28]. Photoelectrons were collected at a take-off angle of 30° with respect to the sample plane. After a Shirley-type background removal, the high resolution spectra were deconvoluted using an in-house least mean squares program. Quantitative analyses were performed using peak areas and elemental sensitivity factors [29].

For adhesion measurements, 2000 Å thick Cu layers were thermally evaporated onto the PFA surfaces in an NRC model 3114 vacuum coater equipped with a shutter and a quartz monitor. The deposition was carried out at a pressure of about 10^{-6} Torr and a deposition rate of around 1 Å/s. Strips of metallized sample (1.3 cm wide) were adhered to a flat metal plate using double-sided-adhesive tape (3M Company, 136C). After compression at 2×10^6 Pa for 20 s, the assembly was fixed to the support of a peel tester (Instrumentors Inc. 3M90, USA). While either metal or PFA film can be peeled off to measure the adhesion strength, the latter was chosen for convenience. Peeling was carried out at 90°, at a speed of 7.6 cm/min. The adhesion strength was averaged over five measurements taken on different strips.

Adhesion was also measured by a scratch test, using a CSEM Micro-Scratch-Tester, under a loading rate of 3 N/min and a scratch speed of 1 cm/min. A diamond indenter, ground to a 120° cone with a spherical apex of 800 μm radius, was used in all measurements. The critical load where the metal delamination events began to occur, L_c, was recorded. The data from five scratch measurements were averaged.

3. RESULTS AND DISCUSSION

First, we evaluated the influence of both plasma treatment time and applied power on the adhesion of Cu to O_2-plasma-treated PFA, the gas flow rate and pressure being

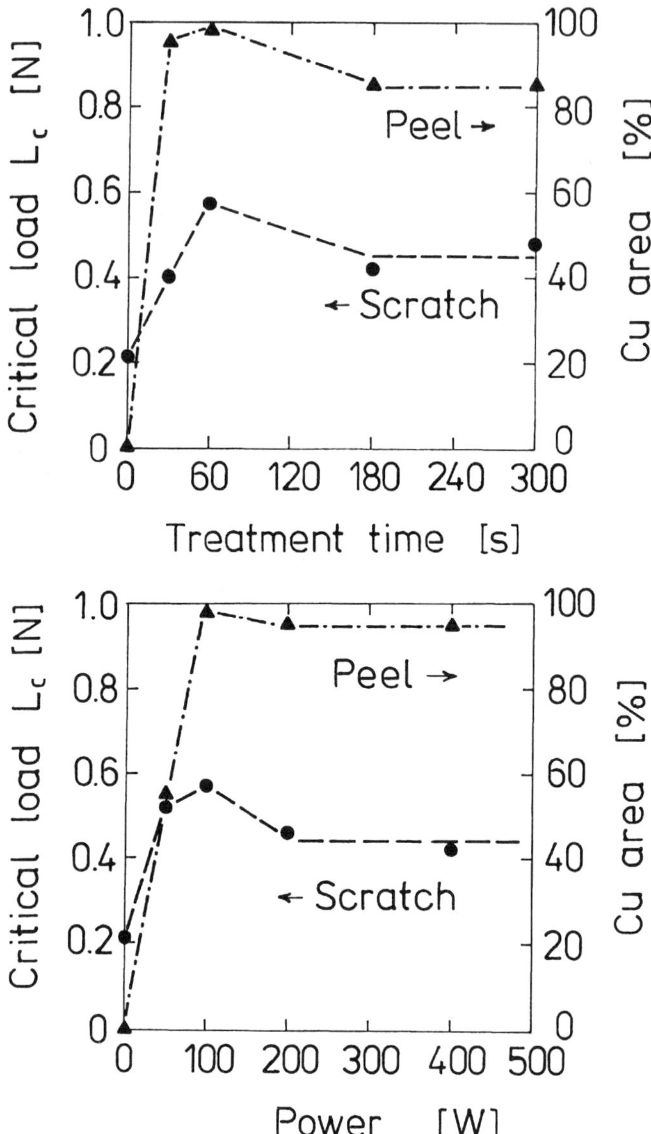

Figure 1. The effect of plasma treatment time (a) and applied power (b) on the adhesion of Cu to O_2-plasma-treated PFA (Cu% represents the areal percentage of Cu which remained on the PFA after peeling).

kept constant (50 sccm and 200 mTorr, respectively). The results are shown in Fig. 1. Here, L_c represents the critical load obtained from scratch tests, while Cu% represents the areal percentage of metal which remained in contact with the PFA after peeling. The adhesion of Cu to the untreated PFA was very weak, and the PFA film could be peeled, easily and totally, from the metal. As noted from Fig. 1, the adhesion was greatly enhanced by plasma treatment, to the extent that, now, only a fraction of the metal film could be separated from the PFA. It should be mentioned that, compared to the bulk material, the thin film of Cu is polycrystalline and the fracture may take place along grain boundaries. The adhesion showed a maximum for the treatment time of around 60 s (Fig. 1a) and near 100 W (Fig. 1b), respectively. The adhesion decreased when either the power or the treatment time was further increased. This behaviour may possibly be explained by the formation of a weak surface layer, caused by plasma-induced degradation [13, 30, 31]. The good correlation between the 90°-peel test and the micro-scratch test indicates that the critical load L_c can be effectively used as a qualitative measure of Cu/PFA adhesion.

The effects of different plasma gases on the enhancement of metal adhesion were compared under the same treatment conditions (i.e. a power of 100 W, a flow rate of 50 sccm, a pressure of 200 mTorr and a treatment time of 60 s). Metal adhesion, evaluated using a scratch test, was found to decrease in the order: $N_2 > O_2 > (N_2+H_2) > (O_2+H_2) > H_2$. The high efficiency of N_2 was confirmed by a 90°-peel test measurement, which does not show any separation of the metal film from the PFA on peeling. On using H_2, the enhancement was small, similar to what had been observed for the autohesion of corona-exposed polyethylene sheets [32].

The substrate surface energy affects the sticking probability and the cluster size of metal particles, both of which have been found to contribute to metal adhesion [15, 16]. Measurement of the contact angle water makes with the surface permits a rapid, qualitative evaluation of polymer surface energy. The lower the water contact angle, the higher the surface energy. In this experiment both advancing (θ_a) and receding (θ_r) water contact angles for plasma-treated PFA were measured; the results are summarized in Table 1. The values for the untreated surface were found to be 112° and 87° for θ_a and θ_r, respectively. They are greatly reduced by plasma treatment, more so for the case of N_2 than for O_2. These results, when compared to the adhesion enhancement shown in Fig. 1, suggest that polar surface groups, introduced by plasma treatment, are a major factor in the enhancement of metal adhesion.

The polymer surface was also analyzed by XPS. The atomic concentration ratios of F, O and N to C for different plasma treatments are listed in Table 2. Plasma treatment produces ablation of fluorine and incorporation of polar groups containing O and N at the polymer surface. The introduction of oxygen, after treatment in H_2, N_2 and $(H_2 + N_2)$ environments, is due to reaction when the sample was removed from the reactor and exposed to air, as mentioned earlier. Among the different gases, H_2 is the most efficient for fluorine removal, and the (N_2+H_2) mixture for surface functional-

Table 1.
Advancing (θ_a) and receding (θ_r) contact angles (in degrees) of water on untreated and plasma-modified Teflon PFA surfaces (treatment conditions: 100 W, 50 sccm, 200 mTorr and 60 s)

	Untreated	O_2-plasma	N_2-plasma
θ_a	112	105	80
θ_r	87	65	46

Table 2.
Atomic concentration ratios of F, O and N to C on Teflon PFA, after different plasma treatments, as determined by XPS (treatment conditions: 100 W, 50 sccm, 200 mTorr and 60 s)

Treatment	F/C	O/C	N/C
Untreated	1.86	0.004	–
O_2	1.78	0.014	–
N_2	1.58	0.031	0.032
$O_2 + H_2$	1.51	0.069	–
$N_2 + H_2$	0.50	0.13	0.19
H_2	0.35	0.064	–

ization, while O_2 shows the weakest performance for both effects. The low reactivity of O_2 plasma with PFA is similar to that observed with Teflon PTFE [18–20].

Adhesion is a complex phenomenon, related to both physical and chemical effects [33]. Under our treatment conditions, no modification in the polymer surface topography was observed by scanning electron microscopy, suggesting that physical effects do not play a major role in the Cu/PFA adhesion.

To ascertain the effect of the metal/polymer interface chemistry on adhesion, submonolayers of Cu were evaporated onto both untreated and ($N_2 + H_2$)-plasma-treated surfaces, and the interfaces were analyzed *in situ* by XPS. The variations of the C1s spectra upon plasma treatment and metal deposition are shown in Fig. 2. Untreated PFA exhibits only a single CF_2 peak at 291.8 eV, the small peaks near 283 eV being due to X-ray satellites [28]. The CF_2 peak is relatively broad because our XPS apparatus is not equipped with a monochromator and the charging effect of the PFA sample is high (about 10 eV). Plasma treatment causes loss of fluorine, leading to a decrease in the CF_2 peak intensity and a concomitant increase in lower binding energy components, with C—H and C—C (near 285 eV) being the most prominent [28]. The high-binding-energy peaks near 292 eV can be attributed to perfluorinated species (CF_3, CF_2, etc.), while the peaks appearing between 291 and 285 eV are due to C with various combinations of N, O and F [18–20, 28]. The formation of CF_3 is due to recombination reactions between F and CF_2 radicals produced upon plasma exposure [21, 23].

The deposition of Cu (120 Å) causes a decrease in the CF_2 peak intensity, due to the shielding effect of metal film, combined with a slight loss of fluorine. A small increase in the peak near 285 eV was observed, probably due to the formation of C—C and Cu—O—C structures [11, 28]. Compared to the untreated surface, all the components in the C1s spectra on the plasma-treated surfaces were reduced in intensity after metal deposition, although the quantity of metal deposited was substantially less (10 Å). This indicates an enhanced sticking probability of Cu to the plasma-treated than to the untreated surface [11, 14].

The single CF_2 peak at the untreated PFA surface appeared at 688.7 eV in the F1s spectrum (Fig. 3a). Its intensity was greatly reduced, and its shape modified, after plasma treatment: a new peak at a lower binding energy (687.3 eV) was added to better fit the spectrum. This new species may tentatively be attributed to C—CF_2 [19].

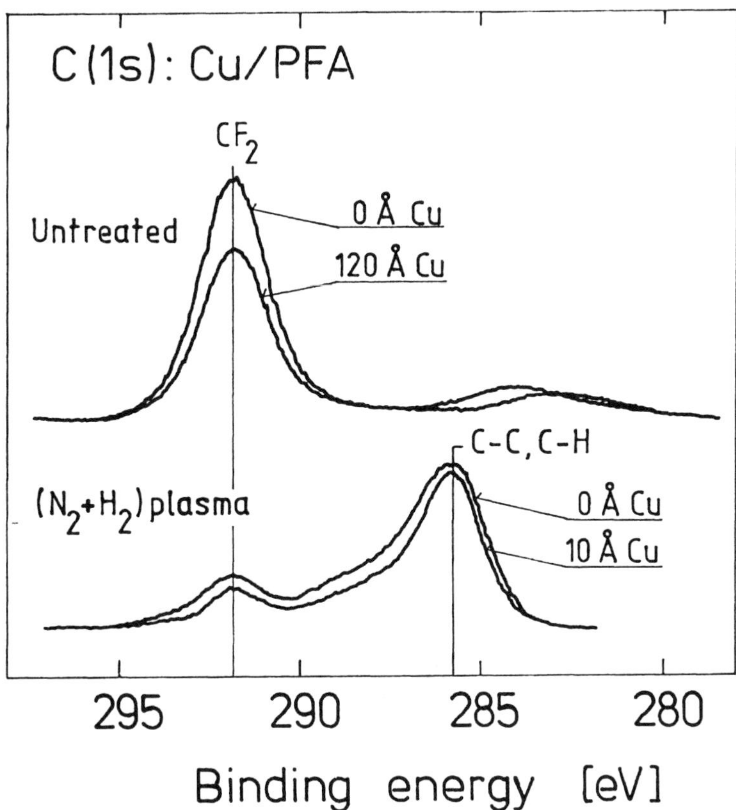

Figure 2. C1s spectra for the untreated and (N_2+H_2)-plasma-treated PFA, before and after Cu deposition (treatment conditions: 100 W, 50 sccm, 200 mTorr and 60 s).

The evaporation of Cu causes a further decrease in the F1s peak intensity (Fig. 3b), more so for the plasma-treated than for the untreated surface. This behaviour may be explained by an increase in the metal sticking probability [11, 14] caused by surface modification upon plasma treatment, giving rise to both a more pronounced shielding effect and a higher surface defluorination. The outline of the F1s peak remained unchanged for both untreated and plasma-treated surfaces, suggesting the absence of formation of metal fluorides, such as CuF_2 ($\sim$ 685 eV, [5, 28]). This excludes the contribution of metal fluorides to metal adhesion.

The N1s spectra of the $(N_2 + H_2)$-plasma-treated sample were also modified by metal deposition, as shown in Fig. 4. Not only was the nitrogen peak intensity greatly reduced, its position was also shifted to a lower binding energy. This result is similar to what was observed for the Ag/polyethylene system [11], and suggests that Cu has interacted with N, probably to form Cu—N bonding. However, due to the complexity of N functionalities [11, 34–37] and the lack of data in the literature, it is difficult to specify the reaction mechanisms.

The oxygen content in untreated PFA is very low and the oxygen peak can be clearly observed only after adding several scans (Fig. 5). The O/C atomic ratio was

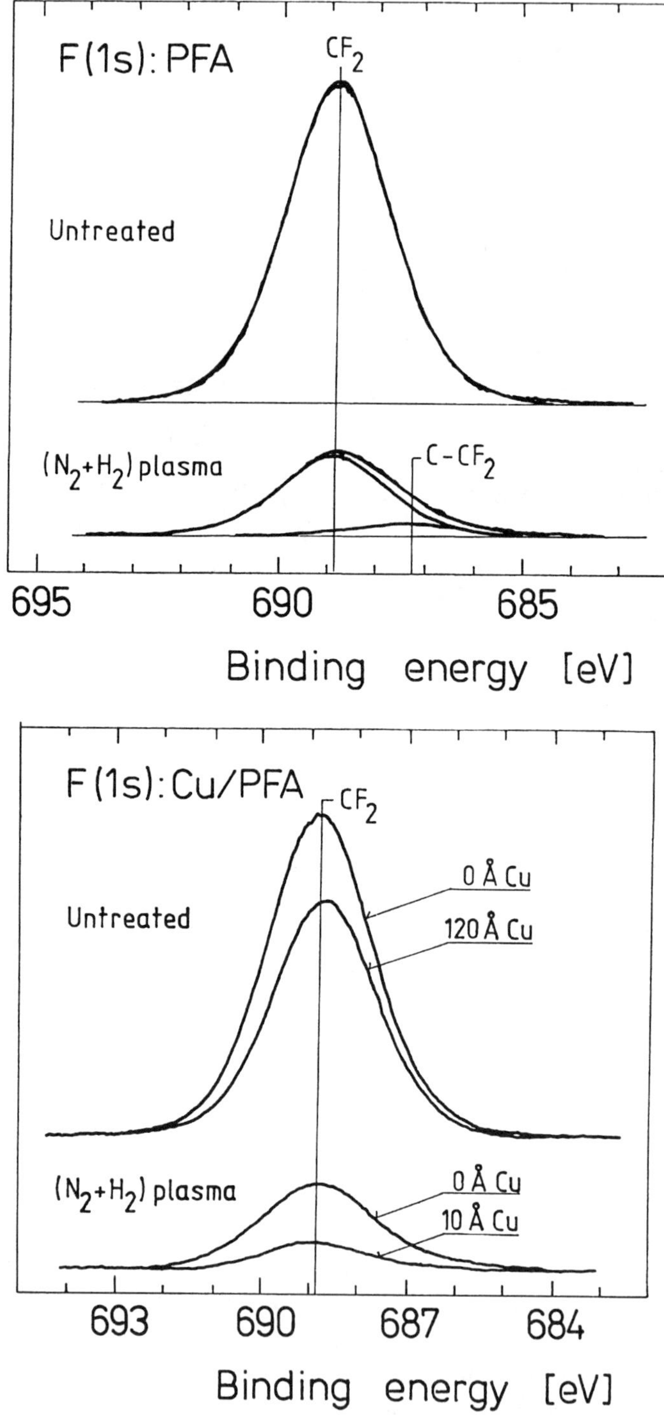

Figure 3. F1s spectra of PFA before and after $(N_2 + H_2)$ plasma treatment (a) and metal deposition (b) (treatment conditions: see Fig. 2).

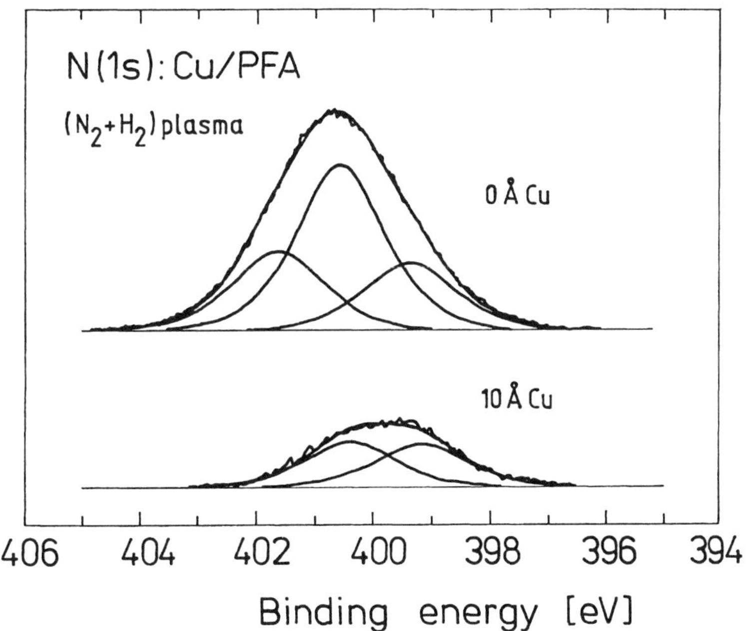

Figure 4. N1s spectra of ($N_2 + H_2$)-plasma-treated PFA, before and after Cu deposition (treatment conditions: see Fig. 2).

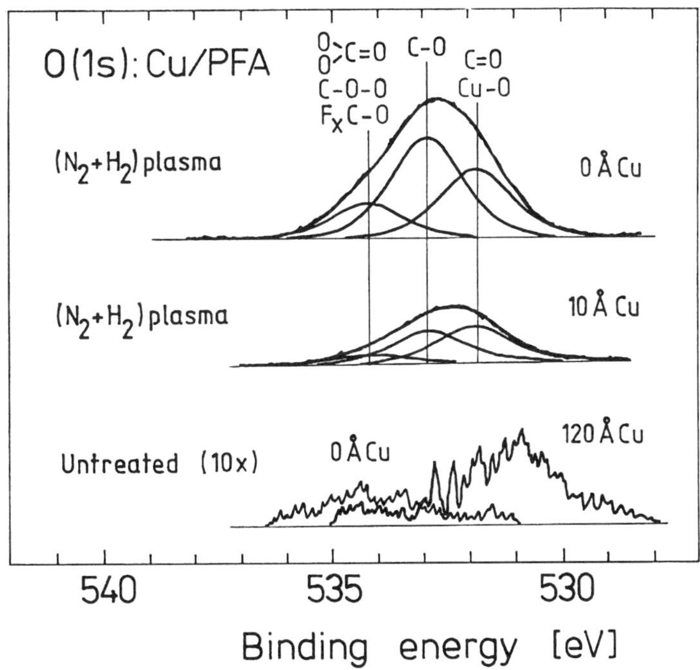

Figure 5. O1s spectra for the untreated and (N_2+H_2)-plasma-treated PFA, before and after Cu deposition (treatment conditions: see Fig. 2).

greatly increased by plasma treatment and the centroid of the peak shifted to a lower binding energy due to a significant loss of fluorine. Curve-fitting revealed three peaks, located at 534.7 eV, 533.6 eV and 532.6 eV which can be attributed, respectively, to $F_xC{-}O$ ($x = 1, 2$; C—O—O and O—CO—O), C—O and C=O groups [21, 28, 34, 37, 38], with C—O being the most prominent.

The oxygen concentration at the untreated surface increased slightly and the peak shifted to a lower binding energy after Cu deposition, due probably to a slight contamination and the formation of Cu—O—C bonds, as observed in Fig. 2. When Cu was deposited onto the plasma-treated surface, the $F_xC{-}O$ (C—O—O and O—CO—O) and C—O peak intensities were reduced much more than the C=O peak intensity, suggesting that Cu has reacted with $F_xC{-}O$ (C—O—O and O—CO—O) and C—O groups to form Cu—O linkages having a binding energy close to that of C=O [6, 39]. The reaction of $F_xC{-}O$ with Cu should be small, due to a reduction in the electron density at the O atoms by the presence of fluorine [40–42]; the loss of the $F_xC{-}O$ (C—O—O and O—CO—O) peak intensity is, thus, due mainly to a loss of C—O—O and O—CO—O, which are more reactive. The absence of the peak below C=O component indicates that the reactivity of C=O is much less than that of C—O—O (O—CO—O) and C—O, and no metal oxide, either CuO (530.3 eV) or Cu_2O (529.6 eV) [28], was formed.

The $Cu2p_{3/2}$ spectra are presented in Fig. 6. Due to an enhanced sticking probability, discussed above, the $Cu2p_{3/2}$ peak intensity is much greater on the plasma-treated, than on the untreated, surfaces, although the nominal thickness of deposited Cu is

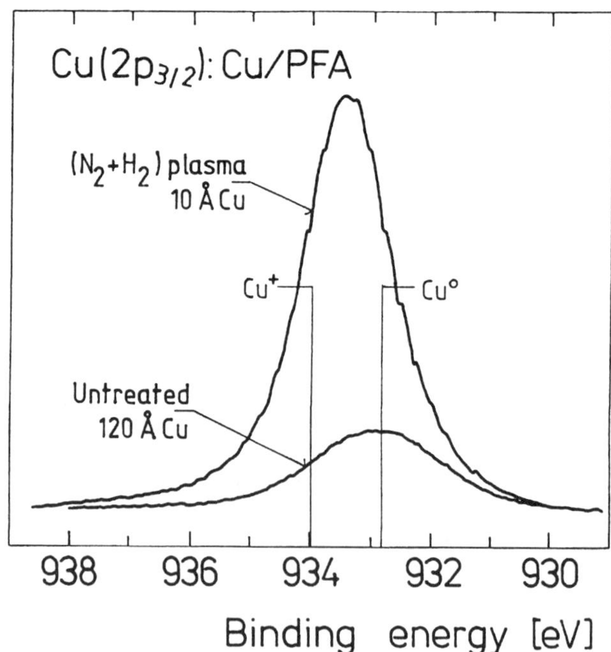

Figure 6. $Cu2p_{3/2}$ spectra for Cu deposition onto untreated and (N_2+H_2)-plasma-treated PFA (treatment conditions: see Fig. 2).

substantially less. This behaviour is similar to that observed in the metallization of polyethylene and polypropylene [11, 14]. The peak is located near 933 eV on the untreated surface, dominated by a pure metallic component [28]. It shifts to a higher binding energy on the plasma-treated surface, indicating that a substantial fraction of the Cu exists in higher valence states [28, 39, 43].

The formation of Cu—O and Cu—N bonds at the interface leads to a higher adhesion of Cu to plasma-treated PFA, similar to what has been observed for other systems [6, 11, 17]. While a good correlation between surface modification and adhesion enhancement was established for O_2 and N_2 plasma treatments, a discrepancy appears when H_2 is used: the surface modification is greater (see Table 2) but the adhesion improvement is less ($O_2 > H_2$, $O_2 > (O_2 + H_2)$, and $N_2 > (N_2 + H_2)$). Apart from the chemical modification of the outermost surface, plasma treatment can cause hardening of the polymer subsurface layers through crosslinking by U.V. radiation, which also appears to affect metal adhesion [13]. However, this effect should be more pronounced in H_2 than in O_2 plasma because of the stronger U.V. radiation of the former [22, 44]. Kim *et al.* [32] observed that the autohesion of polyethylene was greatly enhanced by corona treatment, but it was not improved by ozone exposure, although both caused similar surface modifications. A recent study [45] of the PFA surface, using multiple techniques, indicates that the pure PFA sample does not contain mechanically weak boundary layers; instead, its outermost surface is slightly contaminated during manufacture. It is, therefore, possible that these contaminants are incompletely removed by H_2 plasma treatment and that they are responsible for the lower adhesion enhancement than was obtained with O_2 [13, 46]. Furthermore, the selective attack of the H_2 plasma on fluorine (see Table 2) leads to the formation of a thin defluorinated layer [21], which may be weakly bonded to the bulk. Finally, different mechanisms of oxygen incorporation between O_2- and H_2-plasma-treated surfaces may result in different extents of adhesion improvement, and should also be taken into account.

4. CONCLUSIONS

The adhesion of Cu to Teflon PFA was greatly enhanced by plasma pretreatments. N_2 was found to be the most efficient among the gases used, in the order $N_2 > O_2 > (N_2 + H_2) > (O_2 + H_2) > H_2$. Cleaning, surface wetting, and the formation of chemical linkages at the interface are the major factors affecting metal adhesion. Similar behaviour has been observed for Ag and Au.

Acknowledgements

The Natural Sciences and Engineering Research Council of Canada and the Fonds pour la Formation de Chercheurs et l'Aide à la Recherche of Québec are thanked for financial support. The authors are indebted to Mr P. Leroux for his assistance with the microscratch test measurements.

REFERENCES

1. J. M. Park, L. J. Matienzo and D. F. Spencer, *J. Adhesion Sci. Technol.* **5**, 153 (1991).
2. E. H. A. Granneman, *Thin Solid Films* **228**, 1 (1993).
3. S. L. Vogel and H. Schonhorn, *J. Appl. Polym. Sci.* **23**, 495 (1979).
4. C. A. Chang, Y. K. Kim and A. G. Schrott, *J. Vac. Sci. Technol.* **A8**, 3304 (1990).
5. M. K. Shi, B. Lamontagne, A. Selmani, L. Martinu, E. Sacher, M. R. Wertheimer and A. Yelon, *J. Vac. Sci. Technol.* **A12**, 29 (1994).
6. J. M. Burkstrand, *Appl. Phys. Lett.* **33**, 387 (1978).
7. D. R. Wheeler and S. V. Pepper, *J. Vac. Sci. Technol.* **20**, 442 (1982).
8. P. Bodö and J. E. Sundgren, *J. Appl. Phys.* **60**, 1161 (1986).
9. C. A. Chang, J. E. E. Baglin, A. G. Schrott and K. C. Lin, *Appl. Phys. Lett.* **51**, 103 (1987).
10. A. Celerier and J. Marchet, *Thin Solid Films* **148**, 323 (1987).
11. L. J. Gerenser, *J. Vac. Sci. Technol.* **A6**, 2897 (1988).
12. M. H. Bernier, J. E. Klemberg-Sapieha, L. Martinu and M. R. Wertheimer, *Amer. Chem. Soc. Symp. Ser.* **440**, 147 (1990).
13. E. M. Liston, L. Martinu and M. R. Wertheimer, *J. Adhesion Sci. Technol.* **7**, 1091 (1993).
14. S. Nowak, M. Collaud, G. Dietler, O. M. Küttel and L. Schlapbach, *J. Vac. Sci. Technol.* **A11**, 481 (1993).
15. J. F. Silvain, J. J. Ehrhardt, A. Picco and P. Lutgen, *Amer. Chem. Soc. Symp. Ser.* **440**, 453 (1991).
16. P. Phuku, P. Bertrand and Y. DePuydt, *Thin Solid Films* **200**, 263 (1991).
17. L. J. Gerenser, *J. Vac. Sci. Technol.* **A8**, 3682 (1990).
18. M. A. Golub and T. Wydeven, *Polym. Degrad. Stab.* **22**, 325 (1988).
19. M. A. Golub, T. Wydeven and R. D. Cormia, *Polymer* **30**, 1576 (1989).
20. M. A. Golub, T. Wydeven and R. D. Cormia, *Langmuir* **7**, 1026 (1991).
21. D. T. Clark and D. R. Hutton, *J. Polym. Sci., Polym. Chem. Ed.* **25**, 2643 (1987).
22. G. Smolinsky and M. J. Vasile, *Eur. Polym. J.* **15**, 87 (1979).
23. B. J. Tan, M. Fessehaie and S. L. Suib, *Langmuir* **9**, 740 (1993).
24. C. L. Lasko, B. M. Pesic and D. J. Oliver, *J. Appl. Polym. Sci.* **48**, 1565 (1993).
25. Y. Qin, *J. Appl. Polym. Sci.* **49**, 727 (1993).
26. H. L. Spell and C. P. Christenson, *Tappi* **62**, 77 (1979).
27. J. E. Klemberg-Sapieha, L. Martinu, O. M. Küttel and M. R. Wertheimer, in: *Metallized Plastics 2*, K. L. Mittal (Ed.), pp. 315. Plenum Press, New York (1991).
28. J. F. Moulder, W. F. Stickle, P. E. Sobol and K. D. Bomben, in: *Handbook of X-ray Photoelectron Spectroscopy*, J. Chastain (Ed.), pp. 18, 217–231. Perkin-Elmer Corporation, Eden Prairie, MN (1992).
29. C. D. Wagner, L. E. Davis, M. V. Zeller, J. A. Taylor, R. M. Raymond and L. H. Gale, *Surf. Interface Anal.* **3**, 211 (1981).
30. C. Y. Kim and D. A. I. Goring, *J. Appl. Polym. Sci.* **15**, 1357 (1971).
31. M. Strobel, C. Dunatov, J. M. Strobel, C. S. Lyons, S. J. Perron and M. C. Morgen, *J. Adhesion Sci. Technol.* **3**, 321 (1989).
32. C. Y. Kim, J. Evans and D. A. I. Goring, *J. Appl. Polym. Sci.* **15**, 1365 (1971).
33. K. L. Mittal, *J. Vac. Sci. Technol.* **13**, 19 (1976).
34. D. T. Clark and H. R. Thomas, *J. Polym. Sci., Polym. Chem. Ed.* **16**, 791 (1978).
35. D. T. Clark and A. Harrison, *J. Polym. Sci., Polym. Chem. Ed.* **19**, 1945 (1981).
36. J. E. Klemberg-Sapieha, O. M. Küttel, L. Martinu and M. R. Wertheimer, *J. Vac. Sci. Technol.* **A9**, 2975 (1991).
37. F. D. Egitto, F. Emmi and R. S. Horwath, *J. Vac. Sci. Technol.* **B3**, 893 (1985).
38. S. Noël, L. Boyer and C. Bodin, *J. Vac. Sci. Technol.* **A9**, 32 (1991).
39. A. W. Czanderna, D. E. King and D. Spaulding, *J. Vac. Sci. Technol.* **A9**, 2607 (1991).
40. V. Maurice, K. Takeuchi, M. Salmero and G. A. Somorjai, *Surf. Sci.* **250**, 99 (1991).
41. P. Herrera-Fierro, W. R. Jones, Jr and S. V. Pepper, *J. Vac. Sci. Technol.* **A11**, 354 (1993).
42. B. M. DeKoven and G. F. Meyers, *J. Vac. Sci. Technol.* **A9**, 2570 (1991).

43. A. J. Pertsin and Yu. M. Pashunin, *Appl. Surf. Sci.* **47**, 115 (1991).
44. G. Liebel and R. Bischoff, *Kunststoffe-German Plastics* **4**, 4 (1987).
45. K. Piyakis, E. Sacher, A. Dominque, J. J. Pireaux, G. Leclerc, P. Bertrand and J. B. Lhoest, (in preparation).
46. D. M. Brewis and D. Briggs, *Polymer* **22**, 1 (1981).

Polymer Surface Modification: Relevance to Adhesion, pp. 87–99
K. L. Mittal (Ed.)

Plasma treatment of polymers: the effect of the plasma parameters on the chemical, physical, and morphological states of the polymer surface and on the metal–polymer interface

M. COLLAUD,* P. GROENING, S. NOWAK and L. SCHLAPBACH
Physics Department, University of Fribourg, Pérolles, 1700 Fribourg, Switzerland

Revised version received 12 April 1994

Abstract—The results of the characterization of plasma-treated polypropylene (PP) and polymethylmethacrylate (PMMA) by X-ray photoelectron spectroscopic, atomic force microscopic (AFM), and surface resistivity measurements are presented. Chemical, physical, and morphological modifications of the polymer surface were investigated. The importance of the neutral gas of the electron cyclotron resonance-radio frequency (ECR-RF) plasma treatment is principally demonstrated. Large differences between the effects of noble and reactive gas plasmas were observed by AFM and surface resistivity measurements. However, the maximum sticking coefficient of evaporated Mg on PP is similar for both Ar and N_2 plasma treatments.

Keywords: PP; PMMA; polymer surface; plasma treatment; XPS; AFM; surface resistivity; sticking coefficient; adhesion.

1. INTRODUCTION

Most industrial applications requiring the printing or coating of polymers use some surface pretreatment. However, we are far from understanding the physics and chemistry of the treatment processes and also their limitations. Therefore, the number of research publications in this field has been continuously increasing. Among all the pretreatment methods, plasma techniques have been most commonly investigated for both technological applications and from the view point of basic research [1–3]. Various aspects which are related to adhesion improvement by plasma treatments have been studied: characterization of the plasma mechanisms acting on the polymer surface; chemical, physical, and morphological changes induced by the plasma treatment; the reason for the improvement in macroscopic adhesion; and aging of the surface and of the interface.

*To whom correspondence should be addressed.

The plasma is a complex state of matter that can only be described fully by a characterization of all its constituents (ions, electrons, neutral and excited species, UV and VUV radiation). The plasma equilibrium state is determined by the choice of pressure, gas, potential, and reactor geometry. Any change in these parameters leads to a new equilibrium state of the plasma. Therefore, many studies have been devoted to the understanding of the effects of plasma treatment on polymer surfaces and on interfaces [4–9].

In this paper, we report on polymer surface treatments by an electron cyclotron resonance-radio frequency (ECR-RF) plasma. The main advantage of our low-pressure microwave-RF plasma chamber is that the ECR plasma allows the generation of high densities of charged and excited species at very low pressures (< 0.1 Pa), whereas the RF-power permits the control of the energy of ions striking the sample surface. This study involves several polymers, especially polypropylene (PP) and polymethylmethacrylate (PMMA), and different gases (Ar, He, N_2, O_2, H_2). The surface and interface chemical properties have been investigated by X-ray photoelectron spectroscopy (XPS), whereas the morphology has been determined by an atomic force microscope (AFM). All XPS measurements were done after vacuum transfer of the sample from the plasma chamber to avoid any atmospheric contamination. Surface resistivity measurements provide additional information on the physical modifications induced by the plasma treatments.

2. EXPERIMENTAL

The substrates used in this study were commercial non-crosslinked PMMA foils of 100 μm thickness from Röhm GmbH, and isotactic polypropylene on a base of Propaten HF 24 (ICI). This base material was mixed and extruded with 0.1% Irganox® 1330 (Ciba-Geigy), 0.1% Irgafos® 168 (Ciba-Geigy), 0.05% Ca stearate, and 0.03% DHT-4A (Kyowa Chemical Industries, $Mg_6Al_2(OH)_{16}CO_3 \cdot 4\ H_2O$, a radical scavenger) at 260°C. After cooling, the PP pellets were pressed to plates of 0.8 mm thickness in a heated mold at 230°C. The samples were treated in a low-pressure plasma obtained by ECR with permanent rare-earth magnets at 2.45 GHz microwave frequency. The base pressure in the chamber was 5×10^{-6} Pa. Plasma treatments were performed in the pressure range between 0.04 and 0.5 Pa with N_2, O_2, H_2, Ar, or He gas. The samples can be capacitively coupled to a 13.56 MHz RF-potential, so that a negative DC bias develops on the surface. Contamination due to poor cleaning of the plasma chamber (gas desorption from the walls) never exceeded a few atomic percent as measured by XPS. Details of this ECR plasma experiment have been discussed elsewhere [10]. For comparison, some treatments were also performed with Ar^+ ion bombardment in the preparation chamber of the spectrometer using an Ar^+ sputter gun.

For interface and sticking coefficient studies, Mg films were deposited by thermal evaporation in the preparation chamber of the spectrometer. All the samples used for the sticking coefficient studies were exposed to a defined amount of Mg (13 ML) measured by a quartz crystal microbalance (QCM). Then XPS measurements allow the amount of Mg (atomic percentage) remaining on the surface to be determined. This quantity is a measure of the sticking coefficient of Mg on treated polymers. Thicker films for adhesion measurements were evaporated using an electron beam evaporator in a separate chamber, which was not connected to the plasma chamber through a vacuum line.

XPS studies were performed on a VG ESCALAB 5 spectrometer at a base pressure lower than 10^{-7} Pa with non-monochromatized Mg K_α (1253.6 eV) and Si K_α (1740.0 eV) radiation. Since the plasma chamber was directly connected to the spectrometer by a vacuum line kept at 10^{-5} Pa, all the XPS experiments were done *in situ*. Survey spectra were taken with a 50 eV pass energy, and high-resolution spectra with a 20 eV pass energy. Charging of the sample was corrected by setting the C 1*s* peak at 285.00 eV [11].

The topographic measurements were done on a Digital Instrument Nanoscope III (Santa Barbara, California). The Si_3N_4 levers were gold-coated. The AFM images were made in the constant applied force mode.

The surface resistivity measurements were carried out by a two-wire method, where a voltage is applied and the resulting current is measured [12]. A 617 Keithley electrometer provided the source voltage and the current measurements. Shielded cables were used for the current measurement. Silver electrodes were painted on the polymer surface to allow better contacts. All the experiments were performed in air under controlled humidity [26% relative humidity (RH)]. The resistivity was calculated from a V/I plot and corrected for the geometry of the silver electrodes. The sensitivity threshold of the resistivity is given by the internal resistance of the electrometer and by the geometry of the electrodes, and is 10^{16} Ω.

3. RESULTS AND DISCUSSION

3.1. Chemical modifications

3.1.1. Plasma treatments of polypropylene. The effects (as well as the applications) of the plasma treatments vary depending on the plasma conditions (pressure, treatment time, potential, gas, etc.). For different neutral gases but similar pressures, potentials, and treatment times, plasma treatments induce different states of the surface. Large differences are particularly visible between noble and reactive gases. Figure 1 shows the C 1*s* and valence band (VB) peaks of as-received PP and PP treated with several gases at a low pressure (between 0.04 and 0.1 Pa) for 15–30 s at a floating potential. All the peaks have been normalized to the same intensity to allow a better comparison. The Ar plasma produces a clean hydrocarbon surface. It also induces an increase in the width of the carbon peak, which means a larger number of different bonds in the polymer. The typical PP features in the VB spectra [C—H (C 2*p*) at ~10 eV and C—C (C 2*s*) at ~20 eV], clearly visible on the as-received sample, are mostly lost after these treatments. Previous studies [13] have shown that the VB shapes become less distinguishable for treatments at low pressures or a high RF-potential. This behavior has been attributed to the higher ion energy and flux of this treatment.

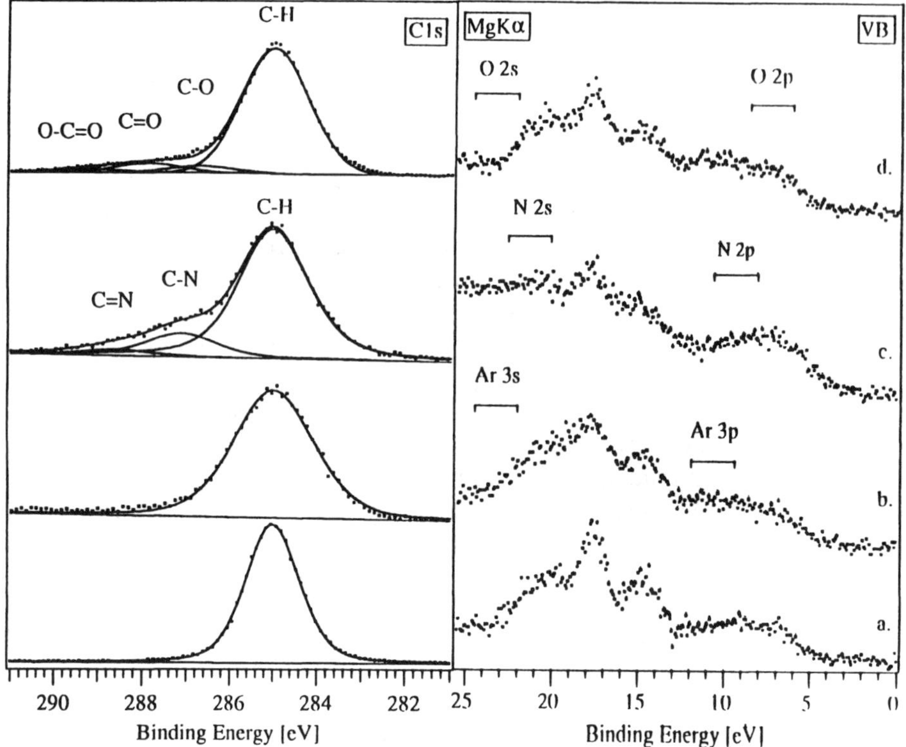

Figure 1. C 1*s* and VB spectra of (a) as-received, (b) Ar-plasma-treated, (c) N_2-plasma-treated, and (d) O_2-plasma-treated PP at low pressures (between 0.04 and 0.1 Pa) and a floating potential. The treatment time was 15–30 s.

The reactive gas plasma treatments (N_2 and O_2) lead to the incorporation of new chemical species, visible in both the C 1*s* and the VB spectra (Figs 1c and 1d). Under these plasma conditions, the incorporation of oxygen or nitrogen is 15–20%. Depending on the pressure and on the treatment time, up to 45% atomic concentration of nitrogen can be incorporated, leading to saturation of the surface [13]. Because of this incorporation and the degradation of the polymer surface by the plasma, the PP features in the VB spectra also partly disappear after reactive gas plasma treatments.

3.1.2. Plasma treatments of polymethylmethacrylate. Survey spectra of as-received, N_2-, O_2-, and Ar-plasma-treated PMMA foils are presented in Fig. 2; high-resolution C 1*s* spectra are displayed in Fig. 3. Both figures show that 10 s of plasma treatment can induce drastic changes on the PMMA surface. The calculated atomic percentage of each species is given in Table 1. The as-received sample contains 26% of oxygen in the ester (C–C=O) form. The C 1*s* peak can be clearly fitted with the four functionalities of the PMMA. The binding energies and the relative peak areas of the functionalities are similar to those given by Beamson and Briggs [11].

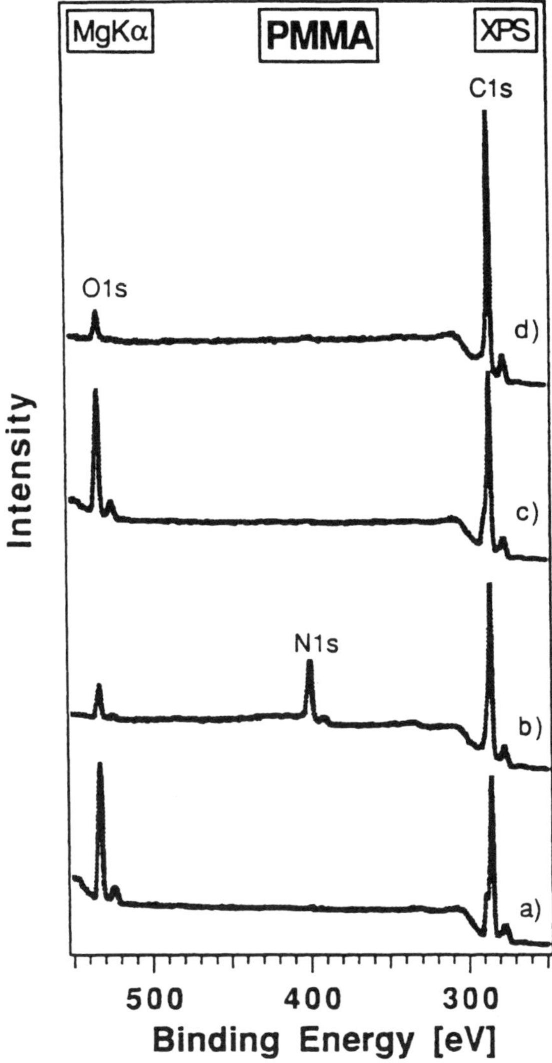

Figure 2. XPS survey spectra of (a) as-received PMMA and PMMA treated for 10 s with (b) N_2 plasma, (c) O_2 plasma, and (d) Ar plasma at low pressures and a floating potential.

Table 1.
Atomic compositions (%) of the PMMA samples after various low-pressure plasma treatments

Treatment	C	O	N
As-received	74	26	—
N_2 plasma	76	6	18
O_2 plasma	79	21	—
Ar plasma	95	5	—

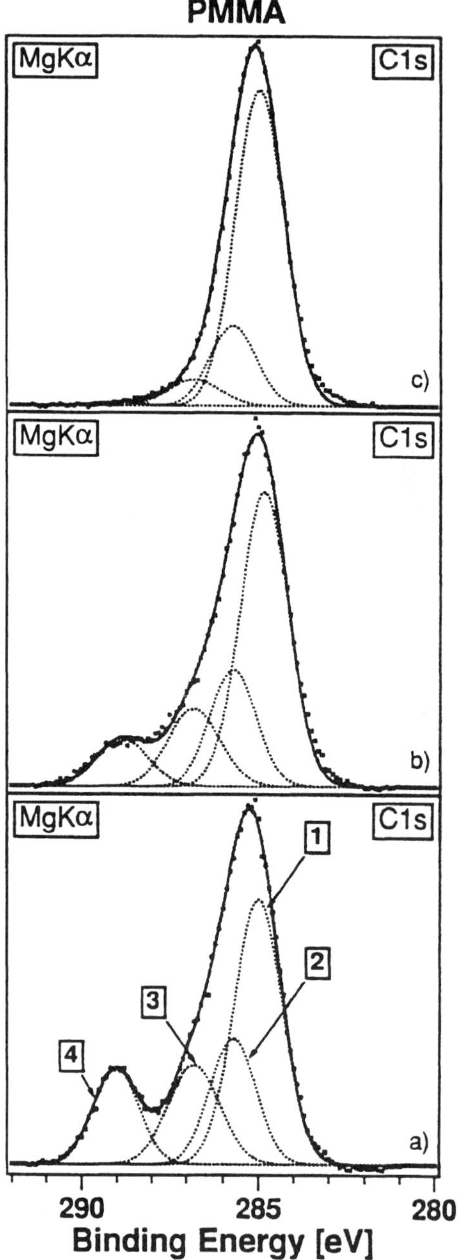

Figure 3. Deconvoluted C 1*s* XPS spectra of (a) as-received, (b) O_2-plasma-treated, and (c) Ar-plasma-treated PMMA at low pressures and a floating potential.

All the plasma treatments, including the O_2-plasma treatment, result in a decrease of the oxygen content, which reveals the destruction of the polar ester groups. The fitted C 1*s* peaks, and particularly the carbon peak of the Ar-plasma-treated sample, show that the C=O bonds are more depleted than the C−O bonds. This means that the plasma first attacks the most polar bonds (C=O) [14].

The reduction of the oxygen content during the N_2 plasma treatment is balanced by the incorporation of nitrogen. The sum of the nitrogen and oxygen atomic concentrations is almost equal to the initial oxygen content. It is clear from the oxygen content and from the fit of the C 1*s* peak that the ester groups are also reduced by the O_2-plasma treatments. Therefore, both reactive gas plasmas lead to a complete degradation of the polymer chains, but the reactive species of the plasma are able to saturate the dangling bonds created in the polymer.

After the Ar-plasma treatment, the oxygen content is almost zero at a depth corresponding to three times the mean free path of the C 1*s* photoelectrons excited by Mg K_α radiation, namely $\sim$8 nm. Therefore carbon−oxygen bonds are depleted by the Ar plasma and a restructuring of the PMMA surface occurs. It is worthwhile noting that if the PMMA ester groups are ejected as volatile compounds, the remaining monomers are similar to polyolefin monomers.

3.2. Morphological modifications

Figure 4 shows AFM measurements made on as-received PP and 135 s plasma-treated PP at low pressures with a RF-potential between −120 and −160 V. The as-received PP surface presents some elongated macroscopic structures (0.1−1 μm). These macroscopic structures can result from macroscopic arrangements of the polymer, but are more likely due to the manufacturing process. Figure 4 shows that the N_2 plasma induces some morphological changes. The surface topography and the roughness change, but the basic morphology of the PP persists. On the contrary, the He plasma leads to a drastic modification of the surface morphology (Fig. 4c). A network of macroscopic chains (0.1 μm in diameter) is present over the whole surface. These changes are due to the sputtering effect of the He plasma.

As mentioned above, the ionized or the atomic nitrogen can react with the dangling bonds of the polymer during the plasma treatment. Therefore, the etching process in a N_2 plasma is different from that in a He plasma, where the He atoms do not react with the surface. The broken polymer chains of a surface subjected to a He plasma will recombine and form new structures. It is also possible that crystalline and amorphous regions of the PP do not react similarly to the plasma etching, or a few monolayers of the polymer are quasi-melted by the plasma treatment.

3.3. Electrical modifications

The surface resistivities of 135 s plasma-treated PP samples at low pressures (between 0.04 and 0.1 Pa) and with a RF-potential are presented in Fig. 5 for several gases. For all treatments, the minimum surface resistivity is reported. The surface resistivity of the as-received PP could not be measured because of the sensitivity of the electrometer (highest measurable resistance = 1×10^{16} Ω) and is therefore taken from literature measurements [15]. The same value of resistivity as that for the as-received PP has been assigned to other non-measurable samples. All the reactive

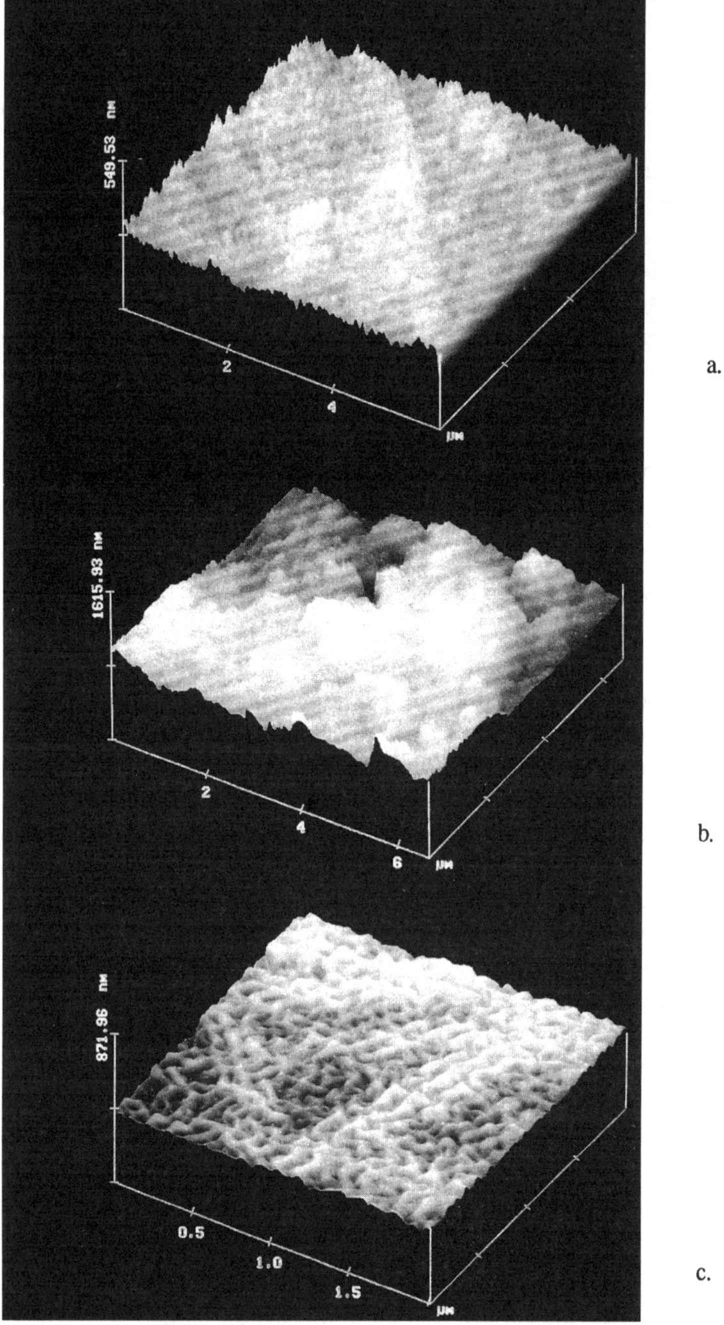

Figure 4. AFM images of (a) as-received PP (5 μm $\times$ 5 μm), (b) N_2-plasma-treated PP at 0.04 Pa (6.5 μm $\times$ 6.5 μm), and (c) He-plasma-treated PP at 0.1 Pa (2 μm $\times$ 2 μm). The treatment time was 135 s and the RF-potential was between -120 and -160 V for both treated samples.

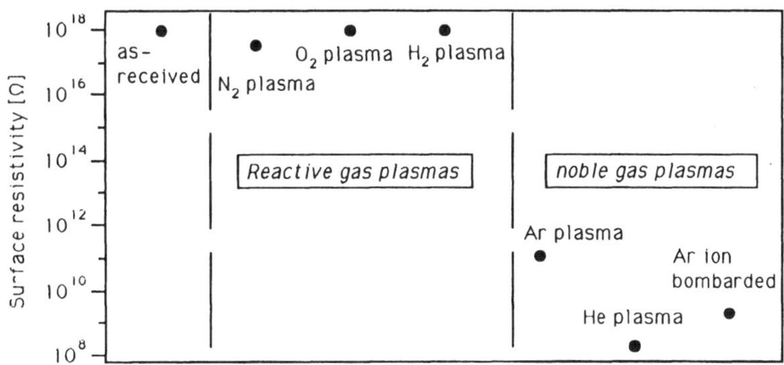

Figure 5. Minimum surface resistivity of PP samples as a function of different treatments. All the plasma treatments were performed for ~2 min at low pressures and with a RF-bias between −100 and −150 V. The Ar-ion-bombarded sample was treated at 2.7 keV and received an ion of 4×10^{16} ions/cm^2.

gas plasmas (N_2, O_2, H_2) lead to almost non-measurable sample resistances and, therefore, to no decrease of the PP surface resistivity. On the contrary, the noble gas plasmas (Ar, He) induce a surface resistivity decrease of 7–10 orders of magnitude. For comparative purposes, the resistivity for an Ar-ion-bombarded sample, 2.7 keV for 25 min, is also reported. The ion dose was 4×10^{16} ions/cm^2. Even with this more highly energetic treatment, the surface resistivity did not reach 1×10^8 Ω.

The resistivity drop depends strongly on the plasma treatment conditions. For both Ar and He plasmas, treatments at a floating potential do not lead to a significant decrease in the PP resistance. Figure 6 shows the dependence of the surface resistivity on the RF-potential for 135 s treatments with the He plasma. The higher the RF-potential, the lower the induced surface resistivity. The application of a RF-bias induces an increase in the energy of ions striking the polymer surface. The ion flux should also be increased by the RF-potential, but to a smaller extent. Therefore, stronger treatments (higher ion energy and flux) result in a lower surface resistivity.

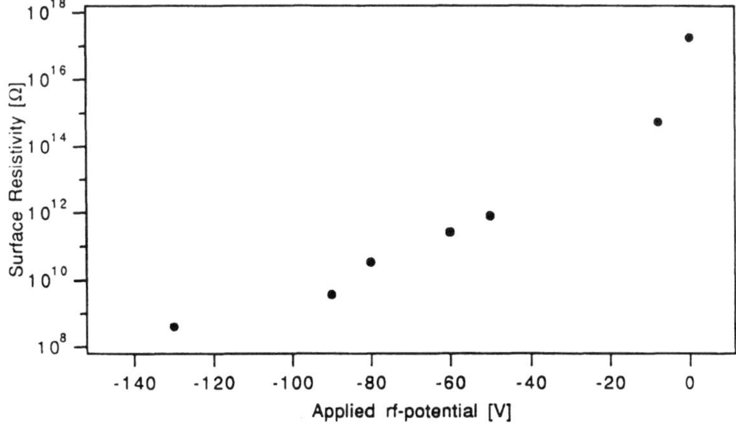

Figure 6. Surface resistivity as a function of the RF-potential for He-plasma-treated PP at 0.1 Pa for 135 s.

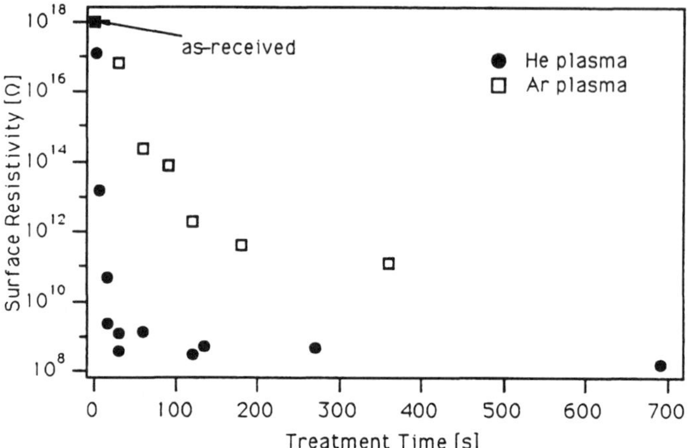

Figure 7. Surface resistivity as a function of the treatment time for Ar- and He-plasma-treated PP samples. The Ar-plasma treatments were performed at 0.04 Pa with a RF-potential of −150 V (30 W). The He-plasma treatments were performed at 0.1 Pa with a RF-potential of −120 V (30 W).

Figure 7 shows the treatment time dependence of the resistivity for Ar- and He-plasma treatments at a low pressure with a RF-potential. 15–30 s is sufficient to induce minimal resistivity for He-plasma treatments, whereas 3–4 min is necessary for Ar-plasma treatments. However, for both plasmas, saturation occurs at long treatment times.

It is well known that the bulk resistivity also decreases by several orders of magnitude after keV or MeV ion irradiation [16, 17]. Usually saturation occurs with increasing dose, i.e. with increasing treatment time for a defined condition. The reason given for this drop in bulk resistivity is that the number of carbon–carbon bonds increases so much that it allows a non-zero conductivity.

In the case of plasma-treated PMMA and PET, noble gas plasmas (Ar, He, Xe) have been shown to eliminate almost all of the oxygen from the surface [14]. The VB spectra of plasma-treated polymer samples also show that the plasma destroys the initial polymeric monomer. Moreover, the VB spectra of PP samples Ar-ion-bombarded with a high energy and a high dose have been found to be similar to those for the highly oriented pyrollitic graphite [18]. Therefore, noble gas plasmas induce scissions of the polymer chains and favor the formation of carbon–carbon bonds. On the other hand, reactive gas plasmas do not allow carbon–carbon bond formation, because the radicals or the dangling bonds in the polymeric chains recombine during the treatment with some plasma species.

3.4. *Sticking coefficient and adhesion*

If a determined amount of metal is evaporated onto each polymer sample and then measured by XPS, the metal content found at the polymer surface is a measure of sticking coefficient of the metal vapor on the various surfaces. Mg has been chosen because its sticking coefficient on the PP surface shows large changes as a function of the plasma treatment. Actually, no Mg atoms at all stick on the as-received PP.

Table 2.
Characterization of the interface and Mg overlayer properties on Ar- and N_2-plasma-treated PP

Sticking coefficient characteristics	Plasma treatments	
	Ar plasma	N_2 plasma
Interfacial component	Metallic Mg only	Mg—N bonds
Maximum measured Mg atomic percentage	80	80
Pressure threshold	0.1 Pa	1 Pa
Optimal treatment time	$\geqslant 30$ s	$= 15$–30 s
Application of a RF-basis	Always improved	Usually improved

Table 2 summarizes the effects of the plasma conditions on the sticking coefficient and on the interface formation [19]. First, after N_2-plasma treatment, interfacial Mg atoms are bonded to nitrogen, whereas in the case of the Ar-plasma treatment, no Mg—C bonds can be observed. Second, the maximum sticking coefficient of Mg on Ar- or N_2-plasma-treated PP is the same for optimal conditions of the treatment. A quantitative estimation of the sticking coefficient of Mg on PP was done by comparison with the sticking coefficient on a clean metal sample. This estimation gave a maximum sticking coefficient of 0.3 ± 0.2 [19]. This sticking coefficient can be obtained for plasma conditions that maximize the ion energy and flux; namely, for lower pressures or greater RF-potentials. Actually, not all the treatments lead to an improvement in the sticking coefficient. For example, there is a pressure threshold (Table 2) above which the sticking coefficient is almost zero. Finally, 30 s was found to be the optimal treatment time for both gases. However, long Ar-plasma treatments (> 10 min) result in the same sticking coefficients as that after a 30-s treatment; long N_2-plasma treatments result in overtreated surfaces and low sticking coefficients.

A simple Scotch tape test was also performed on N_2-plasma-treated PP samples with 3 μm thick aluminum films. This metallization was not done *in situ*, but in a separate electron beam evaporator under identical conditions for all the samples. The as-received PP interacts only weakly with the Al coating, as it can be completely peeled. After 5 s of N_2-plasma treatment, the PP surface leads to good adhesion properties, as the Al film cannot be removed by the same Scotch tape test. The 120 s plasma treatment causes an overtreatment according to the sticking coefficient results, and poor adhesion occurs.

Actually, the sticking coefficient of Al on PP is greater than that of Mg (Al vapor sticks partially on untreated PP). However, our experiments and comparison with literature results [5] allow us to think that both metals show a similar behavior towards plasma treatments.

Therefore, there is good agreement between the XPS measurements of the chemical state of the surface, the sticking coefficient, and the preliminary Scotch tape macroscopic adhesion measurements. AFM and resistivity measurements have to be investigated further to see how they are related to the macroscopic adhesion phenomena.

4. CONCLUSION

Depending on the plasma treatment conditions, treated polymers can present quite different surface properties. Reactive gas plasmas induce the incorporation of new chemical species. The XPS C 1*s* spectra show that the new oxygen or nitrogen species are linked to carbon atoms, whereas the VB spectra show that some degradation of the monomer unit takes place. This incorporation of new species induces small morphological changes and the surface resistivity is similar to that of the as-received sample. On the other hand, the surface deteriorations induced by noble gas plasmas are clearly visible in the C 1*s* and VB spectra and in the decrease of the oxygen content in the PMMA samples. The rearrangement of the broken polymeric chains induces considerable morphological changes and more carbon–carbon bonds. Therefore, the surface resistivity drops by 7–10 orders of magnitude. However, despite these fundamental differences between noble and reactive gas plasma treatments, the maximum sticking coefficients of Mg vapor on PP surfaces after Ar- or N_2-plasma treatments are similar.

Acknowledgements

We thank H. P. Haerri (Ciba-Geigy, Fribourg, Switzerland) for providing us with the PP samples, S. Kasas for the AFM measurements, O. Chauvet for fruitful discussions, and P. Monney for his contribution to this study. This work was supported by program NFP 24 of the Swiss National Science Foundation, Ciba-Geigy, and Alusuisse-Lonza AG.

REFERENCES

1. D. M. Brewis and D. Briggs, *Polymer* **22**, 7 (1980).
2. E. M. Liston, *J. Adhesion* **30**, 199 (1989).
3. (a) K. L. Mittal and J. R. Sausko (Eds), *Metallized Plastics 1: Fundamental and Applied Aspects*. Plenum Press, New York (1989).
   (b) K. L. Mittal (Ed.), *Metallized Plastics 2: Fundamental and Applied Aspects*. Plenum Press, New York (1991).
   (c) K. L. Mittal (Ed.), *Metallized Plastics 3: Fundamental and Applied Aspects*. Plenum Press, New York (1992).
4. L. J. Gerenser, *J. Vac. Sci. Technol.* **A6**, 2897 (1988).
5. L. J. Gerenser, *J. Vac. Sci. Technol.* **A8**, 3682 (1990).
6. R. Foerch and D. H. Hunter, *J. Polym. Sci., Part A Polym. Chem.* **30**, 279 (1992).
7. F. Normand, J. Marec, Ph. Leprince and A. Granier, *Mater. Sci. Eng.* **A139**, 103 (1991).
8. M. Bou, J. M. Martin and Th. Le Mogne, *Appl. Surface Sci.* **47**, 149 (1991).
9. S. Akhter, X. L. Zhou and J. M. White, *Appl. Surface Sci.* **37**, 201 (1989).
10. S. Nowak, P. Gröning, O. M. Küttel, M. Collaud and G. Dietler, *J. Vac. Sci. Technol.* **A10**, 3419 (1992).
11. G. Beamson and D. Briggs, *High Resolution XPS of Organic Polymers*. John Wiley, New York (1992).
12. *Low Level Measurements*. Keithley Instruments, Inc. (1992).
13. M. Collaud, S. Nowak, O. M. Küttel, P. Gröning and L. Schlapbach, *Appl. Surface Sci.* **72**, 19 (1993).
14. P. Gröning, M. Collaud, G. Dietler and L. Schlapbach, *J. Appl. Phys.* **76**, 887 (1994).
15. J. Brandrup and E. H. Immergut (Eds), *Polymer Handbook*. John Wiley, New York (1989).
16. P. Mazzoldi and G. W. Arnold (Eds), *Ion Beam Modification of Insulators*. Elsevier, Amsterdam (1987).

17. L. B. Bridwell, *Solid State Phenomena* **27**, 163 (1992).
18. S. Nowak, M. Collaud, G. Dietler, P. Gröning and L. Schlapbach, *J. Vac. Sci. Technol.* **A11**, 481 (1993).
19. M. Collaud, S. Nowak, O. M. Küttel and L. Schlapbach, *J. Adhesion Sci. Technol.* **8**, 435 (1994).

Polymer Surface Modification: Relevance to Adhesion, pp. 101–120
K. L. Mittal (Ed.)

Interfacial chemistry in Al and Cu metallization of untreated and plasma treated polyethylene and polyethyleneterephthalate

A. RINGENBACH,*,[1] Y. JUGNET[2] and TRAN MINH DUC[1]

[1]*Centre Esca de NanoAnalyse et Technologie de Surface, Université Claude Bernard Lyon 1, 43, Boulevard du 11 Novembre 1918, F69622 Villeurbanne Cedex, France*

[2]*Institut de Recherche sur la Catalyse, CNRS, 2, Avenue Albert Einstein, 69622 Villeurbanne Cedex, France*

Revised version received 27 February 1995

Abstract—The growth of Cu and Al films thermally evaporated onto polyethylene (PE) and polyethyleneterephthalate (PET) surfaces is followed *in situ* by XPS (X-ray Photoelectron Spectroscopy) and XAES (X-ray Auger Electron Spectroscopy) from the early submonolayer stages up to the completion of a metallic film. PE and PET surfaces were metallized first without any preliminary treatment. A second series of metallization experiments were run on the polymer surfaces but pretreated by a remote O_2 microwave plasma (2.45 GHz). These metal films have also been investigated by AFM (Atomic Force Microscopy) in air.

Both metals are shown not to undergo chemical interaction with low surface energy polyolefin such as PE. While an abrupt interface is seen with Al, a diffusion of Cu into the bulk of the polymer is demonstrated. Large size clusters are evidenced by AFM in the initial steps of deposition.

Cu and Al are both shown to react with PET, but not in the same way. In the case of Al, the chemical interaction across the metal/polymer interface proceeds through an electron transfer from the metal toward the ester group O=C—O. With Cu, the chemical interaction is not so clearly evidenced and the Cu is found to diffuse into the PET.

Oxygenated functionalities grafted by O_2 plasma on PE and PET are C—O, C=O, O—C—O, O—C=O, and $O_2C{=}O$. The roughness of the PE and PET surfaces is observed by AFM to increase with the plasma treatment. A metal–CO type complex is clearly observed with Al/treated PE and Cu/treated PET. No chemical interaction was observed at the Cu/treated PE interface.

Keywords: Polymer metallization; polyethylene; polyethyleneterephthalate; copper; aluminium; chromium; X-ray photoelectron spectroscopy; oxygen-plasma treatment; Auger; atomic force microscopy.

1. INTRODUCTION

Metallization of polymer substrates is currently widely investigated because of its various applications in microelectronics, packaging, optics, medical implants, etc.

*To whom correspondence should be addressed.

A knowledge of the chemical bonding between the metallic film and the polymer substrate and that of the microscopic morphology of the interfacial region represents key routes for understanding microscopic mechanisms of adhesion. XPS combined with *in situ* metal deposition, within controlled UHV conditions, has been shown to be a valuable methodology to address the interface problem in metallizing process. Very early, Burkstrand [1] studied by XPS the interfaces of Cu, Ni and Cr grown on various polymers and showed the importance of oxygen in the bond strength. He then suggested that the increasing order of adhesion strength when going from Cu to Ni to Cr, was related to the electronegativity and ability to form oxygen complexes with these metals. Thus the adhesion strength of a metallic thin film on a polymeric substrate will depend, among other factors, on the chemistry across the interface and the strength of chemical interaction between the metal and the polymer [2] (i.e. on the chemical reactivity and nature of both constituents of the interface).

A large number of photoemission studies relative to the metallization of polymers have been performed in the last twenty years, mostly on polyimides [3].

Most metals do not chemically react with PE, a rather inert low surface energy polymer. From a comparative study by XPS and pull strength measurements, Bodö and Sundgren [4] showed that the adhesion was relatively high for Ti, Ni and Cr films, whereas it was low for Al, Cu, Ag, and Au films deposited on PE. The deposition of Cu on PE has also been investigated by Pertsin and Pashunin [5] using XPS studies. Chtaïb *et al.* [6] reported a study of adsorption sites at the Cu/PET interface by XPS and VUV photoemission. Al reacts with PET as observed in various experiments [2, 7] through the formation of Al—O—C type complex. XPS results on the Ag/PET interface [8] suggest a charge transfer between silver and the carbonyl oxygen in PET in the early stages of metallization.

Surface treatment by oxygen plasma is generally used to produce polar oxygenated functionalities in order to enhance the wettability and bondability of the polymer and to improve the adhesion of metallic films. Gerenser [9] has investigated the Ag/plasma treated PE interfaces and correlated the change in chemical bonding to the adhesion strength which increases in the following order: untreated < argon plasma < oxygen plasma < nitrogen plasma. Changes in PE core levels are explained by the formation of Ag—O—C and Ag—N—C species on oxygen and nitrogen plasma treated PE.

Here we focus on the characterization of the chemical bonding across four interfaces formed by cross combinations of an inert (Cu) and a reactive (Al) metal with an inert (aliphatic PE) and an oxygen containing polymer (PET) which is expected to be more reactive. These materials encompass the full range of chemical interaction strengths at metal/polymer interfaces. To investigate the influence of O_2 plasma treatment of the polymer surface, low reactivity Cu and PE appear to be most suitable probes. We have examined the influence of the plasma treatment of PE and PET surfaces upon Cu metallization and of PE surface upon Al and Cu metallization. Our first objective is to explain XPS chemical shifts ΔE_b (binding energy shifts) in terms of the interplay of interfacial bonding, charging and cluster size effects. The second issue is the morphology at interfaces: interdiffusion and growth modes. When no chemical interaction takes place, diffusion of the deposited metal atoms into the polymer can occur. The growth mechanism will depend on the balance between the adsorption energy of the metal on the polymer (E_{ads}) and the cohesion energy of the deposited metal (E_{coh}). When E_{ads} is lower than E_{coh}, a Volmer–Weber (three-dimensional island growth) mechanism is expected, whereas a Franck–van der Merwe (layer-by-layer) mechanism is more probable when $E_{ads} > E_{coh}$. The intermediate Stranski–Krastanov (one or two layer deposit followed by an island growth) mechanism would result after completion of one or two metal layers and a decrease of E_{ads}.

2. EXPERIMENTAL

X-ray photoemission spectra were recorded with a Vacuum Generators ESCA III spectrometer using an unmonochromatized Mg K_α ($h\nu = 1253.6$ eV) radiation. The base pressure was 10^{-10} mbar in the spectrometer. Binding energies were calibrated against the C—H component of C 1*s* set at 285.0 eV. No attempt was made to compensate for the charging effect.

AFM images were obtained using a Nanoscope II equipment from Digital Instruments with the 0.7 μm scanner. All the images were taken at room temperature, in air, using the constant force mode.

The plasma equipment consisted of a 2.45 GHz microwave generator from Raytek. The plasma was created by means of a surface guide in a quartz tube ($\varnothing = 23$ mm) with the distance between the sample and the center of the discharge being fixed at 25 cm, allowing a downstream treatment and thus avoiding electrons and ions impinging on the surfaces of the samples. Plasma treatments were carried out at 50 W in a separate preparation chamber with oxygen pressures of 1 mbar for PE and of 2.5 mbar for PET and flow rates = 250 cc min^{-1} for PE and 1700 cc min^{-1} for PET. Treatment times were 30 min for PE and 25 min for PET. These conditions were determined in previous work and correspond to optimized values obtained for high grafting rates and just before partial degradation [10]. Once treated, the samples were immediately transferred to the preparation chamber of the VG spectrometer.

Cu and Al were deposited in the preparation chamber at a residual pressure in the 10^{-9} mbar range. They were evaporated from a Knudsen type cell previously outgassed for several hours until no contamination was detected by XPS. During metal evaporation the polymer sample was held at approximately 6 cm from the cell extremity. Under these conditions, the evaporation rate is about 0.1–0.2 Å/s.

The low density PE and the biaxially stretched PET were commercial films about 10 μm thick. These samples were cleaned by ultrasonic agitation in trifluorotrichloroethane for a few minutes. No contamination was observed by XPS on these surfaces (no oxygen on the PE samples, no F, and no Cl).

3. RESULTS

3.1. In situ *XPS analysis*

The formation of the metal/polymer interfaces was followed by monitoring the polymer C $1s$ and O $1s$ core levels as well as the Cu $2p$ core level and CuLVV Auger peaks and Al $2p$ as a function of the metal coverage. CuLVV Auger peak is more sensitive to chemical change than the Cu $2p$ photopeak and so is more extensively investigated. The metal coverage was estimated from the atomic ratios Al/C or Cu/C determined from XPS measurements at a detection angle of 45° and corrected for photoionization cross section.

3.2. Cu and Al metallization of untreated PE

Uncovered PE displays (see Fig. 1) a single C $1s$ peak at 285.0 eV with a Full Width at Half Maximum (FWHM) of 1.7 eV. As a function of Cu coverage, no significant change is observed in C $1s$ (Fig. 1), except for a shift towards higher E_b which increases with the amount of deposited Cu and a slight broadening (FWHM = 1.9 eV) observed until Cu/C = 0.15. For larger coverages, the C $1s$ peak recovers its original FWHM. From the very early coverages, the CuLVV peak (Fig. 1) displays all the structural features characteristic of metallic Cu although better resolved when Cu/C increases. The modified Auger parameter α', defined as the summation of the E_b of a core level and the kinetic energy of the corresponding Auger peak, is measured for all coverages at 1851.3 eV (i.e. very close to the bulk Cu value). In the mean time, the Cu $2p$ FWHM decreases continuously from 2.2 eV at very low coverage (Cu/C = 0.06) down to 1.5 eV when Cu/C = 0.85 (i.e. close also to the bulk Cu value). These results indicate that Cu is deposited in metallic form on PE, in agreement with Pertsin and Pashunin [5].

Similarly for Al/PE (see Fig. 2), neither shape modification nor splitting is observed in either C $1s$ or Al $2p$ at any coverage. However, a continuous decrease of both C $1s$ and Al $2p$ binding energies is observed due to a lowering of the charging effect with increasing metal deposition. The C $1s$–Al $2p$ binding energy splitting is constant and equal to 212.1 eV irrespective of Al coverage. After correction for the charging effect, the Al $2p$ peaks appear at the metallic value (E_b = 73.0 eV). Very early, when Al/C $>$ 0.03, the presence of plasmon loss peaks clearly corroborates the metallic nature of the deposited Al [11].

3.3. Cu and Al metallization of untreated PET

The C $1s$ spectrum for clean PET (Fig. 3) is resolved into four components related to the C–C and C–H bonds in the phenyl group (E_b = 285.0 eV), the carbon atoms singly bonded to oxygen (E_b = 286.6 eV), the ester functions (E_b = 289.0 eV) and the 291.7 eV component related to the $\pi \rightarrow \pi^*$ shake-up transition. After deposition of Cu (Fig. 3), we observe a slight shift in all the peaks towards higher E_b, a slight decrease in the relative intensities of the highest E_b components and an overall

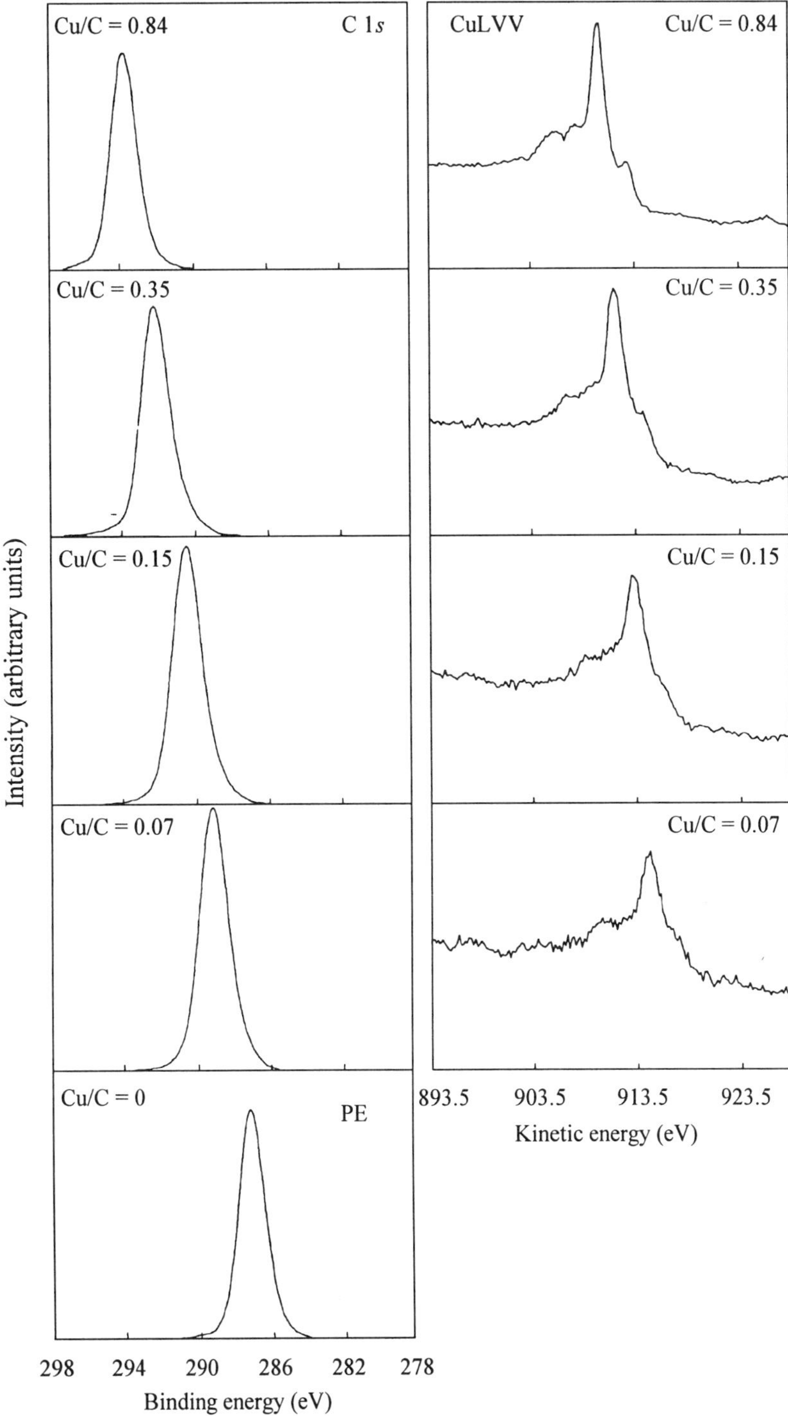

Figure 1. C 1*s* core level spectra and CuLVV X-ray induced Auger peaks measured on Cu/PE interface as a function of Cu coverage.

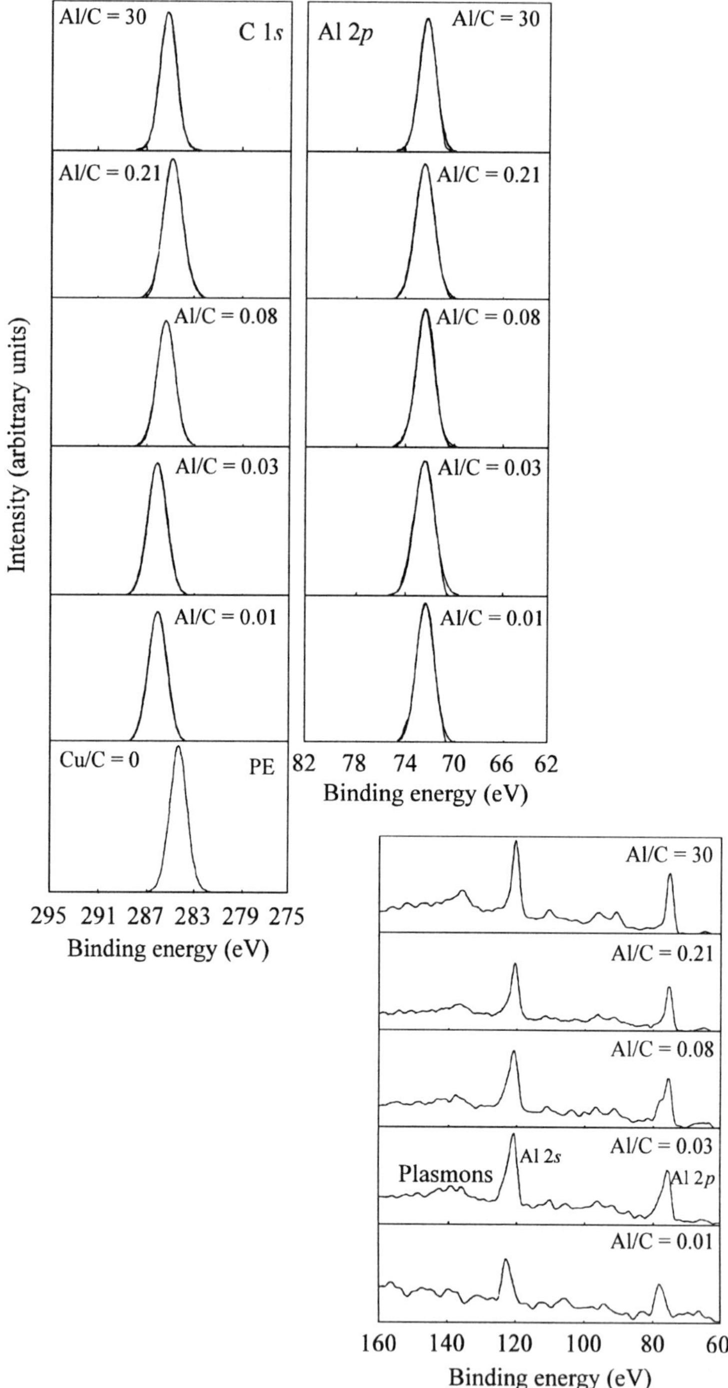

Figure 2. C 1*s* and Al 2*p* core level spectra and characteristic plasmon loss structures measured on Al/PE interface as a function of Al coverage.

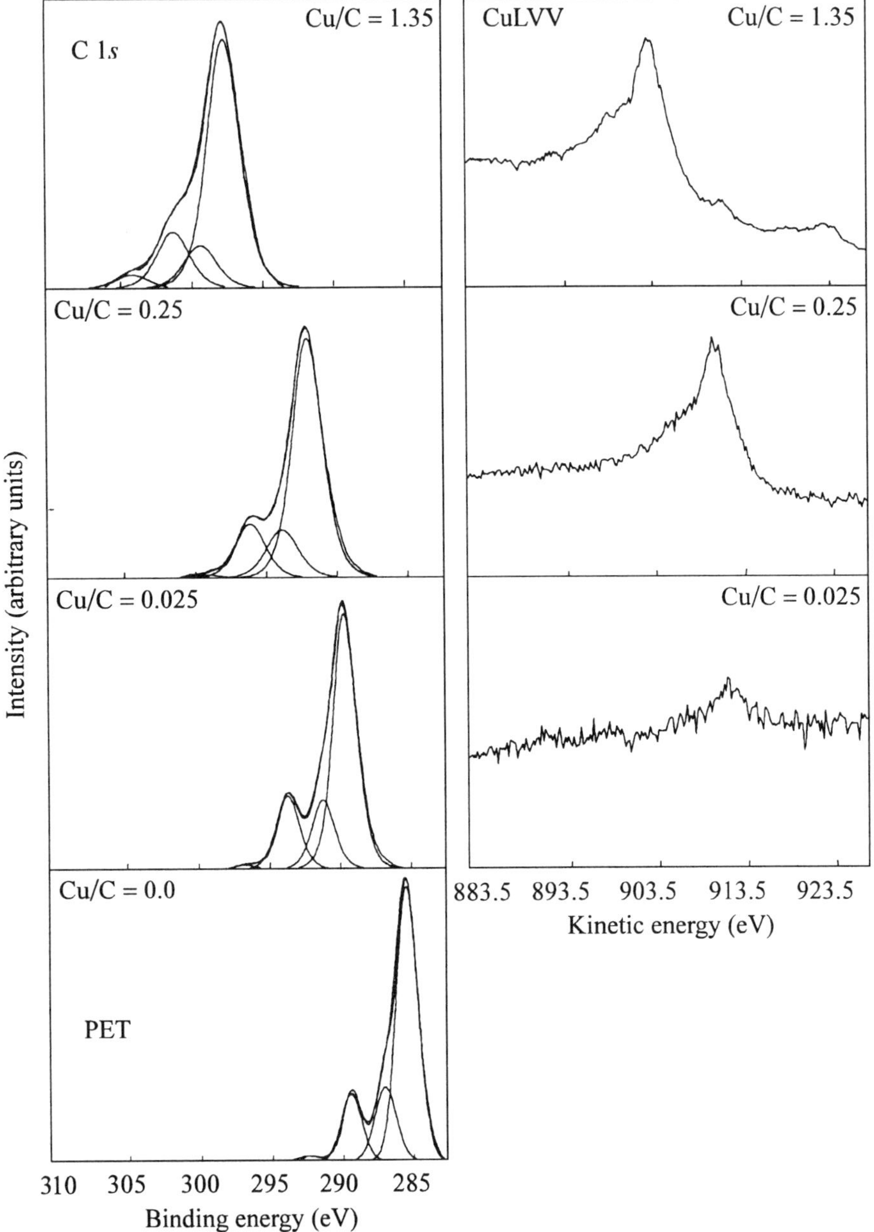

Figure 3. C 1*s* core level spectra and CuLVV induced Auger peaks for Cu/PET interface as a function of Cu coverage.

broadening of each component. We notice here, that the $\pi \rightarrow \pi^*$ shake-up component does not disappear with Cu increasing deposit. This is not the case when Al instead is evaporated on PET (see Fig. 4). The CuLVV peak is broad for any coverage and shifts towards lower kinetic energies with increasing Cu coverage but never

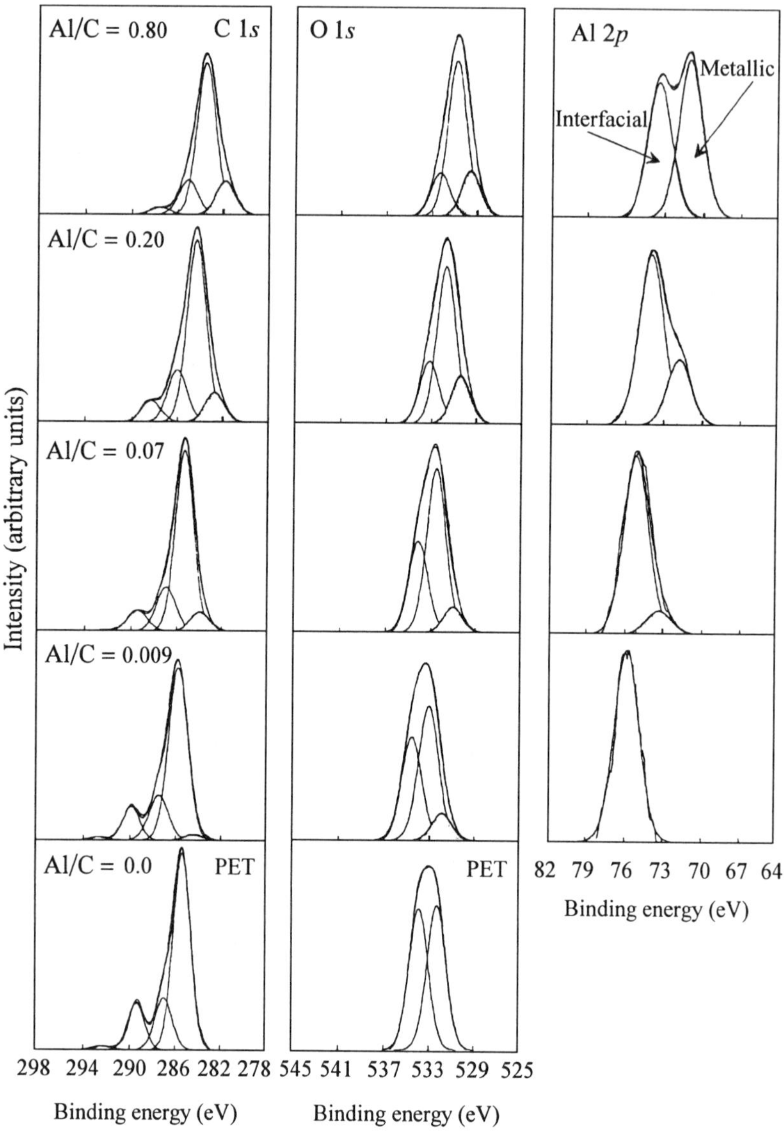

Figure 4. C 1*s*, O 1*s* and Al 2*p* core level spectra for Al/PET interface as a function of Al coverage.

displays the sharp and well resolved structures observed with the Cu/PE interface (see Fig. 1). Furthermore, the α' parameter never reaches the Cu bulk value whatever the Cu coverage. α' starts from 1848.5 eV when Cu/C = 0.02, then increases to 1849.5 and 1849.7 eV for Cu/C = 0.25 and 0.65, respectively, and then decreases to 1848.6 eV when Cu/C = 1.35. The Cu 2*p* FWHM increases continuously from 2.2 eV (Cu/C = 0.02) up to 2.9 eV (Cu/C = 0.83). The large Cu 2*p* FWHM as well as the α' parameter value (α' Cu_2O = 1849.0 eV) indicate a chemical interaction between Cu and PET: π complexation is not expected since the $\pi \rightarrow \pi^*$ shake-up transition does not decrease with increasing Cu coverage. Thus we would rather propose the formation of a Cu–O–C charge transfer complex. However, in contrast to what will be observed later on with Al, no new component, which we would expect following the formation of such a complex, is detected in the C 1*s* signal. This could be the result of a weak interaction between the Cu atoms and the PET that would result in a small shift ΔE_b in the C 1*s* peak. Indeed, Droulas *et al.* [2] observed in the case of metal/polymer interfaces that strongest interaction caused the largest chemical shift. So, in the case of a very weak interaction such as Cu/PET, a very small chemical shift is expected. In such a case, the aromatic and the Cu–O–C interfacial components overlap. The O 1*s* spectra are not reported since no new component has been clearly identified, since it probably overlaps the C=C component.

Cu diffuses into PET, as was already observed with PE. Figure 5 reports the variation of the Cu/C and O/C ratios as a function of the detection angle θ (relative to the surface plane). The Cu/C ratio increases for higher θ, and at grazing, and thus more surface sensitive detection angle, the lower Cu/C ratio is indicative of Cu segregation in the bulk. This diffusion process has also been observed with Cu on polyimides, and has been found to be very dependent on the evaporation flux and temperature [12]. The O/C ratio is characteristic of the polymer: its variation, if any, is a sign of Cu–O interactions.

In contrast with the previous systems, strong modifications of XPS spectra are observed when Al is evaporated on PET (Fig. 4). Even at very low coverages (Al/C = 0.01), we notice a decrease of both O=C–O component and shake-up satellite of the C 1*s* peak. New components appear in, respectively, the C 1*s* spectrum at 283.6 eV, i.e. shifted by 1.4 eV from the main component on the low E_b side, and the O 1*s* spectrum at 531.1 eV on the low E_b side of the main O–C and O=C components located at 533.5 and 531.9 eV [2, 11]. These new components are attributed to C–O–Al components. In the meantime, at very low coverage, the Al 2*p* peak appears as a single component at 75.0 eV and corresponds to an oxidized from of Al. After-while the metallic component starts to increase on the lower binding energy side. Indeed the O 1*s* peak at 531.1 eV and Al 2*p* peak at 75.0 eV are very close to the values usually measured for Al_2O_3 (i.e. O 1*s* at 531.0 eV and Al 2*p* at 74.5 eV).

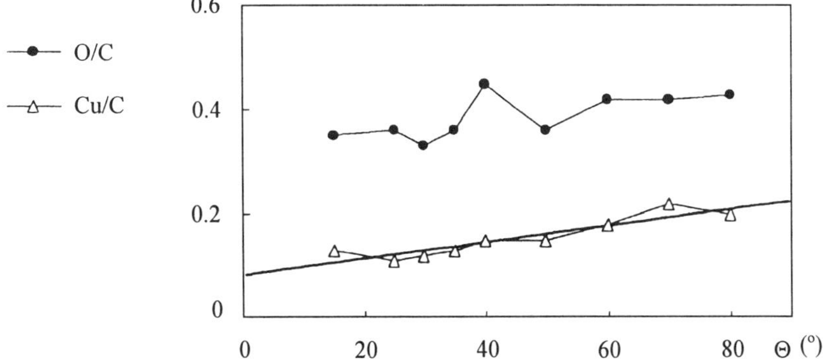

Figure 5. Variation of the Cu/C and O/C atomic ratios (from XPS measurements) as a function of the detection angle for the Cu/PET interface.

The shake-up peak disappears completely for Al/C larger than 0.07, suggesting the occurrence of a second interaction site on the phenyl group. However, no interfacial chemical compound formation has been evidenced by XPS for this site. At Al/C = 0.7 we start having metallic Al [11]. This is evidenced through the appearance of the metallic Al $2p$ component at $E_b = 72.9$ eV and through the appearance of plasmon losses, characteristic of metallic Al as well, for Al/C = 0.5 [11].

3.4. Cu and Al metallization of oxygen plasma treated PE

After O_2 plasma treatment of PE, the C $1s$ spectrum is considerably modified because of the presence of oxygenated species. The C $1s$ peak can be resolved (Fig. 6) into five components shifted toward the high E_b side by 1.4 eV (C–O functionalities), 2.8 eV (C=O functionalities), 4.2 eV (O–C=O functionalities), and 5.2 eV ($O_2C{=}O$ functionalities) from the main aliphatic component. In our experimental plasma treatment conditions, the O/C atomic ratio is 0.23.

In contrast to untreated PE, XPS spectra of Cu/treated PE (Fig. 6) indicate some interaction between the metal and the polymer. However, the C $1s$ spectra skew toward low binding energy, reflecting the contribution of several phenomena: chemical shifts due to plasma oxidation and interaction with the deposited metal atoms, Cu diffusion into the PE bulk, and charging effect. As a result, the C $1s$ profile presents a skewed line shape, and as a result the decomposition of this peak may give inaccurate results.

A very small deposit of Cu induces a large loss of grafted oxygen which is translated into decrease of the highly oxygenated functionalities. For Cu/C = 0.12, the carbonate and ester functionalities vanish and the O/C atomic ratio decreases to 0.13. With further Cu depositions (Cu/C = 0.28), the trend is even more pronounced with the disappearance of the C=O functionalities and a further decrease of the atomic ratio O/C down to 0.1. For larger Cu coverages (Cu/C > 0.3), certain high E_b C $1s$ components are still observed, but oxygen is no longer detected. Considering now the C $1s$ E_b shifts, as previously observed with all the investigated Cu/polymer interfaces, we notice first an overall shift towards higher E_b as a function of Cu coverage up to Cu/C = 0.9. This shift, which reaches a value of 5.4 eV, is followed by a reverse shift towards lower E_b to recover the C $1s$ PET value. In the mean time, we notice an increase of the FWHM (from 1.8 eV when Cu/C = 0 up to 2.4 eV when Cu/C = 0.9) followed by a decrease down to 1.8 eV for higher Cu coverages. The CuLVV Auger peak is also reported in Fig. 6. Although less resolved at low Cu coverages, this peak contains all the structures observed in metallic Cu and for high Cu coverage (Cu/C = 1.95). Similarly to the C $1s$ peak, the CuLVV Auger peak shifts first towards lower kinetic energies and then towards higher kinetic energies for Cu/C > 0.9. The α' parameter is 1850.2, 1850.5, 1851.1, and 1851.2 eV for Cu/C = 0.11, 0.28, 0.9, and 1.95, respectively. The last two values are characteristic of metallic Cu. At the same time, the Cu $2p$ FWHM takes the respective values: 1.9, 2.2. 2.4, and 1.6 eV. At this point, we should remember that such a metallic behavior was observed for much lower Cu coverages on the untreated PE (see Fig. 1).

At very early stage of Al evaporation on plasma treated PE (Al/C = 0.02), as observed with Cu, we notice a rapid loss of O deduced from the C/O ratio and a

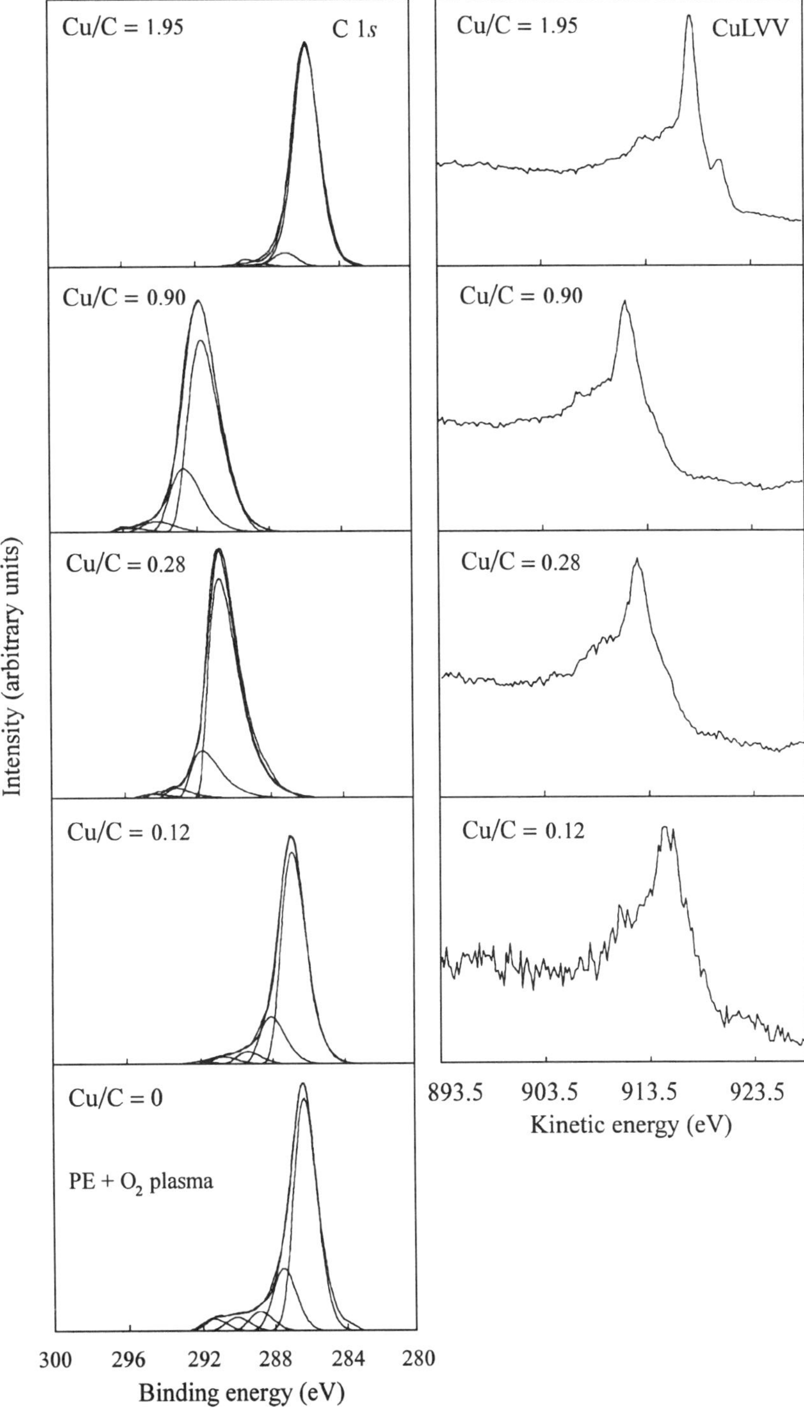

Figure 6. C 1*s* XPS spectra and CuLVV induced Auger peaks for Cu/O_2-plasma-treated PE interface as a function of Cu coverage.

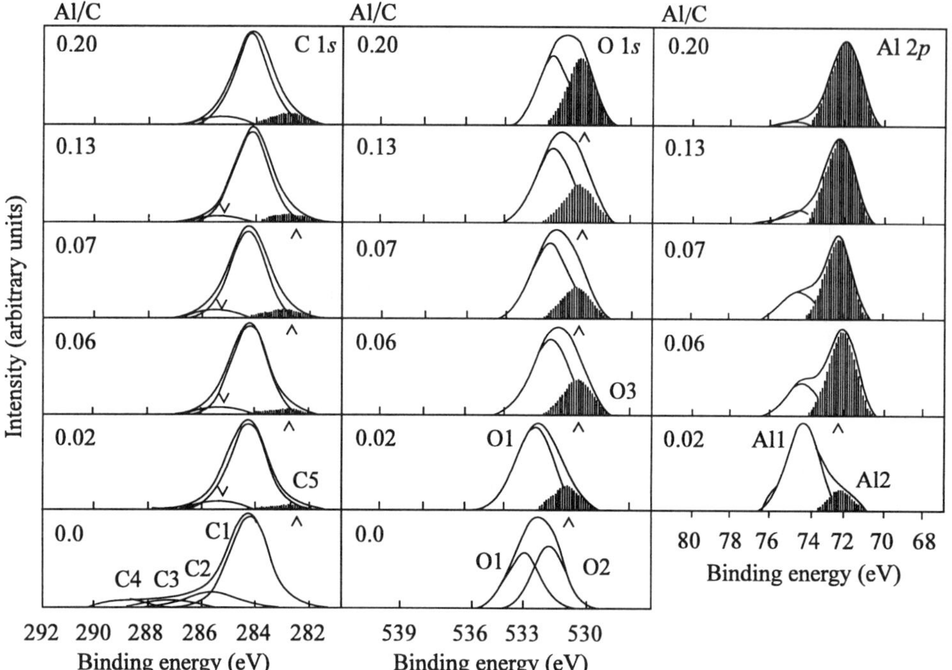

Figure 7. C 1s, O 1s and XPS spectra for Al/O_2-plasma-treated PE interface as a function of Al coverage.

strong attenuation of O–C=O, and C=O functionalities (Fig. 7). Further, as previously quoted for the Al/PET interface, a new C 1s component appears on the low E_b side shifted by 1.4 eV from the main C–H component. At the same time, a new O 1s component appears also at lower E_b (±530.8 eV) together with the appearance of the Al 2p component corresponding to an oxidized Al form at E_b = 74.7 eV (Fig. 7). These components are thus assigned to a C–O–Al type complex. On the Al 2p signal, a second component appears at E_b = 72.9 eV attributed to the metallic form of Al. Plasmon losses are observed very early in the Al 2p peak, for Al/C ratio as low as 0.07. The formation of both interfacial complex and metal occurs simultaneously as observed by XPS until Al/C ratios up to 0.2. Afterwhile, metallic Al is the only observed component. All C 1s, O 1s, and Al 2p peaks shift to lower E_b when the Al coverage increases. To summarize, in contrast to the abrupt interface formed between Al and the untreated PE, the Al/treated PE interface appears to be diffuse with the formation of an Al–O–C complex.

3.5. *Cu metallization of oxygen plasma treated PET*

A large amount of oxygen containing groups are produced on PET after O_2 plasma treatment. The C 1s spectrum of uncoated treated PET (Fig. 8) can then be resolved into five components shifted by 1.8 eV (C–O), 2.8 eV (C=O), 4.2 eV (O=C–O), and 5.2 eV (O_2C=O) from the main aromatic component. The grafting rate determined from the Δ(O/C) uptake is about 55%. The copper evaporation leads to a slight and

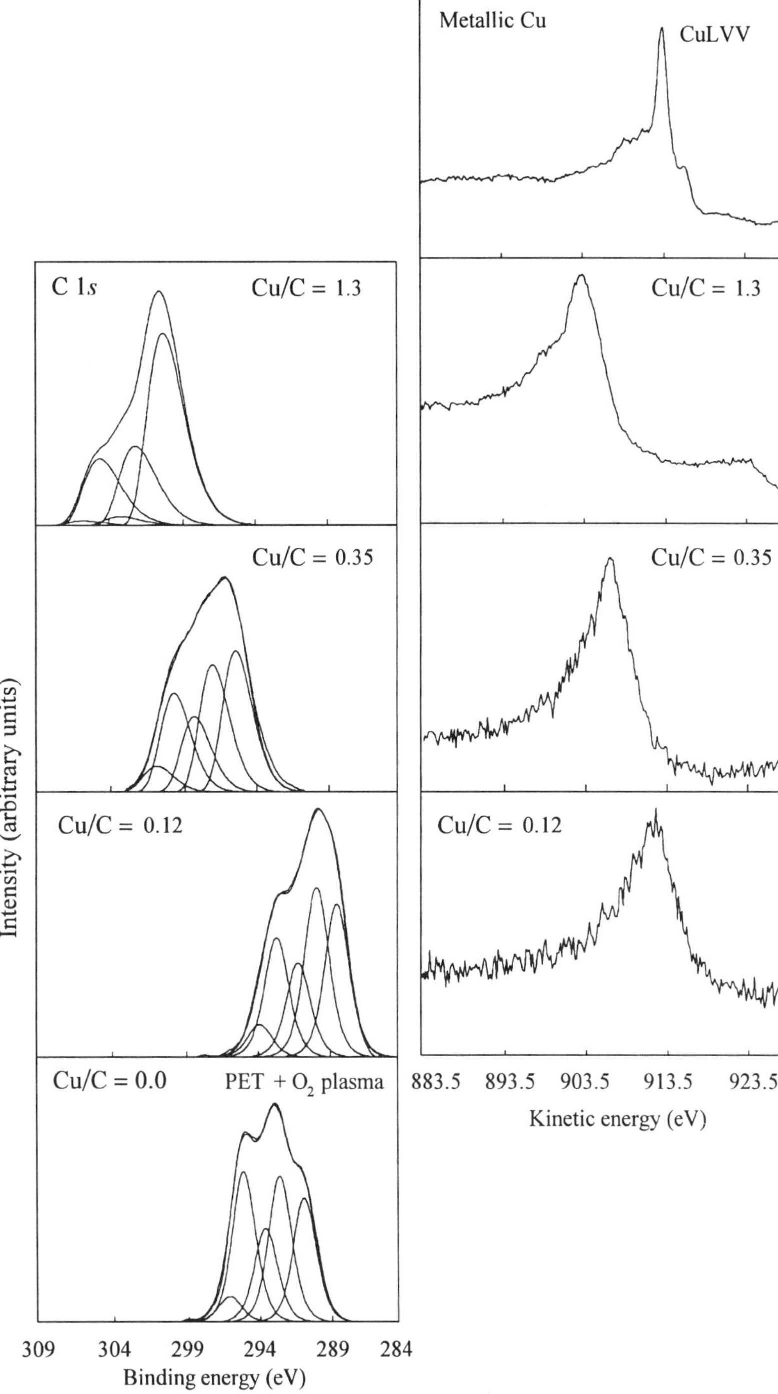

Figure 8. C 1*s* core level spectra and CuLVV induced Auger peaks for Cu/O_2-plasma-treated PET as a function of Cu coverage.

slow decrease of the oxygenated C 1*s* functionalities (Fig. 8). When the Cu/C ratio increases from 0 to 0.12, 0.35 and 1.3, the O/C ratio decreases from 1.1 down to 0.95, 0.8 and 0.55, respectively. In pure PET this ratio is 0.4. At high Cu coverage (Cu/C = 1.3), the various C 1*s* components are very broad, both aliphatic and aromatic, and the C–O and O–C=O components of PET are observed with probably some remaining oxygenated functionalities (C=O and O_2=CO). The CuLVV Auger peak appears as a broad structureless peak (Fig. 8) which is not at all characteristic of metallic copper for all the investigated coverages. The α' Auger parameter is 1848.8 eV for Cu/C = 0.12 and 0.35. It increases to 1850.0 and 1850.4 eV for Cu/C = 0.65 and 1.3, respectively. Here again, we observe an overall shift in all peaks towards higher binding energy. The Cu 2*p* FWHM is quite large ~2.8 – 3.9 eV~ and increases with metal coverage.

3.6. AFM results

Imaging of some of the above surfaces has been carried out in the air by AFM in the contact mode. We are aware that the tip may interact with the polymer surfaces, so only a comparative description of these images will be performed.

Figure 9 presents a set of AFM images of Al/PE (untreated and plasma treated) samples. The mean roughness R_a has been determined for all these surfaces on a scanned (1 μm × 1 μm) are. The images of untreated PE before Al coating (Fig. 9A) and after an Al coating corresponding to a Al/C = 0.22 (Fig. 9B) are displayed. These images have the same z scale and can be directly compared. We notice an increase in the R_a value when going from the uncoated surface (R_a = 5.3 nm) to the coated surface (R_a = 8.1 nm). Similarly R_{max}, the difference in height between the highest and the lowest points on the profile relative to the mean line over the length of the profile, increases from 25.5 nm for the uncoated PE to 35.2 nm after Al deposit. These results indicate a metal deposition in the form of Al islands which do not spread over the PE surface and corresponds to a Volmer–Weber growth.

R_a has been measured on plasma treated PE (Fig. 9C) and after deposition of Al (Al/C = 0.2, Fig. 9D) on that treated surface. After oxygen plasma treatment, PE surface is made rougher, R_a = 8.7 nm. Subsequently, a slight decrease in R_a upon metallization is found. An R_a value of 7.4 nm is measured at Al/C = 0.2.

AFM image of uncovered untreated PET is shown in Fig. 10A as well as the Cu coated surface corresponding to a Cu/C = 0.8 (Fig. 10B) as determined from XPS results. The measured R_a values are, respectively, 2.75 nm and 3.2 nm, with the corresponding R_{max} values being 13.9 and 16.5 nm. A very slight increase in roughness upon metallization is then found. Large structures (mean diameter measured around 100 nm) are observed on the coated surface. These structures are not made of Cu since diffusion of copper into the polymer has been observed by XPS. AFM data rather support the formation of Cu islands which diffuse into the bulk of the polymer and react to form Cu–O–C bonds. The observed structures would be (Cu–O–C + Cu) islands buried in PET.

In Fig. 10, we present the AFM images obtained for O_2 treated PET after Cu deposition. The image (Fig. 10C) obtained for Cu/C = 0.1 displays islands the mean

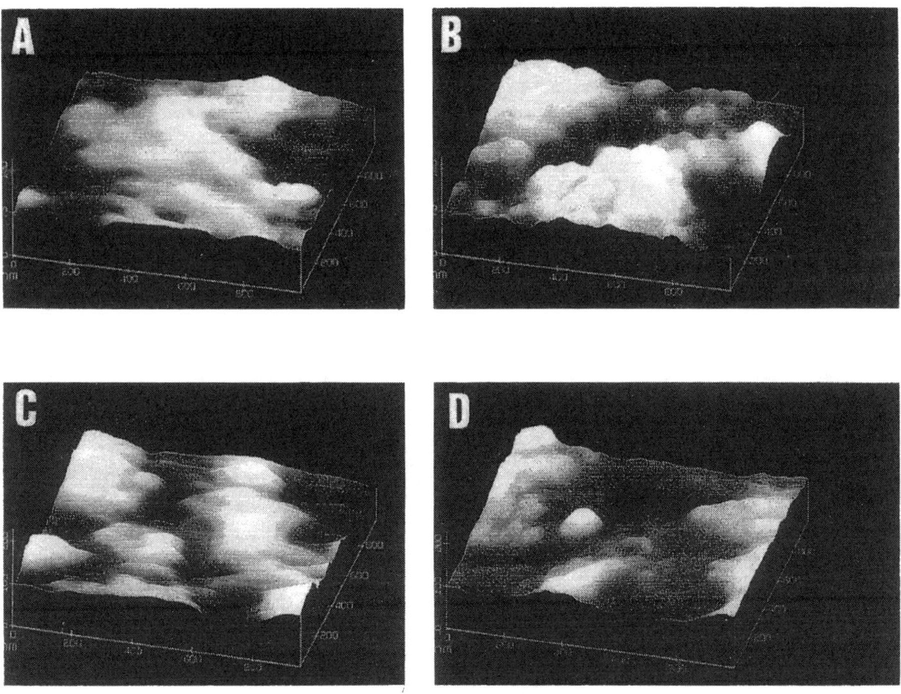

Figure 9. Al/PE (untreated and O_2 plasma treated) interface AFM images: uncoated and untreated PE (A), Al coated untreated PE (Al/C = 0.22) (B), uncoated O_2 plasma treated PE (C), Al coated O_2 plasma treated (Al/C = 0.2) (D).

diameter of which is about 50 nm (R_a = 1.75 nm). The size of these islands is greatly enhanced (mean diameter about 100 nm) for Cu/C = 0.7 (R_a = 9.15 nm, Fig. 10D). At higher coverage, corresponding to a thickness > 10 nm, we observe (Fig. 10E) smaller islands characteristic of metallic copper as deduced from XPS results (R_a = 3.95 nm).

4. DISCUSSION

All these results constitute a set of varying interfaces which we will try now to understand and classify with respect to the metal–polymer bond strength. Indeed, the interpretation of the spectra is made difficult because of the competition between the chemical bonding, the charging and the clustering effects. Both the initial and final states of the photoemitted electrons depend on the physical state of the deposited metal [13]: isolated atoms, clusters or metallic film.

4.1. Unreactive interface: Cu and Al/PE

On untreated PE no chemical bond is formed with both evaporated Cu and Al. Metallic Cu or Al is deposited from the very low coverages without formation of an in-

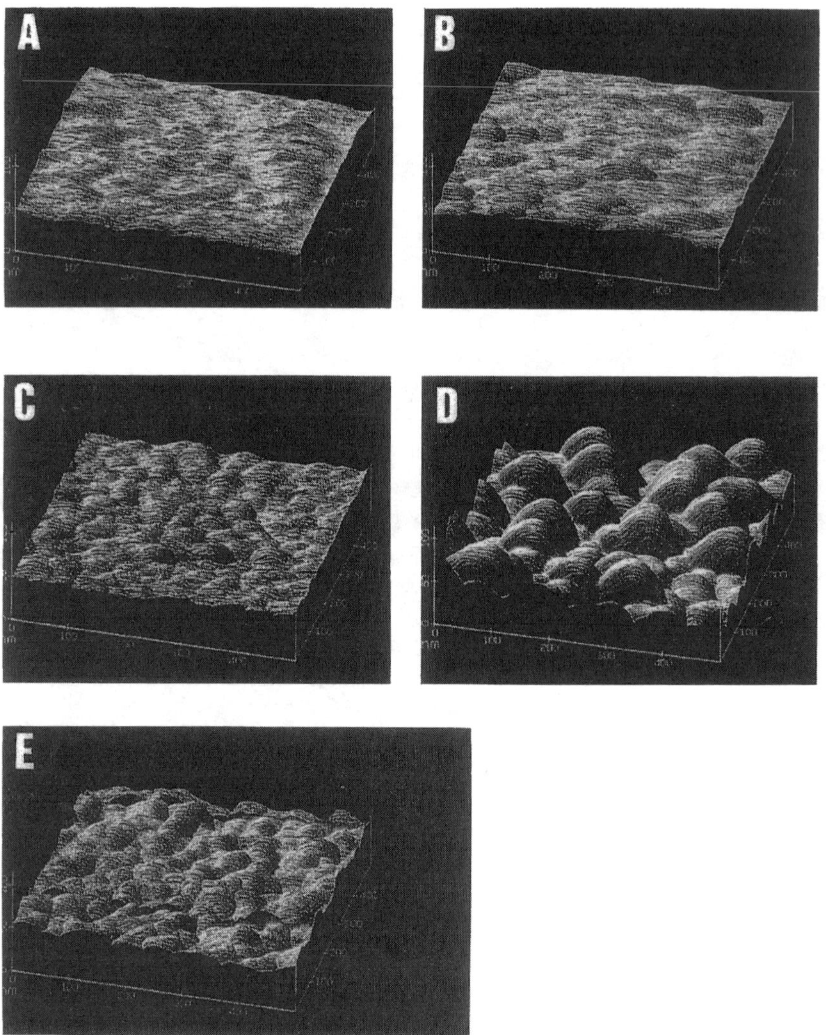

Figure 10. Cu/PET (untreated and O_2 plasma treated) interface AFM images: uncoated and untreated PET (A), Cu coated untreated PET (Cu/C = 0.8) (B), Cu coated O_2 plasma treated PET (Cu/C = 0.1) (C), Cu coated O_2 plasma treated PET (Cu/C = 0.7) (D), Cu coated O_2 plasma treated PET (Cu thickness greater than 10 nm) (E).

terfacial compound. With Al a very abrupt interface is formed. A slight increase in the mean roughness is observed as well as the formation of small metallic clusters. In that case, a Volmer–Weber island growth mechanism is proposed in which the Al adsorption energy is lower than its cohesion energy.

Although no chemical bond is observed between Cu and PE, the Cu/PE interface is not so sharp as Al/PE because of the diffusion of Cu into the polymer. The smaller atomic radius of Cu (1.28 Å) compared to Al (1.43 Å) may be the reason for the difference in behavior between Al and Cu on PE.

For these metal/polymer interfaces with no interfacial chemical bond formation, XPS shifts arise mainly from charging effects. An increase of the charging effect at the very beginning of metallization as observed for Cu and Al on PE is not so surprising. Because of the very high photoionization cross section σ of the core levels of metals (i.e. σAl 2p and σCu 2p) relative to that of the polymer core level (i.e. σC 1s), a considerable amount of photoholes will indeed be produced as soon as metal atoms are deposited onto polymers. These photoholes could not disappear due to the insulating character of the polymer. This excess of positive charges compared to that produced on the uncoated polymers will increase the charging effect when going from uncovered to metallized polymer surfaces, a result which may appear at first glance unrealistic. As a matter of fact, when a bulk-like metallic film is completed, charging effects are then effectively neutralized. Further, we expect charging to be especially enhanced when metal atoms diffuse into the polymer and form smaller clusters embedded in the subsurface region. By comparing Cu/PE and Al/PE interfaces we conclude that charging in the early stage of metallization must be larger for Cu, as indeed is observed experimentally. The effect of cluster size on XPS shift should act in the same direction as charging and so cannot be clearly resolved from charging.

4.2. Reactive interfaces: Cu and Al/PET

We first discuss Al/PET since Al clearly reacts with PET to form an interfacial chemical compound as evidenced by the decrease of the PET C 1s oxygenated functionalities, the appearance of new chemically shifted C 1s and O 1s components and the presence of an oxidized form of Al at very low coverage. Two adsorption sites have been evidenced

The ester function provides a first site for Al adsorption at low coverage which is associated with the formation of an Al–O–C complex. Novis *et al.* [14] had also reported a decrease of the oxygenated functionalities of the PET after Al deposition that they studied by HREELS (High Resolution Electron Energy Loss Spectroscopy).

A second site is the phenyl group with lower adsorption energy since no interfacial compound formation has been evidenced. Metallic Al is deposited once all the adsorption sites are saturated. A very small increase in the mean roughness is observed. Al atoms are not so mobile on PET as on PE owing to their chemical interaction with the oxygenated functionalities of PET. Here, at least at low coverage, we clearly do not observe the formation of clusters. The interaction energy E_{ads} between the metal atoms and the polymer is then larger than the cohesion energy E_{coh} of the metallic layer and we rather suggest the van der Merwe (layer-by-layer) growth mechanism.

With Cu on PET, the situation is not clear out. The interaction is presumably weak and further its observation is made difficult by Cu diffusion into the polymer. The phenyl group does not appear to be a bonding site since no attenuation of the $\pi \rightarrow \pi^*$ shake-up is observed. On the other hand, the Cu α' Auger parameter never reaches the characteristic value of bulk metallic copper (Fig. 11). Cu core levels and Auger peaks are characteristic of Cu^+ rather than Cu° suggesting a weak interaction.

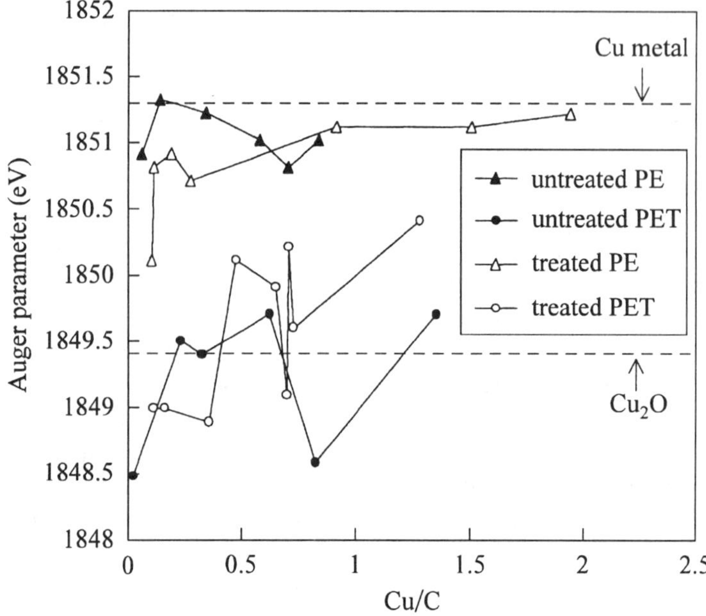

Figure 11. Auger parameter (α') variation as a function of Cu coverage (Cu/C) for Cu deposition on PE, PET and O_2 plasma treated PE and PET. The values of Cu and Cu_2O α' parameters are reported for comparison.

Although not so clearly evidenced as with Al/PET (only a weak attenuation of the oxygenated components of C 1*s* is observed), a Cu–O–C type complex is probably formed. Indeed, the presence of untreacted PET at the surface prevents an easy identification of the chemical species on this emebedded interface. Chtaïb *et al.* [6] reported that at low coverage, Cu binds preferentially to the oxygen from carbonyl and ether groups of PET, and at high coverage, the additional Cu remains metallic giving rise to a three-dimensional island-like growth mode. However they do not mention any Cu diffusion, probably because they worked with very thin PET films. The deposit of Cu on PET leads to in-depth formation of islands whose mean diameter is around 100 nm.

4.3. Role of oxygen plasma treatment

The oxygenated functional groups created by plasma prior to metallization provide adsorption sites for deposited metal atoms. Furthermore, it increases the mean roughness of the polymer and particularly that of the PET.

In contrast to untreated PE, Al interacts with O_2 plasma treated PE. The interaction proceeds through the formation of an Al–O–C complex at low coverage. While metallic state was instantaneously obtained with the untreated PE, it appears later for higher Al coverages when the PE is treated. The mean roughness of the treated PE decreases upon metallization. Due to the chemical interaction between Al and the treated PE, the Al atoms lose their mobility and form small aggregates around the anchoring points or adsorption sites before the formation of a metallic layer.

The effect of O_2 plasma treatment on the reactivity is evidenced in Fig. 11 for Cu/PE and Cu/PET interfaces summarizing the evolution of Auger parameter α' in these interfaces. Cu does not undergo a strong chemical interaction with treated PE and it diffuses into the polymer. However, the metallic layer is observed for larger Cu coverages of the untreated PE. An increase in the R_a is observed as well as the presence of small clusters.

Cu reacts with O_2 plasma treated PET as evidenced by the α' parameter value (Fig. 11). Cu clusters embedded in the polymer are formed, the size of which increases with coverage. Comparing now the images of metallized untreated and O_2 plasma treated PET for comparable coverages ($Cu/C = 0.8$ and $Cu/C = 0.7$, respectively), we notice a large increase in the roughness on the treated PET. The (Cu–O–C + Cu) islands are more clearly evidenced when PET is O_2 plasma treated, indicating that they are less buried inside the polymer. This effect results probably from a depth limited diffusion of Cu when PET is O_2 plasma treated and when more active sites are present at the surface of the polymer.

Moreover we found that the deposition of evaporated metal atoms induces a loss of the oxygen amount incorporated on the polymer surface by the plasma treatment. This loss is more important as the metal is less oxidizable (i.e. Cu evaporated on oxygen plasma treated PE).

It is interesting to note that all the oxygenated functionalities produced by plasma are not equally beneficial for chemical adsorption of Al (or Cu). On the contrary, the heavily oxygenated species are desorbed upon metal atom impact, as they do so with aging in the air.

5. CONCLUSIONS

In situ XPS has been used to study the interfacial chemistry of vacuum deposited Al and Cu on PE and PET surfaces. AFM imaging in the air has provided additional information on the morphology of the interface. The influence of an O_2 plasma on the surface of the polymer prior to metallization has been studied as well.

No interfacial bond is formed when both Cu and Al are evaporated onto PE: metallic state is evidenced at the very early stage of metallization. With Al, a very sharp interface is grown following a Volmer–Weber island-like mechanism, while the Cu/PE interface is more diffused due to Cu diffusion into the polymer and the formation of metallic clusters in the subsurface. XPS shifts arise mainly from charging. AFM images evidence large cluster formation.

Cu and Al both chemically interact with PET to form interfacial compounds before completing a metallic film. With Cu the interaction is weaker, thus Cu diffuses into the bulk of PET. Al interaction proceeds via two adsorption sites. The first site at low coverage is the ester function and the formation of an Al–O–C complex is clearly evidenced from XPS and Auger peaks from substrate and adsorbate as well. The second site is the phenyl group with a π electron transfer too small for inducing detectable XPS shifts. A van der Merwe layer-by-layer growth mechanism is suggested for the Al/PET surface. For Cu, only one site, the ester function, is evidenced, but the Cu–O–C complex has not been detected, first because the interaction is weak and second because the interfacial compound is formed in the subsurface due to Cu diffusion into PET.

The O_2 plasma produces new oxygenated species at the surface of PE and PET with a slight increase in R_a. At low Al coverage on PE, a chemical interface with the formation of an Al–O–C charge transfer complex is observed. This chemical interaction is limited by the desorption of the oxygenated functionalities. No chemical interface is evidenced with Cu deposited on treated PE due presumably to Cu diffusion. Nevertheless, Cu is shown to react with treated PET through the formation of Cu–O–C complexes. The O_2 plasma limits the diffusion depth of Cu into PET by chemically attaching Cu atoms.

REFERENCES

1. J. M. Burkstrand, *J. Appl. Phys.* **52**, 4795 (1981).
2. J. L. Droulas, Y. Jugnet and Tran Minh Duc, in: *Metallized Plastics 3: Fundamental and Applied Aspects*, K. L. Mittal (Ed.), p. 123. Plenum Press, New York (1992).
3. E. Sacher, in: *Metallization of Polymers*, E. Sacher, J. J. Pireaux and S. P. Kowalczyk (Eds), p. 1, ACS Symposium Series No. 440, (1990) and references therein.
4. P. Bodö and J. E. Sundgren, *Surf. Interface Anal.* **9**, 437 (1986).
5. A. J. Pertsin and Yu. M. Pashunin, *Appl. Surf. Sci.* **47**, 115 (1991).
6. M. Chtaïb, J. Ghijsen, J. J. Pireaux, R. Caudano, R. L. Johnson, E. Orti and J. L. Brédas, *Phys. Rev. B* **44**, 10815 (1991).
7. M. Bou, J. M. Martin and T. Lemogne, *Appl. Surf. Sci.* **47**, 149 (1991).
8. L. J. Gerenser, *J. Vac. Sci. Technol.* **A8**, 3682 (1990).
9. L. J. Gerenser, *J. Vac. Sci. Technol.* **A6**, 2897 (1988).
10. C. Magistrini Rotach, PhD Thesis, University of Lyon (1992).
11. J. L. Droulas, PhD Thesis, University of Lyon (1992).
12. R. M. Tromp, F. Legoues and P. S. Ho, *J. Vac. Sci. Technol.* **A3**, 782 (1985).
13. M. G. Mason, *Phys. Rev. B* **27**, 748 (1983).
14. Y. Novis, M. Chtaïb, J. Vohs, J. J. Pireaux, R. Caudano, P. Lutgen and G. Feyder, in: *Metallized Plastics 1: Fundamental and Applied Aspects*, K. L. Mittal and J. R. Susko (Eds), p. 193. Plenum Press, New York (1989).

Polymer Surface Modification: Relevance to Adhesion, pp. 121–136
K. L. Mittal (Ed.)

Barrier properties of plasma-modified polypropylene and polyethyleneterephthalate

J. F. FRIEDRICH,[1,*] L. WIGANT,[1] W. UNGER,[2] A. LIPPITZ,[2] H. WITTRICH,[3] D. PRESCHER,[1] J. ERDMANN,[4] H.-V. GORSLER[4] and L. NICK[5]

[1]*Institut für Angewandte Chemie Berlin-Adlershof, Rudower Chaussee 5, D-12489 Berlin, Germany*
[2]*Bundesanstalt für Materialforschung und -prüfung (BAM), Unter den Eichen 80-85, D-12205 Berlin, Germany*
[3]*Ferdinand-Braun-Institut für Höchstfrequenztechnik, Rudower Chaussee 5, D-12489 Berlin, Germany*
[4]*Ahlbrandt System GmbH, Vogelsbergstraße 45-47, D-36339 Lauterbach, Germany*
[5]*Technische Universität Clausthal-Zellerfeld, Institut für Physikalische Chemie, D-38678 Clausthal-Zellerfeld, Germany*

Revised version received 8 September 1994

Abstract—Plasma treatment changes the solvent absorption and permeation as well as the swelling properties of polymers. Enchanced solvent absorption and swelling are effects of an improved solvent compatibility. The plasma introduces a large number of different groups at the polymer surface depending on the nature of the plasma. Fluorine-containing plasmas can replace hydrogen atoms of the polymer molecule with fluorine atoms. Moreover, fluorine-containing plasma polymer layers can be formed. All these processes reduce the resulting surface free energy, reduce the diffusion length of solvent molecules, and produce a barrier layer. We have studied the formation of solvent barriers by plasma fluorination and by crosslinking by ultraviolet (UV) radiation. Thin foils of polypropylene (PP) and polyethyleneterephthalate (PET) were used as substrates. CF_4, SF_6, and SOF_2 were applied as sources of fluorine atoms. Hexafluoropropene, tetrafluorethylene, and perfluorohexylethylene form plasma polymer layers on the polymer substrates. Test solvents were *n*-pentane, tetrachloroethylene, dimethylsulfoxide, and mixtures of *n*-pentane and methanol. The permeation rate of solvents through plasma-modified polymers was measured gravimetrically. Mass spectrometry was applied to analyze the permeating components of the solvent mixtures. Fluorination of surface layers by plasma-chemical (CF_4, SF_6) means considerably reduces the permeation rate of PP (95% barrier effect) and PET (100%). The preferred permeation of one component of the pentane/methanol mixture is influenced by the polarity of plasma-introduced groups at the polymer surface.

Keywords: Plasma fluorination of polymers; deposition of fluorine-containing plasma polymers; barrier formation against solvent permeation; permeation mechanism of solvent mixtures.

*To whom correspondence should be addressed.

1. INTRODUCTION

A plasma treatment changes the solvent absorption and swelling properties of polymers. In principle, two very different responses of a polymeric material are possible. On the one hand, an improvement in solvent compatibility, accompanied by enhanced solvent absorption and polymer swelling, can be achieved. On the other hand, the formation of a barrier layer, which hinders permeation of the solvent into the polymer, can be the result.

The compatibility properties of polymers can be modified with the help of polar or non-polar groups formed on a polymer surface by plasma treatment. In this way, one can change the relevant thermodynamic surface properties, e.g. the surface free energy. Another relevant parameter is the polymer free volume. It can be varied, for example, by F–H exchange due to exposure to fluorine-containing plasma gases. A third relevant effect is the formation of barriers by crosslinking of the polymeric material due to plasma UV radiation. In this way, one obtains substantially reduced diffusion lengths of solvent molecules. Finally, plasma polymer films deposited on a polymer can seal the substrates against solvent permeation because they are usually characterized by a denser structure.

In this study, we have analyzed the barrier effect in polypropylene (PP) and polyethyleneterephthalate (PET) films occurring after plasma modification or after coating with aliphatic or fluorine-containing plasma polymers and plasma-deposited nickel layers from nickel tetracarbonyl. We used *n*-pentane, toluene, tetrachloroethylene, dimethyl sulfoxide, and mixtures of *n*-pentane and methanol as test solvents.

The polarity of polyolefin surfaces is increased on exposure to SOF_2, $SOCl_2$, CCl_4, O_2, SO_2 and NH_3 plasmas (see for example, [1–4]). We used CF_4 and SF_6 plasmas for fluorination. Plasma surface fluorination of polyolefins has been shown to be effective in perfluorinating the surface region [5–7]. Chemical surface fluorination of polymers is known to retard solvent permeation into polymers [7]. The plasma fluorination of polymers can be done with the help of fluorine-liberating plasma gases, e.g. CF_4, NF_3, XeF_2, BF_3, and SF_6 [8–10].

The introduction of O-containing or SO_x groups at the surface also improves the barrier properties [11]. The presence of a high density of chemically introduced polar oxygen- and fluorine-containing groups should increase the adhesion to metal layers [12–14]. The plasma polymers were deposited on a PP or PET surface by employing C_3F_6, C_8F_{16}, $C_8H_3F_{13}$, C_5H_{12}, and C_8H_{18} monomers (see also [15]).

The plasma modification of polymer substrates influences the resulting barrier properties also by crosslinking or changes in the degree of crystallinity [16, 17]. At longer plasma exposure times, the plasma UV radiation induced such effects in polymer surface layers to some micrometers [18]. Furthermore, the introduction of polar groups in the near-surface layer can also initiate solvent-induced crystallization of the polymer substrate.

2. EXPERIMENTAL

2.1. Plasma modification

The equipment and procedure for DC glow discharge modification of polymers are described in detail elsewhere [19]. Two types of plasma modification were used: the functionalization of the original polymer structure by fluorine, oxygen, OF, or SO_x groups, and the deposition of fluorine-containing plasma polymers.

2.2. Materials used

CF_4 (99.95%, Aldrich), SF_6 (99.75%, Merck), SOF_2 (99.5%), SO_2 (99.9%, Aldrich), and O_2 (99.999%, Air Liquide) were used as the plasma gases, and C_3F_6 (99%, Hoechst), C_8F_{16} (ICI), $C_8H_3F_{13}$ (Hoechst), *n*-pentane, and *n*-octane as plasma polymerizing monomers. The deposited plasma polymers were *ca.* 100 nm thick. Polypropylene (Hoechst) and polyethyleneterephthalate (Mylar, DuPont) biaxially stretched foils of 1.5–40 μm thickness were used. The average molecular weight ($\overline{M}_W$) of PP was 250 000 (size exclusion chromatography), the degree of crystallinity 55–65% (density measurement), and the stretching ratio 5 ($\alpha_=$) and 8–10 ($\alpha_\perp$). The molecular weight ($\overline{M}_W$) of PET was 250 000, the degree of crystallinity (density measurement) 60–65%, and the stretching ratio 3.5 ($\alpha_=$) and 3.5 ($\alpha_\perp$). The permeation tests were carried out at a constant temperature of 40°C because of the strong dependence of permeation on temperature [20]. After the glow discharge treatment, the permeation of *n*-pentane as a model for aliphatic fuel components, of toluene as a model for aromatic substances, and of tetrachloroethylene for halogen-containing solvents was measured. Methanol permeation was studied because of its increasing use in fuels.

2.3. Permeation measurements

The permeation rate was measured gravimetrically using a permeation cell, filled with the test solvent heated up to 40°C, which was closed by the modified polymer film (see Fig. 1a). The weight loss of this permeation cell was measured at intervals of 14 days or longer. Polymer foil deformation by the slightly enhanced pressure within the heated cell was avoided by a metal grid which was fixed to the polymer film.

2.4. Mass spectrometric measurements

Mass spectrometry (see Fig. 1b) allows us to monitor the permeation kinetics of diffusing components from mixtures of solvents (methanol, *n*-pentane, and toluene). The quantitative results of mass spectrometry were in good agreement with the gravimetric data. The mass spectrometer was attached by a capillary tube (diameter 1 mm, length 600 mm) to the permeation cell. The permeated solvent molecules were continuously pumped from the permeation cell to the mass spectrometer using a bypass rotary pump. The cell was kept at 40°C, and the capillary tube at 150°C. The mass spectrometer was a QMS 421 from Balzers (Liechtenstein). The detector was a secondary electron multiplier. A base pressure lower than 5×10^{-7} Pa was established

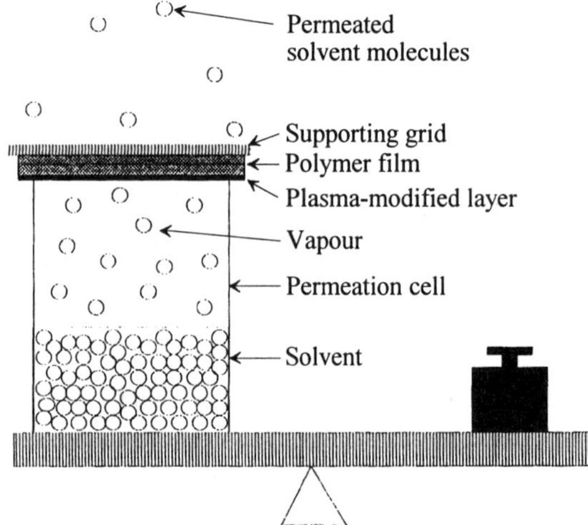

Figure 1a. Permeation cell for the gravimetric measurement of the permeation rate with an analytical balance (40 °C; the polymer film is fixed by a (supporting) metallic grid; 50% of the permeation cell was filled with solvent).

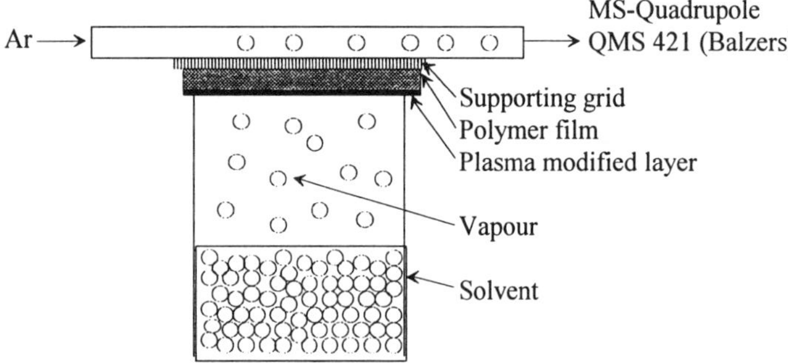

Figure 1b. Permeation cell for mass spectrometric detection of permeated solvent molecules (40 °C; the polymer film is fixed by a (supporting) metallic grid; 50% of the permeation cell was filled with solvent).

within the ionization chamber of the mass spectrometer and the working pressure was kept at 1×10^{-4} Pa. The carrier gas for the transport of solvent molecules into the spectrometer was argon (99.95%, Air Liquide) at a flow rate of 5 l/h.

2.5. Absorption measurements

The absorption of solvents is defined as the increase in mass of the polymer substrates which was gravimetrically measured by using an analytical balance, or a quartz coil microbalance in the solvent-saturated atmosphere at room temperature.

2.6. X-ray photoelectron spectroscopy (XPS) analysis

XPS measurements were carried out using a VG Scientific ESCALAB 200X electron spectrometer. XPS narrow scan spectra were recorded with Mg K_α excitation

(300 W) in the constant retarding ratio mode (CRR 40). The spectrometer energy scale was calibrated by applying the procedure and binding energy (BE) reference data recommended by Anthony and Seah [21].

The PP and PET foils were affixed to the standard VG sample stubs with the help of double-sided adhesive tape. Before recording spectra, the samples were stored within the extended prep-lock (oil free vacuum) of the ESCALAB for degassing at 10^{-5} Pa. Further details are given in [8].

2.7. Atomic force microscopy (AFM) investigation

An atomic force microscope (Ultralever-Omicron) was used in the constant force mode. Unmodified and CF_4 plasma-modified PP films were analysed in air. The swelling of PP was investigated in toluene.

3. RESULTS

3.1. Permeation tests

Different kinds of plasma modification process were tested to obtain data on the resulting permeation of different solvents through PP and PET films. The permeation results of solvents through plasma-modified PP films are summarized in Fig. 2. Fluorinating (CF_4, SF_6) and functionalizing (NH_3, O_2) plasma gases, and plasma polymer-forming fluorocarbon monomers ($CF_2{=}CF{-}CF_3$, C_8F_{16}, $C_8H_3F_{13}$) were tested. The results verify barrier formation in all cases. The highest barrier effects were obtained when simple plasma fluorination was carried out. The formation of a polar polymer surface, characterized by the introduction of N- or O-containing groups (application of NH_3 or O_2 as a plasma gas), shows an unexpected barrier effect against solvent

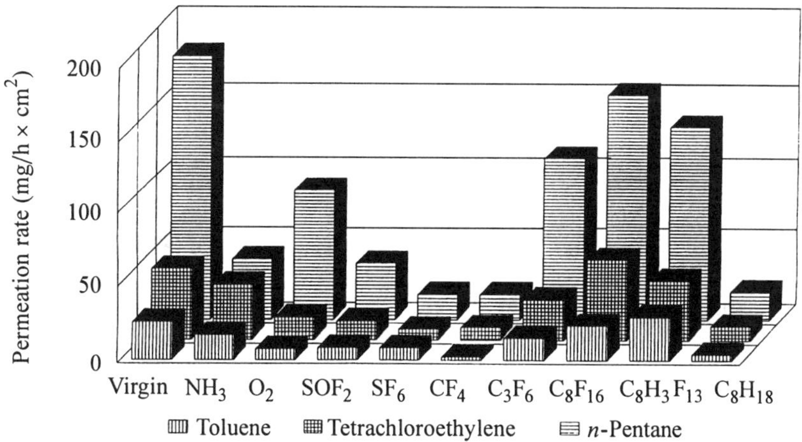

Figure 2. Rates of solvent permeation through plasma-modified PP films (30 μm; 600 s plasma pretreatment).

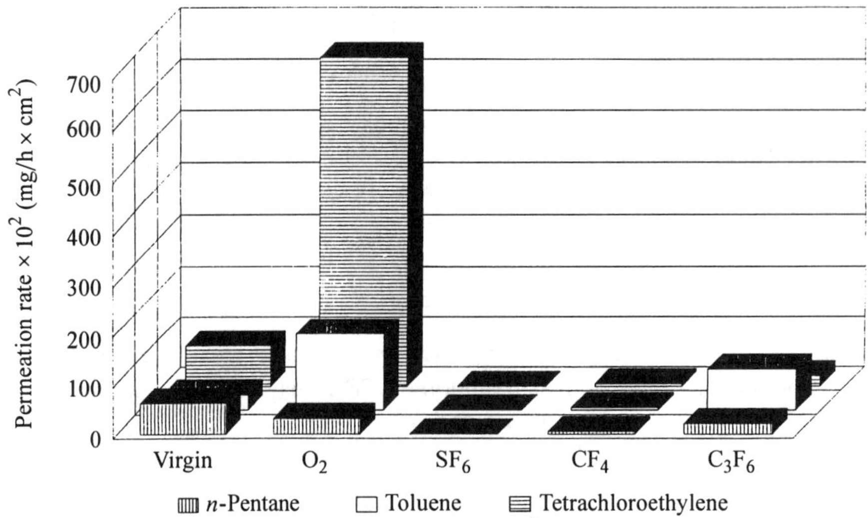

Figure 3. Rates of solvent permeation through plasma-modified PET films (12 μm; 600 s plasma pretreatment; note the change in scale of the Y-axis in comparison with Fig. 2).

permeation. The plasma polymer layers deposited from fluorine-containing monomers show moderate barrier effects (Fig. 2).

In general, PET foils possess a much better barrier behavior than PP against solvent permeation (Fig. 3). A 100% barrier for all the solvents used is reached after fluorination in the SF_6 plasma. This could be confirmed by long-time measurements over 6 months. In contrast, the oxygen plasma enhances strongly the permeation of tetrachlorethylene, while the permeation rate of *n*-pentane decreases.

3.3. Absorption

The absorption of solvents in the polymer foils did not decrease considerably after plasma modification as seen in Fig. 4. Often the absorption was increased by the plasma treatment. Longer plasma treatment (600 s) decreased the absorption. Therefore, a comparison of the solvent absorption data with the solvent permeation results reveals some inconsistencies. It was concluded that a thin surface layer was formed which had a higher affinity for polar solvents. This effect could be due to the formation of oxygen-containing polar groups in this layer, as discussed later. The most probable causes of these oxygen-containing polar groups is the attachment of molecular oxygen to plasma-generated radical sites when the samples are exposed to the laboratory air.

Very low permeation rates and increased absorption resulting after most plasma treatments suggest that the barrier zone is located not at the outermost surface, but deeper in the bulk of the polymer sample. It should be covered by a layer of some micrometers thickness with rather high solvent absorption. In some cases, the compatibility of the plasma-modified layer to the solvent was high enough to dissolve the polymer film. This was observed in absorption tests with CCl_4 plasma-modified polymers and tetrachlorethylene as the solvent.

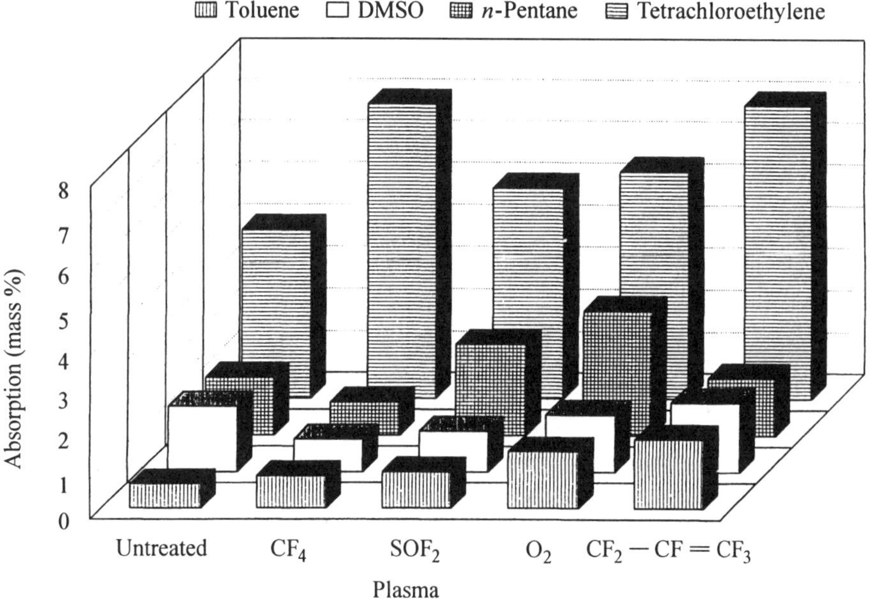

Figure 4. Absorption of different solvents in plasma-modified PP films (600 s plasma pretreatment). DMSO = dimethylsulfoxide.

3.3. Dependence of permeation rates on plasma exposure times

The maximum barrier effect in PP to solvent permeation was reached after a rather long time of exposure to fluorinating or functionalizing plasma. In the case of CF_4 plasma modification, the optimum treatment time was found to be 300–600 s (see Fig. 5). Plasma modification applied for longer than 1000 s did not improve the barrier effect further (Fig. 5). This is due to the degradation of the polymer structure and/or crosslinking processes induced by the UV radiation emitted by the plasma. In the first few seconds of plasma modification, the relevant effects are surface cleaning, changes in molecular orientations, and attachment of functional groups [19]. At such short treatment times we found no effect on the original permeation properties of our samples.

The permeation rate of a mixture of *n*-pentane and methanol (85 : 15) is not so much decreased by a 600 s CF_4 plasma modification as the permeation of the two solvents individually (Fig. 5).

PET is a material with very good solvent barrier properties which can be improved further by plasma fluorination (see Fig. 6). The plasma-fluorinated PET is impermeable to *n*-pentane. The molecular structure of the aromatic polyester allows the penetration of aromatic solvents such as toluene, but only at negligible rates. Similar to the findings for PP, the permeation of *n*-pentane/methanol mixtures through PET is much higher than that of pure *n*-pentane (Fig. 6). After 15 days, the permeation rate decreases, tending to zero values. This phenomenon is interpreted as an effect of methanol absorption blocking the free polymer volume cavities for permeating solvents or as solvent-induced crystallization of the PET.

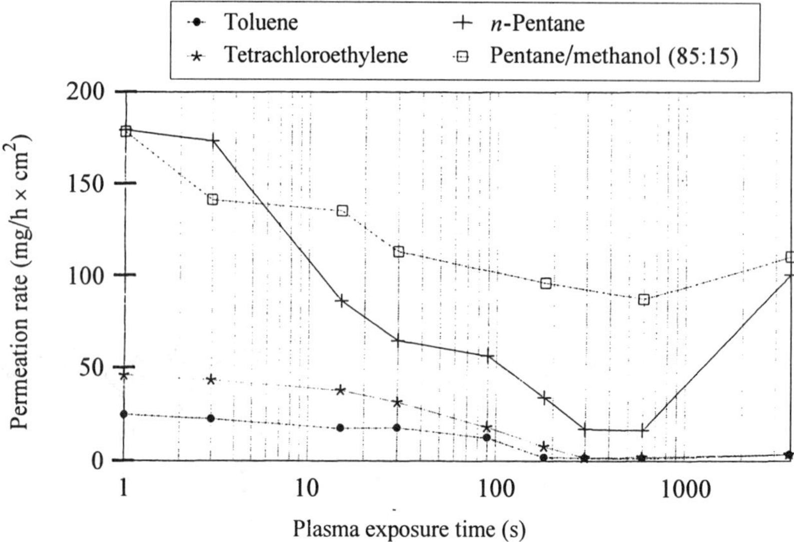

Figure 5. Dependence of the solvent permeation rate through PP on the exposure time to CF_4 plasma (low pressure 6 Pa; DC plasma).

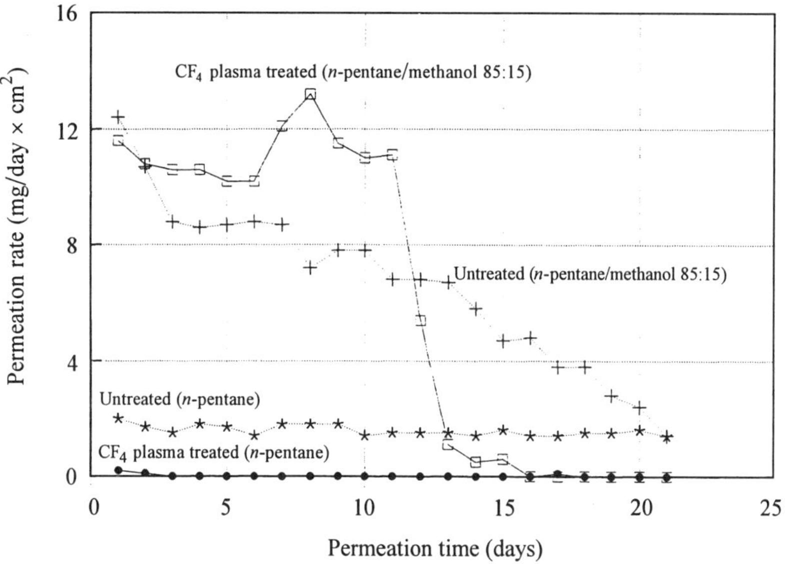

Figure 6. Dependence of the solvent permeation rate through PET on the permeation time (plasma conditions: low pressure 6 Pa; CF_4 DC plasma; note the different scales on the X- and Y-axes in comparison with Fig. 5).

3.4. *Influence of crosslinking*

Argon plasma is known to crosslink polyolefins by its UV radiation [23]. However, it is not clear whether PP is also crosslinked by the argon plasma. Earlier experiments have shown that PP is only weakly crosslinked by the argon plasma [24]. A PP foil modified on exposure to argon plasma for 600 s shows no change in the permeation rate as compared with the unmodified PP.

3.5. Plasma polymer and metallic layers

Fluorine-containing plasma polymer layers improve only slightly the permeation barrier of PP (see Figs 2 and 3). The best results were obtained with the aliphatic hydrocarbon *n*-pentane or *n*-octane as the monomer. The *n*-octane plasma polymer deposition results in a 90% barrier against *n*-pentane permeation which is in the same range as the plasma fluorination.

Nickel metal films of more than 0.5 μm thickness, deposited by a low-pressure plasma process [25], hinder the solvent permeation, but thinner layers are permeable. This is also true for evaporated aluminum-layers.

3.6. Dependence of the permeation rate on the ratio of n-*pentane/methanol in mixtures*

The addition of methanol to *n*-pentane increases the permeation rate of solvents (Fig. 7). A critical mixture at an 85 : 15 pentane/methanol concentration ratio gives a maximum permeation rate. This mixture ratio is of technical relevance because of its use as an alternative fuel. The same effect was observed with plasma fluorinated samples (CF_4 plasma). Methanol itself does not permeate with a significant rate, while *n*-pentane has a minimum permeation rate of *ca.* 15 mg/cm^2 h and the mixture shows a permeation rate of *ca.* 85 mg/cm^2 h as compared with 175 mg/cm^2 h for the unmodified sample. In the case of the unmodified PP foil, the permeation rate of

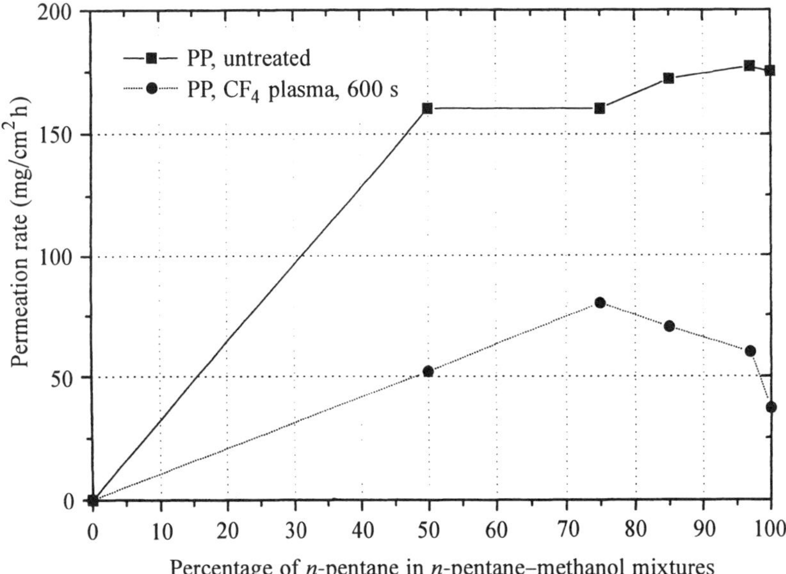

Figure 7. Permeation rate of *n*-pentane/methanol through PP films vs. mixture composition.

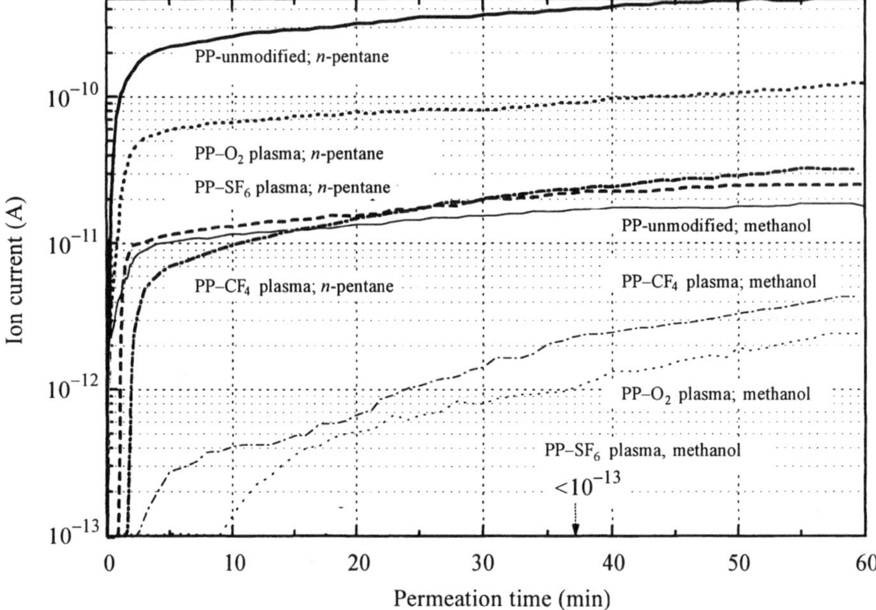

Figure 8. Permeation of the components of an 85 : 15 *n*-pentane/methanol mixture through plasma-modified PP foils measured in terms of ion current (secondary electron multiplier). For mass spectrometric analysis the following characteristic mass numbers were selected: *n*-pentane 43 and methanol 31.

n-pentane (175 mg/cm^2 h) is the same as that for the 85 : 15 mixture of *n*-pentane and methanol.

The conclusion is that methanol swells the plasma-modified PP and 'opens' new cavities for the pentane permeation which are not present when exclusively *n*-pentane was tested. Methanol is absorbed in the surface layer by interaction with the polar groups.

3.7. Mass spectrometric analysis of permeated species

The results of mass spectrometric analysis of the 85 : 15 mixture reveal that *n*-pentane predominantly diffuses through the polymer whereas methanol seems to be hindered (Fig. 8). For the untreated PP, the ratio of the permeation rate of *n*-pentane to that of methanol is *ca.* 28 : 1, and for the 85 : 15 mixture and the CF_4 plasma modified PP (600 s), it is *ca.* 8 : 1. With SF_6 plasma modification (600 s), the highest solvent barrier was observed and no permeation of methanol could be measured. The plasma modification of PP with CCl_4 did not form such a good solvent barrier as the SF_6 or the CF_4 plasma but the so pretreated PP is a good 'selective membrane' because of a permeation ratio of $10^5 : 1$ when the 85 : 15 mixture was used.

3.8. XPS analysis

The chemical state of the polymer surface after CF_4 plasma fluorination was not very different when the PP and PET results are compared. Figure 9 presents the difference spectra of the C 1*s* signals obtained before and after CF_4 plasma exposure

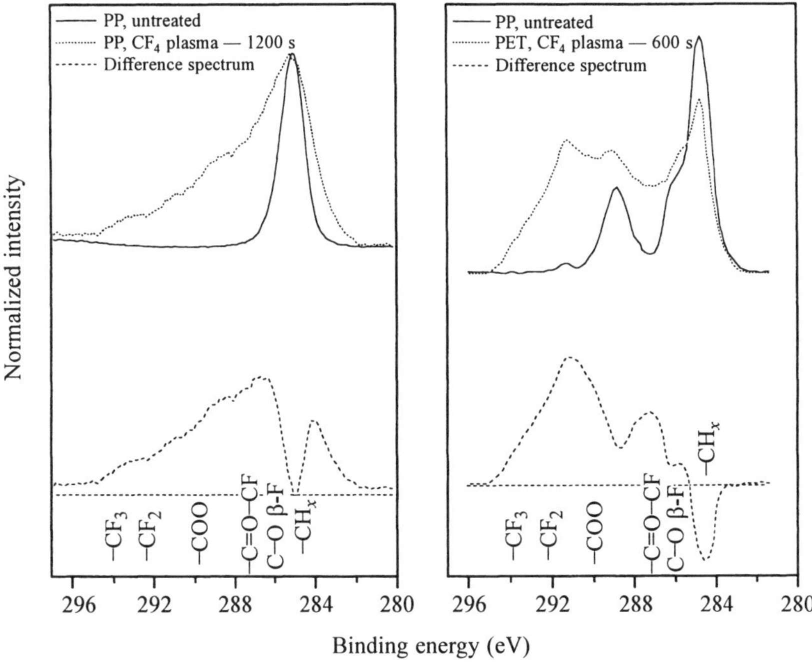

Figure 9. XPS C 1*s* peaks of PP and PET before and after CF_4 plasma exposure.

Table 1.
XPS F/C surface concentration ratios and C 1*s* subpeak intensities (in % of C 1*s* total intensity; estimated error of these results: ±10%) of low pressure (6 Pa) CF_4 DC plasma-treated PP

Exposure time (s)	F/C ratio ±10%	CH_x (BE= 285.0 eV)	C–O– (BE= 286.5 eV)	CF, C=O (BE= 288.0 eV)	COO– (BE= 289.3 eV)	CF_2 (BE= 291.4 eV)	CF_3 (BE= 293.5 eV)
0	—	99	1				
3	2.5	76	16	6	2	—	—
15	30	64	14	13	6	2	1
60	61	69	17	6	5	2	1
180	83	55	21	12	8	2	2
600	150	48	18	10	12	7	5

of PP and PET. Obviously, at the surface of both polymers the same CH_xF_y species were formed. The C 1*s* and F 1*s* peaks reveal the existence of different types of C–F bond and, furthermore, the presence of fluorine anions. A detailed discussion of the plasma fluorination (Table 1) is not possible at this stage of investigation, but it is striking that after CF_4 plasma treatment a high oxygen surface concentration is present (Fig. 10). This oxygen content is not introduced into the surface layer by the plasma fluorination process itself, as revealed by mass spectrometric control. Therefore, post-plasma attachment of oxygen from air is responsible for the formation of oxygen functionalities. The O-containing groups may be responsible for the higher

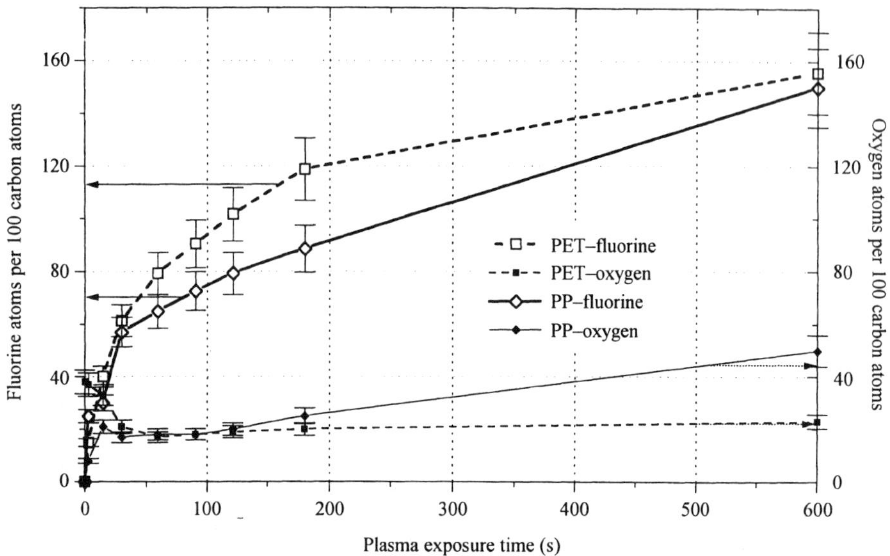

Figure 10. XPS surface concentrations of F and O vs. exposure time to the CF_4 plasma (low pressure; 6 Pa; DC plasma). Note that oxygen results from post-plasma reactions with air and not from reactions with traces of oxygen in the plasma chamber as shown by mass spectrometry.

Table 2.
XPS surface concentrations (in at%) for PP and PET after different plasma treatments [low pressure (6 Pa) DC plasma, 180 s]. The error in the surface concentrations is estimated to be ±10%

	Polymer surface: PP Plasma treatment						
Element	Ar	NH_3	O_2	SO_2	$SOCl_2$	CCl_4	SOF_2
Ar	1.5						
O	0.3	6	28	22	24	4	8
N		21			2		1
S				4	3		
Cl						12	16
	Polymer surface: PET Plasma treatment						
Element	Ar	NH_3	O_2	SO_2	$SOCl_2$	CCl_4	SOF_2
O	36	33	35	28	22	14	35
N		12			3	2	
S				—	4		3
Cl					9	19	
F							9

absorption of (polar) solvents. Under the same plasma treatment conditions (CF_4, DC, 6 Pa), more fluorine atoms were attached to PET in comparison with PP (Fig. 10). The XPS measured concentrations of plasma-introduced elements at polymer surfaces after using non-fluorinating plasma treatments are summarized in Table 2.

Figure 11a. AFM picture of untreated PP immersed in liquid toluene (numbers in nm).

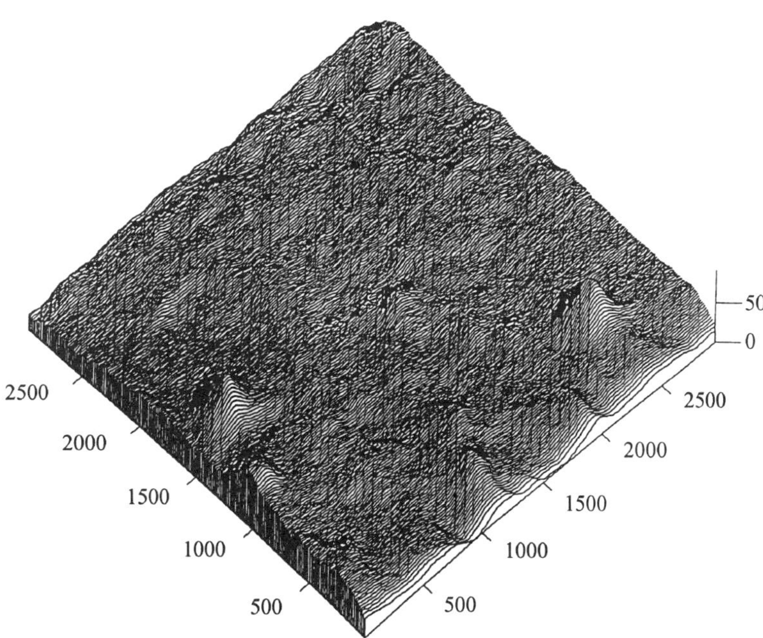

Figure 11b. AFM picture of 600 s CF_4 plasma-modified PP immersed in liquid toluene (numbers in nm).

3.9. AFM surface topography of PP in the presence of toluene

The surface topography of untreated PP in toluene (test solvent) detected by AFM reveals significant differences when it is compared with that for untreated PP in air [19]. The swelling of the polymer surface transforms the surface from a more jagged state into a surface without fine structures (Fig. 11a). The CF_4 plasma-treated PP surface dipped in toluene shows a surface very similar to the picture obtained with untreated PP in an air surrounding (Fig. 11b). Obviously, the plasma treatment results in a picture characterized by an absence of the effects attributed to swelling.

4. DISCUSSION

The barrier and swelling properties of plasma-modified polymers are not well understood because one is not able to explain the effects in terms of one simple model. It seems that the plasma-induced processes, occurring first, i.e. the surface cleaning and functionalization [26], have no significant effect on the permeation properties. On the other hand, the solvent absorption at this state is indeed modified. The absorption of the test solvents used was often slightly increased in relation to the untreated PP. The interpretation is that solvent molecules were absorbed in the surface layer modified by plasma fluorination. The plasma surface modifications result in polar groups as seen by XPS. At these groups, dipole–dipole or hydrogen bonding interaction with polar solvent molecules can occur, thus improving the absorption.

Some radical sites, produced by the plasma process, may pick up oxygen from the ambient air in a post-plasma reaction [22] and cause the formation of oxygen-containing groups at the polymer surface. A low fluorine surface concentration and the presence of some oxygen at the surface create a partially hydrophilic layer, responsible for an intermediate state of high absorption capability.

Longer exposure to the fluorinating plasma gas (300–600 s) results in perfluorination which is characterized by low solvent absorption and permeation. Such long-time plasma fluorination is connected with a high F/C ratio in PP (*ca.* 1.3–1.5 : 1) and PET (1.4–1.7 : 1). The minimum permeation rates, 5% and 7%, respectively, of *n*-pentane through plasma-modified PP in relation to the unmodified PP were reached with SF_6 and CF_4 plasma modification. Plasma exposure longer than 600 s results in degradation processes and the permeation rate again increases.

Perfluorination requires a time of exposure to plasma which should be long enough so that the plasma UV radiation can also play an important role in barrier formation. However, it was found that the argon plasma modification itself could not decrease the permeation rate.

Generally, PET possesses a much lower permeation than PP. For instance, the permeation rate of *n*-pentane/methanol mixtures (85 : 15) through unmodified PET is *ca.* 350 times lower than that for unmodified PP.

As for PP, the permeation of *n*-pentane through PET is increased when a *n*-pentane/methanol mixture is used. The plasma fluorination stops the *n*-pentane and methanol permeation within 15 days by a not well-understood mechanism (Fig. 6). The plasma-fluorinated PET is much more permeable than the untreated material in the first 15 days, but after this time permeation rate of *n*-pentane quickly approaches zero. After this 15-day induction period, PET has a barrier efficiency of 100% for the permeation of *n*-pentane/methanol mixtures examined over a period of 6 months. The mechanism of the abrupt blocking of the permeation of *n*-pentane from the mixture cannot be explained at this level of investigation. One hypothesis is solvent-induced PET crystallization and the adsorption of methanol at polar sites accompanied by a closing of diffusion cavities.

The mass spectrometric results on PP foils support the interpretation of the gravimetric measurements. Detailed information on the permeation of *n*-pentane/methanol mixtures was obtained. It is interesting that the oxygen plasma modification lowers only moderately the pentane permeation, whereas the methanol permeation is strongly decreased.

The swelling of the virgin PP will be decreased by the perfluorination as shown by AFM. Swelling effects may be responsible for the enhanced *n*-pentane permeation from pentane/methanol mixtures by opening new diffusion cavities.

In general, metallic layers should hinder any permeation of solvents. The plasma-chemical induced deposition of nickel from Ni-tetracarbonyl [25]:

$$\mathrm{Ni(CO)_4} \longrightarrow \mathrm{Ni^0} + \text{traces of } (\mathrm{NiO_2} + \mathrm{NiO} + \mathrm{Ni_2O_3})$$

should be a good method for metallizing the inner sides of hollow bodies. However, in practice, a 100 nm thick nickel layer has some pinholes because of the roughness of the substrate and, therefore, the permeation cannot be reduced down to zero. Plasma polymer layer formation results only in slight improvements in barrier properties. All plasma polymers absorb solvents to a great extent. The best results were obtained with aliphatic alkane monomers which provide a barrier up to 90%.

In the case of PET, the high level of fluorination is related to the complete destruction of aromatic rings. The high concentration of $-CF_3$ end-groups may be explained by the attachment of CF_3 species from the plasma or by the occurrence of many polymer chain scissions giving completely fluorinated carbon end-groups.

5. CONCLUSION

The best barrier properties with PP and PET were obtained by plasma fluorination. Plasma treatments with SF_6 or CF_4 result in a 95% or 93% permeation barrier to *n*-pentane for PP. O_2 plasma treatment gives a barrier effect of 85%, and the sulfo-fluorination (SOF_2) of 65%. A very special problem was the permeation of *n*-pentane/methanol mixtures through the investigated polymers. Here, the SF_6 plasma pretreatment is the preferred approach. This special plasma treatment could improve very efficiently the barrier properties of PP to the permeation of the 85 : 15 *n*-pentane/methanol mixture. After CF_4 plasma fluorination of PET, barrier effects of 100% for *n*-pentane and 100% for *n*-pentane/methanol mixtures (after 15 days saturation of the PET with *n*-pentane) were obtained.

Under the plasma conditions used in this study, the deposition of plasma polymer layers from aliphatic hydrocarbons improves the barrier effect up to 99%. Metallic layer deposition was suitable when thick and dense layers could be deposited.

REFERENCES

1. F. Arefi, P. Montazer-Rahmati, V. Andre and J. Amouroux, *J. Appl. Polym. Sci.* **46**, 33 (1990).
2. Y.-M. Tsu, M. C. Kimble and R. E. White, *J. Electrochem. Soc.* **139**, 1913 (1992).
3. P. M. Scott, L. J. Matienzo and S. V. Babu, *J. Vac. Sci. Technol.* **A8**, 2382 (1990).
4. N. Inagaki, S. Tasaka and M. Imai, *J. Appl. Polym. Sci.* **48**, 1963 (1993).
5. H. Schonhorn and R. H. Hansen, *J. Appl. Polym. Sci.* **12**, 123 (1966).
6. L. J. Hayes and D. D. Dixon, *J. Appl. Polym. Sci.* **23**, 1907 (1979).
7. M. Anand, R. E. Cohen and R. F. Baddour, *Polymer* **22**, 361 (1981).
8. G. A. Corbin, R. E. Cohen and R. F. Baddour, *Macromolecules* **18**, 98 (1985).
9. M. Strobel, S. Corn, C. S. Lyons and G. A. Korba, *J. Polym. Sci., Polym. Chem. Ed.* **23**, 1125 (1985).
10. Y. Yagi and A. Pavlath, *J. Appl. Polym. Sci.* **38**, 215 (1984).
11. T. Volkmann and H. Widdecke, *Kunststoffe* **79**, 743 (1989).
12. R. Milker and B. Möller, *Kunststoffe* **82**, 978 (1992).
13. O.-D. Hennemann and G. Krüger, *Kunstsoffberater* **5**, 1 (1992).
14. G. Krüger, *Gummi, Fasern, Kunststoffe* **46**, 292 (1993).
15. H. I. Wang, M. W. Rembold and J. Q. Wang, *J. Appl. Polym. Sci.* **49**, 701 (1993).
16. A. K. Taraija, G. A. Orchard and I. M. Ward, *Plastics, Rubber Composites Processing Applications* **19**, 273 (1993).
17. A. V. Popoola, *J. Appl. Polym. Sci.* **49**, 2115 (1993).
18. J. Friedrich, I. Loeschcke, H. Frommelt, H.-D. Reiner, H. Zimmermann and P. Lutgen, *Polym. Degrad. Stabil.* **31**, 97 (1991).
19. J. Friedrich, W. Unger, A. Lippitz, W. Saur, P. Rohrer, J. Erdmann and H.-V. Gorsler, *J. Adhesion Sci. Technol.* (in press).
20. P. I. Lee, *Polymer* **34**, 2397 (1993).
21. M. T. Anthony and M. P. Seah, *Surface Interface Anal.* **6**, 95 (1984).
22. J. Friedrich, I. Loeschcke and P. Lutgen, *Z. Chem.* **30**, 177 (1990).
23. M. Hudis, *J. Appl. Polym. Sci.* **16**, 2397 (1972).
24. J. Friedrich, Unpublished results (1992).
25. J. Friedrich, H. Wittrich and J. Gähde, *Acta Polymerica* **31**, 59 (1980).
26. J. Friedrich, W. Unger, A. Lippitz, J. Erdmann and H.-V. Gorsler, *Surface Coat. Technol.* (accepted).

Polymer Surface Modification: Relevance to Adhesion, pp. 137–149
K. L. Mittal (Ed.)

Surface biomedical effects of plasma on polyetherurethane

DEJUN LI* and JIE ZHAO

Deparment of Physics, Tianjin Normal University, Tianjin, 300074, PR China

Revised version received 2 March 1995

Abstract—Surface biomedical effects of plasma treatment and plasma polymerization on medical-grade polyetherurethane were studied. N_2 and Ar plasma treatments and hexamethyldisiloxane (HMDS) plasma polymerization were performed at a power of 100 W with exposure times ranging from 1 to 15 min. The results showed that the contact angle of water was decreased from 79° to 62° by N_2 and Ar plasma treatments, and N_2 plasma treatment caused a slight enhancement in anti-coagulability and anti-calcific behavior. HMDS polymerization resulted in a decrease from 79° to 43° in the contact angle and an increase from 30.5 to 37.4 s in the recalcification time. At the same time, the anti-coagulability of polymerized samples for the exposure time of 2–5 min was 2.5 times that of the untreated sample. Results of XPS and ESR analyses showed that HMDS deposited onto the polyetherurethane surface and formed new Si—N bonds, and increased the number of radicals in the sample. XPS analysis also showed that N_2 and Ar plasma treatments broke some of the C—O and C=O bonds at the surface and resulted in oxidation of the surface.

Keywords: Polyetherurethane; wettability; anti-coagulability; anti-calcific behavior; surface modification; plasma treatment; plasma polymerization.

1. INTRODUCTION

At present, polyurethanes are widely used in the medical materials industry for the manufacturing of various implants [1]. Polyurethanes exhibit several properties that make them particularly attractive for biomedical applications. They appear to be well tolerated in implant situations; they can be made durable enough to withstand continuous stress and flexing; and they can be readily synthesized from a variety of starting materials with a wide range of physical and chemical properties. Although they possess good biocompatibility as compared to silicone rubber, certain biomedical behaviors such as blood compatibility and anti-calcific behavior are inferior to those of body organs. It is still very important to improve their surface properties in order to attain the most effective clinical applications.

There are many surface modification techniques currently applied to biomaterials [2, 3]. Among these techniques, plasma treatment is a useful method to modify

*To whom correspondence should be addressed.

the surface properties of polymers. Usually two different processes can be used for the plasma modification of a polymer surface. One is plasma polymerization, i.e. the deposition of a thin, highly crosslinked polymer film on the surface of the material, and the other is the surface modification of the structure of the material itself under the influence of the glow discharge. The properties of a surface exposed to plasma are governed by many parameters including the gas introduced, the substrate properties, the reaction conditions (power, pressure, exposure time), and the reactor geometry. For a more complete description of this technique the reader is referred to [4] and [5]. This technique offers many advantages in industrial applications. In addition to the technological simplicity and cleanliness of this technique, it modifies only surface characteristics without affecting bulk properties. Therefore, if a biomaterial has the desired bulk mechanical properties, but the surface does not exhibit the appropriate biocompatibility, the surface can be modified by the plasma technique. There are some reports on the surface modification of biomaterials by means of the plasma techniques [6–8].

In this work, the effects on the wettability, anti-coagulability, and anti-calcific behavior of medical-grade polyetherurethane of N_2 and Ar plasma treatments and of hexamethyldisiloxane plasma polymerization were investigated. The surface chemical states and oxidation, and radicals were analyzed by means of X-ray photoelectron spectroscopy (XPS) and electron spin resonance (ESR), respectively.

2. EXPERIMENTAL

The medical-grade polyurethane used was polyetherurethane (PEU) film of 0.05 cm thickness. PEU was synthesized using a two-step solution polymerization method. First, polytetramethylene glycol (PTMG) and 4,4′-methylenebis(phenylene isocyanate) (MPI) were solution-polymerized for a reaction time of 1.5 h and a reaction temperature of 60 °C, and then ethylenediamine (ED) was added to extend the chain reaction for a time of 12 h and a temperature of 30 °C. The molar ratio of PTMG, MPI, and ED was 1 : 2.5 : 1.5. The PEU synthesized consisted of an aromatic 'hard' segment (MPI) connected by a polyether 'soft' segment (PTMG); the chain extender was ED. Its structure is shown in Fig. 1. The molecular weight of the polyether soft segment is 1040. The hard and soft segments separate into phases; this phase separation gives the PEU a unique combination of flexibility and strength.

The PEU surface was treated by plasmas of the nonpolymerizable gases N_2 and Ar, or was coated with a plasma polymer layer of polymerizable monomer hexamethyldisiloxane (HMDS). The structure of HMDS is shown in Fig. 2.

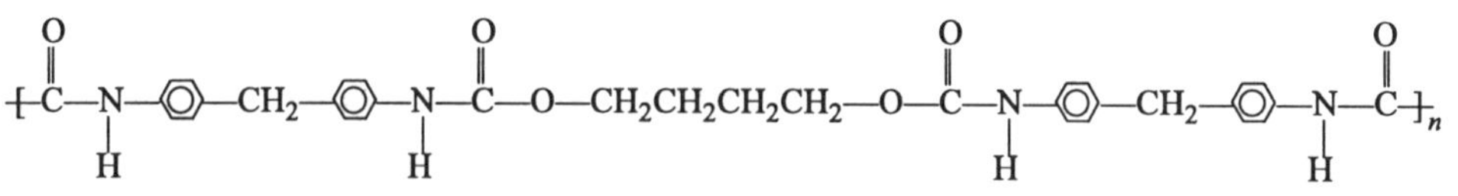

$$\left[\overset{O}{\overset{\|}{C}}-\underset{H}{\underset{|}{N}}-C_6H_4-CH_2-C_6H_4-\underset{H}{\underset{|}{N}}-\overset{O}{\overset{\|}{C}}-O-CH_2CH_2CH_2CH_2-O-\overset{O}{\overset{\|}{C}}-\underset{H}{\underset{|}{N}}-C_6H_4-CH_2-C_6H_4-\underset{H}{\underset{|}{N}}-\overset{O}{\overset{\|}{C}} \right]_n$$

Figure 1. The structure of PEU.

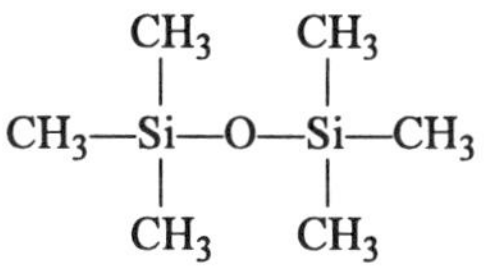

$$
\begin{array}{ccccc}
 & CH_3 & & CH_3 & \\
 & | & & | & \\
CH_3— & Si & —O— & Si & —CH_3 \\
 & | & & | & \\
 & CH_3 & & CH_3 &
\end{array}
$$

Figure 2. The structure of HMDS.

Low-temperature N_2 and Ar plasma treatments were performed using a glow discharge reactor. This apparatus, equipped with a bell-jar-type reaction cell, was a Model GPT-A, manufactured by Central National College, China. The frequency of the RF power source applied was 10 MHz and the glow discharge power was 100 W. Two internal electrodes with an area of 15×15 cm^2 each were placed 4 cm apart from each other in a glass jar. PEU films were placed on a sample holder situated between the two electrodes. The pressure in the glass jar was reduced to 0.3–15 Pa followed by the introduction of N_2 or Ar gas at a flow rate of 10 ml min^{-1}. The time of exposure to N_2 or Ar plasma varied from 1 to 15 min. The samples were stored in dry air immediately after the treatments.

HMDS plasma polymerization was carried out using a bell-jar-type plasma deposition reactor (PD-2, Samco Corp., Japan). The frequency applied was 13.5 MHz and the glow discharge power was 100 W. Two internal electrodes with an area of 15×15 cm^2 for the upper electrode and 20×20 cm^2 for the lower one were placed 2.5 cm apart in a glass jar. The pressure in the glass jar was reduced to 0.3 Pa followed by introduction of the HMDS monomer at a monomer flow rate of 10 ml min^{-1}. The pressure in the bell jar was kept at about 10 Pa after monomer introduction. The exposure time varied from 1 to 15 min. The samples were stored in dry air immediately after polymerization.

The wettability of the samples was investigated by measuring the contact angle of distilled water by depositing a drop on the surface exposed to plasma and compared with the untreated (control) surface using a contact angle analyzer (FACFCA-D, Samco Corp., Japan). The anti-coagulability of the samples was investigated at 37°C by measuring the coagulation time for the surface in contact with 1 ml of rabbit's blood using a water-bath orbital shaker (Certomat-WK, Germany). The anti-calcific behavior of the samples was investigated by measuring the recalcification time, i.e. the time interval from introducing 0.1 ml of $CaCl_2$ solution into a test-tube containing 0.1 ml of rabbit's plasma and the sample to the time for the appearance of white particles in the plasma. These three properties were determined 1 day after the plasma treatments and HMDS polymerization.

An ES-750 spectrometer manufactured by Shimadzu Corp., Japan was employed to carry out the XPS measurements on the plasma-treated and plasma-polymerized samples in order to investigate the surface chemical states and atomic concentrations of C, N, O, and Si. Sputter etching was performed using a 3.5 keV Ar^+ beam. Electron ejection from the samples was induced by 10 kV, 25 mA Mg K_α X-ray radiation at a pass energy of 35.75 eV, a pass speed of 0.1 eV s^{-1}, and a time constant of 0.3 s. The pressure in the sample chamber was kept below 10^{-6} Pa.

The measurement of radicals in the plasma-treated and plasma-polymerized samples was performed using the scan magnetic field method by means of ESR (JES-3BX,

JEOL Corp., Japan). In order to obtain electron spin resonance (resonance condition is described by $h\nu = g\beta H$, where β is the Bohr magneton and h is the Planck constant), microwaves were produced by a microwave generator at a microwave frequency of 9.4×10^9 Hz and a wavelength of 3.2 cm^{-1}. The time constant was 0.1 s and the magnification was 50 db in the analysis.

3. RESULTS AND DISCUSSION

3.1. N_2 and Ar plasma treatments

3.1.1. Contact angle change. Figure 3 shows the contact angle of water for N_2 and Ar plasma-treated samples as a function of the exposure time at a power of 100 W. Each point is the average of five readings (±2°). The decrease in contact angle due to the plasma treatment depends on the time of exposure to plasma. The contact angle on the untreated sample is approximately 79°, and the N_2 and Ar plasma treatments cause a gradual decrease with the exposure time. It is clear from this result that the N_2 and Ar plasma treatments increase the wettability of the surface.

The decrease in the contact angle caused by the plasma treatments is also dependent on the nature of the plasma gas. N_2 plasma treatment causes the contact angle to decrease gradually as the exposure time increases; the angle becomes constant after 10 min and its value is 62°. A similar result was obtained when the samples were treated with Ar plasma; however, the contact angle becomes constant after an exposure time of 2 min and its value is 65°. In any case, the enhancement of the surface wettability due to the plasma treatments depends on both the exposure time and the nature of the plasma gas.

3.1.2. Coagulation time and recalcification time changes. Figure 4 shows how the coagulation time changes with the time of exposure to N_2 and Ar at a power of 100 W.

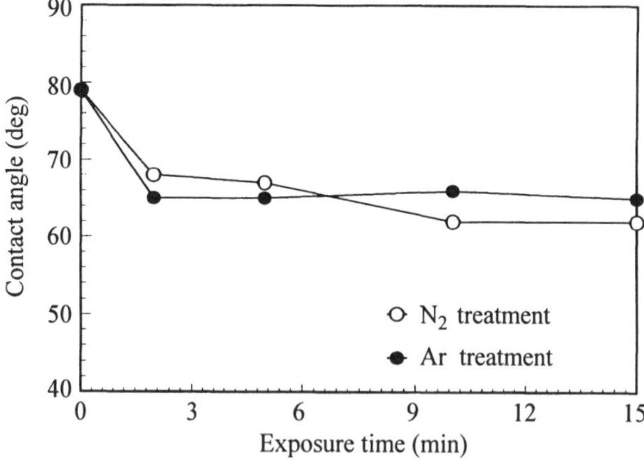

Figure 3. Contact angle vs. plasma exposure time at 100 W.

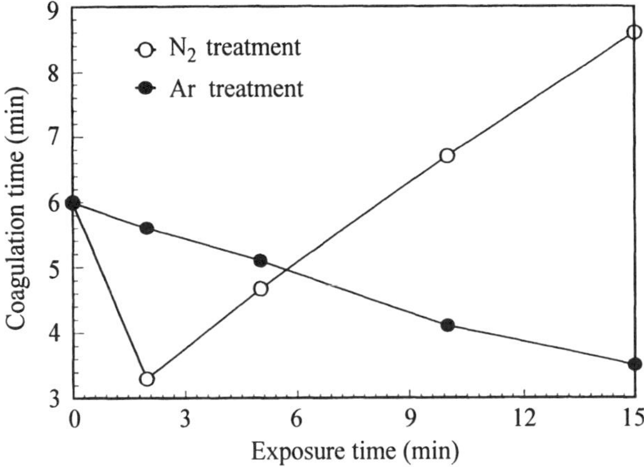

Figure 4. Coagulation time vs. plasma exposure time at 100 W.

Each value in Fig. 4 is the average of five measurements ($\pm$0.2 min). The coagulation time on the untreated sample is approximately 6 min. N_2 plasma treatment results in a slight decrease in its value for an exposure time up to 2 min. The coagulation time, however, shows a tendency to increase linearly as the exposure time increases; its value increases gradually from 3.2 to 8 min within an exposure time of 2–15 min. This result means that the anti-coagulability of the PEU surface is enhanced gradually with the time of exposure to N_2 plasma. On the contrary, Ar plasma treatment leads to a linear decrease in anti-coagulability tendency with the exposure time; its value decreases gradually from 6 to 3.5 min.

Figure 5 shows the change in recalcification time with the time of exposure to N_2 and Ar plasmas at 100 W. Each value is the average of five measurements ($\pm$1.5 s). The recalcification time on the untreated sample is 30.5 s. N_2 plasma treatment results in a decrease in its value for an exposure time up to 2 min, reaching a value of 24.5 s.

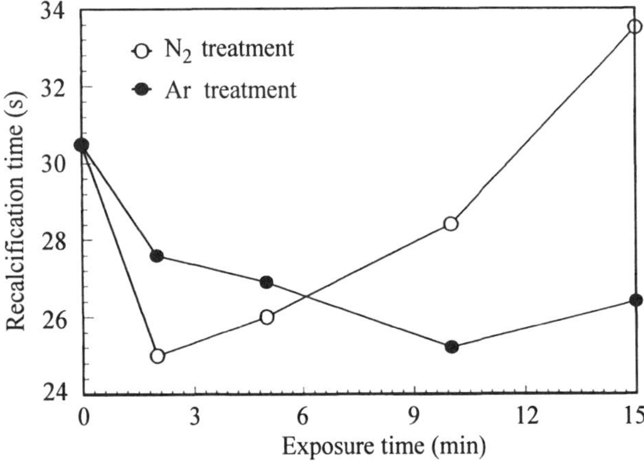

Figure 5. Recalcification time vs. plasma exposure time at 100 W.

However, the recalcification time increases gradually to 33 s after the exposure time of 2 min, and shows a tendency to increase gradually as the exposure time increases, which indicates that the N_2 plasma treatment leads to a slight enhancement in the anti-calcific behavior at the longest exposure time. Compared with the N_2 plasma treatment, the Ar plasma treatment results show a gradual decrease in the anti-calcific behavior as the exposure time increases.

It is clear from the results presented that the anti-coagulability and anti-calcific behavior changes are dependent on the exposure time and the nature of the plasma gas. N_2 plasma treatment can lead to a slight enhancement in both biomedical behaviors at the longest exposure time, although no enhancement in these behaviors is observed for the samples treated by Ar plasma.

3.1.3. Surface chemical states, oxidation, and radical analysis. In order to investigate surface chemical states and oxidation, XPS analysis was performed. The binding energies of the C_{1s}, N_{1s}, and O_{1s} peaks on the untreated sample surface were 284.5, 399.9, and 532.4 eV, respectively, and these values did not change after the plasma treatments. However, the form of the C_{1s} peaks changed after the surface was exposed to plasma. Figure 6 presents XPS spectra of the C_{1s} peaks at binding energies ranging from 280 to 292 eV for N_2 and Ar plasma-treated and untreated samples. The spectrum for the untreated sample shows three peaks at binding energies of 284.5, 286.2, and 289 eV, which correspond to C—H, C—O, and C=O bonds. The intensity of the peaks corresponding to C—O and C=O bonds decreases after N_2 plasma treatment for 15 min, whereas the C_{1s} signal (284.5 eV) does not change, which means that the N_2 plasma treatment broke some of the C—O and C=O bonds. The spectrum of the Ar plasma-treated sample for an exposure time of 15 min is similar to that of the N_2 plasma-treated sample, except that the decrease in the intensity of the peaks

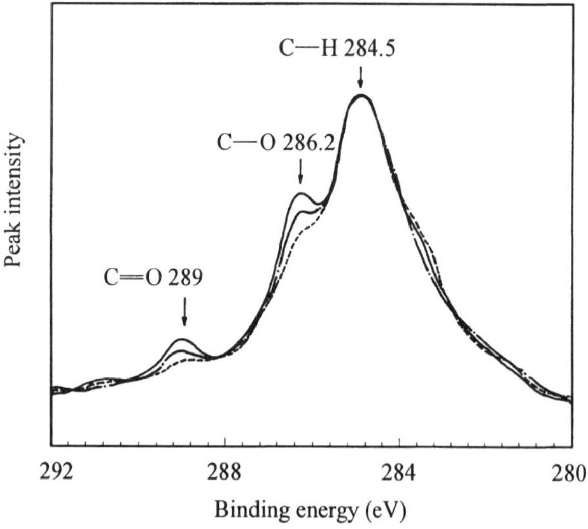

Figure 6. C_{1s} XPS spectra for untreated sample (—); Ar plasma-treated sample, exposure time 15 min (—·—); and N_2 plasma-treated sample, exposure time 15 min (- - - - - -).

Table 1.
Atomic concentrations (%) in the surface region of PEU due to plasma treatments

Sample	Exposure time (min)	Atomic concentration (%)		
		C_{1s}	N_{1s}	O_{1s}
Untreated	0	76.78	1.62	21.60
Ar plasma treated	15	73.10	0.67	26.23
N_2 plasma treated	15	70.13	3.60	26.17

corresponding to the C—O and C=O bonds is not evident. No appreciable change in the XPS spectra of the N_{1s} and O_{1s} peaks was observed for N_2 and Ar plasma-treated samples.

As is well known, polymer surfaces undergo oxidation when exposed to air after exposure to an inert gas plasma. The oxidation of the surface will affect the biomedical behavior of polymer surfaces, because PEU samples treated by the N_2 and Ar plasmas were stored in dry air for about 24 h until the measurements, leading to the oxidation of the surface. In order to clarify this result, atomic concentrations (%) in the surface region due to plasma treatments were measured using XPS analysis. Table 1 presents the results of the atomic concentration measurements. As can be seen, a certain amount of oxygen atoms is introduced in the surface region of the samples treated by N_2 plasma, which is the result of oxidation. It is seen that nitrogen is introduced in the surface region when the sample is treated with N_2 plasma. However, the amount is smaller than that of O atoms. In addition, the carbon content decreases in the surface region due to the breaking of C—O and C=O bonds. The increase in O and N contents and the decrease in C content will lead to a change in the biomedical behavior. For comparison, the results for Ar plasma are also given in Table 1. As can be seen, there is similar oxidation of the surface. However, Ar plasma treatment decreases the N content in the surface, and causes a smaller decrease in C content than N_2 plasma treatment does.

No evident increase in the number of radicals in the samples treated by N_2 and Ar plasmas for an exposure time of 15 min was observed by ESR analysis.

3.2. HMDS plasma polymerization

3.2.1. Contact angle change. Figure 7 shows the change in contact angle upon exposing the PEU samples to HMDS plasma at a power of 100 W. HMDS plasma polymerization results in a greater decrease in contact angle for an exposure time up to 2 min and reaches a value of 43°. On longer exposure time, however, the contact angle increases rapidly from 43° to 84°. A plasma exposure for about 10 min is required for the surface to attain a contact angle to a high constant value. It is clear from the result presented that polymerization for a shorter time will lead to a lower contact angle, i.e. wettability. This result means that the optimum exposure time to HMDS plasma for highest wettability is approximately 2–5 min. As compared to the N_2 and Ar plasma treatments, the HMDS plasma polymerization leads to a significant enhancement of wettability.

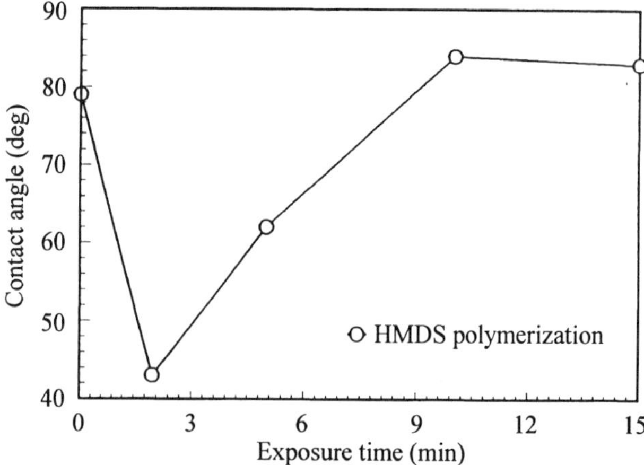

Figure 7. Contact angle vs. plasma exposure time at 100 W.

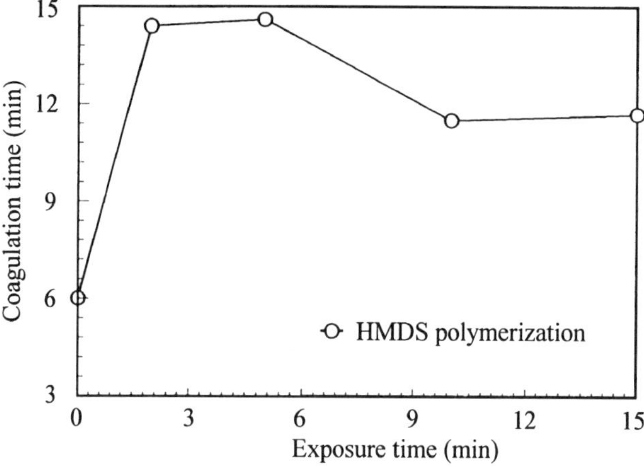

Figure 8. Coagulation time vs. plasma exposure time at 100 W.

3.2.2. Coagulation time and recalcification time changes. Figures 8 and 9 show the coagulation time and the recalcification time for HMDS plasma-polymerized samples as a function of the exposure time at a power of 100 W. Figure 8 indicates that HMDS plasma polymerization leads to a considerable enhancement of the anti-coagulability. For an exposure time of 5 min, the coagulation time increases from 6 to 14.6 min. After 5 min, the coagulation time decreases slightly and reaches a plateau for an exposure time of 10 min. This means that the shorter time (2–5 min) of polymerization is the optimum choice for a significant improvement in the surface anti-coagulability, which is of great value to clinical application. At the same time, Fig. 9 shows that HMDS plasma polymerization leads to a significant increase in the recalcification time. Its value increases from 30.5 to 37.4 s for an exposure time of 5 min, which means that the anti-calcific behavior of PEU is improved significantly. The recalcification time, however, decreases gradually after an exposure time

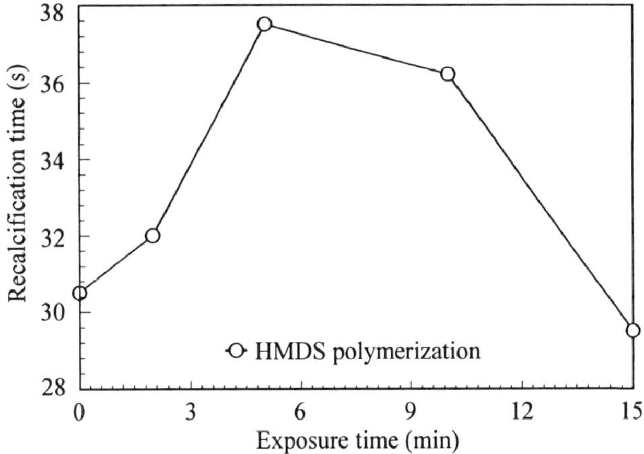

Figure 9. Recalcification time vs. plasma exposure time at 100 W.

of 10 min. This result means that the optimum exposure time for improving surface anti-calcific behavior is 5–10 min.

On the whole, HMDS plasma polymerization can result in a more effective improvement in the anti-coagulability and anti-calcific behavior than can N_2 and Ar plasma treatments.

3.2.3. Surface chemical states and radical analysis. It is clear from the results presented that HMDS plasma polymerization can cause more effective improvements in the surface biomedical behavior than the N_2 and Ar plasma treatments. We think that the introduction of HMDS to the surface is the main reason for these changes.

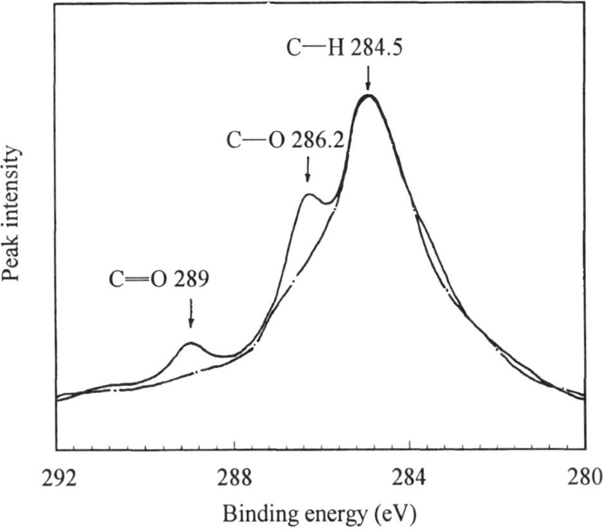

Figure 10. C_{1s} XPS spectra for untreated sample (—) and HMDS plasma-polymerized sample, exposure time 5 min (—·—).

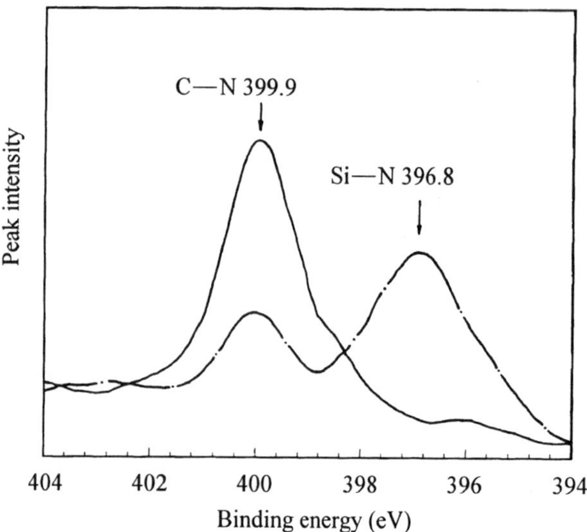

Figure 11. N_{1s} XPS spectra for untreated sample (—) and HMDS plasma-polymerized sample, exposure time 5 min (— · —).

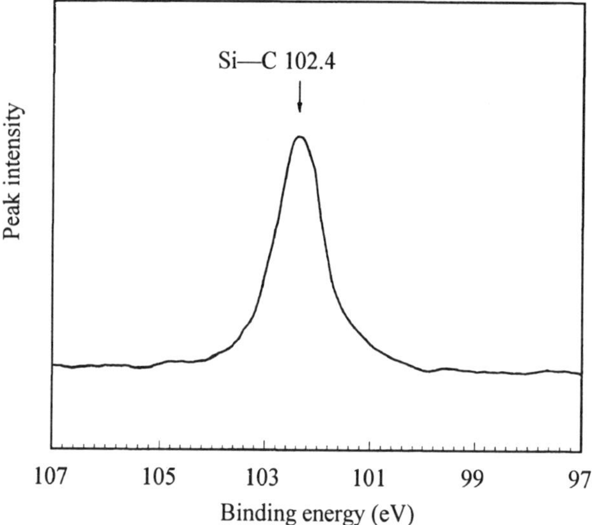

Figure 12. Si_{2p} XPS spectra for HMDS plasma-polymerized sample, exposure time 5 min.

The hypothesis that it is the deposition of HMDS on the surface of the PEU which is responsible for these significant improvements in biomedical behavior was fully confirmed by XPS studies. Figures 10, 11, and 12 present characteristic C_{1s}, N_{1s}, and Si_{2p} XPS signals for the samples (the same as those used for the biomedical behavior studies) with plasma-polymerized HMDS. In these cases, XPS not only confirmed the presence of Si atoms at the surface, but also enabled a distinction between different Si-containing groups. The binding energy range of 280–292 eV corresponds to C_{1s} electrons. It is seen that the two peaks at the binding energies of 286.2 and 289 eV corresponding to C—O and C=O bonds, respectively, disappear after HMDS plasma polymerization for 5 min. This is because the deposition of HMDS covers the surface and the XPS signals are from only C_{1s} electrons of HMDS deposited on the surface. Figure 11 presents XPS spectra of N_{1s} peaks at binding energies ranging from 394 to 404 eV. The spectrum for the untreated sample shows a peak at 399.9 eV corresponding to C—N bonds. HMDS plasma polymerization leads to a large decrease in this peak. At the same time, a new peak at a binding energy of 396.8 eV emerges; this peak corresponds to Si—N bonds. This result means that Si and C atoms of HMDS deposited on the surface constitute new chemical bonds with some N atoms of PEU. Further evidence supporting the occurrence of HMDS deposition onto the PEU surface is obtained by measuring the Si_{2p} signals. No Si atoms exist on the surface of the untreated sample. However, Fig. 12 indicates that the sample with plasma-deposited HMDS for 5 min shows a new peak at a binding energy of 102.4 eV; this peak corresponds to Si—C bonds, which is a reflection of the Si—C bonds of HMDS. The Si concentration (%) in the surface region due to HMDS plasma polymerization is 19.8, which is also evidence of HMDS deposition onto the PEU surface. These results mean that the HMDS plasma polymerization really takes place and that the HMDS is located at the sample surface.

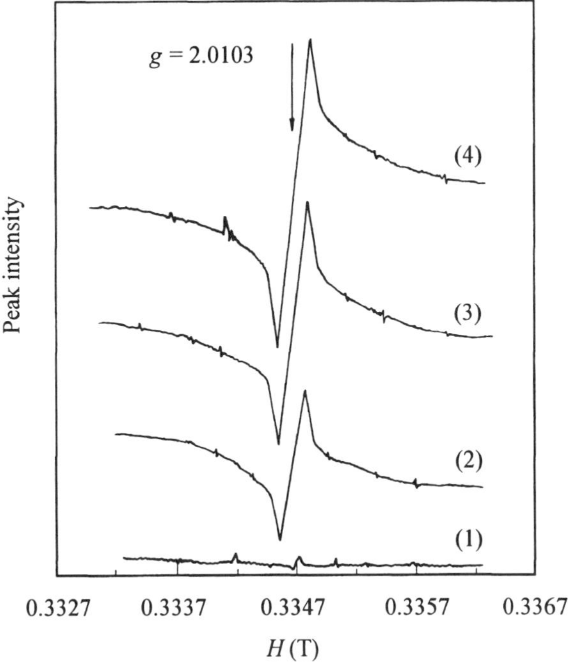

Figure 13. ESR spectra (1) untreated sample; (2) HMDS plasma-polymerized sample, exposure time 2 min; (3) HMDS plasma-polymerized sample, exposure time 5 min; and (4) HMDS plasma-polymerized sample, exposure time 10 min.

In order to estimate further the number of radicals in the samples, ESR analysis was performed. Figure 13 gives the ESR spectra of samples with HMDS plasma polymer as compared with the untreated sample at a magnetic field intensity (H) range of 0.3327–0.3367 T. It was observed that the amplitude of the resonant peak ($g = 2.0100$) at $H = 0.3347$ T increases dramatically with the exposure time. As the resonant peak intensity is a direct measurement of the radical number, this result indicates that the number of radicals increases with the time of exposure to HMDS plasma, which is also the reason why the surface biomedical behavior is affected.

4. CONCLUSIONS

A study of the surface biomedical behavior was made for N_2 and Ar plasma treatments and for HMDS plasma polymerization on medical-grade PEU. The following conclusions can be drawn:

(1) The enhancement in the wettability of plasma-treated samples was dependent on the exposure time and the nature of the plasma gas. HMDS plasma polymerization leads to a significant enhancement of the wettability for an exposure time of 2–5 min.

(2) N_2 plasma treatment resulted in a slight increase and a gradual increase in anti-coagulability and anti-calcific behavior. Compared with the plasma treatments, the coagulation time of samples with HMDS plasma polymer could be enhanced about 2.5 times *vis-à-vis* the untreated sample when the exposure time was 2–5 min. The anti-calcific behavior was significantly improved by HMDS plasma polymerization for an exposure time of 5–10 min.

(3) Plasma treatments broke some C—O and C=O bonds in the surface and resulted in the oxidation of the surface, which caused a change in surface properties.

(4) Plasma polymerization resulted in HMDS deposition onto the PEU surface and formed new Si—N bonds and increased the number of radicals in the samples, which is the main reason why the biomedical behavior is improved significantly by plasma polymerization.

Acknowledgement

This project was supported by Chinese Young Natural Science Foundation.

REFERENCES

1. M. B. Lelah and S. L. Cooper, *Polymerization in Medicine*. CRC Press, Boca Raton, FL (1986).
2. R. Barbucci, A. Magnani, A. Albanese and F. Tempesti, *Int. J. Artif. Organs* **14**, 499–507 (1991).
3. Dejun Li, Jie Zhao, Hanqing Gu, Mozhu Lu, Fuqing Ding and Qiqing Zhang, *Nucl. Instrum. Methods* **B82**, 57–62 (1993).
4. T. Yasuda, M. Gazicki and H. Yasuda, *J. Appl. Polym. Sci., Appl. Polym. Symp.* **38**, 201–214 (1984).
5. H. Yasuda, *Plasma Polymerization*. Academic Press, New York (1985).

6. B. D. Ratner, *J. Biomater. Sci. Polym. Ed.* **4**, 3–11 (1992).
7. T. G. Grasel and S. L. Cooper, *J. Biomed. Mater. Res.* **23**, 311–338 (1989).
8. A. Z. Okkema and S. L. Cooper, *Biomaterials* **12**, 668–676 (1991).

Part 2

Laser Surface Modification Techniques

Polymer Surface Modification: Relevance to Adhesion, pp. 153–184
K. L. Mittal (Ed.)

Modification of polymers with UV excimer radiation from lasers and lamps

JUN-YING ZHANG,[1] HILMAR ESROM,[2,*] ULRICH KOGELSCHATZ[3] and GERHARD EMIG[4]

[1] *Institut für Chemische Technik, Universität Karlsruhe, 75128 Karlsruhe, Germany*
[2] *Fachhochschule für Technik Mannheim, Institute of Technology Mannheim, 68163 Mannheim, Germany*
[3] *Asea Brown Boveri AG, Corporate Research, CH-5405 Baden, Switzerland*
[4] *Lehrstuhl für Technische Chemie I, Universität Erlangen-Nürnberg, 91058 Erlangen, Germany*

Revised version received 30 April 1994

Abstract—Photochemical dry etching and surface modification of various polymers, e.g. polymethylmethacrylate (PMMA), polyimide (PI), polyethyleneterephthalate (PET) and polytetrafluoroethylene (PTFE) were investigated with coherent and incoherent excimer UV sources. Ablation rates of PMMA were measured as a function of laser fluence and laser pulse at the wavelength $\lambda = 248$ nm (KrF^*). Decomposition and etch rates of PMMA and PI were determined as a function of UV intensity and exposure time at three different wavelengths $\lambda = 172$ nm (Xe_2^*), $\lambda = 222$ nm ($KrCl^*$) and $\lambda = 308$ nm ($XeCl^*$). The transmittance of the polymeric films was determined with a UV-spectrophotometer after different exposure times. The morphology of the exposed polymers was investigated with scanning electron microscopy (SEM). The gaseous products occurring during UV exposure were measured using mass spectrometry (MS). Chemical surface changes of the photoetched PMMA were determined by X-ray photoelectron spectroscopy (XPS). The mechanism of the photo-oxidation process of PMMA is discussed. The etching of PMMA can be explained as a result of extensive photo-oxidation. The results are compared with those obtained from mercury lamp and excimer laser experiments. Good adhesion of electrolessly deposited metal layers was achieved by irradiation of the polymeric surfaces from incoherent UV source before depositing the metal layer.

Keywords: Modification of polymers; adhesion of metal layers; UV excimer sources (laser and lamp); photochemical dry etching; photo-oxidation; PMMA; PI; PET; ablation of polymers.

1. INTRODUCTION

The discovery in 1982 of ultraviolet (UV)-induced removal of thin layers of organic polymers with intense radiation of excimer lasers [1, 2] stimulated scientific interest in the physics and chemistry of this surface modification process. The ablative

*To whom correspondence should be addressed.

photo-decomposition (APD) process renders important potential applications in microelectronics, optics, packaging technology and surgery. For example, polyimides and fluoropolymers offer desirable properties for microcircuit fabrication [3]. PMMA is widely used as a photoresist in lithographic processing [4]. Applications of metallized plastics in the automotive industry and in the fields of electromagnetic shielding and semiconductor manufacturing and packaging are widespread [3, 8, 9, 24].

Today, controlled modification of polymeric properties induced by the interaction of energetic particle beams, photons and plasmas has become a field of considerable technical importance. It can be used to improve adhesion properties of coatings as well as wettability and printability of polymers by changing the morphology and chemical properties of the polymer surface. Area-selective processing of polymeric materials by UV irradiation with excimer lasers has become a growing field of applied research [5]. A considerable amount of theoretical and experimental work has been done to understand and to describe the observed phenomena. A wide variety of polymers (PMMA, PI, PET, PTFE, PE (polyethylene), PC (polycarbonate), etc.) can be ablated by UV laser radiation. Most of the etching studies were performed using pulsed excimer lasers. Table 1 summarizes some results currently available for the etching of polymers.

Several models have been suggested to describe the ablative photo-decomposition by ultraviolet radiation [15, 16]. Many investigations have been performed to detect the species which are involved in the ablation process by using emission spectrometry [17], laser induced fluorescence [18], IR- and UV-spectroscopy [19], gas chromatography and mass spectrometry [20, 44], X-ray-photoelectron spectroscopy (XPS) [21], etc. As a result, photochemical and photothermal processes and also combinations of these processes have been identified as principal mechanisms for polymeric material removal.

A characteristic feature of a polymeric surface treated with excimer lasers is that significant etching occurs only if the incident pulse fluence exceeds a well-defined threshold fluence. Its value depends on the wavelength of the UV radiation and is related to the absorption coefficient of the polymer. The relatively small cross-section of excimer laser beams, together with the value of the relative high threshold energy, make it difficult to treat large areas (several $cm^2 - m^2$) of polymeric surfaces efficiently.

Recently, it has been demonstrated that novel incoherent UV excimer sources [6, 7] can be very useful in large area surface processing, e.g. metallization of plastic materials [8–10], photochemical dry etching of polymers [22, 23, 47], photo-deposition of silicon dioxide films [51], silicon nitride films [52] and semiconductors [53, 54].

In this paper, we report results of a systematic study concerning decomposition and dry etching of polymers (PMMA, PI, PET, PTFE) using incoherent excimer UV radiation at $\lambda = 172$ nm (Xe_2^*), $\lambda = 222$ nm ($KrCl^*$) and $\lambda = 308$ nm ($XeCl^*$). Gaseous decomposition products during exposure of the polymeric surface were determined by mass spectrometry (MS). The surface changes of photoetched polymers were studied by scanning electron microscopy (SEM) and X-ray photoelectron spectroscopy (XPS).

Table 1.
UV-induced etching of polymers

Polymer	Ambient	Light sources (λ)	Remarks	References
PET	Air Vacuum	ArF^* (193 nm)	1.2 μm/pulse (0.37 J/cm^2)	[1, 2]
	Vacuum	ArF^* (193 nm), low pressure mercury lamp (185, 254 nm)	XPS, SEM	[25]
PI	Air	ArF^*, KrF^*, $XeCl^*$ (193, 248, 308 nm)	Etch rate, UV, model	[16, 18, 26, 27]
		$XeCl^*$ (308 nm)	IR, SEM	[28]
	Vacuum	F_2^*, ArF^*, $XeCl^*$ (157, 193, 308 nm)	MS	[29]
	Air Vacuum	KrF^*, $XeCl^*$, XeF^* (248, 308, 351 nm)	Threshold fluence, UV, IR GC/MS	[20]
PMMA	Air	KrF^* (248 nm)	Etch rate, model, incubation	[30]
		ArF^*, KrF^*, $XeCl^*$ (193, 248, 308 nm)	Etch rate, SEM, LIF, model	[14, 31, 32, 27]
		$XeCL^*$ (308 nm)	Etch rate, SEM, model	[26]
		Ar ion laser (300–330, 350–380 nm)	Etch rate, SEM	[33]
		ArF^*, KrF^*, $XeCl^*$, XeF^* (193, 248, 308, 351 nm)	Etch rate, UV, SEM	[34]
PMMA, PET, PI	Air	ArF^* (193 nm), low pressure mercury lamp (185, 254 nm)	Etch rate, XPS, SEM	[13]
PMMA, PI PET, PTFE	Vacuum Air	Xe_2^*, $KrCl^*$, $XeCl^*$ (172, 222, 308 nm)	Etch rate, MS UV, SEM, XPS	[22, 23, 47]
PTFE (Teflon)	Air	KrF^* (248 nm)	Etch rate, SEM	[35, 36]
		ArF^*, KrF^* (193, 248 nm)	Etch rate, IR, UV, SEM	[37]
α-MePS, PMMA	Air	ArF^* (193 nm)	Etch rate, XPS	[38]
PC, PET, PS	Air Vacuum	ArF^*, KrF^* (193, 248 nm), low pressure mercury lamp (185, 254 nm)	Etch rate	[39]
PE, PET, PI	Helium	F_2^* (157 nm)	Threshold fluence	[40]
Mylar, PI, PMMA	Air, Helium Vacuum	ArF^*, KrF^*, XeF^* (193, 248, 351 nm)	Threshold fluence, SEM	[17]
PPQ, PI, PS PC, PET	Air Vacum	ArF^*, KrF^* (193, 248 nm), low pressure mercury lamp (185, 254 nm)	Etch rate, UV	[41]

PS: polystyrene, PPQ: poly(phenylquinoxaline), α-MePS: poly(methylstyrene), GC: gas chromatography, LIF: laser induced fluorescence.

The mechanism of the photo-oxidation of PMMA caused by incoherent excimer radiation is discussed and compared with excimer laser experiments. A model describing the mechanism of the photo-oxidation is proposed.

2. EXPERIMENTAL

The polymeric samples used were commercially available thick samples of PMMA (Röhm, Darmstadt), PI (Du Pont; Kapton^{TM}), PET (BASF) and PTFE. Spin coatings of PMMA (Röhm, Darmstadt) on quartz substrates (Suprasil I) were prepared by using a PMMA/trichloroethylene solution. The decomposition of thin PMMA films (thickness: 0.6 μm) was studied by measuring the transmission spectra of the films after different exposure times with a UV spectrophotometer (Beckman). A stock solution of polyimide (Merck) was diluted in the appropriate solvents (Merck, Selectiplast HTR D-2). This polyimide solution was also spin-coated on quartz substrates (Suprasil I) for transmittance measurements.

The samples were irradiated in air with a commercial KrF* excimer laser (Lambda Physik EMG 103, $\lambda = 248$ nm). The experiments were performed with the unfocused beam of the excimer laser or a focused beam (line focus) using a cylindrical quartz lens ($f = 158$ mm). The energy density of the UV laser radiation was measured using a joulemeter (Gen Tec PRJ-D, ED500).

Polymeric surfaces were also irradiated with incoherent excimer sources from a dielectric barrier discharge (silent discharge) operating in pure xenon, or gas mixtures of krypton/chlorine and xenon/chlorine providing intense narrow-band radiation at $\lambda = 172$ nm (Xe_2^*), $\lambda = 222$ nm (KrCl*) and $\lambda = 308$ nm (XeCl*), respectively. The discharge was initiated in an annular gap between coaxial quartz tubes [7]. The radial width of the annular discharge gap was about 0.4 cm. The excimer UV source was operated at frequencies of about 20 kHz and voltage amplitudes up to 20 kV. More details about this special excimer UV source and the experimental parameters can be found in the literature [6, 7, 22, 23]. Irradiation of the polymers was carried out in a small exposure reactor [9] that could be evacuated and flooded with air or argon. The distance between the samples and the excimer lamp was 1 cm. The structuring of the polymer surfaces with excimer lamps was performed by irradiation of the surfaces through the openings of a contact metal mask. The UV intensity was measured using a vacuum phototube (R765, Hamamatsu).

The etch depth and the roughness of the exposed polymers were measured with a stylus profilometer (Dektak). The morphology of the irradiated polymeric films was investigated by scanning electron microscopy (SEM) (JEOL JXA-733).

Gaseous products during incoherent excimer irradiation ($\lambda = 222$ nm) were determined by mass spectrometry (Dataquad, CIT-Alcatel) at different exposure times and at different pressure levels in the exposure reactor.

The chemical changes in the UV exposed PMMA surface layer were investigated using X-ray photoelectron spectroscopy (Model ESCALAB MKII, Fisons Instruments) by measurement of the intensity of the C$-$C, C$-$O, and O$-$C=O bonds after irradiation with the KrCl* excimer lamp ($\lambda = 222$ nm).

3. RESULTS AND DISCUSSION

3.1. Etching polymers with excimer UV sources

Figure 1 shows the depth profile achieved by scanning across non-irradiated and irradiated areas of PMMA using the KrF* laser at different fluences and using one single pulse. At a fluence of 3.1 J/cm^2, an elevated region of a height of about 5 μm was observed in the 0.6 $\times$ 2 mm^2 exposed area instead of the expected removal of PMMA by ablation. At 3.5 J/cm^2, ablation of PMMA started and a narrow groove formed because of high fluence at the middle of the focus width. Higher fluences caused an enlargement of the groove. A fluence value of 4 J/cm^2 led to an ablation depth of about 4.3 μm. Elevated ridges appeared at both sides of the grooves at high fluence. These ridges were due to low intensity at the sides of the laser line focus.

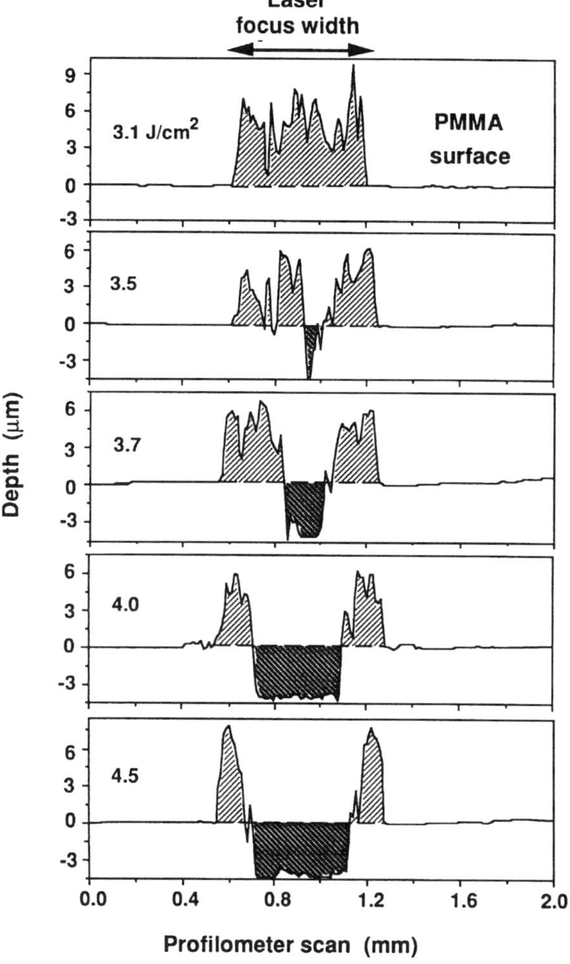

Figure 1. Depth profile of the ablation region of PMMA at different fluences (λ = 248 nm).

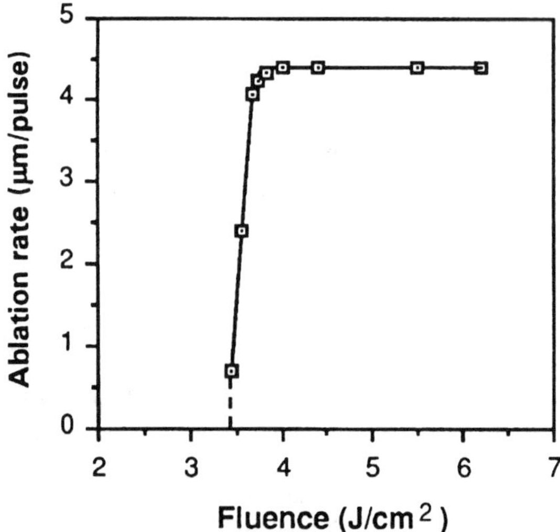

Figure 2. Ablation rate of PMMA as a function of laser fluence per pulse ($\lambda = 248$ nm).

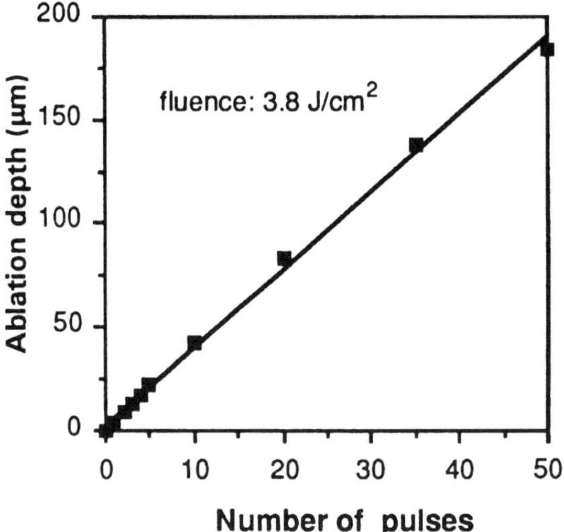

Figure 3. Ablation thickness of PMMA as a function of number of laser pulses ($\lambda = 248$ nm, fluence: 3.8 J/cm^2).

Ablation rates were measured for different fluences (Fig. 2). Up to 4 J/cm^2 the ablation rate increased with the fluence and saturated at higher fluences. The depth of the groove in PMMA shown in Fig. 3 is dependent upon the number of laser pulses at a fluence of 3.8 J/cm^2 per pulse. It increased linearly with the total number of pulses.

Profiles of grooves produced with different exposure times using the KrCL* excimer lamp at an intensity of 30 mW/cm^2 are shown in Fig. 4. As can be seen in Fig. 4 the walls of the etched groove were practically vertical with good edge sharpness.

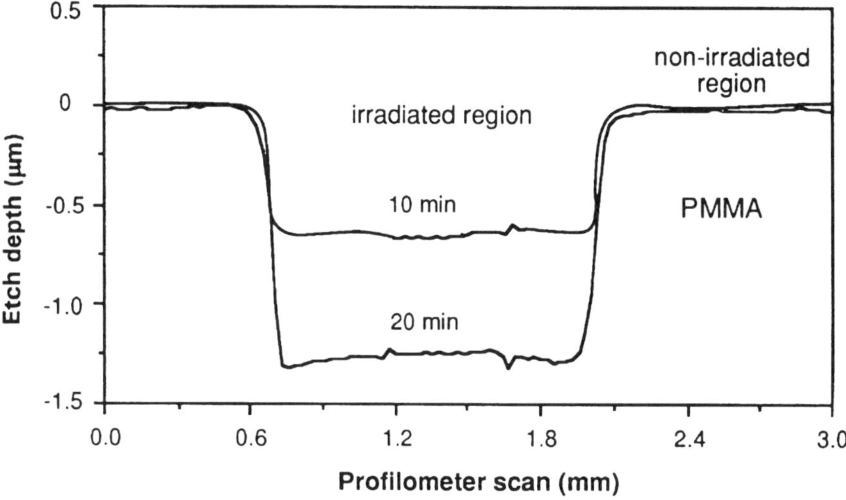

Figure 4. Depth profile of the irradiated region of PMMA with excimer lamp (exposure time: 10 min and 20 min, UV intensity: 30 mW/cm^2, wavelength: $\lambda = 222$ nm).

Absence of warping and redeposited particles around the edges of the groove and sharp edge contours indicate that photolysis is playing a major role in the removal of PMMA using incoherent excimer UV radiation.

In experiments on the etching of polymers with excimer lasers [5] a wavelength dependence of the etch rate was observed and attributed to the changing absorption properties of the etched polymers. This tendency is also found in the case of photoetching of PMMA and PI with incoherent excimer radiation. Figures 5 and 6 show the etch depth of PMMA and PI as a function of the exposure time at different wavelengths $\lambda = 172$ nm, $\lambda = 222$ nm and $\lambda = 308$ nm (30 mW/cm^2). Due to the stronger absorption of PMMA and PI in the VUV region, the highest etch rates 0.15 $\mu\text{m} \cdot \text{min}^{-1}$ (for PMMA) and 0.1 $\mu\text{m} \cdot \text{min}^{-1}$ (for PI) were achieved using radiation at the shortest wavelength $\lambda = 172$ nm (30 mW/cm^2).

In Fig. 7 the etch rate of PMMA is shown as a function of the UV intensity at the three wavelengths ($\lambda = 172$ nm, $\lambda = 222$ nm and $\lambda = 308$ nm). Figure 8 shows etch rate of polyimide as a function of the UV intensity at the three wavelengths indicating that the etch depth of PI as well is a linear function of UV intensity at all wavelengths. Contrary to the experiment with excimer lasers [5], no flattening of the etch curves is observed at high UV intensities. Consequently, an enhancement of the etch rate by higher photon fluxes of more powerful excimer lamps seems possible.

The etching of PMMA, PI and PET in air at atmospheric pressure using UV radiation from a low pressure mercury lamp at $\lambda = 254$ nm (UV intensity: 12 mW/cm^2) and $\lambda = 185$ nm (UV intensity: 2.5 mW/cm^2) was investigated by Srinivasan and Lazare [13]. The etch rate of PMMA was determined to be about 0.007 $\mu\text{m} \cdot \text{min}^{-1}$. The etching mechanism, supported by ESCA investigations, was interpreted as a photo-oxidative etching of the polymeric surface. Using incoherent UV excimer radiation at $\lambda = 172$ nm and $\lambda = 222$ nm, we estimated even lower etch rates in the case of PMMA treated in air at atmospheric pressure.

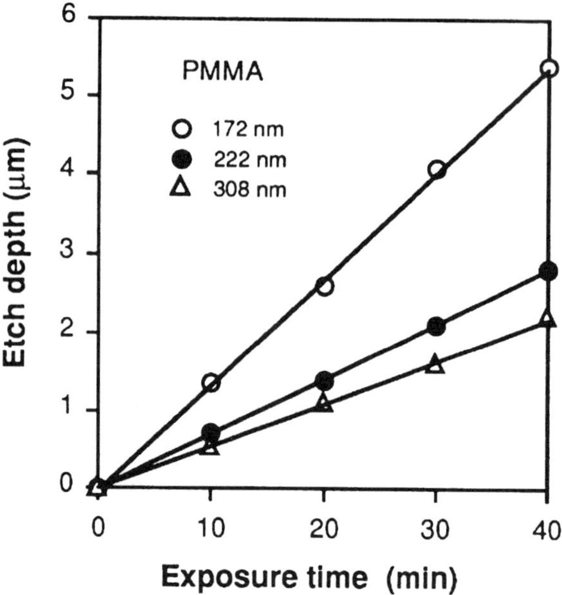

Figure 5. Etch depth (μm) of PMMA as a function of exposure time (min) at different wavelengths: $\lambda = 172$ nm, $\lambda = 222$ nm and $\lambda = 308$ nm; $p = 1$ mbar.

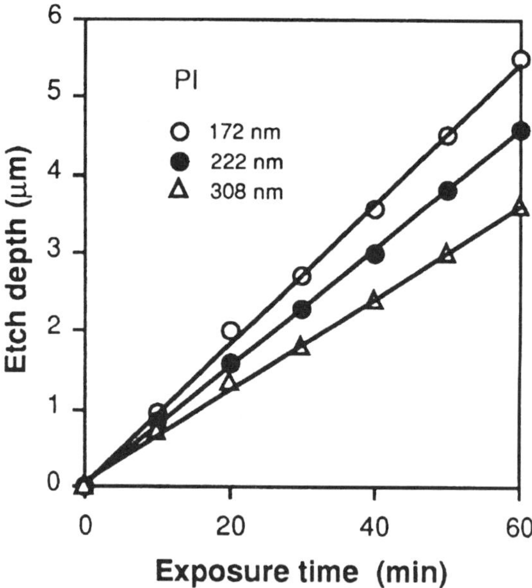

Figure 6. Etch depth (μm) of PI as a function of exposure time (min) at different wavelengths: $\lambda = 172$ nm, $\lambda = 222$ nm and $\lambda = 308$ nm; $p = 1$ mbar.

However, by performing the irradiation at a low background pressure of 1 mbar the etch rate of PMMA could be enhanced in our experiments by a factor of > 20 (see Fig. 9). If the background pressure is further lowered to 0.1 mbar the etch rate

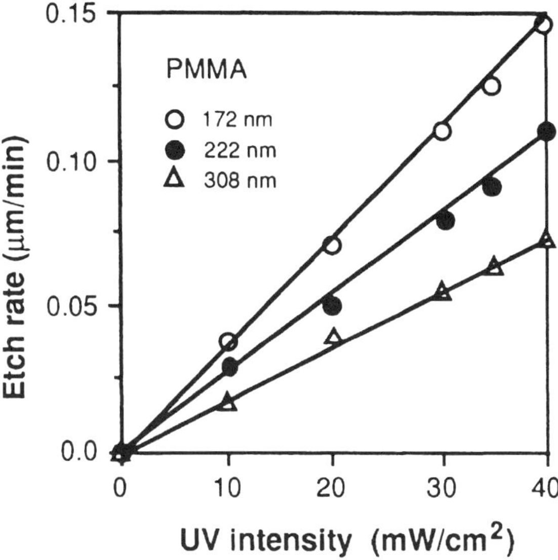

Figure 7. Etch rate of PMMA as a function of UV intensity at different wavelengths: λ = 172 nm, λ = 222 nm and λ = 308 nm; p = 1 mbar.

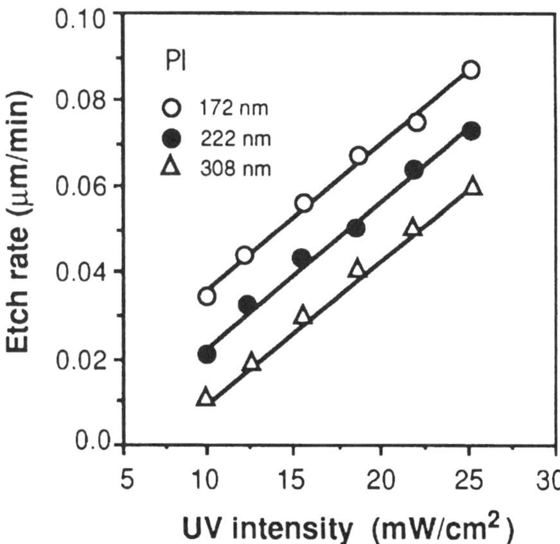

Figure 8. Etch rate of polyimide as a function of UV intensity at different wavelengths: λ = 172 nm, λ = 222 nm and λ = 308 nm; p = 1 mbar.

drops again to less than 0.01 $\mu m \cdot min^{-1}$ (Fig. 9). A tentative explanation for the enhanced etch rates at 1 mbar could be that the etching process involves photolytic oxidation processes relying on the presence of some oxygen. At very low pressures, photo-oxidation is reduced drastically due to the lack of oxygen in the background gas. If the pressure is too high, the removal of the volatile decomposition products is

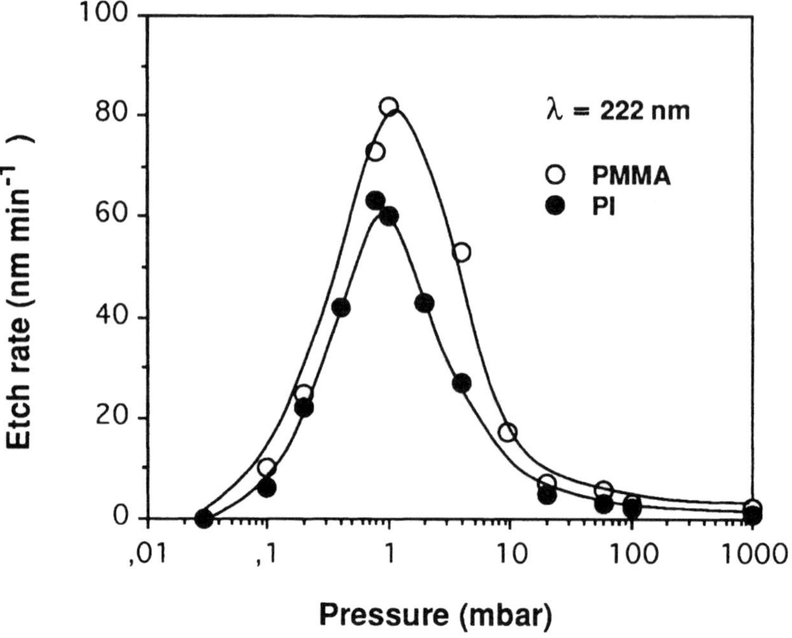

Figure 9. Plot of the etch rate as a function of background pressure with incoherent excimer UV radiation for PI and PMMA (wavelength $\lambda = 222$ nm, UV intensity 20 mW/cm^2).

Table 2.
Photoetching of polymers with conventional mercury lamps and incoherent UV lamps

UV lamp	pressure	Etch rate: μm/min PMMA	PI	PET
Low pressure mercury lamp (185 nm, 254 nm)	1 bar air	0.007	$<$ 0.001	0.001
	$< 10^{-2}$ mbar	0	0	0
Excimer lamp (172 nm)	1 mbar air	0.15	0.1	0.04
'Enhancement factor'		$>$ 20	100	40

hampered. At an intermediate pressure of about 1 mbar, apparently optimal conditions exist for the photo-oxidation of the polymeric surface and its decomposition resulting from photolytic chain scission. The most active species are most likely the free radicals O (^{1}D) and O (^{3}P), as well as excited molecular oxygen species, and perhaps some ozone [42].

In our experiments at low background pressure (1 mbar), we found that using the incoherent excimer UV source, other polymers, especially PI and PET, can be etched with rates higher by a factor from 20 to 100, compared to the rates 0.001 $\mu\text{m} \cdot \text{min}^{-1}$

achieved with conventional mercury lamps (see Table 2) [13, 41, 43]. Surprisingly high etch rate of about 1 $\mu m \cdot min^{-1}$ in the case of Teflon can be achieved by using excimer lamps [47], too. Other authors indicated that these polymers cannot be efficiently ablated at low pressures with mercury lamps. Perhaps a remark about the temporal behaviour of the light emission of the different UV sources should be made. While the intensity of low-pressure mercury lamps is modulated with twice the line frequency, the excimer UV lamps emit bursts of extremely short UV pulses [7]. The duration of individual UV pulses is approximately 10 ns, and the repetition frequency of the bursts was 40 kHz.

3.2. Morphology of the exposed areas of polymeric films with excimer UV sources

Polymeric surfaces are often smooth and chemically inert. Excimer laser-induced etching can lead to alteration of the surface morphology. Wettability, adhesion, chemical, physical and optical properties might be controlled in this way [43]. The roughness of the UV-irradiated area can also be changed by suitable choice of the laser parameters (pulse energy, wavelength, pulse rate, number of pulses, etc.) depending on the properties of the material to be etched. Srinivasan *et al.* [14] demonstrated that in the case of PMMA, the characteristic scale size of the structures of the etched surface depends on the wavelength of the excimer laser radiation. In contrast to the experiments with 193 nm laser pulses, the morphology of the etched surface with $\lambda = 248$ nm laser pulses appears as bubbles due to the high thermal component in the etching mechanism. At $\lambda = 193$ nm, the etching process proceeds mainly by photochemical decomposition, producing a very smooth surface without any structure.

This systematic study was started after obtaining good adhesion of electrolessly deposited metals (copper, nickel, gold, chromium) to a variety of polymers (PI, PMMA, PET, PTFE, PP, ABS (acryl-butadiene-styrene)). This fundamental knowledge of surface modification of polymers with incoherent UV from excimer lamps was previously non-existent. The key to good adhesion in this process is the excellent possibility of controlling the degree of micro-roughness of the polymeric surfaces in a reliable manner. As already found by other authors [50] micro-roughness is a very important parameter in obtaining good adhesion.

Good adhesion of copper, nickel, gold and chromium layers (thickness 1 μm) on a number of polymers (PI, PMMA, PET, PTFE, PP, ABS) was produced with a new electroless deposition method previously described in our papers [9, 48]. No Scotch tape test failed in our investigations of the adhesion of metal layers using an optimal set of parameters such as the time of pre-irradiation, UV power, background pressure of air in the exposure apparatus, and parameters of the metal deposition process [9, 49].

Figure 10 shows the morphology of PMMA treated with an excimer laser after different laser pulses ($\lambda = 248$ nm, fluence: 3 J/cm^2). The morphology of the etched region after a different number of 248 nm laser pulses shows structures like bubbles. This indicates the high thermal component in the etching process. The edge of ablated PMMA can be seen in the scanning micrograph of Fig. 11 at high fluence of 4.5 J/cm^2. In the middle, a smooth structure without bubbles appears (Fig. 11a).

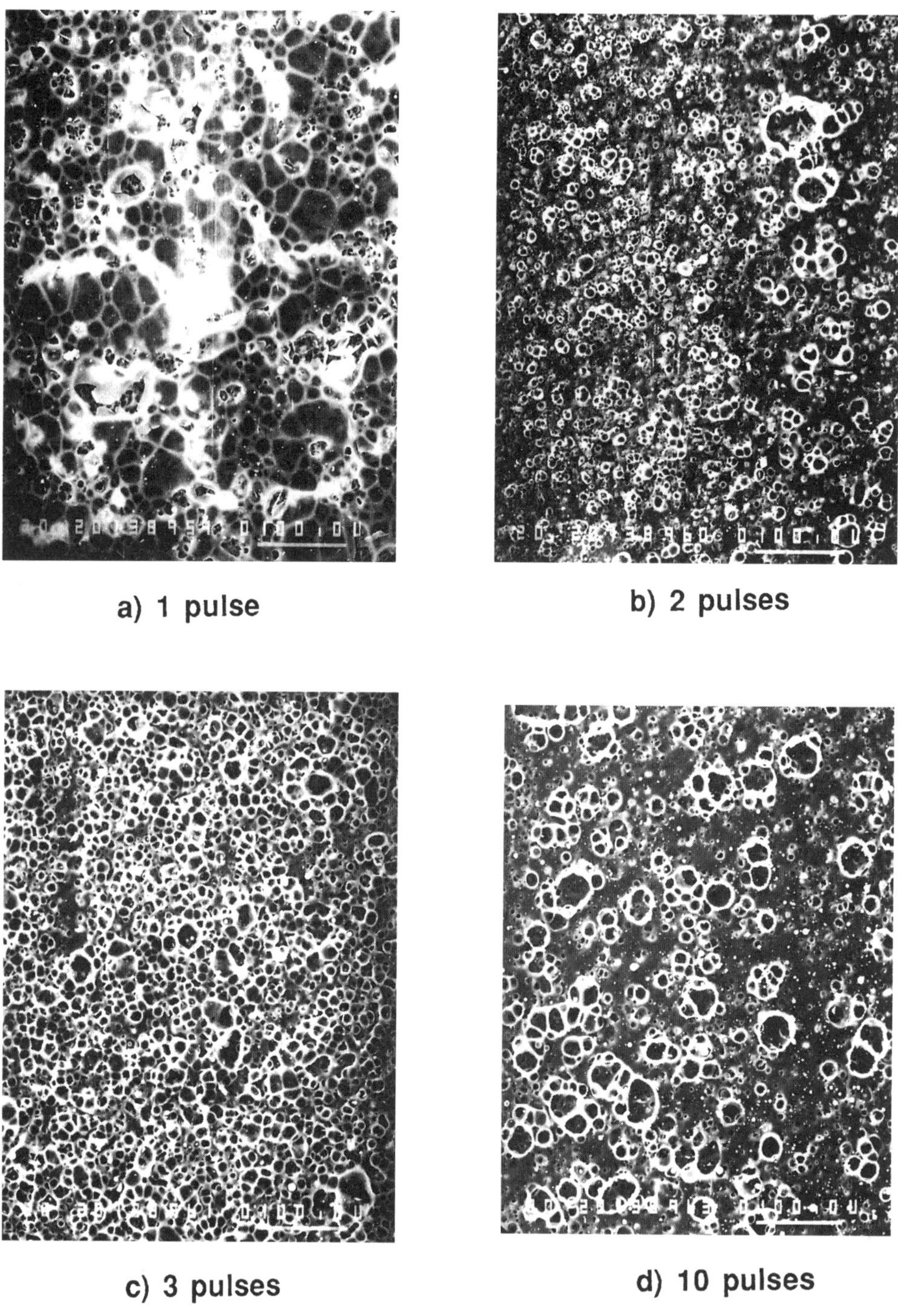

Figure 10. Morphology of PMMA with excimer laser treatment at different laser pulses, ($\lambda = 248$ nm, fluence 3 J/cm^2, scale bar; 100 μm).

Figure 11. Excimer laser-induced structure in PMMA (λ = 248 nm, fluence: 4.5 J/cm^2, 20 pulses); a) whole structure, scale bar; 1000 μm; b) left edge, scale bar; 100 μm; c) right edge, scale bar; 100 μm.

But the edges of the PMMA structure still consist of bubbles due to melting during laser heating (Figs 11b, c).

In Fig. 12 the typical surface roughness observed in the etched region of PMMA after irradiation with the excimer lamp at $\lambda = 222$ nm after different exposure times (0–40 min) is shown. Compared to the laser experiments with 248 nm pulses, the characteristic scale size of the roughness features is smaller. Therefore, our experiments support the hypothesis that the etching process with incoherent excimer radiation is mainly of photolytical nature. No thermal damage such as cracking, melting or bubble formation was observed in and around the etched region. Further experiments showed that the degree of surface roughness can be controlled by exposure time and UV intensity. Figure 13 shows the measured surface roughness (mean peak-to-valley distance) of PMMA as a function of exposure time after excimer lamp irradiation ($\lambda = 222$ nm, $p = 1$ mbar). The roughness of PMMA increased with the esposure time. Even after 50 min exposure time the roughness continued to increase but at a slower rate. Area-selective etching of PMMA was achieved with excimer lamp irradiation using metal contact masks. Figure 14 shows a scanning electron microphotograph of the surface of a PMMA sample which had been exposed for 10 min at $\lambda = 222$ nm (30 mW/cm^2). The edges of the PMMA structure are remarkably sharp without showing obvious thermal effects such as bubbles and melting formation. In summary, we observed characteristic differences in the morphology of polymeric surfaces irradiated with excimer lasers and excimer lamps. Pulsed excimer laser etching of polymers is mainly a pyrolytic process. On the contrary, excimer lamp etching is predominantly a photolytic process.

Ablation of polymers with excimer lasers is characterized by the following observations:

(1) At fluences below a typical wavelength-dependent threshold fluence, practically no ablation of polymers is observed. Because of the required threshold fluence, polymers can be ablated only in relatively small, localized areas.

(2) The etch depth per pulse can be as high as several μm due to high laser intensity. Saturation of the etch rate with increasing laser fluence is observed.

(3) The morphology of the irradiated region and the edge of the ablated region show bubble formation, melting and redeposited material. This indicates a strong thermal component in the etching process.

Etching of polymers with incoherent excimer lamps is characterized by the following observations:

(1) No threshold fluence could be detected in our excimer lamp experiments.

(2) Considerably higher etch rates are achievable as compared to the etch rates reported for conventional mercury lamps. The etch depth increases linearly with UV intensity.

(3) The morphology of the irradiated region and the edge of the ablated region show no indications of thermal effects such as bubbles or melting. The etching of polymers with excimer lamps appears to be mainly photolytic in nature.

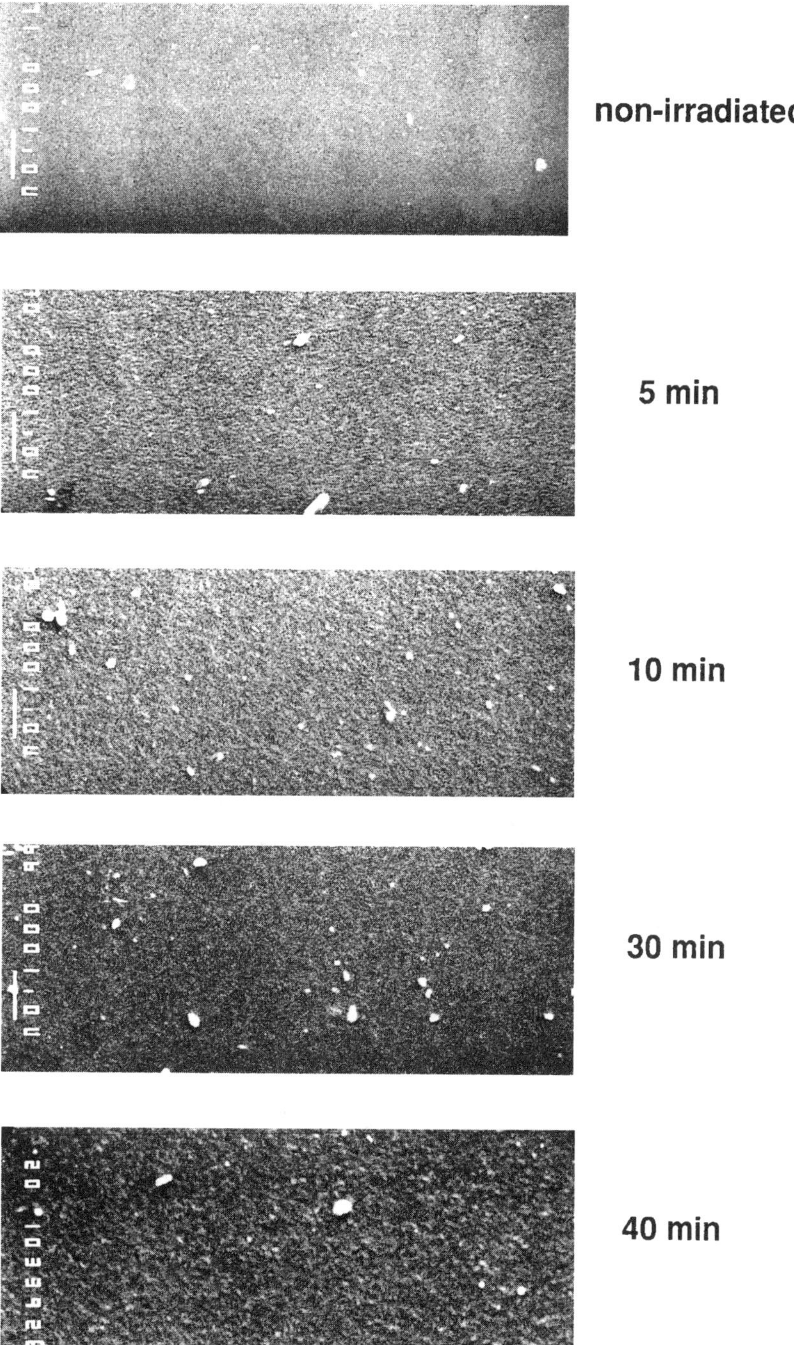

Figure 12. Morphology change of PMMA with excimer lamp radiation after different exposure times ($\lambda = 222$ nm, 30 mW/cm^2, scale bar; 1 μm).

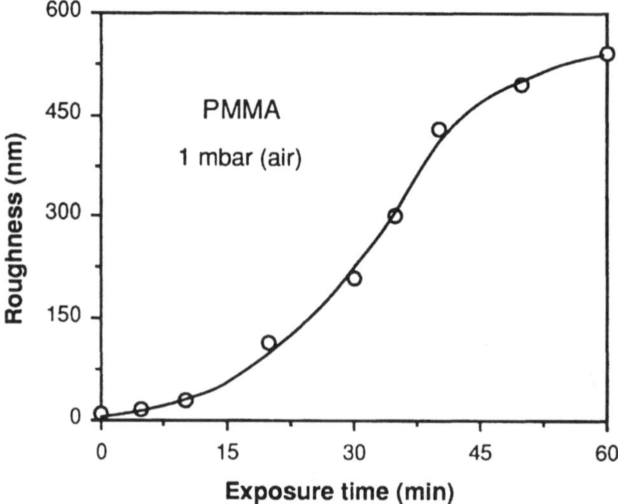

Figure 13. Surface roughness of PMMA as a function of exposure time after excimer lamp irradiation ($\lambda = 222$ nm, 30 mW/cm^2; $p = 1$ mbar).

Figure 14. Structure of PMMA after treatment with the excimer UV lamp ($\lambda = 222$ nm, 10 min, 30 mW/cm^2, scale bar; 1000 μm).

3.3. Mechanisms of the photo-etching process of polymers

Many investigations have been performed in order to understand the mechanisms of photo-etching of polymers with excimer lasers. It was concluded that the process of ablation is mainly photochemical in nature if ArF* ($\lambda = 193$ nm) laser pulses were used, but that both photochemical and thermal mechanisms are operative at longer laser wavelengths (KrF*, $\lambda = 248$ nm; XeCl*, $\lambda = 308$ nm; and XeF*, $\lambda = 351$ nm).

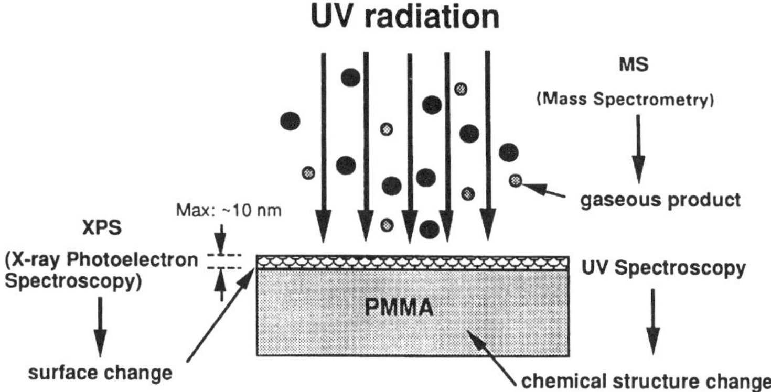

Figure 15. Properties of PMMA analyzed using UV, IR, MS and XPS.

In order to understand the mechanisms of the photo-etching process of polymers with incoherent excimer UV radiation in our experiments, it was necessary to investigate the gaseous decomposition products, the chemical structural changes and the surface composition of polymers. Figure 15 shows the various analytical techniques used for the investigation of the photo-etching mechanism of the polymers in this work. The chemical structural changes of the polymeric surface under investigation after irradiation were determined by UV spectroscopy (UV). The gaseous products formed by UV excimer lamp etching of the polymer were detected by using mass spectrometry (MS). X-ray photoelectron spectrometry (XPS) was used to study the surface composition of the polymer after UV excimer lamp irradiation.

3.3.1. Transmittance change during UV irradiation. Figure 16 shows the change of the UV transmission of thin PMMA films during UV irradiation (20 mW/cm^2) at the wavelength $\lambda = 222$ nm (KrCl*) in air. The initial decrease in the transmission at $\lambda = 240$–400 nm is presumably caused by the formation of aldehyde groups or conjugated double bonds [11]. At very long exposure times (> 60 min), the transmittance increases due to etching of the material. PMMA exhibits yellowing on exposure to ultraviolet light under vacuum due to the formation of conjugated double bonds in the polymer.

Contrary to the behavior of PMMA decomposition, the UV transmission (at $\lambda =$ 250 nm) of thin PI film increases immediately after irradiation (20 mW/cm^2) at $\lambda = 222$ nm (KrCl*) due to ablation of PI (Fig. 17).

Figure 18 shows the UV transmission of thin PMMA films (thickness: 0.6 μm) and thin PI films (thickness: 0.2 μm) during UV irradiation (20 mW/cm^2, $\lambda = 222$ nm, KrCl*) at the different background pressures of 1 mbar and 0.01 mbar. At the background pressure of 1 mbar, the UV irradiation rapidly leads to a decrease of UV transmission of PMMA due to the formation of aldehyde groups or conjugated double bonds [11]. After 5 min exposure time the UV transmission of PMMA increases due to photo-oxidative etching of the thin PMMA film. At a low background pressure of 0.01 mbar, the UV transmission of thin PMMA films continues to decrease even after very long exposure times (> 3 h). In this case only the chemical structure

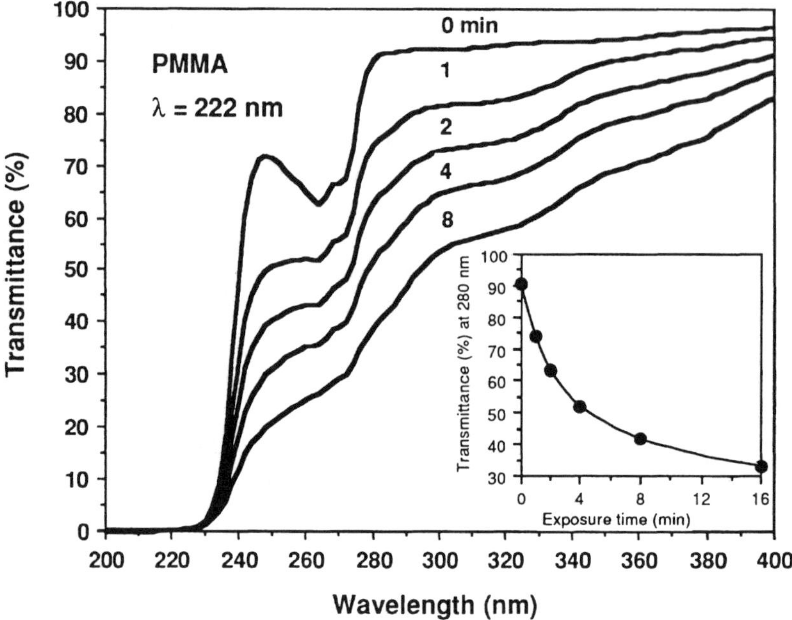

Figure 16. Transmittance of a thin PMMA film (0.6 μm) on quartz as a function of wavelength after different exposure times (min) with incoherent excimer UV radiation ($\lambda = 222$ nm, 20 mW/cm^2) at atmospheric pressure and the transmittance at $\lambda_{tr} = 280$ nm as a function of exposure time.

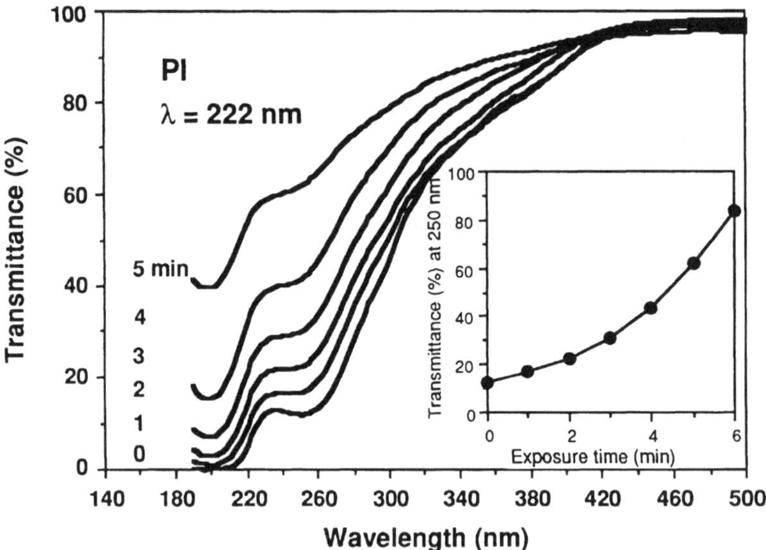

Figure 17. Transmittance of a PI film on quartz as a function of wavelength after different exposure times (min) with incoherent excimer UV radiation ($\lambda = 222$ nm, 20 mW/cm^2) and the transmittance at $\lambda_{tr} = 250$ nm as a function of exposure time.

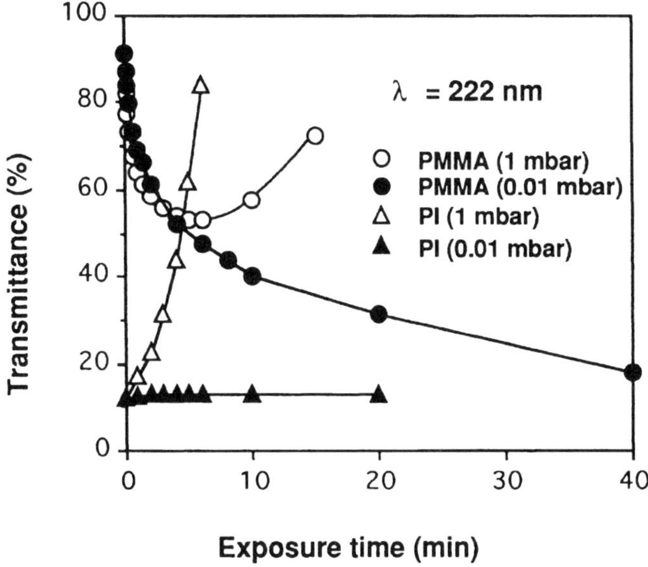

Figure 18. Transmittance of thin PI films (0.2 μm, $\lambda_{tr} = 250$ nm) and thin PMMA films (0.6 μm, $\lambda_{tr} = 280$ nm) on quartz after different exposure times (min) with incoherent excimer UV radiation at $\lambda = 222$ nm (20 mW/cm^2).

of PMMA changes and no etching of PMMA occurs. The results are in agreement with the measurement of etch rates (Fig. 9). Contrary to the behaviour of PMMA, the UV transmission of thin PI films increases immediately after irradiation due to etching of thin PI films at 1 mbar. At a low background pressure of 0.01 mbar, no change in UV transmission was observed during the irradiation of PI. Therefore, no etching or decomposition of PI during the UV irradiation occurs at the latter pressure.

These important experimental results can be explained as follows: after short exposure times the chemical constitution of PMMA has changed, while no etching is observable. These changes in the chemical structure of PMMA have also been observed by Küper and Stuke [19] in the infrared spectrum after subjecting the samples to excimer laser pulses at $\lambda = 248$ nm. They reported that after laser exposure unsaturated aliphatic compounds exist, which can be clearly identified by their absorption at 3078 cm^{-1} and 1643 cm^{-1}. The repeating structural unit of the PMMA is as follows:

$$\left[\, \underset{\substack{|\\ O\!=\!C-OCH_3}}{\overset{\substack{CH_3\\|}}{C}} - CH_2 \,\right]$$

PMMA

These units form unsaturated compounds from saturated aliphatic esters by irradiation-induced removal of a H atom and/or ester group. Unsaturated compounds are known to absorb at longer UV wavelengths than their saturated derivatives. Therefore, a broad UV absorption between $\lambda = 240$ nm and $\lambda = 400$ nm is observed after a short exposure time in our excimer lamp experiments. In comparison with the chemical structure of PMMA, PI is an aromatic derivative with repeating structural unit as follows:

Polyimide

The absorption of UV photons by PI leads to breaking of the weak bonds in the imide groups. It will most probably result in initial loss of carbon monoxide, followed by the loss of other small products such as CO_2, H_2O, HCN and benzene [20]. Therefore, UV exposure leads to the increase of transmission of thin PI film immediately.

As mentioned above, a wavelength dependence of the degradation rate of PMMA and PI in the etch rate experiment was observed in laser experiments. The similar results with respect to UV transmittance change of polymer given in Figs 19 and 20 were obtained with radiation from excimer lamps at $\lambda = 172$ nm (Xe_2^*), $\lambda = 222$ nm ($KrCl^*$) and $\lambda = 308$ nm ($XeCl^*$) at a constant background pressure ($p = 20$ mbar in the case of PMMA and $p = 10$ mbar in the case of PI) in the exposure reactor.

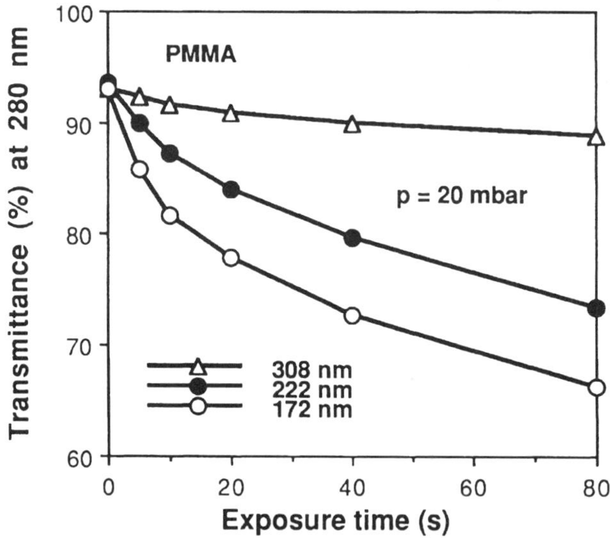

Figure 19. Transmittance at $\lambda = 280$ nm of thin PMMA films (0.6 μm) on quartz after different exposure times (s) with incoherent excimer UV radiation at different wavelengths: $\lambda = 172$ nm, $\lambda = 222$ nm and $\lambda = 308$ nm (25 mW/cm^2).

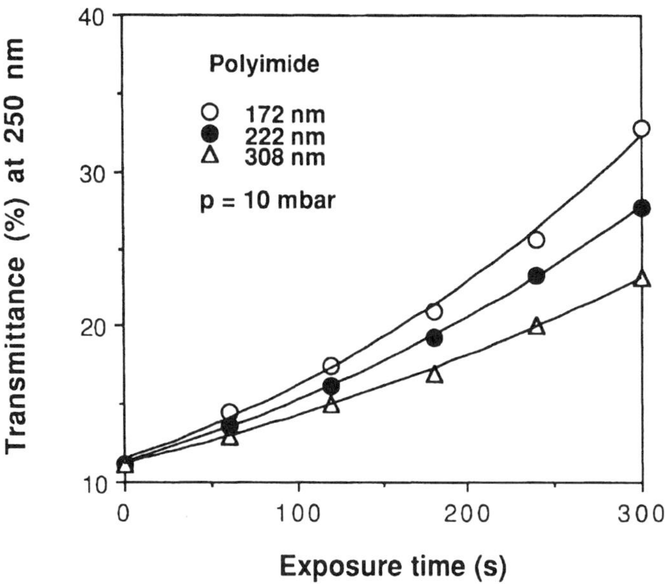

Figure 20. Transmittance at $\lambda = 250$ nm of thin PI films (0.2 μm) on quartz after different exposure times (s) with incoherent excimer UV radiation different wavelengths: $\lambda = 172$ nm, $\lambda = 222$ nm and $\lambda = 308$ nm (20 mW/cm^2).

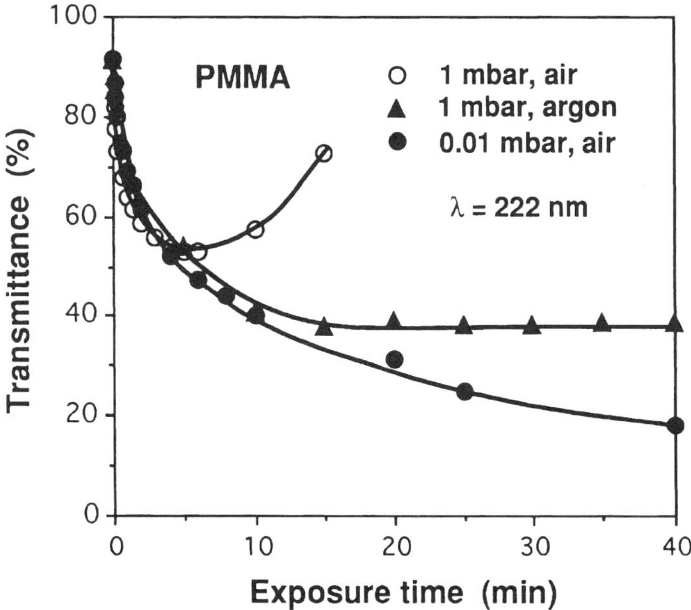

Figure 21. Transmittance of thin PMMA films (0.6 μm, $\lambda_{tr} = 280$ nm) on quartz after different exposure times (min) with incoherent excimer UV radiation ($\lambda = 222$ nm, 20 mW/cm^2) at different background pressures in air atmosphere and argon.

The different decomposition rates correlate with the absorption coefficients at the wavelengths of the UV radiation. The highest decomposition rate was achieved at $\lambda = 172$ nm due to the higher absorption of PMMA and PI in the vacuum ultraviolet range $\lambda < 200$ nm [12]. This supports again the wavelength dependent nature of the etching of polymers by UV irradiation (Figs 5 and 6).

In order to verify that oxygen plays an important role in the etching of PMMA with an incoherent excimer lamp, experiments were performed in argon (instead of air) at the same pressure $p = 1$ mbar (see Fig. 21). In this case during the first 5 min of exposure time the UV transmission of PMMA decreases in a similar way as under atmospheric pressure air. At longer exposure time (> 5 min) no increase of UV transmission was observed. This indicates that only the chemical structure of PMMA changes and no etching of PMMA occurs in the absence of oxygen, similar to the case of the very low pressure of $p = 0.01$ mbar in air. Therefore, we conclude that the etching process with incoherent excimer radiation is mainly a UV-induced photo-oxidation relying on the presence of some oxygen.

3.3.2. Gaseous products determined by mass spectrometry. A knowledge of the composition of the products ablated from the surface of a polymer is of key importance in determining the chemical mechanisms of the process. The gaseous species produced during UV ($\lambda = 213$ nm) and VUV ($\lambda = 121.6$ nm) irradiation of PMMA by using mass spectrometry were investigated by Ueno *et al.* [44]. The results indicated direct main-chain scission induced by deep VUV irradiation of PMMA, as opposed to side-chain scission caused by UV irradiation of PMMA.

Figure 22 shows a typical mass spectrum of the gaseous etching products of PMMA after 15 min irradiation with incoherent excimer UV lamp at 0.1 mbar in air ($\lambda = 222$ nm). UV irradiation causes an increase of various mass peaks, such as, $m/e = 2$ (H_2), 12 (C^+), 15 (CH_3 from CH_4), 18 (H_2O), 26, 27 (CHCH, CH_2CH from CH_3CH_3), 28 (CO), 29, 30 (CHO and CH_2O) and 44 (CO_2). As a result H_2, CH_4, H_2O, CH_3CH_3, CO, CH_2O and CO_2 were positively identified as decomposition

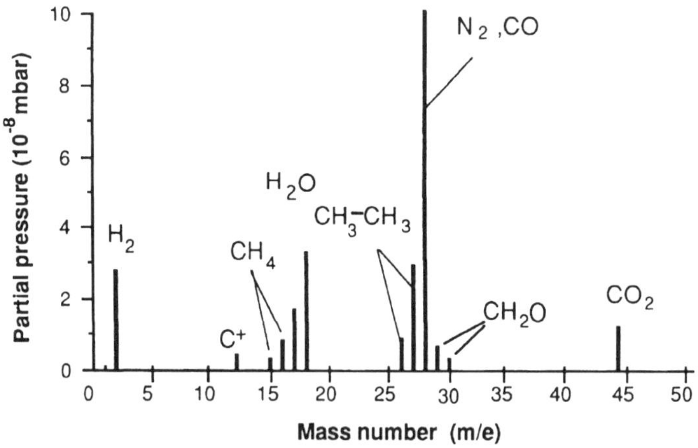

Figure 22. Mass spectrum of PMMA after 15 min exposure time at $p = 0.1$ mbar in air.

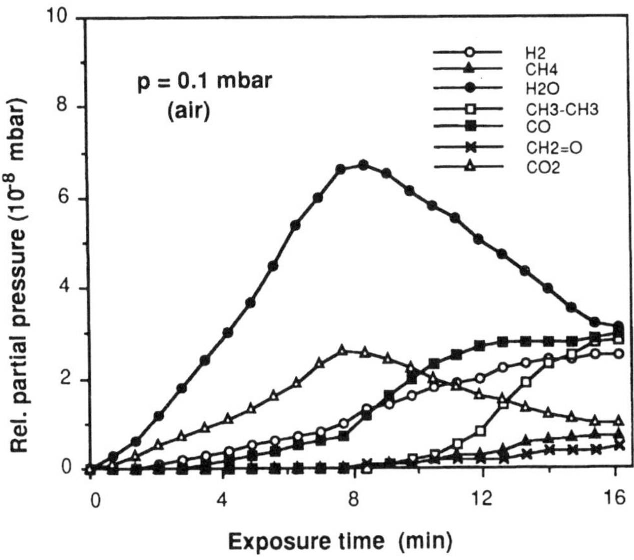

Figure 23. Relative partial pressure of gaseoüs products from PMMA as a function of exposure time ($p = 0.1$ mbar air).

products. The gaseous etching products of PMMA as a function of exposure time during UV irradiation are shown in Fig. 23. At exposure times < 8 min the gaseous products CO_2 and H_2O increase with exposure time. However, in the case of this low air pressure (0.1 mbar) the gaseous products CO_2 and H_2O decrease at longer exposure times due to the lack of oxygen partial pressure in the exposure chamber. Other decomposition products (H_2, CH_4, CH_3CH_3, CO and CH_2O) increase slowly with exposure time for the first 8 min. At long exposure times the concentrations of the decomposition products tend to approach equilibrium conditions.

Figure 24 shows the gaseous etching products of PMMA as a function of exposure time during UV irradiation at 1 mbar in air. In this case only H_2O, CO_2 and H_2 were observed. These etching products increase during the whole exposure time. Even at large exposure times the decomposition products show no saturation. After 6 min exposure time the slope of etching products is higher than that during the first 6 min of exposure time because of the exclusive chemical decomposition of PMMA at short exposure times and the additional photo-oxidative etching of PMMA starting after 6 min of exposure time. Further etching products such as CH_4, CH_3CH_3, CO and CH_2O, which react rapidly with enough oxygen at 1 mbar of air to form CO_2 and H_2O, cannot be observed in this case. This is also in good agreement with UV transmission and etch rate measurements showing that higher etch rates are achieved at 1 mbar in air.

At the low pressure of $p = 0.0001$ mbar the gaseous etch products of PMMA as a function of exposure time during UV irradiation are shown in Fig. 25. The concentration of etch products is small due to chemical decomposition of PMMA. At long exposure times the decomposition products will tend to saturate. As can be seen from Fig. 25, CO is the main etch product.

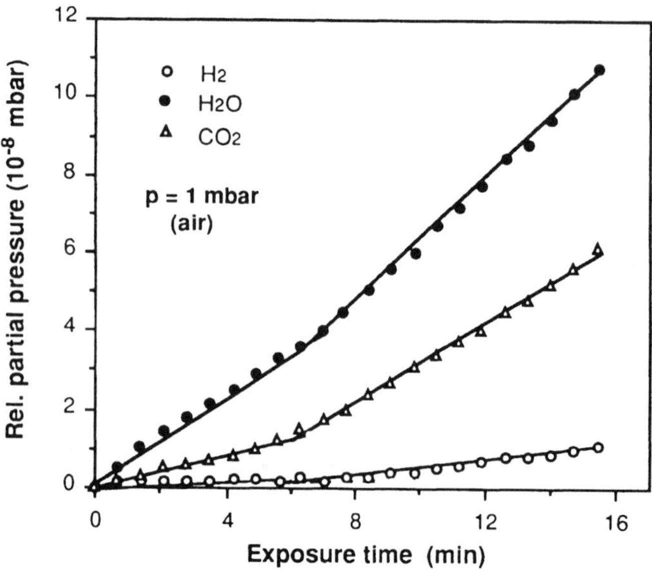

Figure 24. Relative partial pressure of gaseoüs products from PMMA as a function of exposure time ($p = 1$ mbar air).

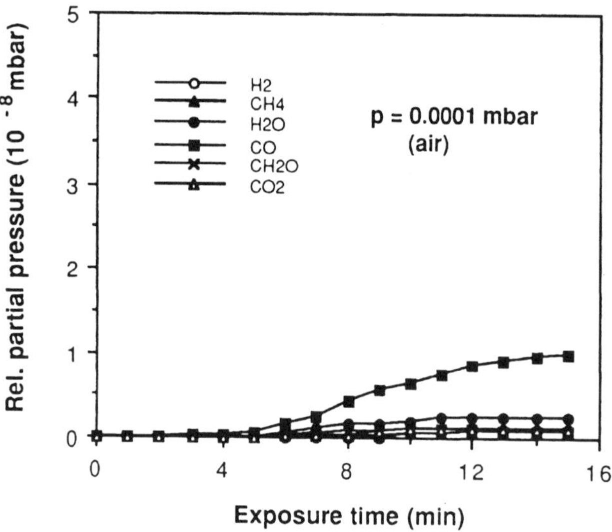

Figure 25. Relative partial pressure of gaseoüs products from PMMA as a function of exposure time ($p = 0.0001$ mbar air).

Figure 26 shows the consumption of oxygen at different pressures. Oxygen decreases rapidly with increasing exposure time at different pressures. Oxygen consumption is nearly zero after 15 min of exposure time. CO_2 first increases up to maximum and then decreases with the further increase of exposure time. If more oxygen is present this will lead to a longer exposure time before the CO_2 maximum is reached.

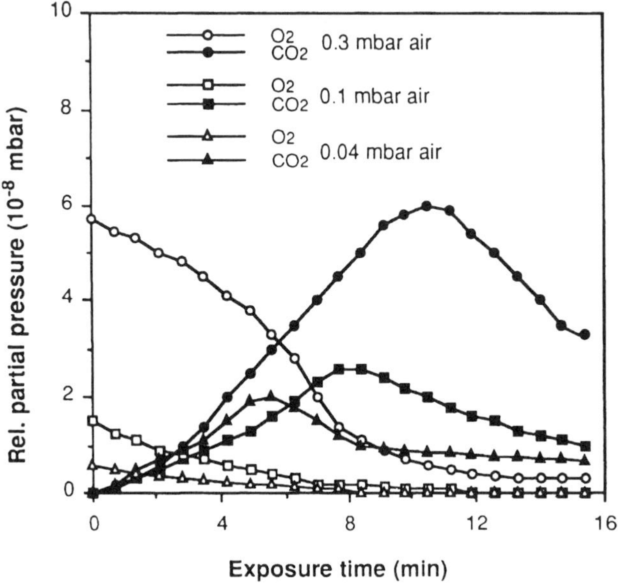

Figure 26. Relative partial pressure of O_2 and CO_2 as a function of exposure time at different air pressures.

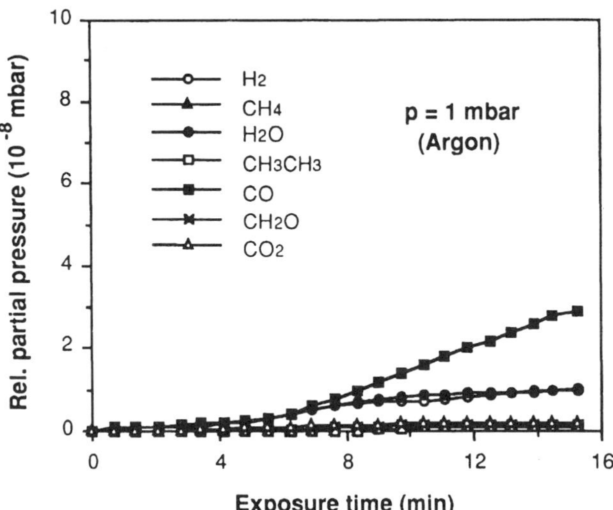

Figure 27. Relative partial pressure of gaseoüs products from PMMA as a function of exposure time ($p = 1$ mbar argon).

At a pressure of $p = 1$ mbar in argon, the gaseous products of PMMA as a function of exposure time during UV irradiation are shown in Fig. 27. The concentration of the etch products H_2O, CO_2 and H_2 are much smaller than those found at 1 mbar in air due to the lack of oxygen, in this case resulting in only chemical decomposition.

Other decomposition products such as CH_4, CH_3CH_3, CO and CH_2O which cannot react to form CO_2 and H_2O because of the lack of oxygen, can be observed. All decomposition products will tend to reach equilibrium conditions at long exposure times. Comparison of Figs 27 and 24 indicate that oxygen plays an important role in the etching of PMMA. This result is in good agreement with UV measurements (see Fig. 21) showing that there is only chemical decomposition of PMMA under the lack of oxygen and that no etching of PMMA occurs. Therefore, we conclude that the presence of oxygen plays a major role in the etching of PMMA. In the presence of oxygen ($p = 1$ mbar air) etching of PMMA occurs efficiently. Conversely, with the lack of oxygen ($p = 1$ mbar argon or $p = 0.0001$ mbar air) only chemical decomposition of PMMA is observed.

3.3.3. Changes of C—C, C—O, and O—C=O bonds investigated by XPS. With XPS analysis, carbon–carbon and carbon-oxygen bonds were detected at the surface of UV irradiated PMMA samples. This provides information on the chemical composition of the exposed surface to a depth of about 10 nm. Figure 28 shows the C 1s spectra of the initial PMMA surface and of the UV irradiated surface after different exposure times using the excimer lamp. The C 1s peak consists of three components corresponding to the three resolvable carbon species in the PMMA: (i) carbon bonded only to H or other carbon atoms (C—C bond, 284.6 eV), (ii) carbon singly bonded to one oxygen atom (C—O bond, 286.2 eV), and (iii) the ester carbonyl (O—C=O bond, 288.5 eV). The surface composition and relative intensities of C 1s components after UV exposure

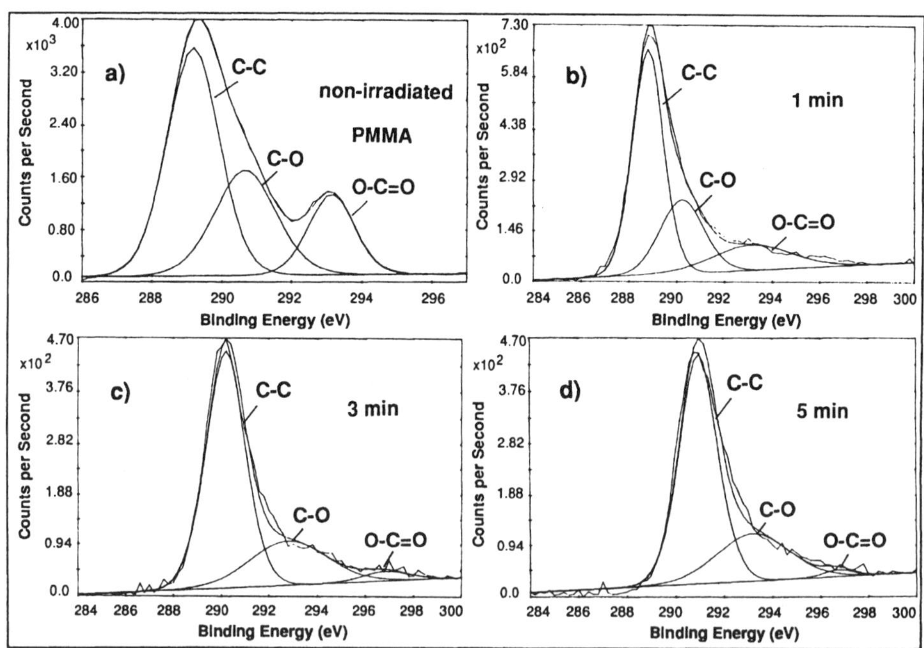

Figure 28. XPS spectrum of non-irradiated PMMA (a) and change in XPS spectra of PMMA during UV irradiation (b–d).

are different from those of the virgin sample, especially the O—C=O bond. After 5 min irradiation the intensity of the O—C=O bond of the PMMA surface decreases to a value of 0.01 starting from 0.17 for the initial PMMA surface.

Figure 29 shows the components of C 1s as a function of the exposure time at 1 mbar air. The intensity of the O—C=O bond decreases with increasing exposure time up to 5 min, and after a distinct minimum, this component increases with the further increase of exposure time. The initial decrease is due to the decomposition of PMMA. After 5 min, the removal of PMMA begins, layer by layer. Therefore, the intensity of the O—C=O bond increases again with the further increase of exposure time. The intensity of the O—C bond changes slowly during the first 10 min. After 15 min the intensity increases rapidly due to photo-oxidative etching of PMMA. This is in good agreement with the UV measurements (Fig. 18). The initial decrease of transmission and the intensity of the O—C=O bond of PMMA is due exclusively to the chemical decomposition of PMMA. On the contrary, after 5 min the increase of transmission and intensity of the O—C=O bond at the PMMA surface is due to photo-oxidative etching of PMMA.

The components of C 1s as a function of exposure time at the low pressure $p = 0.01$ mbar air are shown in Fig. 30. In this case, because of lack of oxygen the components of O—C=O and C—O bonds always decrease even after long exposure times corresponding to chemical decomposition of PMMA. The intensity of the O—C=O bond changes at different pressures in air and argon is shown in Fig. 31. After 5 min at 1 mbar air the intensity of the O—C=O bond increases due to the beginning of photo-oxidative etching of PMMA. At 1 mbar argon and 0.01 mbar air, even after large exposure time no increase of the intensity of the O—C=O bond was observed because of lack of oxygen, which is again in good agreement with the UV measurement (Fig. 21). No significant etching of PMMA is observed in the absence of oxygen.

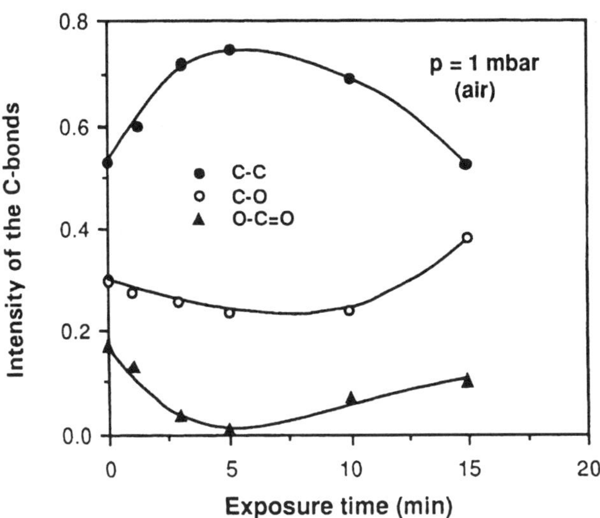

Figure 29. Components of the different carbon bonds of PMMA versus exposure time ($p = 1$ mbar air).

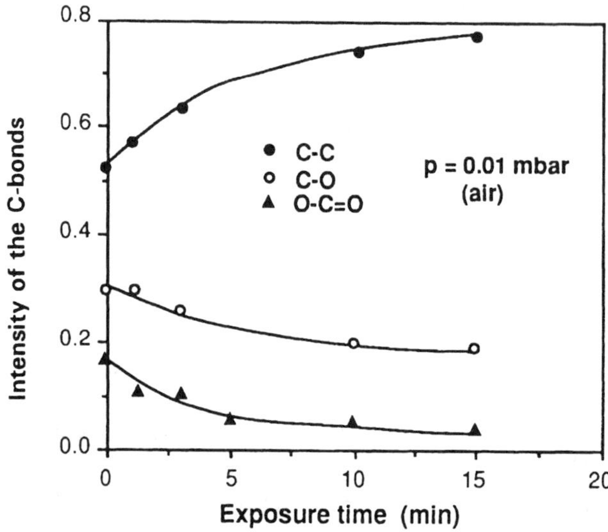

Figure 30. Components of the different carbon bonds of PMMA versus exposure time ($p = 0.01$ mbar air).

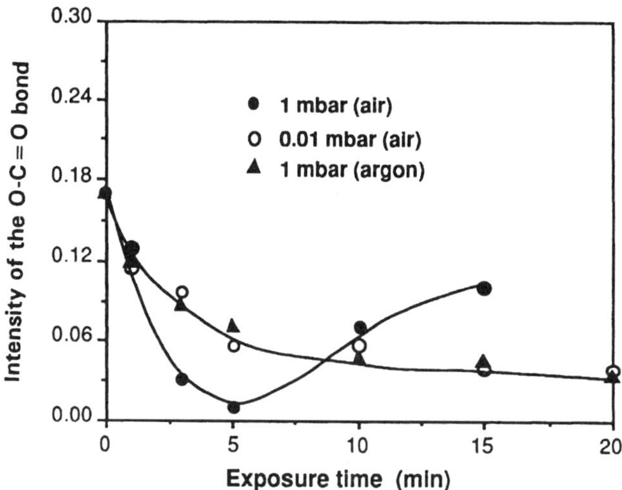

Figure 31. Intensity of the O—C=O bond of PMMA versus exposure time ($p = 0.01$ mbar air, $p = 1$ mbar air, and $p = 1$ mbar argon).

3.3.4. Mechanisms of etching of polymethylmethacrylate (PMMA). Figure 32 shows the three photo-degradative reaction channels for the ester side groups of PMMA under the influence of UV irradiation. Referring to the spectroscopic investigation of Gupta *et al.* [45] and of Küper and Stuke [19, 46] (UV, IR, ESR, GC, MS, and NMR spectroscopy) the formation of methyl formate $HCOOCH_3$ (path: k_3) is the major reaction path in the case of laser irradiation ($\lambda = 266$ nm and $\lambda = 248$ nm). On the contrary, in our mass spectroscopic investigations, no $HCOOCH_3$ was observed as decomposition product using excimer UV lamps. Therefore, the photo-degradation of

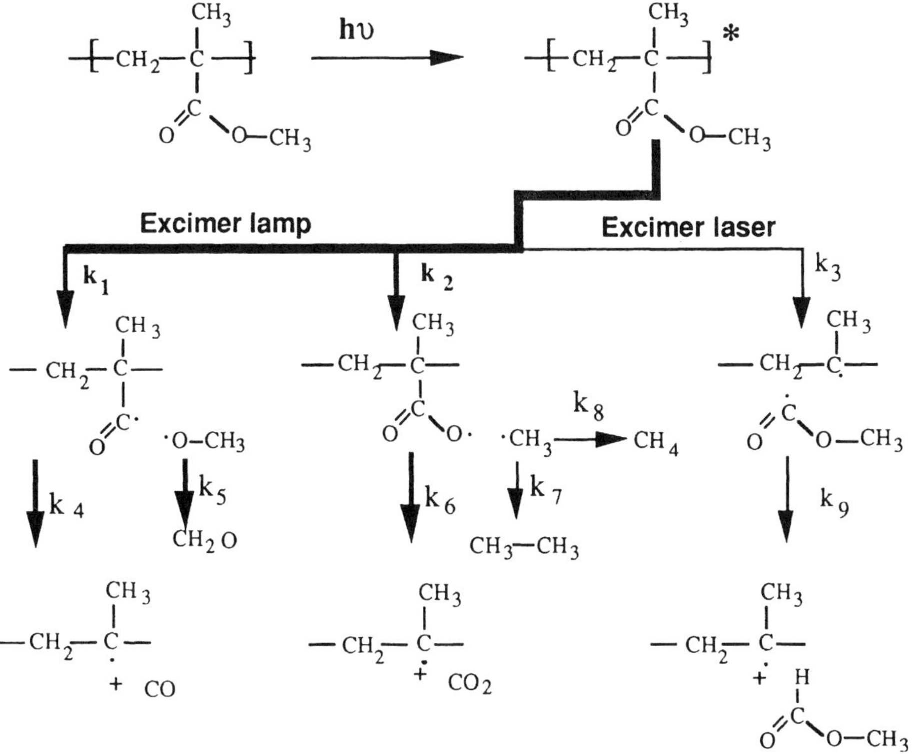

Figure 32. Photodegradation scheme of PMMA with excimer UV irradiation.

the ester side groups of PMMA with excimer UV lamp radiation occurs via the paths k_1 and k_2 (Fig. 32). At low air pressure $p < 0.01$ mbar and $p = 1$ mbar argon instead of air, CO was identified as the major product. Therefore, in the absence of oxygen, path k_1 dominates in the photo-degradation of PMMA with excimer UV lamps.

Oxygen strongly influences the photochemical process of PMMA, which may be postulated as in Fig. 33. In the presence of oxygen, radicals on the polymer chain formed via paths k_1 and k_2 can react further with O_2 to yield the gaseous products H_2, H_2O, CO, CO_2, CH_4, etc. (see Fig. 33). On the other hand, if no oxygen is present, the radicals of the polymer chain stabilize to conjugated double bonds on the polymeric surface (see Fig. 33). These conjugated double bonds on the polymer can be verified from the following experimental results: PMMA exhibits yellowing after UV irradiation in the absence of oxygen; photolysis of PMMA results in an increase in the ultraviolet absorption between $\lambda = 260$ nm and $\lambda = 400$ nm; after UV irradiation, unsaturated aliphatic compounds exist, which can be clearly identified by their absorption at 3078 cm^{-1} and 1643 cm^{-1}.

From the above results, we conclude that the photo-degradation of PMMA with excimer UV lamps follows the paths k_1 and k_2 (Fig. 32). If no oxygen is present, the path k_1 dominates and then the radicals of the polymer chain stabilize to conjugated double bonds on the polymeric surface, thus preventing further chemical changes

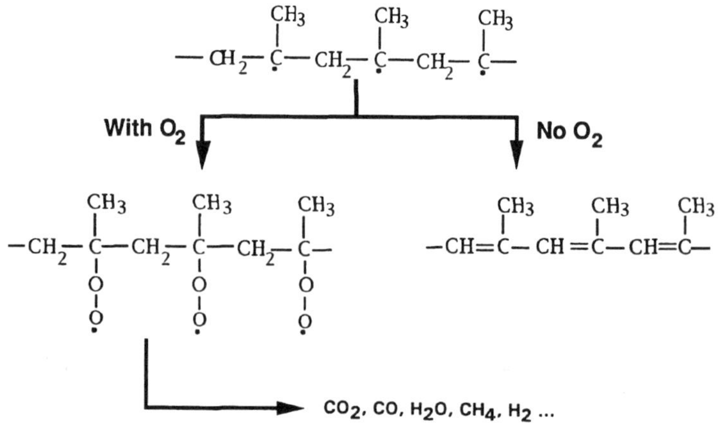

Figure 33. Influence of oxygen in photo-oxidation of radicals formed in the PMMA chain with excimer UV irradiation.

(see Fig. 33). Therefore, oxygen plays an important role in the photo-oxidative etching of PMMA with excimer UV lamps, resulting in a photo-oxidative decomposition process.

4. CONCLUSIONS

We have demonstrated that surface modification and area-selective dry etching of polymers is not only possible with excimer lasers but also with novel incoherent excimer UV sources. The reaction mechanisms of the UV induced material removal were studied in detail for PMMA. Provided that irradiation is performed at a reduced pressure of about 1 mbar in air, surprisingly high etch rates by using excimer lamps can be achieved, as compared to etch rates resulting from the use of conventional UV lamps. Efficient photo-decomposition of the polymeric material with excimer lamps depends mainly on photo-oxidative processes which, in the case of optimal parameters, can lead to volatile end products. Contrary to the process in PMMA using focused excimer lasers which produced particles on the interior and exterior surfaces of the exposed regions, the process using excimer lamps together with contact metal masks always resulted in sharp etch contours and relatively smooth etched surfaces without cracks or bubbles. The complete absence of redeposited particles in the neighbourhood of the treated area of all investigated polymers (PMMA, PI, PET, PTFE) may be a remarkable advantage of excimer lamp treatment, especially for semiconductor manufacturing. Good adhesion of coatings on a variety of polymers promises to be a tremendous break-through in the excimer lamp technology. Since it is quite conceivable that large areas of some cm^2 or even m^2 can also be exposed to UV radiation from excimer lamps, surface modification and microstructuring of large polymeric surfaces is expected to become an important industrial process.

Acknowledgements

The authors would like to thank Dr J. Demny, U. Feller, and B. Heiliger, ABB Heidelberg, for SEM and XPS results.

REFERENCES

1. R. Srinivasan and V. Mayne-Banton, *Appl. Phys. Lett.* **41**, 576–578 (1982).
2. R. Srinivasan and W. J. Leigh, *J. Am. Chem. Soc.* **104**, 6784–6785 (1982).
3. A. T. Barfknecht, J. P. Partridge, C. J. Chen and C.-Y. Li (Eds), *Advanced Electronic Packaging Materials, Mater. Res. Symp. Soc. Proc.* **167** (1990).
4. I. Higashikawa, M. Nonaka, T. Sato, M. Nakase, S. Ito, K. Horioka and Y. Horiike, *Mater. Res. Soc. Symp. Proc.* **101**, 3–12 (1990).
5. R. Srinivasan and B. Braren, in: *Lasers in Polymer Science and Technology: Applications III*, J.-P. Fouassier and J. F. Rabek (Eds). CRC Press, Boca Raton, Florida (1989).
6. (a) B. Eliasson and U. Kogelschatz, *Appl. Phys.* **B46**, 299–303 (1988).
   (b) B. Gellert and U. Kogelschatz, *Appl. Phys.* **B52**, 14–21 (1991).
   (c) U. Kogelschatz, *Appl. Surf. Sci.* **54**, 410–423 (1992).
7. U. Kogelschatz, *Pure Appl. Chem.* **62**, 1667–1674 (1990).
8. H. Esrom and U. Kogelschatz, *Appl. Surf. Sci.* **54**, 440–444 (1992).
9. H. Esrom, J. Demny and U. Kogelschatz, *Chemtronics* **4**, 202–208 (1989).
10. H. Esrom and U. Kogelschatz, *Appl. Surf. Sci.* **46**, 158–162 (1990).
11. A. R. Shultz, *J. Phys. Chem.* **65**, 967–972 (1961).
12. H. R. Philipp, H. S. Cole, Y. S. Liu and T. A. Sitnik, *Appl. Phys. Lett.* **48**, 192–194 (1986).
13. R. Srinivasan and S. Lazare, *Polymer* **26**, 1297–1300 (1985).
14. R. Srinivasan, B. Braren, R. W. Dreyfus, L. Hadel and D. E. Seeger, *J. Opt. Soc. Am.* **B3**, 785–791 (1986).
15. B. J. Garrison and R. Srinivasan, *Appl. Phys. Lett.* **44**, 849 (1984).
16. R. Srinivasan, E. Sutcliffe and B. Braren, *Appl. Phys. Lett.* **51**, 1285–1287 (1987).
17. G. Koren and J. T. C. Yeh, *J. Appl. Phys.* **56**, 2120–2126 (1984).
18. R. Srinivasan, B. Braren and R. W. Dreyfus, *J. Appl. Phys.* **61**, 372–376 (1987).
19. S. Küper and M. Stuke, *Appl. Phys.* **A49**, 211 (1989).
20. J. H. Brannon, J. R. Lankard, A. I. Baise, F. Burns and J. Kaufman, *J. Appl. Phys.* **58**, 2036–2043 (1985).
21. F. Kokai, H. Saito and T. Fujioka, *J. Appl. Phys.* **66**, 3252–3255 (1989).
22. H. Esrom, J.-Y. Zhang and U. Kogelschatz, *Mater. Res. Soc. Symp. Proc.* **236**, 39–45 (1992).
23. J.-Y. Zhang, H. Esrom, U. Kogelschatz and G. Emig, *Appl. Surf. Sci.* **69**, 299–304 (1993).
24. (a) K. L. Mittal and J. R. Susko (Eds), *Metallized Plastics 1: Fundamental and Applied Aspects*. Plenum Press, New York (1988).
   (b) K. L. Mittal (Ed.), *Metallized Plastics 2: Fundamental and Applied Aspects*. Plenum Press, New York (1991).
   (c) K. L. Mittal (Ed.), *Metallized Plastics 3: Fundamental and Applied Aspects*. Plenum Press, New York (1992).
25. S. Lazare and R. Srinivasan, *J. Phys. Chem.* **90**, 2124–2131 (1986).
26. R. Srinivasan, in: *Interfaces under Laser Irradiation*, L. D. Laude, D. Bäuerle and M. Wautelet (Eds), pp. 359–370. Martinus Nijhoff, Dordrecht (1987).
27. E. Sutcliffe and R. Srinivasan, *J. Appl. Phys.* **60**, 3315–3322 (1986).
28. S. Mihailov and W. Duley, *Laser Beam Surface Treating and Coating, SPIE Vol. 957*, 111–118 (1988).
29. C. E. Otis, *Appl. Phys.* **B49**, 455–461 (1989).
30. S. Küper and M. Stuke, *Appl. Phys.* **B44**, 199–204 (1987).
31. R. Srinivasan, B. Braren, D. E. Seeger and R. W. Dreyfus, *Macromolecules* **19**, 916–921 (1986).

32. R. Srinivasan, *J. Vac. Sci. Technol.* **B1**, 923 (1983).
33. R. Srinivasan, *J. Appl. Phys.* **70**, 7588–7593 (1991).
34. M. Bolle, K. Luther, J. Troe, J. Ihlemann and H. Gerhardt, *Appl. Surf. Sci.* **46**, 279–283 (1990).
35. J. H. Brannon, D. Scholl and E. Kay, *Appl. Phys.* **A52**, 160–166 (1991).
36. S. Küper and M. Stuke, *Appl. Phys. Lett.* **54**, 4 (1988).
37. H. Hiraoka and S. Lazare, *Appl. Surf. Sci.* **46**, 342–347 (1990).
38. M. C. Burrell, Y. S. Liu and H. S. Cole, *J. Vac. Sci. Technol.* **A4**, 2459–2462 (1986).
39. S. Lazare and V. Granier, *J. Appl. Phys.* **63**, 2110–2115 (1988).
40. P. E. Dyer and J. Sidhu, *J. Opt. Soc. Am.* **B3**, 792–795 (1986).
41. M. Bolle and S. Lazare, *Appl. Surf. Sci.* **54**, 471–476 (1992).
42. B. Eliasson and U. Kogelschatz, *Ozone Sci. & Eng.* **13**, 365 (1991).
43. S. Lazare, P. D. Hoh, J. M. Baker and R. Srinivasan, *J. Am. Chem. Soc.* **106**, 4288–4290 (1984).
44. N. Ueno, T. Mitsuhata, K. Sugita and K. Tanaka, *ACS Symposium Series* **412**, 424–436 (1989).
45. A. Gupta, R. Liang, F. D. Tsay and J. Moacanin, *Macromolecules* **13**, 1696–1700 (1980).
46. S. Küper, Ablation mit UV-Laserlicht, Thesis. Georg-August-Universität Göttingen, Germany (1989).
47. J.-Y. Zhang, Thesis. Universität Karlsruhe, Germany (1993).
48. H. Esrom and U. Kogelschatz, *Thin Solid Films* **218**, 231–246 (1992).
49. H. Esrom and G. Wahl, *Chemtronics* **4**, 216–223 (1989).
50. M. Paunovic and I. Ohno (Eds), *Electroless Deposition of Metals and Alloys*. Vol. 88-12 of the Proceedings of the Electrochemical Society, Pennington, NJ, USA (1988).
51. P. Bergonzo, U. Kogelschatz and I. W. Boyd, *Appl. Surf. Sci.* **69**, 393–397 (1993).
52. P. Bergonzo and I. W. Boyd, *Appl. Phys. Lett.* **63**, 1757–1759 (1993).
53. (a) F. Kessler, H.-D. Mohring and G. H. Bauer, in: *9th International Symposium on Plasma Chemistry*, pp. 1383–1388. Pugnochiuso, Italy (1989).
    (b) F. Kessler and G. H. Bauer, *Appl. Surf. Sci.* **54**, 430–434 (1992).
54. C. Manfredotti, F. Fizzotti, M. Boero and G. Piatti, *Appl. Surf. Sci.* **69**, 127 (1993).

Polymer Surface Modification: Relevance to Adhesion, pp. 185–197
K. L. Mittal (Ed.)

Photolytic surface modification of polymers with UV-laser radiation

J. BREUER,* S. METEV and G. SEPOLD
BIAS, Bremen Institute for Applied Beam Technology, Klagenfurter Str. 2, 28359 Bremen, Germany

Revised version received 20 July 1994

Abstract—Several polymers are stable towards aggressive media. This is advantageous in general, but on the other hand problems arise if such polymers are to be adhered or coated. This limits their use in certain technical applications. To improve the surface properties of polymers, mechanical, wet-chemical, or plasma treatments are applied. Most of these methods, however, are not applicable to every polymer or are limited in practical use because of the requirement of special treatment conditions. This is the reason why improvement of the existing technologies and a search for alternative methods are subjects of current research. One relatively new way to alter the surface properties of polymers is through the use of excimer lasers. In the present paper recent results are presented concerning the interaction of UV-laser radiation with polymers and resulting photolytical modification and activation of their surfaces. On the basis of such activation, a significant enhancement of the adhesive bond strength between a polymer and an adhesive (in this case epoxy resin) has been achieved.

Keywords: Surface activation; adhesion; carbonization; carbonyl group; epoxy resin; excimer laser; FTIR-spectroscopy; laser parameter dependence; surface modification; morphology change; oxidation; physico-chemical treatment; polypropylene; pretreatment; ripple structure; XPS-measurement.

1. INTRODUCTION

A widespread joining method in industrial applications of polymers is adhesive bonding. With this technique satisfactory joints can be produced. However, because of the inherent low surface energy of polymers, it is difficult to bond these materials. Only when appropriate pretreatment techniques are applied is the bond strength sufficient.

Nowadays, mechanical, wet chemical, plasma or corona treatments are industrially applied for such pretreatment [1]. However, because of certain disadvantages of these techniques, for example the environmental considerations or poor controllability of the process, alternative methods for pretreatment of polymers are looked for. The properties of excimer lasers with wavelengths in the near and deep UV regions

*To whom correspondence should be addressed.

(351–157 nm) and very high spectral density of the laser radiation make it possible to controllably alter the physico-chemical structure of polymeric materials photolytically. These structural changes could enhance the activity of the surface and it has been shown that better adhesion strengths could be achieved [2]. Under certain conditions the action of UV-laser radiation on polymers may also result in ablation [3] or may cause certain morphological changes [4]. The actual investigations deal with the fundamental laser-initiated processes and finding the appropriate treatment conditions.

The aim of the UV-laser assisted pretreatment of polymers by photolytical modification of the surface is the enhancement of its adhesion properties. It has already been shown that the irradiation of polypropylene (PP) with UV-laser light in an oxidizing environment results in the oxidation of its surface [2, 5]:

$$
-\underset{\substack{|\\CH_3}}{CH}-CH_2-\underset{\substack{|\\CH_3}}{CH}-CH_2-\underset{\substack{|\\CH_3}}{CH}- \xrightarrow{h\nu,\,O_2,\,\ldots}
\begin{cases}
-\underset{\substack{|\\CH_3}}{CH}-CH_2-\underset{\substack{\|\\O}}{C}-CH_2-\underset{\substack{|\\CH_3}}{CH}- + CH_3\cdot \\
-\underset{\substack{|\\CH_3}}{CH}-CH_2-\underset{\substack{|\\CH_3}}{C}=O + CH_2\cdot \\
-\underset{\substack{|\\CH_3}}{CH}-CH_2-\overset{\substack{OH\\|}}{\underset{\substack{|\\CH_3}}{C}}-CH_2-\underset{\substack{|\\CH_3}}{CH}- + -\underset{\substack{|\\CH_3}}{CH}-
\end{cases}
\tag{1}
$$

The chemical structures of UV-irradiated PP (1) show that carbonyl- (−C=O) or hydroxyl- (−OH) groups are formed. It is known that the presence of these strongly polar groups on a polymer surface can positively influence at least the physisorption and hence the adhesion [6, 7, 8]. In the present investigation the main interest was in the creation of carbonyl groups during photolytical UV-laser treatment of PP and its relation to the adhesion properties of the material.

2. EXPERIMENTAL

In the experiments an excimer laser (Lambda Physik LPX120i) with wavelengths 308, 248, and 193 nm was used to irradiate PP foils 40 μm thick (manufactured by Goodfellow). The samples were cleaned in methyl ethyl ketone and placed in a chamber which allowed control of the environmental conditions, Fig. 1.

The laser radiation was guided through a quartz lens and a quartz window onto the sample surface in the chamber. The gas atmosphere inside the chamber was oxygen, helium or air. The laser energy density was varied in the range from 0.05 to about 2 J/cm^2 with a spot size of 5×5 mm^2 on the PP-surface. The morphology of the irradiated samples was studied by microscopic techniques and chemical structure changes were investigated mainly by FTIR-spectroscopy.

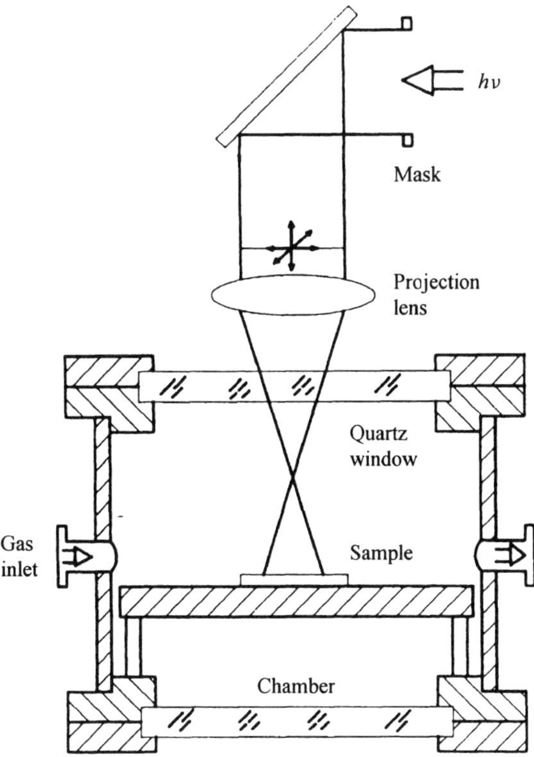

Figure 1. Experimental setup for irradiation investigations with excimer lasers under specific environmental conditions.

3. RESULTS AND DISCUSSION

3.1. Influence of different experimental conditions

Irradiation of PP at high energy density (>0.5 J/cm^2) leads to its ablation. In this case by applying FTIR-spectroscopy no chemical changes could be detected in the wavelength range used. At energy densities <0.5 J/cm^2 some significant chemical changes (formation of C=O groups) could be detected, and a dependence on the laser parameters and on the surrounding atmosphere was found. In this treatment regime no damage to the surface could be observed. However, with increasing irradiation dose (e.g. with increasing number of pulses) specific morphological changes of the surface such as microcraters, cracks or ripples appear depending on the experimental conditions. In the following the dependence of the C=O-concentration of the irradiated polymer on some experimental parameters is shown.

Figure 2 shows the dependence of the IR spectra of PP samples irradiated in O_2 atmosphere on the laser energy density. With increasing energy density and fixed other conditions the carbonyl concentration increases to a maximum value and then drops as the energy density continues to increase (Fig. 2, on the right). High energy density leads to a direct fragmentation of the polymeric chains or — especially at longer wavelengths — to partial melting. The dependence of the carbonyl concentration on

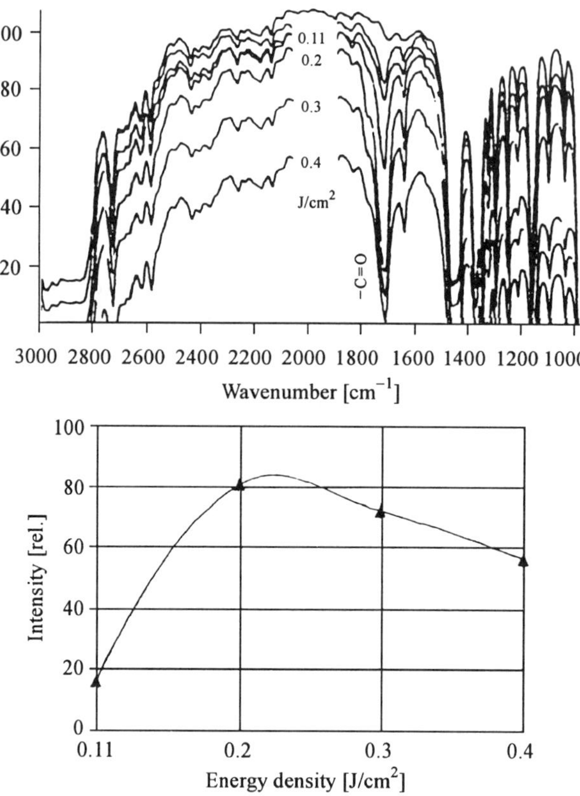

Figure 2. FTIR-spectra of PP-foils irradiated with excimer laser light of different energy densities; $\lambda =$ 248 nm, 10 Hz, 3000 pulses, energy density varies in the range 0.1–0.4 J/cm^2. The lower diagram shows relative intensity of the carbonyl absorption band in relation to the energy density.

the surrounding atmosphere and on the laser wavelength for different pulse numbers is presented in Fig. 3.

In general all spectra show increasing changes with increasing pulse numbers. The results also show a definite dependence on the wavelength. Maximum concentration of C=O-bonds was obtained at 193 nm. Slightly less changes were found at 248 nm and significantly less changes were indicated at 308 nm. Irradiation with an IR-laser (10.6 μm) leads to almost negligible carbonyl formation, which appears only when the energy density was considerably enhanced. Taking into account the strongly different photon energies of the respective laser sources this behaviour of the C=O-formation suggests a dominating photochemical mechanism of the oxidation process.

The atmosphere influences the treatment process also. Carbonyls were not detected at any wavelength when the irradiation was performed in the helium atmosphere. In air the concentration was less than that in oxygen. This shows that heterogeneous reactions with the surrounding gas atmosphere occur. In order to investigate thermal effects of the laser radiation on the oxidation process, irradiation experiments at different pulse repetition rates were performed, with other conditions being the same. The FTIR-analysis of the irradiated zones shows a small increase of the carbonyl con-

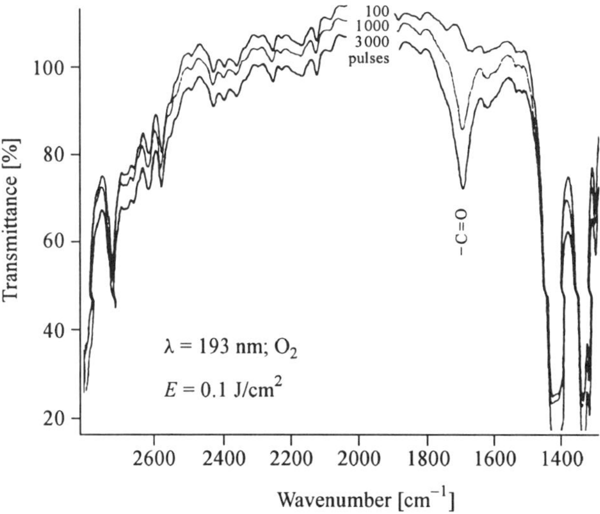

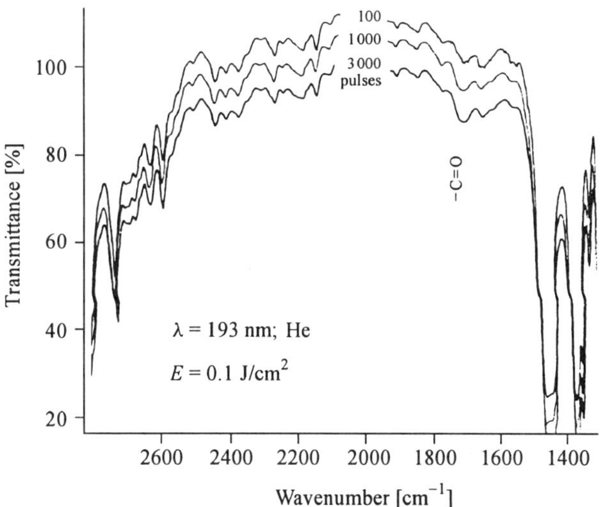

Figure 3. FTIR-spectra of PP-foils irradiated with excimer laser light of different wavelengths and in different atmospheres; pulse freq.: 10 Hz (other conditions are given in the diagrams).

centration at higher repetition rates (Fig. 4). This relatively slight dependence of the carbonyl concentration on the pulse repetition rate points out the photolytical character of the laser process. Further increase of the repetition rate leads to material damage and with this the C=O-concentration decreases. Also bonds in the polymer structure will be broken (depolymerisation) and consequently volatile products are formed. The spectra in Fig. 4 are the result of PP-irradiation with excimer laser light of 248 nm wavelength. Similar behaviour is observed when 193 nm wavelength is used (Fig. 5).

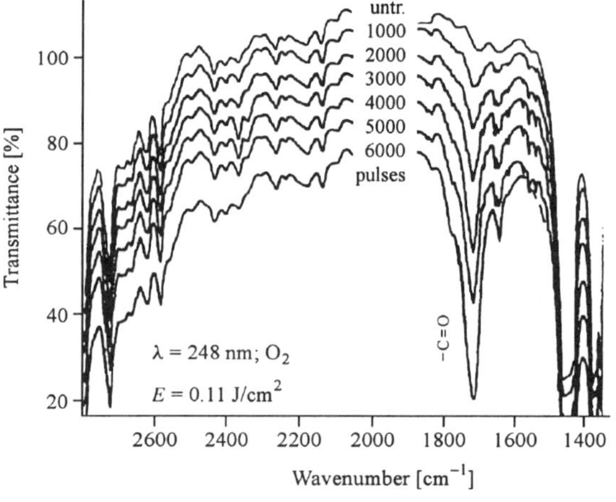

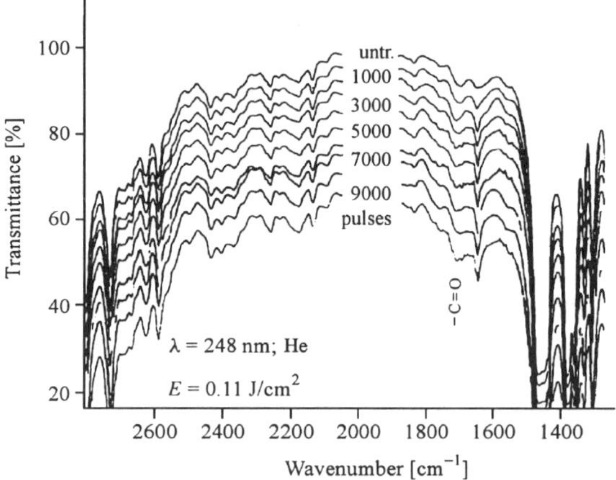

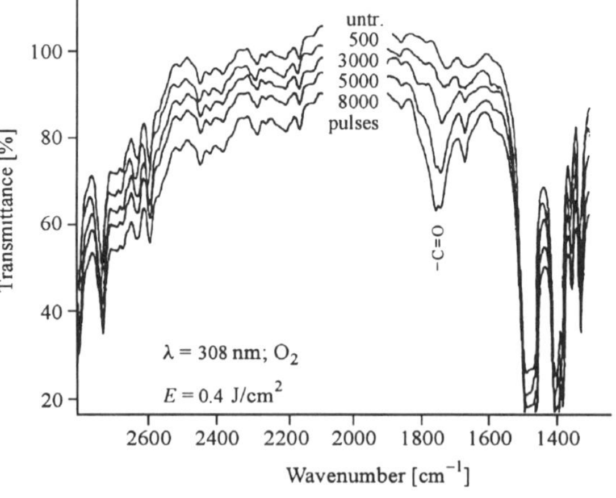

Figure 3. (Continued).

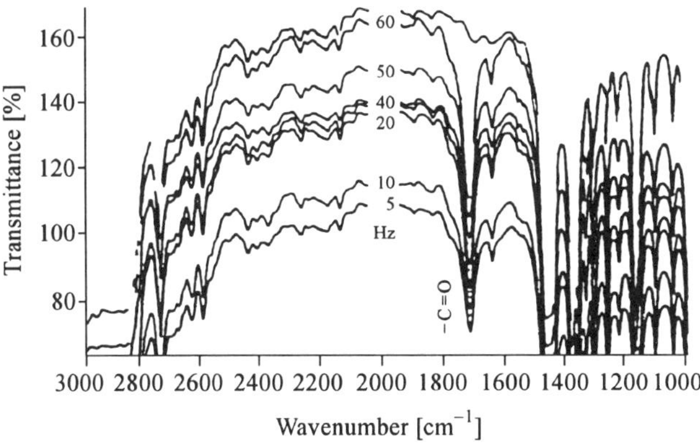

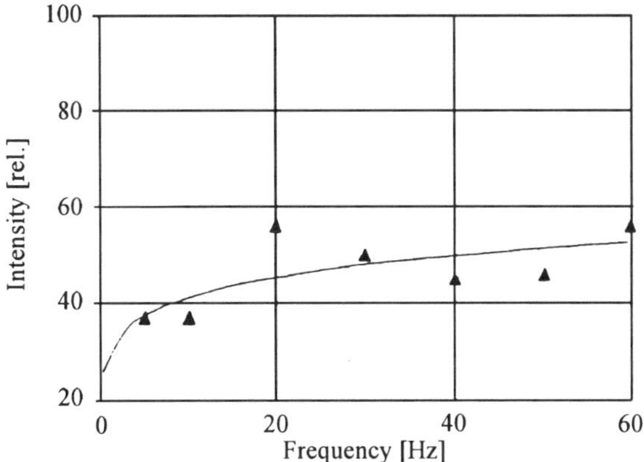

Figure 4. FTIR-spectra of PP-foils irradiated with excimer laser light of different repetition rates; the frequency varies in the range from 1 to 60 Hz; $\lambda = 248$ nm; $E = 0.1$ J/cm^2; $n = 4000$ pulses; atmosphere: oxygen. The lower diagram shows relative intensity of the carbonyl absorption band in relation to the frequency.

There is an increase in the carbonyl concentration when the energy density or the repetition rate is increased until a maximum is reached. It should be pointed out here that there is hardly any further increase when the pulse number is enhanced.

One analysis was also made with XPS (Fig. 6). The spectrum of the treated sample shows an additional O1s-peak as compared to the untreated one. Fit procedures for the C1s-peak indicated that carbonyl bondings were formed. Measurements with different take-off angles showed that these were formed in the uppermost surface layers. This also is an indication of heterogeneous reactions with the environmental gas.

3.2. Influence of the pretreatment on adhesion behaviour

The influence of the laser treatment on the adhesion behaviour was examined by peel-off tests. With this simple method only qualitative estimates were possible, but nevertheless some trends could be shown. The irradiated area of the PP-foil where C=O-bonds were detected was covered with liquid epoxy resin. After epoxy

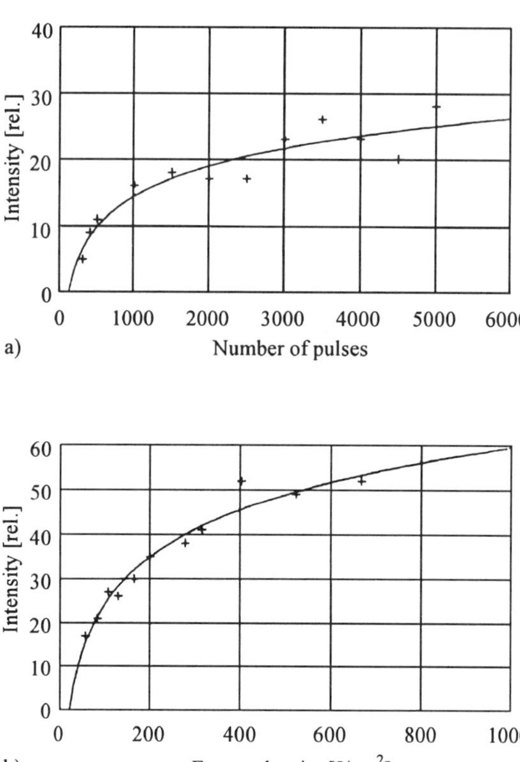

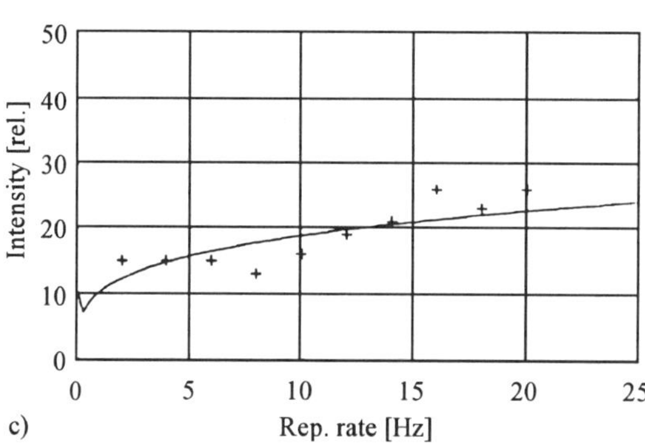

Figure 5. Relative intensities of the carbonyl absorption band in PP-foils (FTIR-measurements) irradiated with 193 nm excimer laser light under different conditions: (a) influence of number of pulses ($E = 0.1\ \mathrm{J/cm^2}$), (b) influence of energy density ($n = 3000$), (c) influence of frequency ($E = 0.1\ \mathrm{J/cm^2}$, $n = 3000$).

hardening, the foil was peeled off the epoxy mass. The peeled epoxy interface was investigated by scanning electron microscopy (Fig. 7).

The SEM-micrographs show that most of the PP-foil remains on the epoxy when 1000 pulses of 248 nm wavelength were applied. After 3000 pulses the foil remains completely on the resin within the laser treated area because the area of destruction

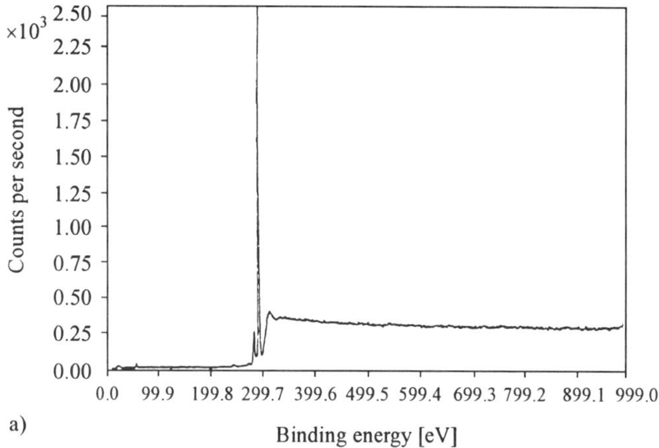

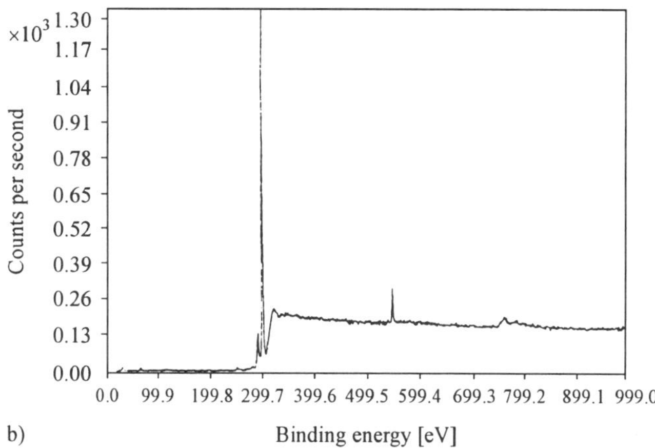

Figure 6. XPS-measurements on UV-excimer laser irradiated PP-foils: $\lambda = 248$ nm; $E = 0.12$ J/cm^2; $f = 10$ Hz; $n = 800$; (a) untreated, (b) excimer laser pretreated.

is where probably the PP-foil was irradiated with scattered laser light. In any case, unirradiated zones of the foil do not adhere to the epoxy surface. This result shows that irradiation of PP with 248 nm positively influences the adhesion and that the application of an increasing number of pulses does not necessarily further enhance the adhesion properties of the foils.

Foils irradiated with 308 nm exhibited a somewhat different behaviour. First, treatment of the PP-foil with 1000 to 2000 pulses resulted also in some PP remaining on the epoxy surface (but less than observed in the case of 248 nm). With increasing number of pulses the adhesion dropped down to the complete removal of the PP-foil from the epoxy, although the carbonyl concentration in the irradiated area was enhanced as indicated in the FTIR-spectra. This effect could probably be due to some competing processes as, for example, damage to the polymer structure and weakening of its surface layer on one hand and melting effects with rearrangement of radicals to

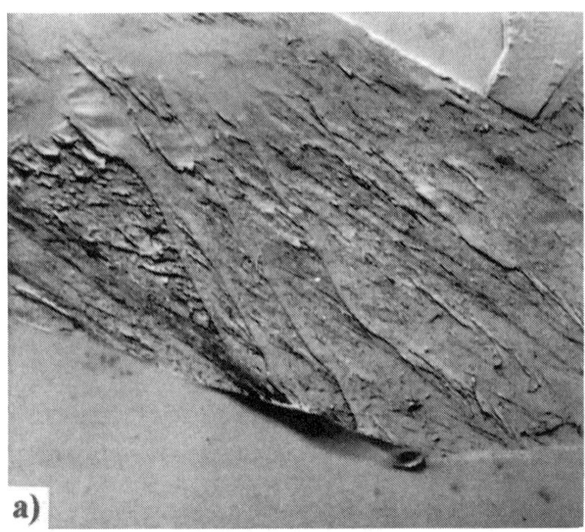

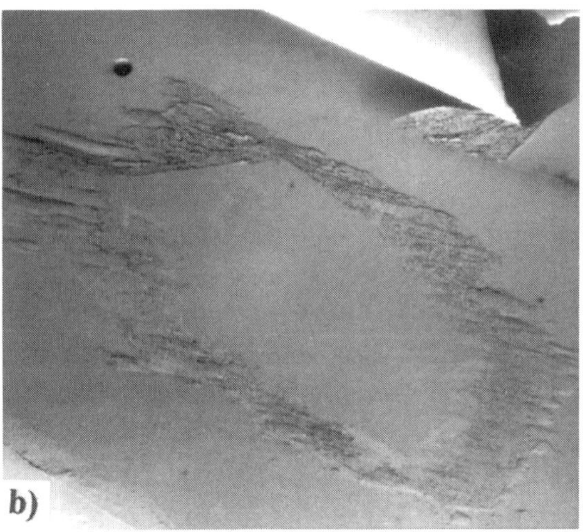

Figure 7. SEM-photographs of epoxy-PP-foil interfaces after partial removal of the foil which was irradiated with 248 nm UV-laser light with different pulse numbers ($E = 0.1$ J/cm^2; $f = 10$ Hz); (a) 1000 pulses, (b) 2000 pulses, (c) 3000 pulses.

stable products that have the low surface energy of the origin material, on the other hand.

3.3. Influence of the pretreatment on morphology

Under certain irradiation conditions chemical and also morphological changes have been observed. At low energy density these changes appear with increasing irradiation dose (pulse number) and differ from the typical irradiation damage observed at high

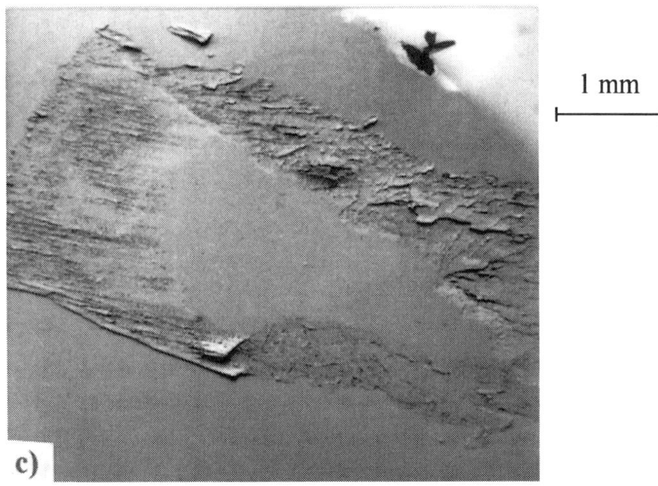

Figure 7. (Continued).

laser energy. The morphological changes also depend on the wavelength and the surrounding atmosphere.

Irradiation with 248 nm wavelength in oxygen leads to the appearance of black spots on the surface as well as in the bulk. The spot density and spot size increase with increasing number of pulses. Irradiation in helium leads to less pronounced changes and, in addition, a slight yellowing is observed. These results are believed to be due to carbonization of the polymer by its partial photolytical dissociation under the laser action. This assumption is supported by the fact that at sufficiently high irradiation doses microscopic defects such as small craters and cracks appear on the surface. They are probably due to gas formation (for example, H_2, CH_n) in the dissociation process. At high irradiation doses the gas pressure in the bulk damages the surface layers at some weak points, causing the creation of the mentioned defects.

Morphological changes of a completely different character have been observed after irradiation with shorter wavelength (193 nm), Fig. 8. At high irradiation doses a rippled pattern is observed. The ripples produced in He-atmosphere (Fig. 8b) have one-sixth the frequency of the ripples produced in oxygen (Fig. 8a). The orientation of the ripples does not depend on the experimental conditions, but seems to depend only on the orientation of the macromolecules in the polymer dictated by the manufacturing process. The morphological changes have some similarities with the ones observed when stretched polymers were irradiated with excimer laser light [8, 9]. The influence of the atmosphere could be interpreted in terms of different chemical modifications of the surface in different atmospheres, leading to different surface energies for the polymer foil and hence to different periodicity of the rippled morphological structure. Further investigations are needed to fully explain these phenomena.

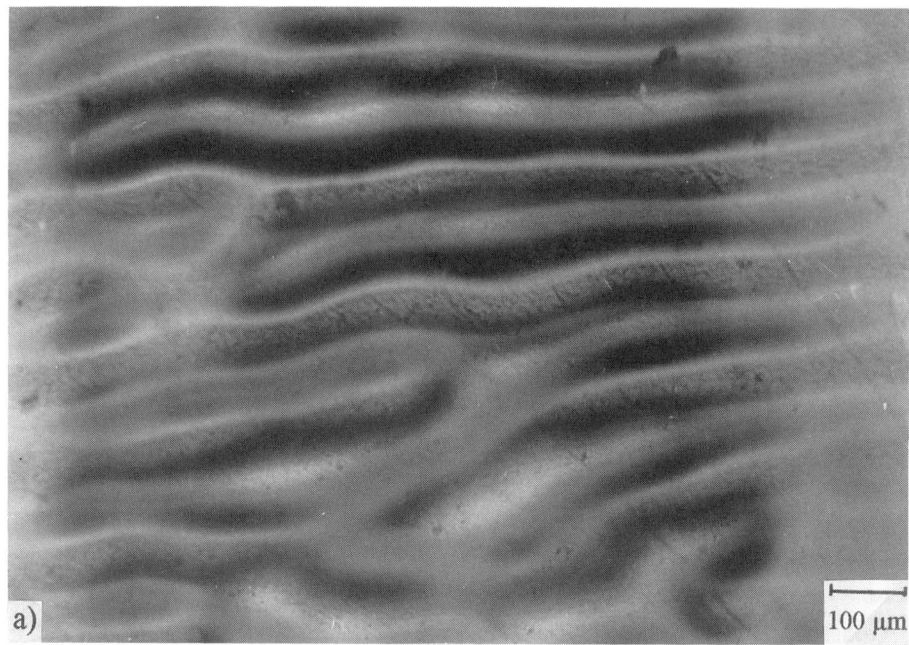

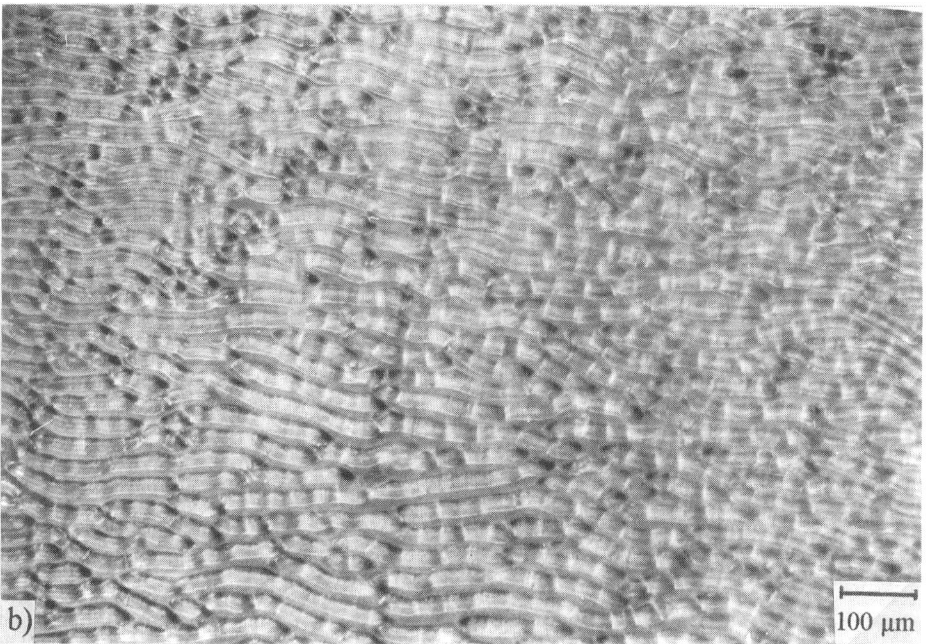

Figure 8. Microscopic views (polarized light) through irradiated PP-foils; atmosphere: (a) oxygen; (b) helium. $\lambda = 193$ nm; $E = 0.1$ J/cm^2; $f = 10$ Hz; $n = 4000$.

4. CONCLUSIONS

Experiments on irradiation of polypropylene with UV-excimer lasers in an oxidizing environment show that it is possible to initiate photolytically the formation of carbonyl groups on the polymer surface. Peel-off tests with epoxy resin have shown qualitatively a significant enhancement of the adhesion strength in the irradiated zones when appropriate conditions are applied.

Currently, intensive investigations of the pretreatment mechanisms and their dependence on the irradiation conditions are being carried out in order to find the main influencing parameters and thresholds as well as to control and optimize the treatment process.

Acknowledgement

This work was supported by the 'Deutsche Forschungsgemeinschaft' (DFG) which the authors gratefully acknowledge. The authors are indebted to Dr Schlett of 'Fraunhofer-Institut für angewandte Materialforschung' (IFAM) in Bremen, for the XPS-measurements.

REFERENCES

1. D. Satas, *Plastics Finishing and Decoration.* Van Nostrand Reinhold, New York (1986).
2. J. Breuer, S. Metev, G. Sepold, O. D. Hennemann and W. Kollek, *Appl. Surf. Sci.* **46**, 336–341 (1990).
3. R. Srinivasan and B. Braren, in: *Lasers in Polymer Science and Technology: Applications*, J. P. Fouassier and J. F. Rabek (Eds), Vol. 3, pp. 133–179. CRC Press, Cleveland, OH (1989).
4. W. Kesting, D. Knittel and E. Schollmeyer, *Die Angew. Makromol. Chem.* **191**, 145–161 (1991).
5. H. H. G. Jellinek, *Degradation and Stabilization of Polymers*, Vol. 1, Ch. 3. Elsevier, Amsterdam (1983).
6. S. Wu, *Polymer Interface and Adhesion.* Marcel Dekker, New York (1982).
7. C. Bischof and W. Possart, *Adhäsion.* Academie-Verlag, Berlin (1982).
8. A. W. Adamson, *Physical Chemistry of Surfaces.* Wiley & Sons, New York (1990).
9. E. Arenholz, V. Svorcik, T. Kefer, J. Heitz and D. Bäuerle, *Appl. Phys. A* **53**, 330–331 (1991).
10. T. Bahners and E. Schollmeyer, *J. Appl. Phys.* **66**, 1884 (1989).

Polymer Surface Modification: Relevance to Adhesion, pp. 199–212
K. L. Mittal (Ed.)

Laser-induced adhesion enhancement of polymer composites and metal alloys

A. BUCHMAN,[1] H. DODIUK,[1] M. ROTEL[2] and J. ZAHAVI[2,*]

[1]*Materials and Processes Department, RAFAEL, PO Box 2250, Haifa 31021, Israel*
[2]*Israel Institute of Metals, Technion, Israel Institute of Technology, Haifa 32000, Israel*

Revised version received 25 May 1994

Abstract—Proper surface treatment of an adherend is among the decisive factors concerning the final quality and durability of an adhesive joint. Various surface treatments are applied to plastic and metal adherends; among them are abrasive treatment, chemical treatment, and plasma etching. An alternative method is presented here which utilizes an excimer UV laser as a new technique for preadhesion surface treatment. Experimental results indicated that preadhesion laser surface treatment improved significantly the shear strength of adhesively bonded aluminum joints as compared with untreated or anodized substrates. Laser treatment also improved the adhesion strength of polycarbonate, polyetherimide, and composite adherends as compared with untreated or simply abraded substrates. Optimal laser treatment parameters (intensity, repetition rate, and number of pulses) depend on the substrate material and its chemical nature. As a result of laser treatment, the mode of failure changed from interfacial to cohesive as long as ablation did not occur. The adhesion enhancement resulted from morphological changes of the adherend surface caused by the laser irradiation as revealed by SEM (scanning electron microscopy), and from chemical modification and surface cleaning as indicated by ESCA (electron spectroscopy for chemical analysis) and FTIR (Fourier transform infrared spectroscopy). It can be concluded that the excimer laser has potential as a precise, clean, ecologically favorable, and simple preadhesion treatment for a wide variety of materials.

Keywords: Adhesion; surface treatment; excimer laser; surface chemistry; morphology.

1. INTRODUCTION

Adhesives are often required when joining materials in order to make the best use of their properties. Effective surface treatment methods for the adherends are among the important factors assuring bond performance and durability. Many treatments have been devised for preparing the surfaces of materials for adhesive bonding, coating, and the like [1]. The general purpose of these preparation procedures is to modify the original surface of the adherend material (a) to promote the development of interfacial

*To whom correspondence should be addressed.

bonds to the adhesives, (b) to enhance the environmental resistance to moisture and humidity effects, (c) to eliminate weak boundary layers and contaminations at the surface, (d) to improve wetting, and (e) to increase surface roughness. Various materials (metals, plastics, composites) require different surface treatments, some examples of which are detailed below.

Present processes for prebond surface preparation of thermoplastic and thermoplastic composite adherends involve the use of abrasive, welding, chemical, or plasma treatments, which are to a certain extent destructive, poorly controllable, and introduce undesirable changes in the morphology and composition of the surface such as cracks, uncontrolled pitting, and contamination [2].

The prebond surface treatments which are commonly used for metals such as aluminum are [3] chromate conversion coating, chromic acid anodization (with or without sealing), sulfuric acid anodization (with or without sealing), phosphoric acid anodization (PAA) and chromic sulfuric etch (FPL — Forest Product Laboratory). All of these treatments involve the use of acids (sulfuric, nitric, hydrochloric), strong bases, or hexavalent chromium compounds. New OSHA (Occupational Safety and Health Administration) and EPA (Environmental Protection Agency) regulations ban such chemicals in industrial operations.

In recent years, the accelerated development of ultraviolet (UV) laser systems has led to an increasing tendency to use these devices in various process applications. The main uses include cleaning of surfaces, precise cutting of materials, photopolymerization, ablation, coating in microelectronic circuits, and surgical applications. The outstanding feature of UV laser radiation is its ability to break chemical bonds or cause chemical reactions on the irradiated surface due to photochemical effects in addition to a thermal heating effect. The range of energy of UV lasers is within the bond energy of organic chemical bonds. The shorter the laser wavelength applied, the higher the photon energy achieved. For 193 nm, the photon energy is 6.4 eV, which can break up most organic bonds. The ability to use specific coherent energies can result in specific chemical reactions such as segregation, crosslinking, or radical formation which are otherwise difficult to obtain. The change of wavelength can also determine the amount of heat effect compared with the photochemical effect. Because of these effects, the laser irradiation may provide a new method for surface preadhesion treatment.

The laser treatment provides a clean and rather simple method of surface preparation and reduces the extent of damage to the treated surfaces. It cleans the surface from contaminations, modifies the chemical composition, and increases roughness. The modification of the chemistry and the morphology of the thin surface layer by the laser irradiation does not affect the bulk properties.

In this paper, the application of an ArF excimer laser (193 nm) for the surface treatment of thermoplastic (Ultem® and Lexan®) [4, 7], a carbon fiber-reinforced thermoplastic (PEEK — polyetheretherketone) composite, aluminum [5], and sealed anodic aluminum coating [6] is demonstrated. Recently the effect of laser treatment on adhesive bonding was also investigated on polypropylene [8, 9] and fluorocarbon [10]. Thus, a simple new technology is presented for the treatment of a variety of materials based on changing laser parameters according to the type of material to be treated.

2. EXPERIMENTAL

2.1. Laser treatment

The laser used during the course of this investigation was an excimer laser 201 MSC, a product of Lambda Physik, Germany. The laser parameters (intensity, repetition rate, energy density, area, and number of pulses) could be varied. All experiments were conducted at ambient temperature and in an air environment.

2.2. Materials

Table 1 lists the adherends and adhesives tested during the course of the investigation.

2.3. Testing

The various adherends were treated with laser UV by varying the following parameters: intensity, repetition rate, energy density, and number of pulses. The optimal laser treatment for each material was defined by achieving the maximum shear strength of the corresponding bonded joint. Joints were prepared by laser treatment of the adherends followed by bonding with the adhesives as listed in Table 1. The adhesive layer was 0.1 mm thick. Joint properties were determined using single lap shear (SLS) tests according to ASTM D-1002-72. Specimens were tested 4 days after preparation in an Instron machine, model 1185, at a crosshead speed of 20 mm/min. The mode of failure was determined visually to be either interfacial or cohesive (within the adhesive). Surface morphology following laser treatment and following shear fracture

Table 1.
Adherends and adhesives

Adherend	Commercial name (producer)	Adhesive/(producer) curing conditions
Polycarbonate	Lexan 9023-112 (General Electric)	Two-component polyurethane (PU) (Hexcel, USA), 7 days at RT
Polyetherimide	Ultem 1000 (General Electric)	Two-component polyurethane (PU) (Hexcel, USA), 7 days at RT
PEEK composite (polyether-etherketone reinforced with continuous carbon fibers)	APC-2/AS-4 (ICI)	Structural epoxy adhesive film FM 300 2K (American Cyanamid), 2 h at 120°C, 2.7 atm
Aluminum	2024-T351	Rubber-modified epoxy adhesive (RAFAEL, Israel), 7 days at RT [11]
	2024-T351	Structural epoxy adhesive film FM73 (American Cyanamid), 2 h at 127°C, 2.4 atm
Sealed anodized aluminum	2024-T351 treated with chromic acid anodization	Rubber-modified epoxy adhesive (RAFAEL, Israel), 7 days at RT

RT = Room temperature.

was analysed by means of a scanning electron microscope (SEM), model JSM-840, JEOL, Japan. The surfaces were coated with a sputtered Au/Pd layer prior to analysis. Chemical changes at the laser-irradiated surfaces and at the fracture surfaces were investigated by means of Fourier transform infrared (FTIR) spectroscopy using a Nicolet 5DX spectrophotometer in an external specular mode, equipped with a horizontal stage in near-to-normal incidence and a gold-coated mirror as reference.

Surfaces of the laser-treated compared with untreated adherends were studied for chemical changes due to laser irradiation by electron spectroscopy for chemical analysis (ESCA), model PHI 555 spectrometer with an Al K_α X-ray source, at 10 kV, 40 mA, and pressure of 3×10^{-8} Torr.

Two types of reference were used for each set of experiments. For the thermoplastics and composite thermoplastics, the adherends were untreated and/or abraded with SiC (36 mesh). For the aluminum and sealed anodized aluminum adherends, the references used were untreated aluminum and/or unsealed chromic acid anodized according to MIL B 8625.

3. RESULTS AND DISCUSSION

Table 2 summarizes the optimal laser treatment parameters for each of the adherends tested. As can be seen from the results, different conditions are required for the various adherends tested.

Table 3 summarizes the maximum joint shear strength achieved for the various laser-treated bonded adherends when applying the optimal laser preadhesion treatment. The results indicate that the ArF excimer laser treatment is effective for all the different adherends tested, providing better or similar joint strength as compared with both references (untreated or conventionally treated).

The results in Table 3 show that improvements of the joint shear strength of 200–600% were achieved for laser-treated adherends as compared with untreated ones, and of 100–200% compared with conventionally treated ones. The best results were achieved for the thermoplastic composite materials.

Table 2.
Optimal laser parameters for the treatment of various adherends

Adherend	Laser energy (J/P cm^2)	Repetition rate (Hz)	No. of pulses
Polycarbonate	0.08	10	12
Polyetherimide	0.08	10	200
PEEK composite	0.19	5	100
	or 1	5	10
Aluminum alloy	0.19	30	2000
Sealed anodized aluminum alloy	0.8	30	1000
	or 1.9	30	100

Table 3.
Joint strengths and failure modes of the various bonded joints

Adherend/adhesive	Surface treatment		
	Untreated	Conventional[c]	Laser-treated
	SLS strength[a] (MPa) (failure mode)[b]	SLS strength[a] (MPa) (failure mode)[b]	SLS strength[a] (MPa) (failure mode)[b]
Polycarbonate/PU	3.5 (A)	5.0 (M) SiC abraded	7.5 (C)
Polyetherimide/PU	2.5 (A)	5.0 (M) SiC abraded	5.5 (C)
PEEK composite (APC2/AS4)/ structural epoxy	6.1 (A)	14.7 (M) SiC abraded	27.8 (M)
Aluminum alloy/ toughened epoxy	2.0 (A)	10.2 (C) unsealed anodized	14.3 (C)
Aluminum alloy/ structural epoxy	12.8 (C)	42.9 (C) unsealed anodized	33.0 (C)
Sealed anodized aluminum/toughened epoxy	4.5 (A)	10.2 (C) unsealed anodized	11.0 (C)

[a] $\pm 5\%$ standard deviation.
[b] A = Interfacial; M = mixed; C = cohesive.
[c] Conventional — optimal surface treatment achieved chemically or mechanically.

Visual inspection of the failure surfaces shows clearly that the laser treatment causes the mode of failure to change from interfacial in a non-laser-treated adherend to mostly cohesive at optimal laser operating conditions, indicating that the interfacial adhesion was significantly improved.

SEM micrographs of the laser-treated adherends revealed morphological changes depending on the nature of the adherend material, the laser energy, and the number of pulses. The thermoplastic adherend surfaces exhibited conic and rounded granules spread all over the surface (Figs 1 and 2). The reinforced PEEK composite exhibited the same granules accompanied by partial exposure of the fibers (Fig. 3) at laser energies of 0.1–0.2 J/P cm^2 at 100 or more pulses. At higher laser energies (1 J/P cm^2 and above), the laser-irradiated surface was smooth with randomly spread cracks (Fig. 4). At 50 pulses and more, severe damage to the matrix was observed.

SEM micrographs of the Al adherend after laser treatment showed no morphological changes at low laser energies (0.18–0.2 J/P cm^2). Increasing laser energy revealed a fine microstructure of the treated surface demonstrating arrays of cracks about 1 μm wide, small holes, and segregation (0.7 J/P cm^2) (Fig. 5).

Irradiation of the sealed anodized aluminum specimens at low energy densities (0.2 J/P cm^2) showed no change in surface morphology even after 1000 pulses. At 0.7 J/P cm^2, changes in morphology included open bubbles, resulting probably from

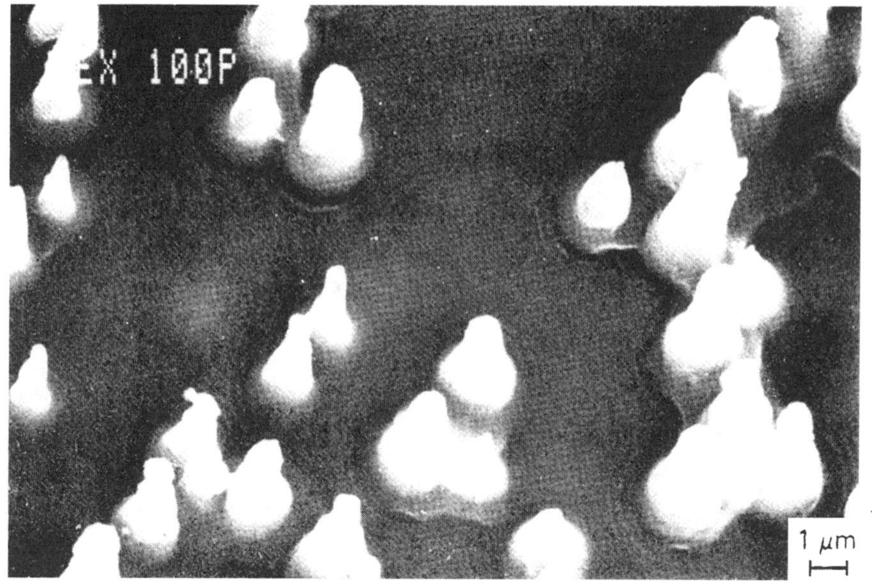

Figure 1. SEM micrograph of the top view of polycarbonate treated with excimer laser irradiation (0.085 J/P cm^2, 100 pulses).

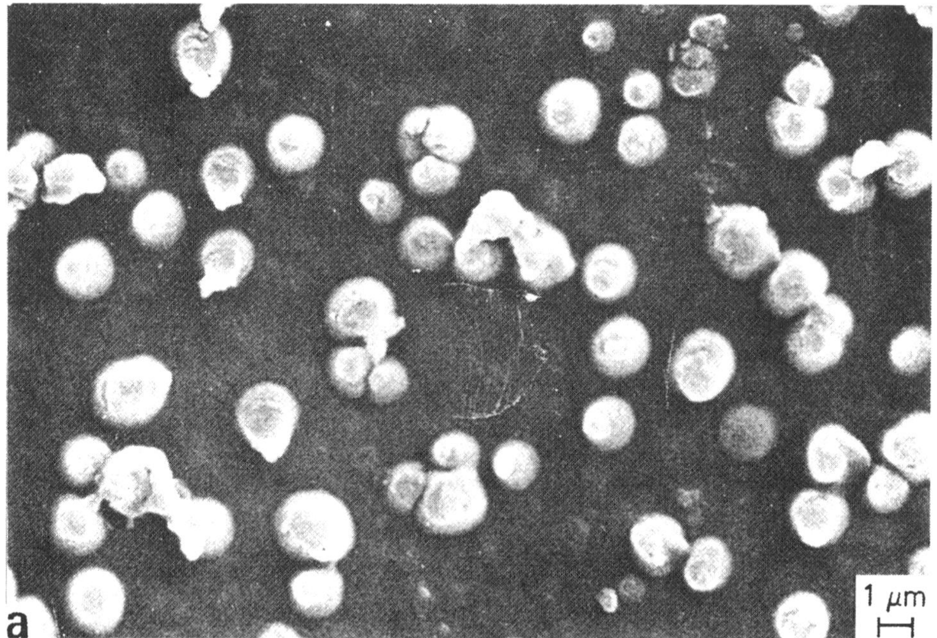

Figure 2. SEM micrograph of polyetherimide treated with excimer laser irradiation (0.085 J/P cm^2, 200 pulses).

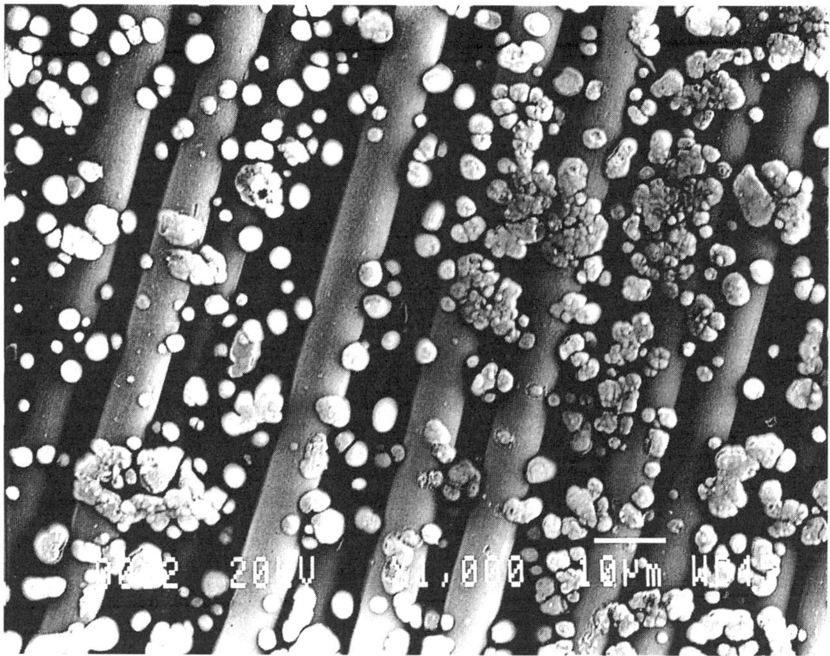

Figure 3. SEM micrograph of PEEK composite treated with low-energy excimer laser (0.18 J/P cm^2, 100 pulses).

Figure 4. SEM micrograph of PEEK composite treated with high-energy excimer laser (1 J/P cm^2, 10 pulses).

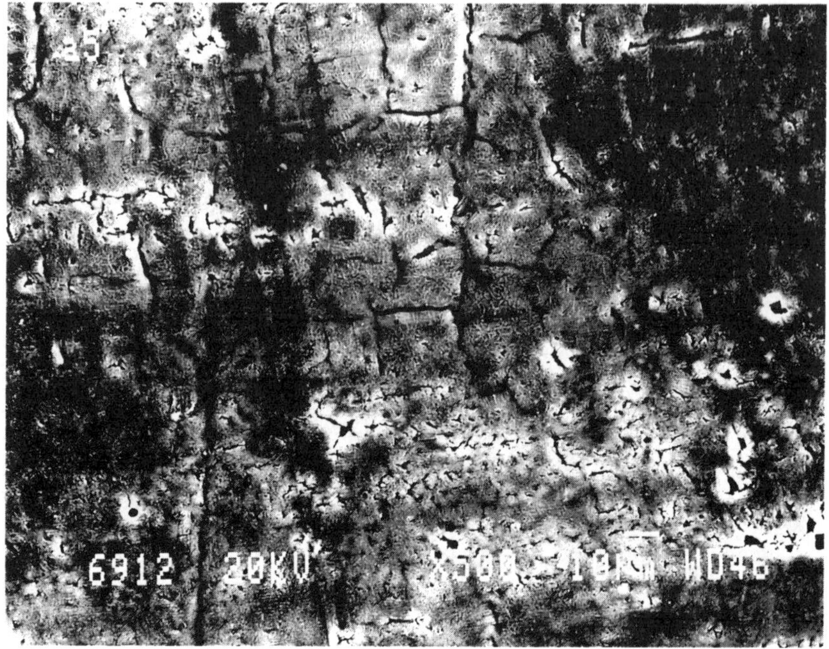

Figure 5. SEM micrograph of Al 2024-T3 aluminum treated with excimer laser (0.7 J/P cm^2, 200 pulses).

water removal. Some spherical droplets of Al_2O_3 due to splashing (caused by laser ablation of Al_2O_3) and cracks could be observed (Fig. 6). All the morphological features observed after laser treatment indicate an increase in surface roughness which contributes to improved mechanical interlocking and increased bondable surface area.

The enhanced adhesion due to mechanical interlocking is clearly revealed in the SEM micrographs at nearly all UV laser treatments and all adherends.

A far better adhesion (partially cohesive failure) can be observed in laser-treated samples compared with the interfacial failure in the case of the untreated adherends. Adhesive can be found on each adherend showing a replica of the granules or bubbles formed during laser treatment. Figures 7 and 8 represent the micrographs showing the replica formed during failure for the thermoplastic and anodized Al adherends. At a higher laser intensity or number of pulses, a weakening of the treated layer of the adherend is observed due to ablation revealing loosely connected granules at the thermoplastic interface, deformed exposed fibers of the composite, and molten dendritic areas of the Al and anodized Al surface.

In addition to the pronounced morphological modification, chemical changes also take place following UV laser treatment of the adherends. These changes are evidenced by FTIR and ESCA. An increase in surface polarity of the polycarbonate occurs due to the scissioning of carbonate bonds to form hydroxyls and carboxyls [4, 7]. These changes probably increase the surface reactivity. A more energetic treatment than the optimum causes massive degradation, even in the aromatic part of the skeletal

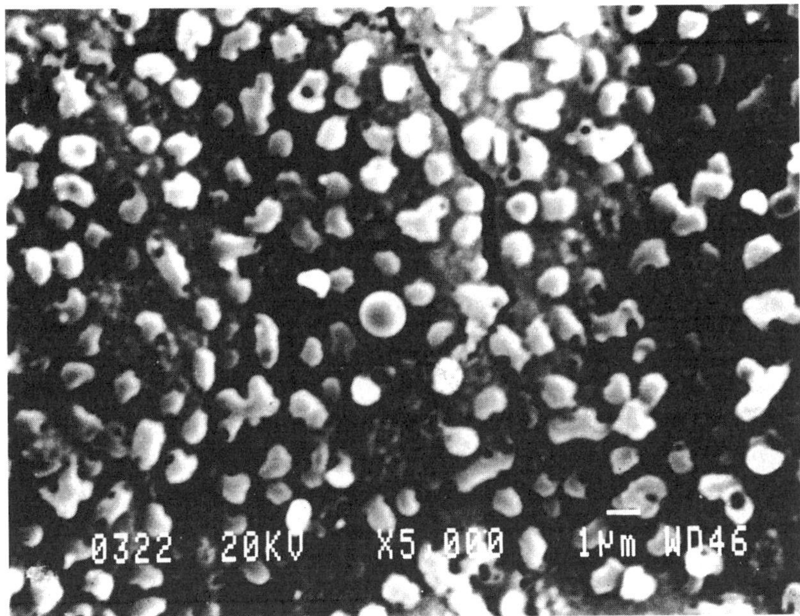

Figure 6. SEM micrograph of sealed anodized aluminum treated with excimer laser (0.7 J/P cm^2, 10 pulses).

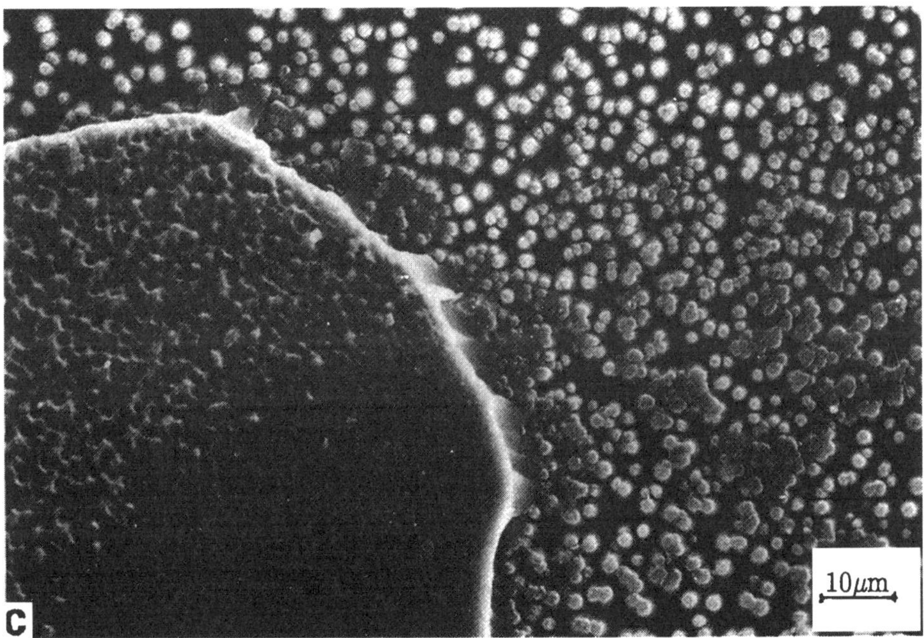

Figure 7. SEM micrographs of fractured surfaces (cohesive failure within the adhesive) of joints consisting of polyetherimide adherends treated with excimer laser radiation (0.085 J/P cm^2, 200 pulses) and bonded with polyurethanes.

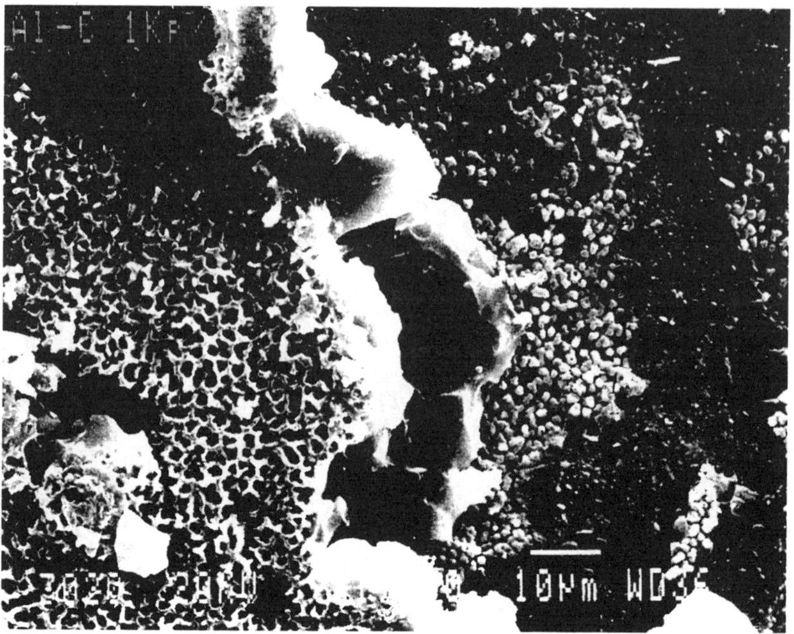

Figure 8. SEM micrograph of fractured surfaces of joints consisting of sealed anodized Al adherends treated with excimer laser radiation (0.8 J/P cm^2, 1000 pulses) and bonded with epoxy adhesive.

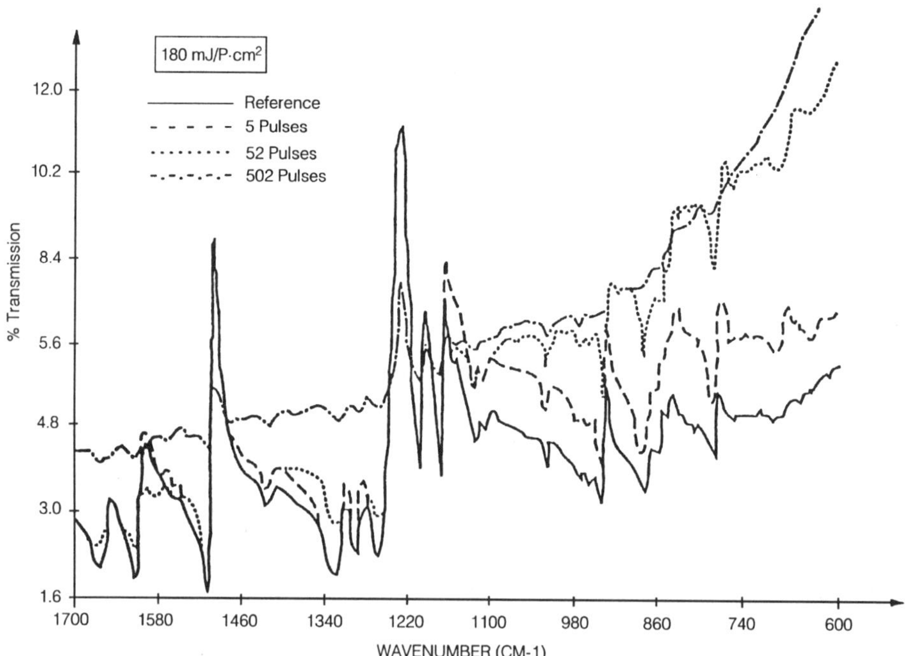

Figure 9. FTIR spectra of carbon-reinforced PEEK composite before and after laser treatment.

chain. This degradation results in the weakening of the outer layer as observed by SEM and probably causes a reduction in the adhesion level.

FTIR spectral analysis of polyetherimide revealed small chemical changes which occur mainly at the imide ring, to form polyamic acid, which improves the surface chemical activity [4, 7]. FTIR spectral analysis of the PEEK composite revealed the formation of hydroxyls and aldehydic groups from the broken main chains and oxidation to carbonate bonds, which probably modify the adherend surface due to optimal laser treatment. Higher laser energies resulted in a total peak reduction due to degradation and ablation (Fig. 9).

FTIR spectra of anodized Al revealed the removal of adsorbed and bound water molecules from the anodic film and dehydration of the hydroxide to Al oxide [6]. The growth of a new Al oxide layer on the laser-cleaned surface was observed by FTIR spetroscopy in the case of the pure Al alloy [5].

ESCA analysis was performed on untreated adherends and on adherends after laser treatment, and the atomic concentration at the surface were determined. As can be clearly seen in Figs 10 and 11, laser treatment evaporates the contaminants (Si, Ca, Na, and N) from the polycarbonate adherend surface and the contaminants (Mg and Si) from the PEEK composite adherend surface, leaving these cleaner. This reveals another advantage of the laser treatment, which is surface cleaning before

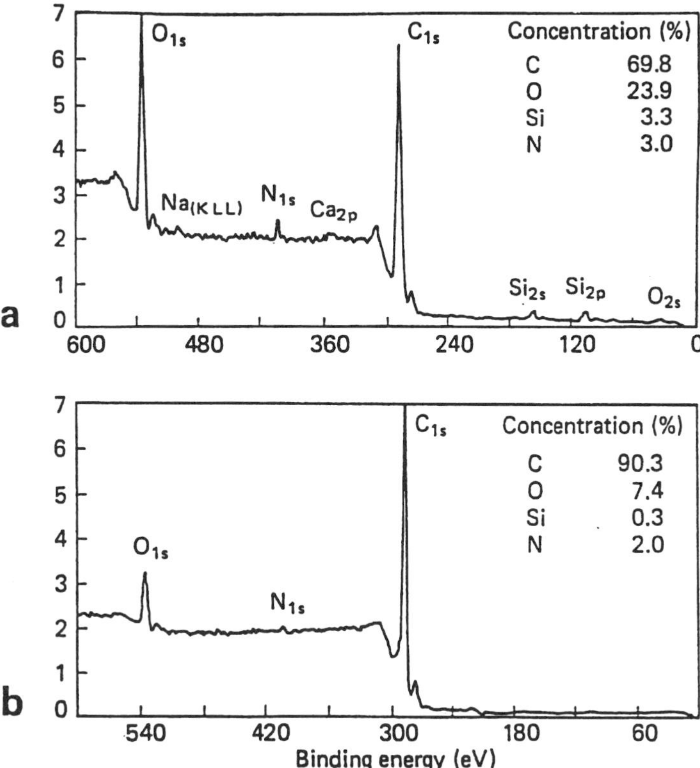

Figure 10. ESCA spectra of polyetherimide before (a) and after (b) laser treatment.

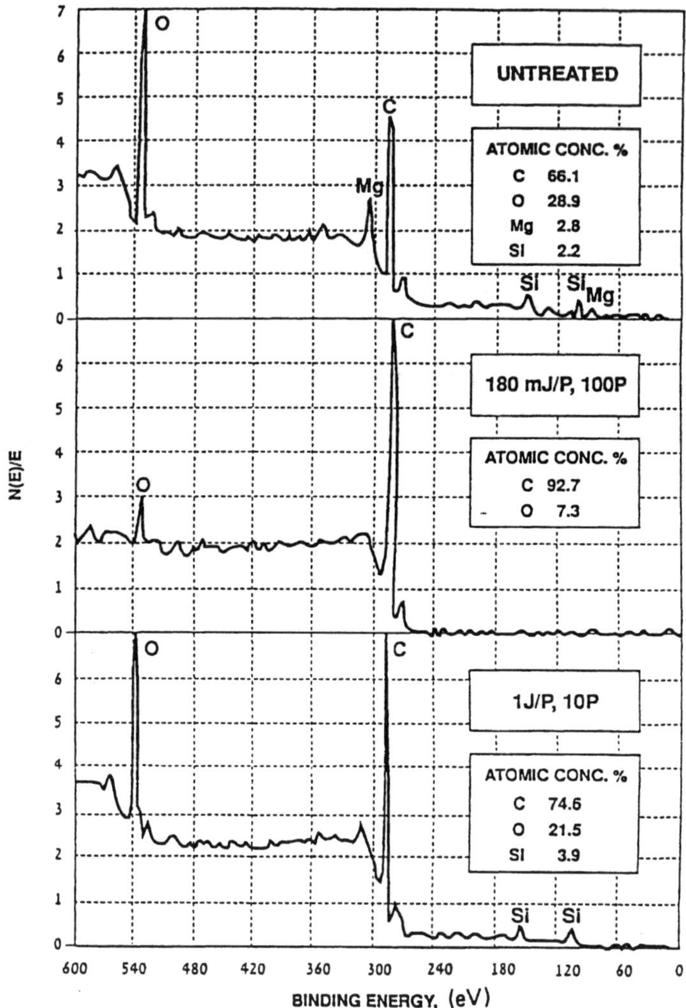

Figure 11. ESCA spectra of carbon-reinforced PEEK composite before and after laser treatment.

bonding, a basic requirement for a strong and durable bond. The same effect was observed for all the adherend materials tested.

ESCA analysis of C_{1s} and O_{1s} spectra for all the adherends tested showed changes as a result of the laser treatment, confirming the FTIR spectral results of surface modification. In laser-treated polycarbonate, the peaks assigned to the carbonate group decreased while the peaks assigned to the hydroxyl and carboxyl groups increased compared with the untreated adherend [7]. The new chemical groups are more active and thus improve the adhesion level.

In the laser-treated PEEK composite, the peaks assigned to the etheric group nearly vanish and the peaks from the carbon fibers increase due to fiber exposure (Fig. 12). The carbonyl peak becomes the major peak due to laser treatment, confirming the modification effect of the surface.

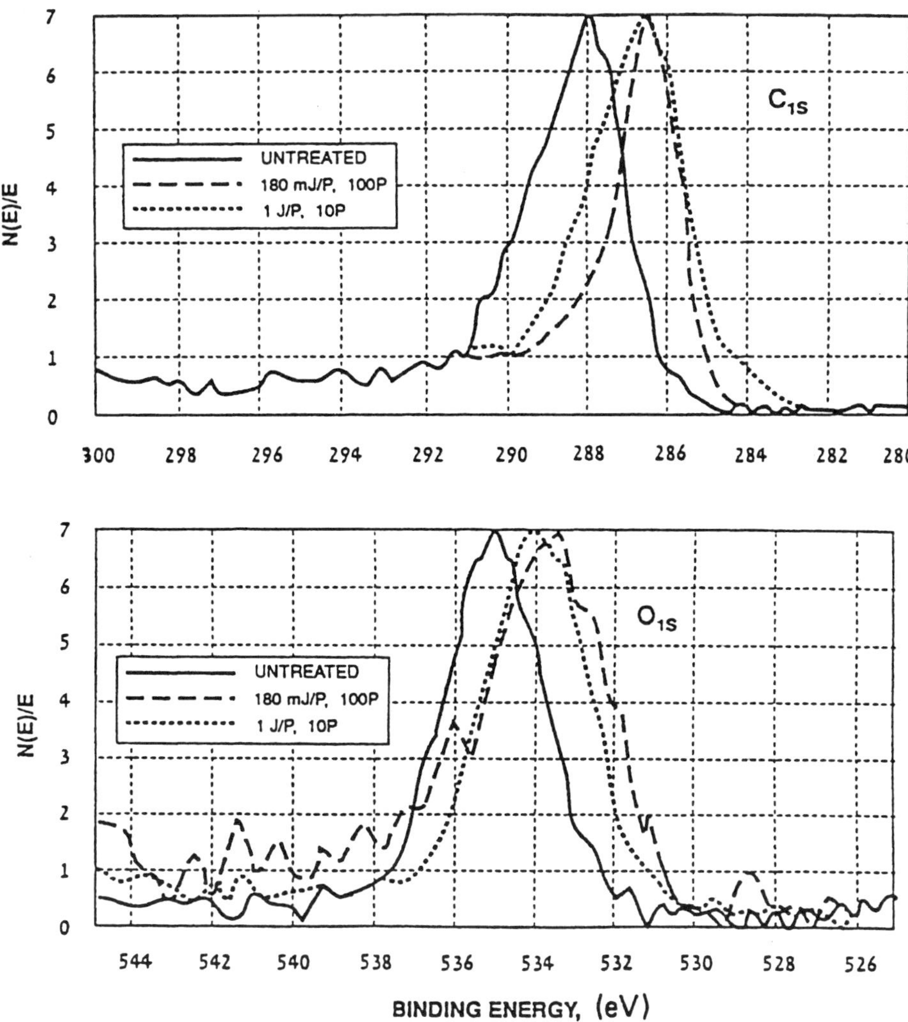

Figure 12. Comparison of C_{1s} and O_{1s} ESCA spectra of carbon-reinforced PEEK composite before and after laser treatment.

4. CONCLUSIONS

The use of an ArF excimer as a preadhesion surface treatment for various substrates (plastic, composite, and metal) has been investigated. Bonded joints with laser-treated adherends had higher lap shear strengths than untreated, abraded, or chemically treated adherends. The morphology was modified to a rougher surface with improved mechanical interlocking and increased bondable surface area.

A higher laser intensity than the optimum caused various phenomena indicating ablation. FTIR and ESCA spectroscopy indicated chemical modifications of the surface following laser treatment and the elimination of contaminants and natural oxide layers, which enhanced the joint strength. Under optimal excimer laser treatment, the mode of failure became cohesive or mixed in nature. Increased surface roughness, cleaning, and chemical modification of the laser-treated adherends are all responsible for and contribute to the higher adhesive joint strength. The optimal laser conditions are related to the adherend material's bonding energy. The higher the bonding energy, the higher the laser energy needed for effective surface treatment (metals > composites > polymers). This mechanism should be further studied, modelled, and verified.

It can be concluded from the present study that the ArF excimer laser is an effective, clean, and precise method for surface preadhesion treatment of various kinds of materials compared with conventional chemical or abrasive methods. For the best results, the laser irradiation intensity, dose, and operating atmosphere should be optimized and durability should be further studied.

Acknowledgements

We would like to thank the USAF, Dr T. J. Reinhart, and Dr Y. Gefen for supporting this research.

REFERENCES

1. J. R. J. Wingfield, *Int. J. Adhesion Adhesives* **13**, 151 (1993).
2. P. Davies, C. Courty, N. Xanthopoulos and H.-J. Mathieu, *J. Mater. Sci. Lett.* **10**, 335 (1991).
3. J. D. Venables, *J. Mater. Sci.* **19**, 2431 (1984).
4. A. Buchman, H. Dodiuk, M. Rotel and J. Zahavi, *Int. J. Adhesion Adhesives* **11**, 144 (1991).
5. H. Dodiuk, A. Buchman, M. Rotel, J. Zahavi, S. Kenig and T. J. Reinhart, *J. Adhesion* **41**, 93 (1993).
6. Z. Gendler, A. Rosen, M. Bamberger, M. Rotel, J. Zahavi, A. Buchman and H. Dodiuk, *J. Mater. Sci.* **29**, 1521 (1994).
7. E. Würzberg, A. Buchman, E. Zylberstein, Y. Holdengraber and H. Dodiuk, *Int. J. Adhesion Adhesives* **10**, 254 (1990).
8. L. Dorn and W. Wahono, *6th Int. Symp. of Swiss Bonding*, Bazel, p. 87 (May 1992).
9. M. Murahara and M. Okoshi, *J. Photopolym. Sci. Technol.* **6**, 379 (1993).
10. J. Breuer, S. Metev, G. Sepold, O.-D. Hennemann, H. Kollek and G. Krüger, *Appl. Surface Sci.* **46**, 336 (1990).
11. H. Dodiuk, U.S. Patent No. 4841010 (20 June 1985).

Polymer Surface Modification: Relevance to Adhesion, pp. 213–221
K. L. Mittal (Ed.)

Excimer laser-induced photochemical modification and adhesion improvement of a fluororesin surface

M. MURAHARA[1,*] and K. TOYODA[2]

[1]*Faculty of Engineering, Tokai University, 1117 Kitakaname, Hiratsuka, Kanagawa 259-12, Japan*
[2]*Riken, The Institute of Physical and Chemical Research, 2-1 Hirosawa, Wako, Saitama 351-01, Japan*

Revised version received 26 August 1995

Abstract—Modification of a selective area of a fluororesin surface was accomplished by using ArF excimer laser radiation and a boron complex with oleophilic or hydrophilic functional groups. The chemical stability of fluororesin is attributed to the presence of C—F bonds. The F atoms were abstracted by B atoms selectively from the area irradiated with excimer laser radiation and were replaced with the desired functional groups. In this modification, $B(CH_3)_3$ and $B(OH)_3$ were used: a boron compound with methyl groups to generate an oleophilic surface, and one with hydroxyl groups to generate a hydrophilic surface. As a result, the resin surface exposed to ArF laser radiation becomes oleophilic or hydrophilic. Both samples were bonded to stainless steel plates with an epoxy bonding agent and the tensile shear strength was 1.2×10^7 Pa in both cases.

Keywords: Fluororesin; excimer laser; defluorination; oleophilic; hydrophilic; photochemical modification; epoxy bonding agent; shear strength.

1. INTRODUCTION

Fluororesin (Teflon) is a chemically very stable material; it repels both oil and water and has poor adhesion. Therefore, surface modification of the resin is required. The following modification methods are commonly used: irradiation with plasma or creation of a rough surface by laser ablation, and chemical methods such as mixed ammonia/water solution with metallic sodium, and the complex compound of naphthalene–tetrahydrofuran solution with metallic sodium [1–4]. In the case of physical methods to make a surface rough, these defects may cause a stress concentration. On the other hand, although chemical methods are not risky in that respect, their usage is very restricted because of the possibility of damaging the environment. In the future, the use of poisonous chemicals will be limited. Thus, a new modification method is needed.

*To whom correspondence should be addressed.

Therefore, we have developed a photochemical modification method which uses excimer laser radiation [5, 6]. By this method, the surface can be modified to be hydrophilic [6, 7] or oleophilic [9, 10]. If the fluororesin surface is photochemically modified to be epoxy-compatible, a high adhesion strength will be obtained with the epoxy bonding agent [11].

2. PHOTOCHEMICAL REACTIONS

A fluororesin is composed of C—F bonds and the bond energy is 128 kcal/mol. ArF excimer laser radiation has a photon energy of 147 kcal, which is higher than the C—F bond energy. Also, fluororesin has a significant optical absorption at the ArF laser wavelength of 193 nm. Therefore, the C—F bonds can be broken directly by the ArF laser photons. Boron atoms are employed to prevent the recombination reaction of fluorine atoms with carbon atoms of the surface [9, 10]. The reason why these atoms are employed is that the B—F bond energy is higher than the C—F bond energy. In particular, boron atoms have a bond energy with fluorine atoms of 184 kcal/mol; the B—F bond energy is higher than the photon energy of ArF laser radiation. Accordingly, B—F bonds are not photodissociated under ArF laser radiation. For these reasons, the defluorination reaction proceeds effectively. On the other hand, the side-chains of epoxy resins are composed of CH_3 and OH functional groups, as shown in Fig. 1.

If the fluorine atoms at the surface are replaced by CH_3 or OH functional groups, the fluororesin can be modified to be epoxy-compatible. Thus, chemical bonding of the modified surface with the epoxy is possible without a primer. To make the fluororesin epoxy-compatible, $B(CH_3)_3$ gas, or orthoboric acid $B(OH)_3$ or H_3BO_3 aqueous solution, which are compounds of boron with CH_3 or OH groups, is used. $B(CH_3)_3$ and H_3BO_3 in the presence of fluororesin are photodecomposed by the ArF laser radiation as follows:

$$B(CH_3)_3 + [-\underset{\displaystyle F}{\overset{\displaystyle F}{C}}-\underset{\displaystyle F}{\overset{\displaystyle F}{C}}-]_n + h\nu\ (193\ \text{nm}) \longrightarrow BF_3 + [-\underset{\displaystyle F}{\overset{\displaystyle CH_3}{C}}-\underset{\displaystyle F}{\overset{\displaystyle CH_3}{C}}-]_n \tag{1}$$

$$\begin{aligned} &H_3BO_3 + 2\,H_2O + [-\underset{\displaystyle F}{\overset{\displaystyle F}{C}}-\underset{\displaystyle F}{\overset{\displaystyle F}{C}}-]_n + h\nu\ (193\ \text{nm}) \longrightarrow B(OH)_4^- + H^+ + H_2O + [-\underset{\displaystyle F}{\overset{\displaystyle F}{C}}-\underset{\displaystyle F}{\overset{\displaystyle F}{C}}-]_n \\ &\longrightarrow HBF_4 + H_2O + [-\underset{\displaystyle F}{\overset{\displaystyle OH}{C}}-\underset{\displaystyle F}{\overset{\displaystyle OH}{C}}-]_n \longrightarrow HBF_3(OH) + [-\underset{\displaystyle F}{\overset{\displaystyle OH}{C}}-\underset{\displaystyle F}{\overset{\displaystyle OH}{C}}-]_n + HF \end{aligned} \tag{2}$$

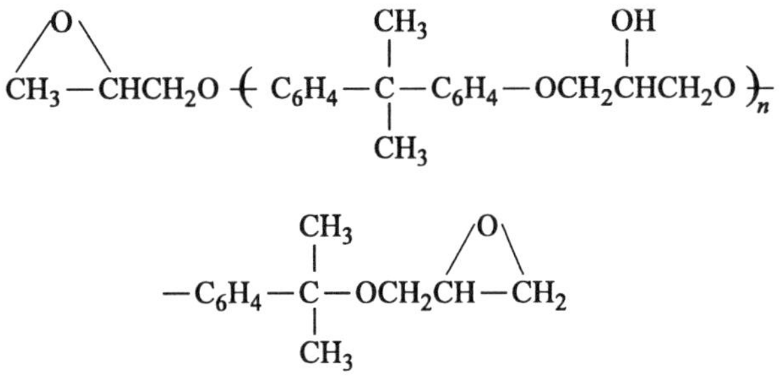

$$CH_3-\overset{\overset{O}{\frown}}{CHCH_2}O \left(C_6H_4-\underset{CH_3}{\overset{CH_3}{C}}-C_6H_4-OCH_2\overset{OH}{C}HCH_2O \right)_n$$

$$-C_6H_4-\underset{CH_3}{\overset{CH_3}{C}}-OCH_2\overset{O}{\overbrace{CH-CH_2}}$$

Figure 1. Structural formula of an epoxy resin.

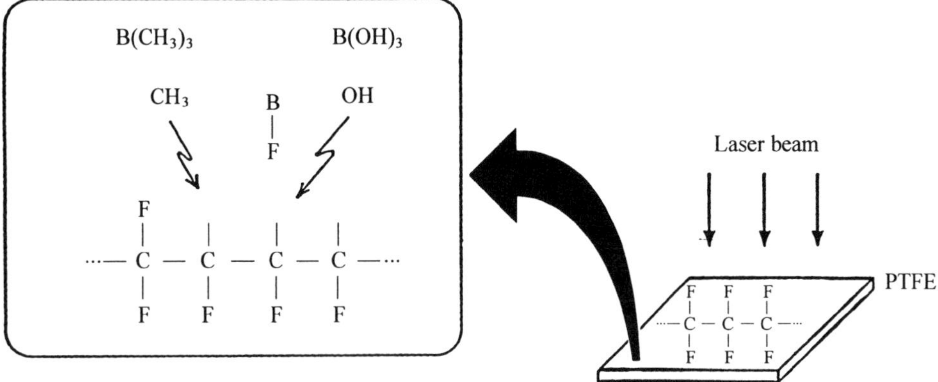

Figure 2. Principle of the photochemical reaction providing replacement of F atoms by CH_3 or OH functional groups.

As a result, the surface is defluorinated by the photodissociated boron atoms; the mechanism for photochemical substitution of the CH_3 or OH functional groups for B atoms is shown in Fig. 2. As a result, an ArF laser-irradiated part of the fluororesin surface can be modified selectively.

3. EXPERIMENTAL METHOD

3.1. Experimental set-up for CH_3 functional group substitution

A schematic diagram of the experimental set-up is shown in Fig. 3(a) [12]. The fluororesin film was placed in the reaction chamber. After evacuation, the chamber was filled with gaseous $B(CH_3)_3$ at a pressure of 40 Torr. Then the ArF laser irradiated the fluororesin surface at a laser fluence of 0–50 mJ/cm^2.

3.2. Experimental set-up for OH functional group substitution

A schematic diagram of the new surface modification method, which can be carried out in the open air, is shown in Fig. 3(b). A fused silica glass plate was placed

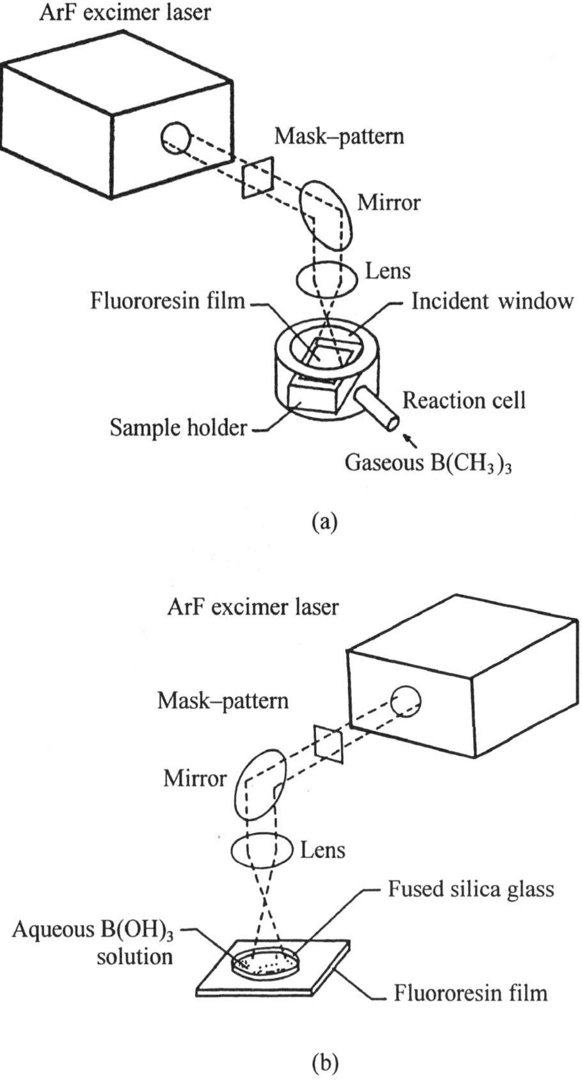

Figure 3. Schematic diagram of the experimental set-up: (a) using gaseous $B(CH_3)_3$, and (b) using aqueous $B(OH)_3$ solution.

on the fluororesin surface, and aqueous orthoboric acid solution was infiltrated into the thin gap between the glass plate and the fluororesin surface by the capillary phenomenon [12].

In this set-up, the solution layer was about 50 μm thick and the orthoboric acid concentration in the water was approximately 1.8%. In the selective modification, low-fluence patterned ArF excimer laser radiation was projected onto the fluororesin surface. In this way, the laser beam irradiated the sample surface only; penetration of the beam into the solution was prevented. The ArF laser radiation irradiated the sample surface through the fused silica glass plate at a fluence of 0–25 mJ/cm^2.

4. RESULTS AND DISCUSSION

4.1. Defluorination of the fluororesin surface

To investigate the defluorination of the modified sample surface, X-ray photoelectron spectroscopy (XPS) analysis of the PTFE samples with and without surface modification was carried out. Figure 4a shows the variation of the F 1*s* peak along the direction of the depth. Depth profiles were obtained by etching the surface by bombarding with an Ar ion beam. The F 1*s* (690 eV) peak due to the $F_2{-}CF_2$ fluororesin structure was minimum at zero etching time, corresponding to the minimum number of fluorine atoms at the surface. By etching repeatedly in the depth direction, the peak amplitude reached asymptotically that of the bulk material [10]. Figure 4b shows the dependence of the F 1*s* intensity as a function of the laser shot number. The peak in the non-irradiated surface is due to the $CF_2{-}CF_2$ fluororesin structure. With increasing laser shot number from 0 to 500 or 1000, the F 1*s* peak gradually decreased. At 3000 laser shots, the F 1*s* peak almost disappeared [12]. Evidently the concentration of fluorine atoms decreases with ArF laser radiation in the presence of boron atoms.

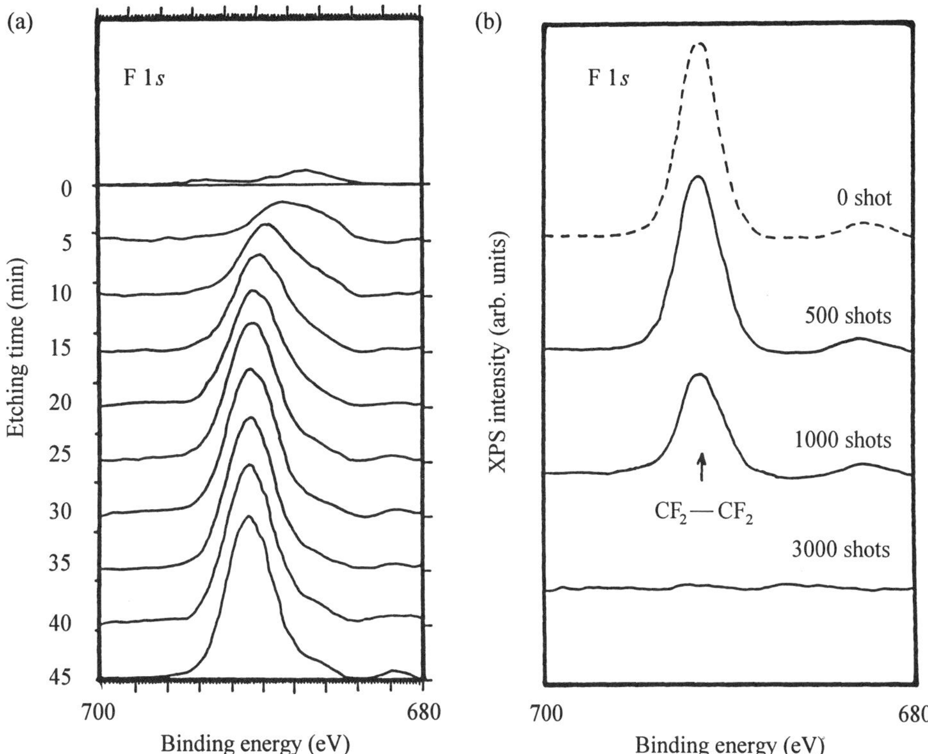

Figure 4. XPS spectra of F 1*s* (690 eV). (a) Variation of the F 1*s* peak in the direction of the depth; (b) the F 1*s* peaks of the modified surfaces at four different laser shot numbers.

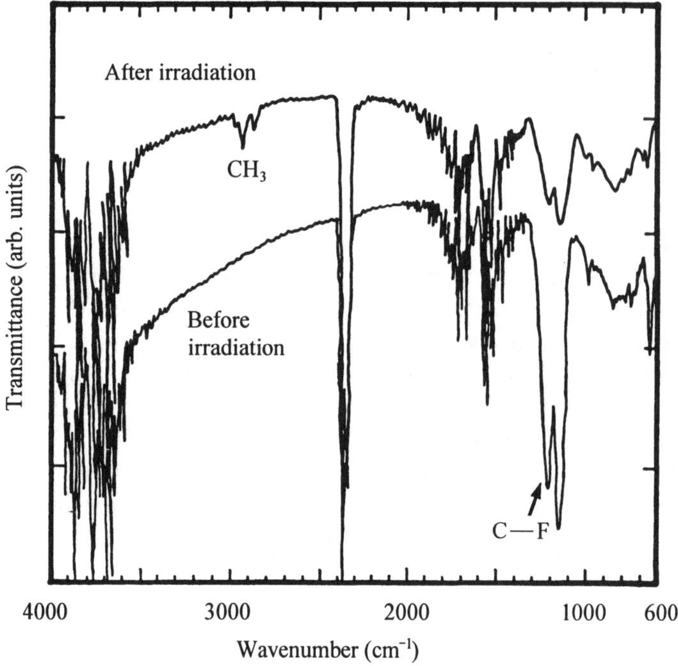

Figure 5. ATR–FTIR spectra of the fluororesin surface before and after laser irradiation.

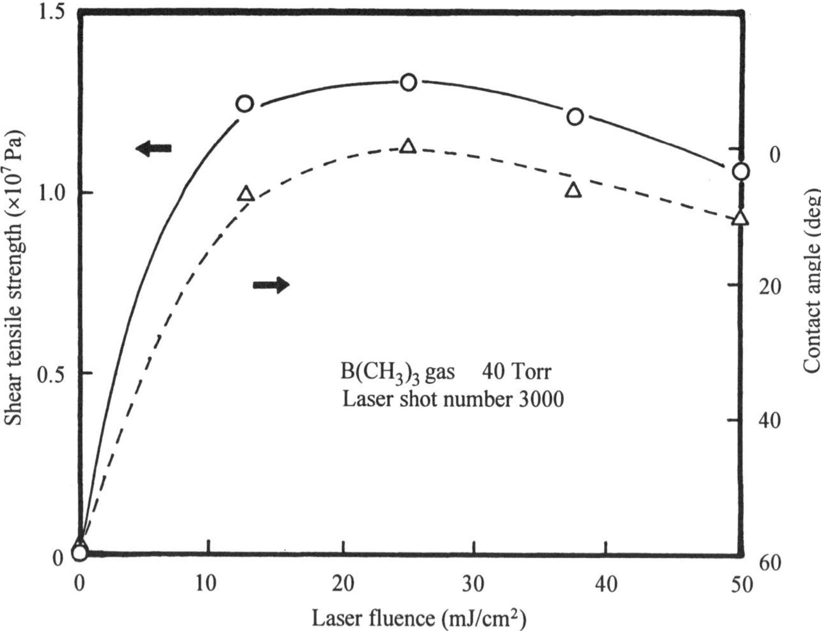

Figure 6. Shear tensile strength and contact angle with benzene as a function of the laser fluence. The specimens were modified in a gaseous $B(CH_3)_3$ atmosphere of 40 Torr.

4.2. Substitution of CH_3 functional groups and evaluation of the oleophilic property

The patterned ArF laser beam was projected onto the fluororesin film surface in a gaseous $B(CH_3)_3$ atmosphere. Figure 5 shows the ATR–FTIR spectra of the fluororesin surface before and after ArF laser irradiation in the $B(CH_3)_3$ atmosphere. The dominant new peaks in the irradiated sample were observed between 3100 and 2700 cm^{-1}; these are well known as the peaks of the stretching vibration mode of CH_3 [10]. The modified surface showed affinity for oil. To evaluate the oleophilic property of the modified surface, the contact angle with benzene was measured as a function of the laser fluence, as shown in Fig. 6. The contact angle of the non-irradiated surface was about 57 deg. With increasing laser fluence, the contact angle gradually decreased. With 25 mJ/cm^2 fluence, the angle became about 0 deg. However, the contact angle increased at a fluence of over 25 mJ/cm^2 because the substituted CH_3 groups were dissociated by the high laser fluence radiation. The modified fluororesin was bonded to a stainless steel plate of about 1 mm thickness degreased with methyl alcohol, with epoxy resin glue. In general, the adhesion strength between fluororesin and epoxy resin is less than 1.92×10^4 Pa. The shear tensile strength was examined and is shown as a function of the laser fluence in Fig. 6 [11, 12]. The shear tensile strength increased with increasing laser fluence. With 25 mJ/cm^2 fluence, the strength was 1.27×10^7 Pa at a shot number of 3000. However, the strength decreased at a fluence of over 25 mJ/cm^2. The shear tensile strength and contact angle show a similar tendency as the laser fluence increases.

4.3. Substitution of OH functional groups and evaluation of the hydrophilic property

Aqueous orthoboric acid solution is very safe compared with the case of using gaseous $B(CH_3)_3$. Moreover, the fluororesin can be modified in open air by using the aqueous orthoboric acid solution. Figure 7 shows the ATR–FTIR spectrum of the photomodified fluororesin surface. The dominant peak in the irradiated sample is observed in the region of 3300 cm^{-1}. A new absorption band appears after photomodification, derived from the OH stretching mode. The IR analysis showed that OH functional groups were substituted on the fluororesin surface. The hydrophilic properties of the photomodified surface were evaluated by measurements of the contact angles with water. The minimum contact angle of about 0 deg was obtained at a fluence of 25 mJ/cm^2, as shown in Fig. 8a. The non-irradiated fluororesin surface has a contact angle of about 120 deg.

As shown in Fig. 8b, the modified fluororesin was bonded to stainless steel and the shear tensile strength was measured. The experimental conditions for this surface treatment were a boric acid concentration of 1.8%, a laser fluence of 25 mJ/cm^2, and a shot number of 3000. The epoxy resin glue was cured for 12 h at room temperature. The maximum shear tensile strength was 1.08×10^7 Pa at a fluence of 25 mJ/cm^2 or more.

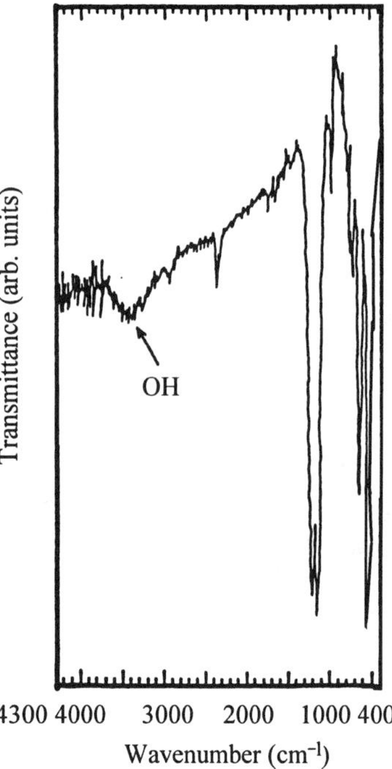

Figure 7. ATR–FTIR spectra of the fluororesin surface after laser irradiation.

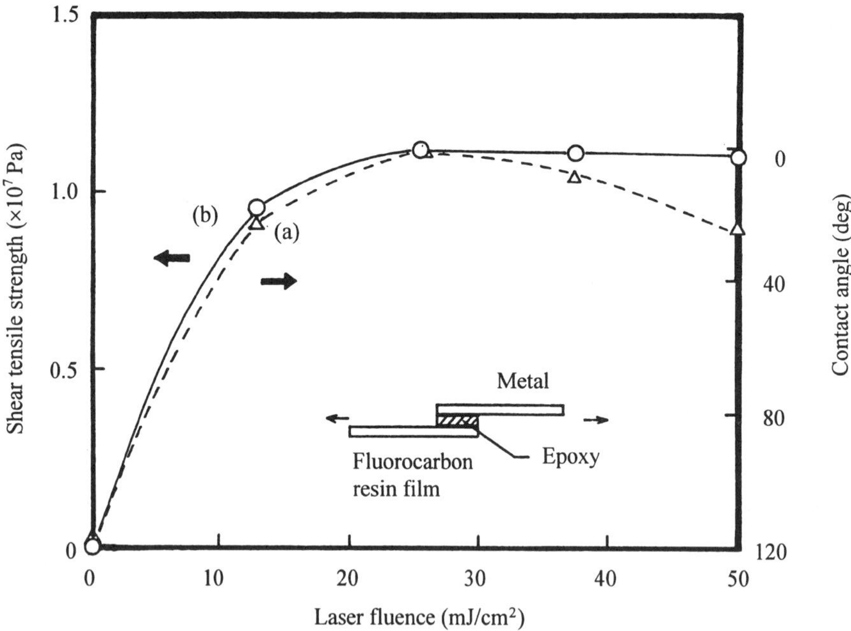

Figure 8. Shear tensile strength and contact angle with water as a function of the laser fluence. The specimens were modified in an aqueous solution containing 1.8% orthoboric acid.

5. CONCLUSION

A good bond strength between a fluororesin and stainless steel using an epoxy resin-based bonding agent was achieved by ArF excimer laser radiation. The bond strength was more than 650 times that of the untreated fluororesin. In this study, B atoms were employed for the defluorination reaction, and the F atoms at the fluororesin surface are replaced with CH_3 or OH functional groups. A noteworthy finding was the development of a new surface modification process in the open air, in which the dangerous gas is replaced by the safe aqueous orthoboric acid solution. This new safe surface modification process should be of great industrial appeal.

REFERENCES

1. J. R. Hall and C. A. L. Westerdahl *et al.*, *J. Appl. Polym. Sci.* **16**, 1465 (1972).
2. R. H. Hansen and H. Schonhorn, *J. Polym. Sci.* **B-4**, 203 (1966).
3. E. R. Nelson, T. J. Kilduff and A. A. Benderly, *Ind. Eng. Chem.* **50**, 329 (1958).
4. L. Soukup, *Int. Polym. Sci. Technol.* **5**, 19 (1978).
5. T. Sato, N. Takahashi, M. Murahara, K. Toyoda and S. Namba, *Riken Symp. Proc. Laser Sci.* **9**, 55 (1986) (in Japanese).
6. Lambda Highlights, No. 43, 1 (January 1994).
7. M. Murahara and M. Okoshi, in: *Proceedings of CLEO '91*, CPDP32-1, 634 (1991).
8. M. Okoshi, M. Murahara and K. Toyoda, *Mater. Res. Soc. Symp. Proc.* **201**, 451 (1991).
9. M. Okoshi, M. Murahara and K. Toyoda, *Mater. Res. Soc. Symp. Proc.* **158**, 33 (1990).
10. M. Okoshi, M. Murahara and K. Toyoda, *J. Mater. Res.* **7**, 1912 (1992).
11. M. Murahara and M. Okoshi, *J. Polym. Sci. Technol.* **6**, 379 (1993).
12. M. Okoshi, T. Miyokawa, H. Kashiura and M. Murahara, *Mater. Res. Soc. Symp. Proc.* **334**, 365 (1994).

Polymer Surface Modification: Relevance to Adhesion, pp. 223–229
K. L. Mittal (Ed.)

Photochemical surface modification of polypropylene for adhesion enhancement by using an excimer laser

M. MURAHARA* and M. OKOSHI

Faculty of Engineering, Tokai University, 1117 Kitakaname, Hiratsuka, Kanagawa 259-12, Japan

Revised version received 26 August 1995

Abstract—Polypropylene (PP) surface in water was photochemically modified to render it hydrophilic using ArF excimer laser radiation. The chemical stability of PP is attributed to the C—H and C—H_3 bonds present. Thus, it is considered that H atoms are selectively pulled out from the area irradiated with ArF excimer laser light and are replaced with OH functional groups in the presence of water. In this treatment, the irradiated sample becomes hydrophilic with enhanced adhesion properties. The experimental conditions for this surface treatment were ArF laser fluence of 12.5 mJ/cm^2 and a shot number of 10 000. The treated PP and stainless steel were bonded with epoxy adhesive and the tensile shear strength was 46 kg/cm^2.

Keywords: Polypropylene; excimer laser; photochemical modification; dehydrogenation; epoxy adhesive.

1. INTRODUCTION

Polypropylene (PP) is a chemically stable material and is the lightest of all the plastics, with a specific gravity of 0.9. Moreover, it has a high tensile strength and elasticity. Therefore, it is used widely for engineering plastics. However, this material shows poor dyeing and adhesion characteristics because of its hydrophobic nature. Therefore, its surface modification is needed to enhance these characteristics.

In general, the surface modification of PP has been carried out by chemical or physical treatment methods. In the chemical method, potassium bichromate/concentrated sulfuric acid solution is usually employed to produce C—OH, >C=O, and —SO_3H groups at the surface [1–4]. On the other hand, corona discharge treatment is physical and simply produces surface roughness [5]. However, there is the possibility of damaging the environment with chemical treatment with the potassium bichromate/concentrated sulfuric acid solution; thus, its use will be limited in the future. The plasma method requires an exhaust system; thus, a simple modification process is needed. We have previously reported on the excimer laser-induced photochemical surface modification of Teflon and polyethylene [6, 7]. In this study, we

*To whom correspondence should be addressed.

develop a new method of surface modification in the open air which uses a safe water environment [8].

2. PHOTOCHEMICAL MODIFICATION

The chemical stability of PP is attributed to the presence of C—H bonds with bond energy 110 kcal/mol. The photon energy of the ArF excimer laser radiation (193 nm) is 147 kcal, which is higher than the dissociation energy of the C—H bond. Polypropylene also has an optical absorption band at the ArF laser wavelength of 193 nm. Therefore, the $C—CH_2$ or C—H bonds will be dissociated directly by ArF laser photons. Simultaneously, the H atoms in the $C—CH_2$ and C—H bonds will be dehydrogenated by the photoexcited H atoms of the alternative groups. The reason why these atoms are employed is that the H—H bond with 104.2 kcal/mol bond energy is higher than the C—H bond energy. Although the H—H bond energy is less than the photon energy of ArF laser radiation, the absorption wavelength of the H_2 gas generated by this photochemical reaction is 111 nm or smaller, as compared to the ArF laser wavelength of 193 nm. Therefore, H—H bonds are not photodissociated with ArF laser radiation, which means that the dehydrogenation reaction of the PP surface is carried out effectively [8].

On the other hand, the OH functional group is important in reacting chemically with the epoxy adhesive agent. We have demonstrated that the Teflon surface can be rendered hydrophilic, oleophilic or conductive in an atmosphere of gas or aqueous chemical solution [6–10]. We have now applied this new surface modification technique to PP in the open air in tap water. The water has an optical absorption at the ArF laser wavelength of 193 nm; the H_2O is photodecomposed by the ArF laser radiation on the PP surface as follows:

$$H_2O + [-\underset{\substack{|\\H}}{\overset{\substack{H\\|}}{C}}-\underset{\substack{|\\H}}{\overset{\substack{CH_3\\|}}{C}}-]_n + h\nu\ (193\ \text{nm}) \longrightarrow$$

$$\longrightarrow H + OH + [-\underset{\substack{|\\H}}{\overset{\substack{H\\|}}{C}}-\underset{\substack{|\\H}}{\overset{\substack{CH_3\\|}}{C}}-]_n \longrightarrow [-\underset{\substack{|\\H}}{\overset{\substack{OH\\|}}{C}}-\underset{\substack{|\\H}}{\overset{\substack{CH_2-OH\\|}}{C}}-]_n + H_2$$

The PP surface is effectively dehydrogenated by the photodissociated hydrogen atoms, and the hydrogen atoms of the PP surface are replaced by OH groups produced by the photodissociation of H_2O. Consequently, the PP surface is photomodified to a hydrophilic surface only in the photoirradiated areas.

3. EXPERIMENTAL METHOD

A schematic diagram of the new surface modification method which can be carried out in the open air is shown in Fig. 1. The PP surface was covered with a fused

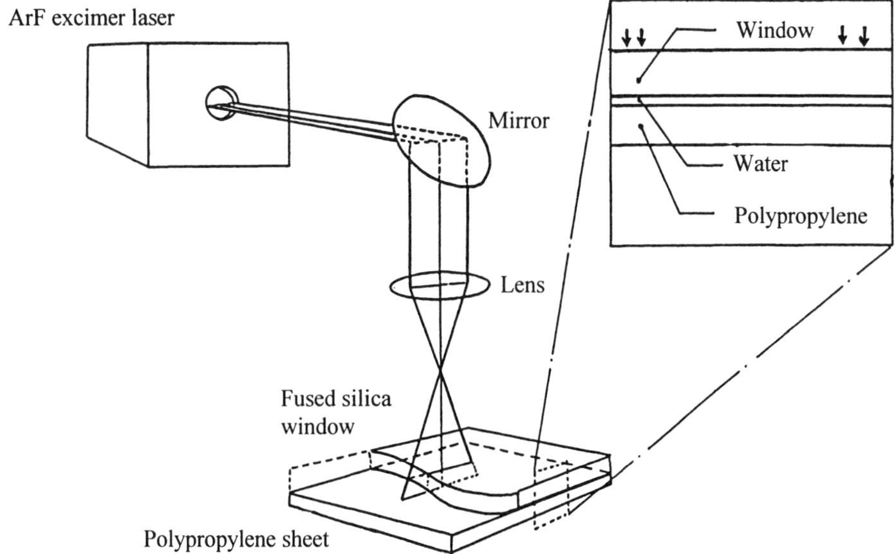

Figure 1. Schematic diagram of the experimental set-up.

silica glass plate; tap water of pH 7.1 was poured into the thin gap between the glass plate and the PP surface [8, 11]. Low-fluence ArF excimer laser light irradiated the PP surface through the fused silica glass plate at a fluence of 12.5 mJ/cm^2 and a shot number of 0–10 000.

4. RESULTS AND DISCUSSION

4.1. Substitution of OH functional groups

To investigate the presence of OH groups on the photomodified PP surface, X-ray photoelectron spectroscopy (XPS) was used, as shown in Fig. 2. This figure shows the O 1*s* spectra as a function of the etching time. The O 1*s* peak was maximum at zero etching time and the etch rate was about 10 Å/min.

By etching repeatedly with argon ions in the depth direction, the peak reached that of the bulk material. Oxygen was not detected on the untreated sample surface except for the adsorbed oxygen molecules, but a new oxygen peak was detected on the treated sample surface. Figure 3 shows the O 1*s* concentration as a function of the laser shot number. The concentration of O 1*s* increased with the ArF laser shot number in the presence of the water on the sample surface.

To analyze the extremely thin modified surface layers, the IR–ATR method was employed. The dominant peaks in the irradiated sample were observed at the center of 3300 cm^{-1}. A new absorption band appeared after photomodification, originating from the OH stretching mode. The IR analysis indicated, that OH functional groups were present on the PP surface. These findings indicate that the OH functional groups

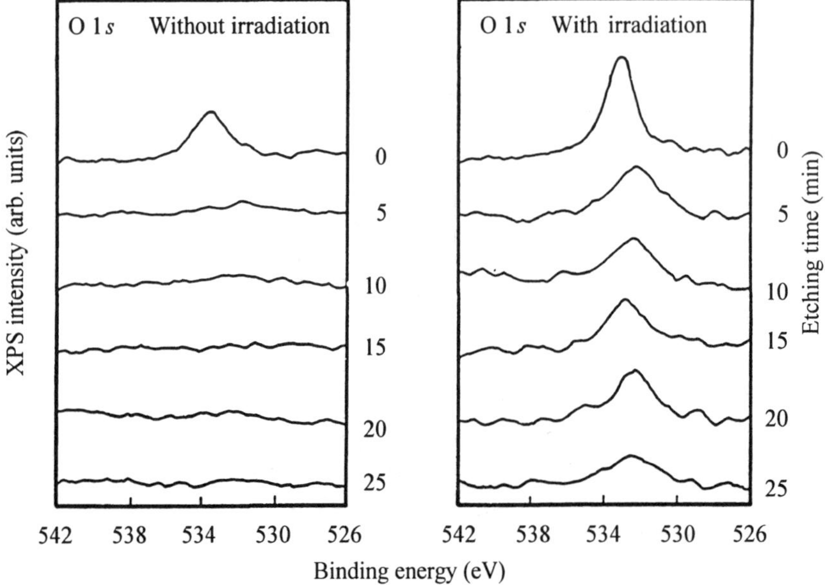

Figure 2. O 1*s* XPS spectra of the polypropylene surface with and without laser irradiation.

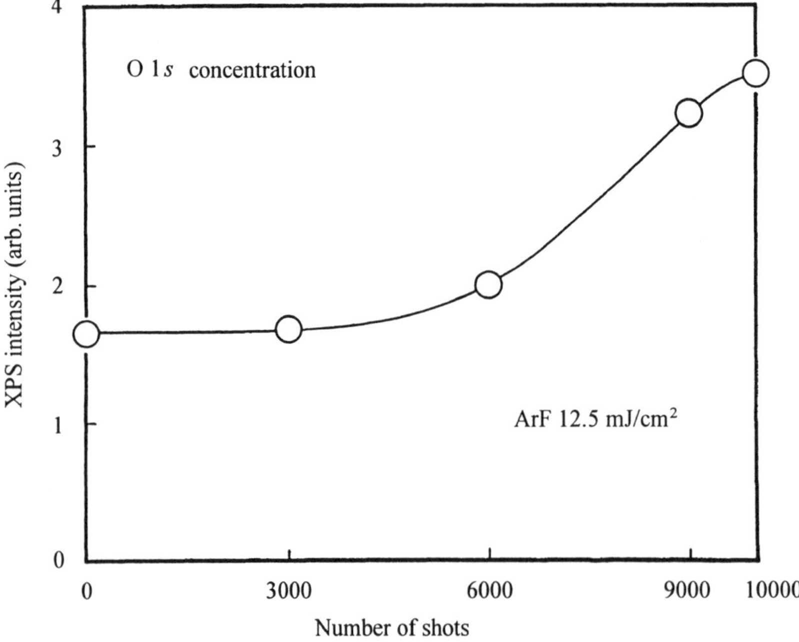

Figure 3. Laser shot number dependence of the O 1*s* concentration in the polypropylene surface.

photodissociated from H_2O replaced the H atoms of the PP surface: thus, the sample surface was modified to be hydrophilic.

4.2. Evaluation of the hydrophilic property

To investigate the hydrophilic property of the modified sample surface, the contact angle on the photomodified surface with water was measured. Figure 4 shows the contact angle as a function of the laser shot number. The contact angle on the non-irradiated PP surface was about 93 deg. In the case of water, the contact angle gradually decreased with increasing laser shot at a laser fluence of 12.5 mJ/cm^2. The contact angle reached about 65 deg at 10 000 shots.

Because of the presence of OH groups, the photomodified surface also showed good adhesion [7]. The modified PP surface and a stainless steel plate were bonded with an epoxy-based adhesive (Araldite standard). The experimental conditions were as follows: laser fluence, 12.5 mJ/cm^2; laser shot number, 10 000; curing condition of the epoxy resin, 24 h at room temperature. Figure 5 shows the relationship between the tensile shear strength and the displacement of the adhesion plane with PP and the epoxy bonding agent.

The tensile shear strength of the treated sample was 46 kg/cm^2. The tensile shear strength as a function of the laser shot number is also shown in Fig. 4. The tensile shear strength for untreated PP and epoxy resin is less than 6 kg/cm^2. As the laser shot number increased, the strength gradually increased and was 45 kg/cm^2 at a shot number of 10 000. This dependence corresponds closely to the dependence of the contact angle on the laser shot number. Also, the increase in tensile shear strength depends on the increase in O 1*s* concentration, as shown in Fig. 4. The photomodified PP surface was observed by a scanning electron microscope, and the surface was the same as the non-irradiated surface. Thus, the adhesion property of the sample was

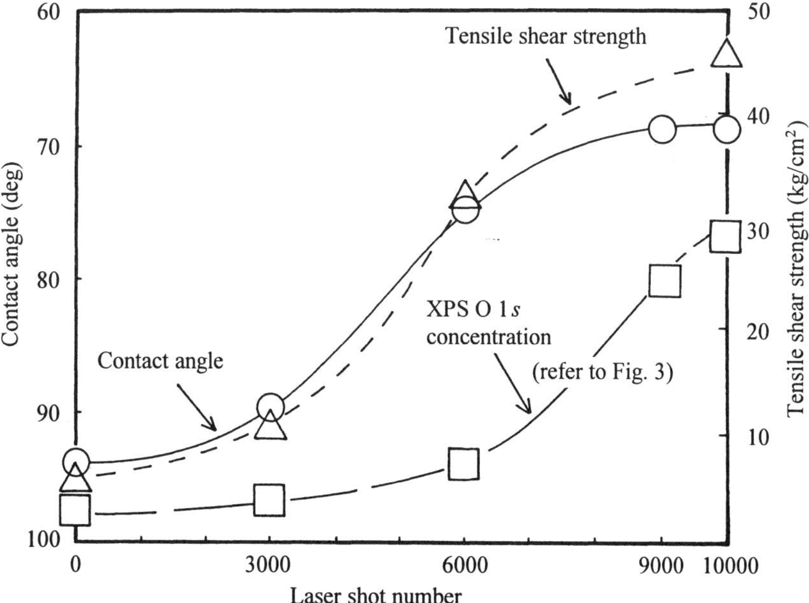

Figure 4. Laser shot number dependence of the contact angle with water, the tensile shear strength, and the O 1*s* concentration.

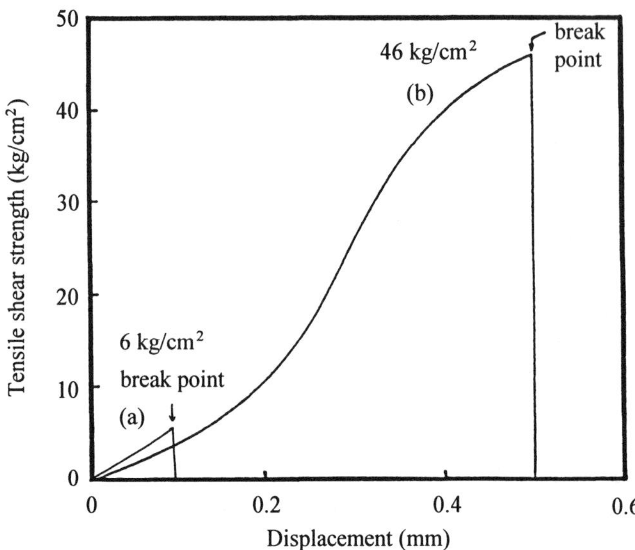

Figure 5. Relationship between the tensile shear strength and the displacement of the adhesion plane with polypropylene (PP) and the epoxy adhesive agent. (a) Untreated PP; (b) treated PP.

not caused by the anchoring effect due to surface roughness, but was due to chemical interaction between the OH groups and the epoxy adhesive.

5. CONCLUSION

The PP surface was photochemically modified using ArF excimer laser radiation and tap water. In this study, H atoms which were photodissociated from the water were employed for the dehydrogenation reaction. On the other hand, the OH functional groups photodissociated from H_2O replaced the H atoms of the PP surface. The hydrophilic property of the photomodified surface was evaluated by contact angle measurement with water; the contact angle reached a minimum at ArF laser irradiation of the optimum fluence and shot number. The hydrophilic property originated from the presence of OH groups. The treated PP was successfully bonded with an epoxy adhesive, and the tensile shear strength was seven times higher than that of an untreated sample. This increase in adhesion is not due to surface roughness, but results from chemical interaction between the OH groups and the epoxy adhesive.

REFERENCES

1. I. A. Abu-Isa, *Polym. Plast. Technol. Eng.* **2**, 29 (1973).
2. P. Blais, D. J. Carlsson, G. W. Csullog and D. M. Wiles, *J. Colloid Interface Sci.* **47**, 636 (1974).
3. D. Briggs, D. M. Brewis and M. B. Konieczko, *J. Mater. Sci.* **11**, 1270 (1976).
4. S. Wu, *Polymer Interface and Adhesion*, p. 295. Marcel Dekker, New York (1982).
5. M. Stradal and D. A. I. Goring, *Polym. Eng. Sci.* **17**, 38 (1977).
6. M. Okoshi, M. Murahara and K. Toyoda, *J. Mater. Res.* **7**, 1912 (1992).

7. M. Okoshi and M. Murahara, *J. Photopolym. Sci. Technol.* **7**, 381 (1994).
8. M. Murahara, Extended Abstract, 54th Autumn Meeting, The Japan Society of Applied Physics, 2, 568 (1993) (in Japanese).
9. Lambda Highlights (Lambda Physik GmbH, Germany), No. 43, 1 (January 1994).
10. M. Murahara and M. Okoshi, *J. Photopolym. Sci. Technol.* **6**, 379 (1993).
11. M. Okoshi, M. Miyokawa, H. Kashiura, K. Toyoda and M. Murahara, *Mater. Res. Soc. Symp. Proc.* **334**, 365 (1964).

Part 3
Other/Miscellaneous Surface Modification Techniques

Polymer Surface Modification: Relevance to Adhesion, pp. 233–251
K. L. Mittal (Ed.)

A comparison of gas-phase methods of modifying polymer surfaces

MARK STROBEL,[1,*] MARY JANE WALZAK,[2] JOSEPHINE M. HILL,[2] AMY LIN,[2] ELIZABETH KARBASHEWSKI[3,†] and CHRISTOPHER S. LYONS[1]

[1] *3M Company, 3M Center, Building 208-1, St. Paul, MN 55144, USA*
[2] *Surface Science Western, Western Science Centre, The University of Western Ontario, London, Ontario N6A 5B7, Canada*
[3] *3M Canada, Post Office Box 57571, London, Ontario N6A 4T1, Canada*

Revised version received 29 July 1994

Abstract—Oxidation is the most common surface modification of polymers. This paper presents a comparison of five gas-phase surface oxidation processes: corona discharge, flame, remote air plasma, ozone, and combined UV/ozone treatments. Well-characterized biaxially oriented films of polypropylene and poly(ethylene terephthalate) were treated by each of the five techniques. The surface-treated films were then analyzed by X-ray photoelectron spectroscopy (XPS or ESCA), contact-angle measurements, and Fourier-transform IR (FTIR) spectroscopy.

Corona, flame, and remote-plasma processes rapidly oxidize polymer surfaces, attaining XPS O/C atomic ratios on polypropylene of greater than 0.10 in less than 0.5 s. In contrast, the various UV/ozone treatments require orders of magnitude greater exposure time to reach the same levels of surface oxidation. While corona treatment and flame treatment are well known as efficient means of oxidizing polymer surfaces, the ability of plasma treatments to rapidly oxidize polymers is not as widely appreciated.

Of the treatments studied, flame treatment appears to be the 'shallowest'; that is, the oxygen incorporated by the treatment is most concentrated near the outer surface of the film. Corona and plasma treatments appear to penetrate somewhat deeper into the polymers. At the other extreme, the UV/ozone treatments reach farther into the bulk of the polymers.

Keywords: Surface modification; plasma; corona; flame; ozone; UV; polypropylene; poly(ethylene terephthalate); contact angle; XPS.

1. INTRODUCTION

Many polymeric materials require some type of surface modification prior to commercial use. Surface oxidation is the most common modification imparted to polymers.

---

*To whom correspondence should be addressed.

†Present address: Novacor Chemicals Ltd., Technical Centre, 3620 32nd Street N.E., Calgary, Alberta T1Y 6G7, Canada.

Oxidized polymer surfaces tend to have improved wetting and adhesion properties when compared with unmodified polymers. Over the years, a number of solution-based and gas-phase processes were developed to oxidize polymer surfaces. The gas-phase surface oxidation processes include corona discharge, flame, ozone, UV/ozone, and glow-discharge plasma.

Of these processes, corona is by far the most widely used. Despite extensive industrial use for many years, corona treatment remained poorly understood. However, in the past ten years, a reasonably clear understanding of the effects of corona treatment has finally emerged. Many fundamental and applied aspects of corona treatment are discussed in the publications by Briggs and co-workers [1–3], Gerenser and co-workers [4, 5], and Strobel *et al.* [6–8].

Along with corona-discharge processes, flame treatment was first developed in the 1950s to improve the wetting and adhesion properties of polyolefin films. Although originally developed to treat films, in the past 25 years gas flaming has been primarily used to modify paperboard or relatively thick polyolefin materials, such as blow-molded bottles [1]. Currently, however, flame treatment is receiving renewed industrial interest as an option for modifying polymer films. While flame has been studied much less than corona, the flame treatment of polyolefins is discussed in the papers of Briggs and co-workers [1, 9], Garbassi *et al.* [10, 11], Sutherland *et al.* [12, 13], and Papirer *et al.* [14, 15].

Although plasma treatment has probably been the most extensively studied surface-modification technique for many years [16], plasmas are used in industry to a far lesser extent than corona treatment or flame treatment. The principal reason for this seemingly unusual situation is that, while small laboratory plasma systems are relatively easy to construct and operate, industrial-scale equipment is quite complex and costly. Although plasma treatment has the potential to be used as a high-speed film treatment [17], there are few incentives to replace corona or flame treatments with a plasma process. Nonetheless, as discussed in a number of recent papers [18–21], the potential for obtaining unique surface modifications by plasma treatment is widely recognized. Plasmas are used in some industries for the batch (non-continuous) treatment of high-value-added materials.

Ozone-generating UV light has been used for many years to clean organic contaminants from various surfaces [22]. UV/ozone processes are now generally accepted as an effective method of cleaning surfaces used in the semiconductor industry [23]. A number of researchers have examined the individual and combined effects of UV light and ozone on the surfaces of polyethylene (PE) [24], polypropylene (PP) [25], and poly(ethylene terephthalate) (PET) [26, 27]. In all of these studies the treatment times for the ozonation of the polymers were lengthy as compared with the treatment times used in our study. In the UV/ozone treatment of polymers, the major reactive species are ozone and atomic oxygen. Ozone is constantly formed and destroyed by the action of 183 nm and 254 nm wavelength light, respectively, with both processes involving free atomic oxygen. Our studies [28] have shown that photooxidation in the presence of ozone is more effective than oxidation by ozone or UV light alone. These effects can depend greatly on the UV-absorbing characteristics of the polymers.

There have been few direct comparisons of the various gas-phase surface-oxidation techniques. Briggs [1] has comprehensively reviewed both corona and flame treatments, but provided no direct comparative data. Using PE substrates, Foerch and co-workers [29, 30] attempted the direct comparison of remote plasma treatment with both corona and UV/ozone. Novis *et al.* [31] compared the UV and corona treatments of PET films, while Onyiriuka *et al.* [32] conducted a brief study comparing the corona and oxygen-plasma treatments of polystyrene (PS) petri dishes.

In this paper we present a comparison of five gas-phase surface oxidation processes: corona, flame, remote air plasma, ozone only, and combined UV/ozone treatments. In all cases, the treatments were conducted using well-specified process equipment representative of the types of systems that might be used in industry. Well-characterized biaxially oriented films of PP and PET were treated by each of the five techniques. For this study, we attempted to bring each polymer surface to a specific oxidation level. The process conditions needed to reach this oxidation level were then compared.

2. EXPERIMENTAL

Thermally extruded, biaxially oriented PP and PET films were the substrates used in this study. The 0.04 mm thick PP film was produced using a homopolymer resin with a weight-average molecular weight of 1.9×10^5 and polydispersity index of 6.0. The base resin contained 500–1000 ppm each of an inorganic acid scavenger and a high-molecular-weight phenolic antioxidant. Biaxially oriented film was produced from this PP resin on a tenter-frame film line. The cast web was quenched at 45 °C prior to orientation. The machine-direction (MD) draw ratio was 5.2 : 1, while the transverse-direction (TD) draw ratio was 9 : 1.

The 0.1 mm thick PET film was produced using a homopolymer resin with a weight-average molecular weight of 25 000 and a degree of polymerization of 100. The PET base resin contained *ca.* 300 ppm of residual metal catalyst and 50 ppm of an organometallic thermal stabilizer. Biaxially oriented film was produced from this PET resin on a tenter-frame film line. The draw ratios were 3.4 : 1 in the MD and 4.6 : 1 in the TD. The finished PET film had a glass transition temperature of 83 °C.

As discussed in a previous paper [8], the additives contained in these PP and PET films are *not* surface active; that is, they are *not* detectable by ESCA or static secondary ion mass spectrometry (SSIMS) on either untreated or surface-treated polymers. Considerably different results might be obtained in process comparisons performed using films that contain significant quantities of surface-active additives.

2.1. Corona

Corona treatment was performed using an industrial corona system built by Sherman Treaters Ltd., (Thame, United Kingdom) and equipped with an ENI Power Systems Inc. (Rochester, New York) Model RS-48 (4 kW) power supply. A schematic diagram of the experimental apparatus has been published previously [6]. During treatment, the 30 cm wide substrates were in contact with a 25 cm diameter nickel-plated aluminum ground roll in the so-called 'bare-roll' configuration. The powered electrode consisted of three 15 cm diameter, 35 cm wide aluminum cylinders covered with 2.4 mm thick silicone-rubber sleeves. The powered electrodes were rotated at 60 rpm to prevent excessive thermal and electrical stresses on the dielectric covering. The electrode gap was 1.8 mm. The treater housing was continually flushed with *ca.* 400 litres/min of dry air at 25 °C to maintain a relative humidity in the discharge region of $5 \pm 5\%$.

The net power dissipated in the corona was measured with a directional power meter incorporated into the ENI supply. The ENI power supply is equipped with a feedback circuit that varies the output frequency to maintain optimal impedance matching (minimum reflected power). Thus, the discharge frequency varies somewhat as the corona conditions change. In this study, the discharge frequency was typically 35 kHz.

The normalized energy of the corona treatment was calculated from the net power and the film velocity:

$$\text{normalized energy,} \quad E = P/wv,$$

where P is the net power (in W), w is the electrode width (in cm), and v is the film velocity (in cm/s). Typical units are J/cm^2. In this study, we used corona energies of 0.17 and 1.7 J/cm^2, which correspond to film velocities of 100 m/min and 10 m/min, respectively, at a net power of 1000 W. The exposure times of the films to the corona were 0.05 s at the 0.17 J/cm^2 energy and 0.5 s at the 1.7 J/cm^2 energy. A normalized energy of 1.7 J/cm^2 is near the upper limit of commercially viable corona treatments.

2.2. Flame

Flame treatment was performed using an industrial flaming system built by 3M Company. A schematic diagram of the flame treater is shown in Fig. 1. During treatment, one side of the 30 cm wide polymer film was exposed to a laminar premixed natural gas:air flame while the backside was cooled by contact with a 25 cm diameter water-cooled aluminum roll held at 25–30 °C. To insure intimate contact between the substrate and the chill roll, a 10 cm diameter rubber-covered nip-roll was used. The burner, supplied by Aerogen Ltd. (Alton, United Kingdom), was a 30 cm × 1 cm stainless steel ribbon mounted in a cast-iron housing. The distance between the upermost surface of the ribbon burner and the polymer film was 13 mm. Ignition of the flame was accomplished by an electric spark.

Dust-filtered, 20–25 °C compressed air with a relative humidity of <5% was premixed with natural gas in a venturi mixer located *ca.* 10 m upstream of the burner. The natural gas used for this study had a specific gravity of 0.577 and a heat content of 37.7 kJ/litre (1018 Btu/ft^3). The exact amount of dry air necessary for complete combustion of this natural gas, the 'stoichiometric ratio', was 9.6 volumes per volume of natural gas, or, in other words, a 9.6 : 1 air : fuel ratio.

The flame power is readily quantified and is simply a function of the volume of natural gas burned per unit time. The input of thermal energy (heat) to the film is a complex function of the flame power, the exposure time of the film to the flame,

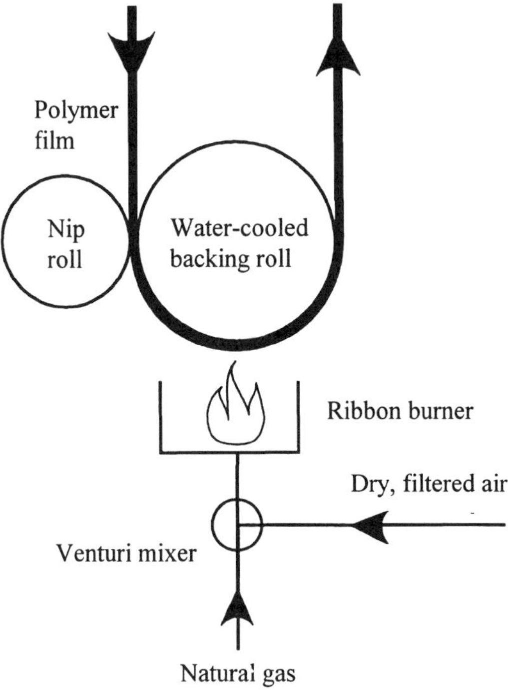

Figure 1. Schematic diagram of the flame treater (not drawn to scale).

and the burner-to-film gap. The input of energy to the film is not easily determined because the heat generated by the flame is dissipated into the chill roll, the burner housing, and the air surrounding the flame as well as into the film itself. This makes the comparison of corona and flame energies of dubious value.

Excess oxygen is generally recommended for the flame treatment of polymer films [1, 9, 12, 13]. For this study, we used a flame power of 15 kW (50 000 Btu/h), a film velocity of 125 m/min, and an air : fuel ratio of 10.4 : 1, which corresponds to about 1.5% excess oxygen. The film was exposed to the visible flame for *ca.* 0.04 s.

2.3. Remote plasma

The plasma was produced inside an Evenson cavity connected to a microwave generator operated at 2.45 GHz. Both the cavity and the generator were manufactured by Opthos (Rockville, Maryland). A schematic diagram of the plasma system is shown in Fig. 2. The microwaves were applied to air as it passed through a 1 cm diameter quartz plasma-drift tube. Below the drift tube is the reactor, which consists of a six-way stainless steel cross flanked by two sets of three differentially pumped chambers separated by a series of PTFE disks. The disks have slits cut into them that are slightly larger in cross-sectional dimensions than the polymer films to be treated. The pumpdown chambers were evacuated by a 65 m^3/h SOGEVAC rotary-vane pump, while a 25 m^3/h TRIVAC rotary-vane pump was used to further reduce the pressure in the plasma reaction vessel. During treatment, 1 cm wide films passed from ambient

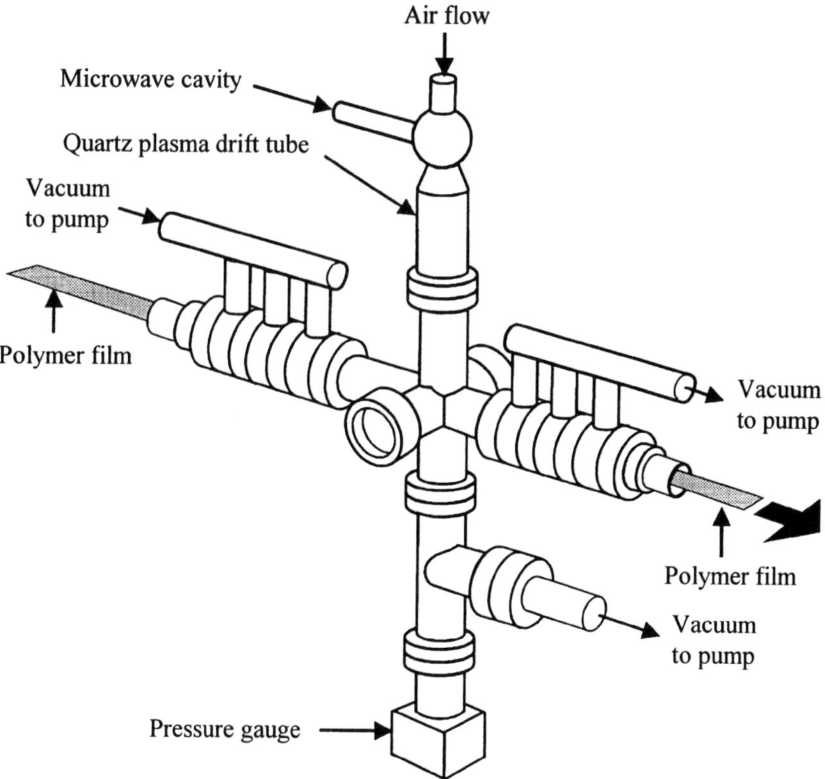

Figure 2. Schematic diagram of the remote-plasma treater (not drawn to scale).

conditions into the inlet seal, through the reaction zone, and then exited the second set of seals. The configuration of the plasma reactor causes the film to be directly exposed to the UV radiation generated by the microwave plasma. The ultimate pressure in the plasma reaction vessel was 3 torr. The pressure in the reactor was measured with a Pirani gauge, while the flow of air into the reactor was controlled and measured with an MKS Inc. mass flow controller.

PP and PET films were exposed to the reactive species from an air plasma generated 10 cm above the film surface. The plasma was produced by the application of 40 W of microwave power to an air (RH $< 5\%$) stream flowing at 1 litre/min. The pressure in the reaction zone was 7 torr. The film velocity was 10 m/min, which corresponds to an exposure time of the films to the remote air plasma of *ca.* 0.1 s. The normalized energy of the plasma treatment was *ca.* 1 J/cm^2.

2.4. Ozone and UV/ozone

A schematic diagram of the UV/ozone system is shown in Fig. 3. Because of the longer times needed to modify the polymers using UV/ozone, a non-continuous batch reactor was used to treat discrete pieces of film. The UV light source was an ozone-generating mercury-vapor grid lamp manufactured by BHK Inc. (Pomona, California).

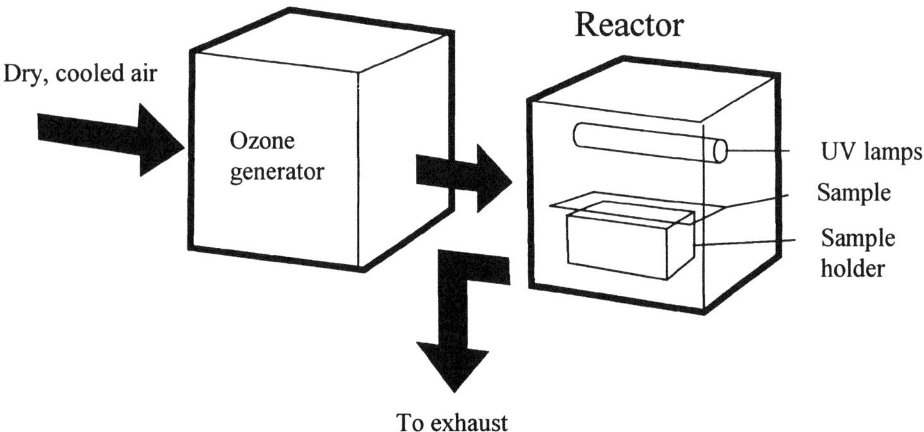

Figure 3. Schematic diagram of the UV/ozone reactor (not drawn to scale).

The radiation intensity at a wavelength of 254 nm was 12–15 mW/cm^2 at a distance of 2.54 cm from the lamp. A Pillar Technologies (Milwaukee, Wisconsin) ozone generator was used to produce supplemental ozone. Film samples (12 cm × 13 cm) were treated on a shelf mounted 1.5 cm from the UV lamp within a sealed 15 cm diameter stainless steel reactor. A cooling coil immersed in an ice/brine bath was used to chill the incoming air prior to entry into the ozone generator, or prior to entry into the reactor if the ozone generator was not used. The flow rates of <1% RH air were 1 litre/min when the UV lamp was used and 2 litres/min when the ozone generator was used alone. The air flow was controlled by an MKS Inc. mass flow controller. The gas inlet was situated so that the air flowed between the sample surface and the UV lamp. The concentration of ozone at the sample surface was measured using a Resonance Ltd. (Alliston, Ontario) Model UVTRANS2 ozone transmissiometer. The accuracy of the ozone concentration measurements was ±10%.

The reaction conditions for the various combinations of UV and ozone were:

(1) UV/air

The PP and PET film samples were exposed to UV for 10 min and 5 min, respectively, in the presence of 1 litre/min of dry flowing air. The concentration of ozone near the film surface was *ca.* 0.3 mg/litre. All of this ozone was formed by the action of the UV light on the molecular oxygen in the air.

(2) UV/air plus ozone

The PP and PET film samples were exposed to UV for 10 min and 5 min, respectively, in the presence of 1 litre/min of dry flowing air containing ozone generated by the Pillar Technologies ozone generator. The concentration of ozone near the film surface was *ca.* 14 mg/litre.

(3) Ozone only

The PP and PET samples were exposed to ozone in 2 litres/min of dry flowing air in the absence of UV for 30 min and 50 min, respectively. The concentration of ozone near the film surface was *ca.* 11 mg/litre. Even though the concentration of ozone in the ozone-only condition was less than with the UV/air plus ozone, the samples were actually exposed to more ozone per unit time in the ozone-only condition because the flow rate of ozone-laden air was higher.

2.5. Analysis

The surface-treated PP and PET samples were analyzed by X-ray photoelectron spectroscopy (XPS or ESCA), contact-angle measurements, and Fourier-transform IR (FTIR) spectroscopy. ESCA spectra were obtained on both a Hewlett-Packard Model 5950B spectrometer and a Surface Science Laboratories SSX-100 spectrometer. Both spectrometers use a monochromatic Al K_α photon source. Spectra were taken at electron take-off angles with respect to the surface of 18°, 37–38°, and 68° to generate a non-destructive depth profile of the surface region. The SSX-100 spectrometer was focused to an X-ray spot size of 600 μm with a pass energy of 150 eV. Sample charging was controlled using the flood gun/screen technique [33]. Spectra were referenced with respect to the 285.0 eV carbon 1s level observed for hydrocarbon. Based on previous studies of corona-treated PP and PET films [8], we estimate that the typical standard deviation of the O/C atomic ratios obtained from ESCA is ± 0.02 for treated PP samples and ± 0.03 for treated PET films. Repetitive analysis of films treated using the other surface-modification techniques described in this study confirm these expected standard deviations. Note that these standard deviations indicate the reproducibility of both the treatments and the ESCA technique. Because PET is an oxygen-containing polymer, the oxygen uptake from surface treatment is expressed as 'ΔO/C', the difference between the ESCA O/C atomic ratio of a treated sample and the O/C ratio of untreated PET. This method of presenting the ESCA data for PET enables easier comparisons to the PP experiments. The ΔO/C values are simply the increase in the amount of oxygen detected by ESCA.

Measurements of the advancing and receding contact angles in air of deionized, filtered water were made using the sessile-drop method on a Ramé-Hart contact-angle goniometer. The surface tension of the water was measured at 72.6 mN/m using a Cahn DCA-322 microbalance. Measurements of the advancing drop were carried out 15–30 s after placement of the drop, while receding measurements were made 15–30 s after the receding liquid front became stationary. Typical standard deviations for the contact-angle measurements were 2–3°.

Contact angles are generally considered to be affected by both changes in surface chemistry and changes in surface topography [34, 35]. Although the untreated films used in this study were not perfectly smooth, none of the surface treatments appeared to *alter* the degree of surface roughness as observed using scanning electron microscopy at a resolution of <50 nm, corresponding to a magnification of ×30 000. Surface topographical changes having dimensions of <100 nm should not significantly affect contact-angle measurements [35]. Therefore, we attribute all changes in contact angle solely to chemical modifications of the polymer surface.

The advancing water contact angle is most sensitive to the low-energy (unmodified) components of the surface, while the receding angle is more sensitive to the high-energy, oxidized groups introduced by the surface treatments [35]. Therefore, the receding contact angle is actually the measurement most characteristic of the modified component of the surface. In addition, in many industrial coating processes, liquids are physically forced to wet a film surface; whether or not the coating continues to uniformly wet the film during the drying or curing of the coating can depend upon the wetting characteristics of the substrate that are probed by the receding contact angle measurement. Therefore, it is important to examine both the advancing and receding contact angles on all surface-modified materials.

The infrared spectra were collected using a Perkin-Elmer Model 2000 Fourier-transform Infrared Spectrometer (FTIR) equipped with a Spectra Tech Baseline Horizontal Attenuated Total Reflectance attachment (ATR). This ATR accessory uses a 45° zinc selenide crystal as the contact crystal. The sample compartment was purged with <5% RH air to maintain stable atmospheric conditions. The data are the result of at least nine co-added scans with a resolution of 4 cm^{-1}.

The depth of penetration of FTIR-ATR spectroscopy depends on a number of factors including the angle of incidence, the refractive index of the crystal, and the refractive index of the sample. The depth of penetration also depends on the wavelength of the incident beam; at higher wavenumbers, the depth of penetration is much less than that at lower wavenumbers. Using PET with a refractive index of 1.64 as an example, the depth of penetration is 0.91 μm at 4000 cm^{-1}, 2.2 μm at 1666 cm^{-1}, and 5.6 μm at 650 cm^{-1}. Using a refractive index of 1.55 for the PP film, the depth of penetration at 1666 cm^{-1} is 1.38 μm.

Samples of modified and unmodified PET were analyzed by FTIR-ATR spectroscopy. We emphasized examination of the region 1590–1880 cm^{-1}, where effects on the carbon–oxygen double bond would be seen. Unfortunately, strong absorbances in this region from the bulk PET polymer masked any changes that may have occurred at the surface after treatment.

The PP was much better suited to analysis by FTIR spectroscopy. For these spectra, the region 1530–1880 cm^{-1} was examined for evidence of oxidation in the form of carbon–oxygen double bonds. We used the band-ratio technique to quantify the effects of treatment. This technique, which has been used extensively in both transmission and ATR infrared spectral analyses, involves integrating the area under the peak of interest and then dividing that area by the area under a reference peak. This technique enables the analyst to correct for variations in sample size and amount of contact with the crystal. When using a band-ratio technique, the reference peak chosen should be unaffected by the surface modification and should occur at a wavelength in close proximity to the peak of interest so that the sampling depths are similar. In our IR studies of modified PP, we chose the signal arising from a combination of bending and scissoring motions in the carbon–hydrogen bonds as our reference peak. Although it is possible that surface treatment causes changes to these bonds, the carbon–hydrogen bond is the most suitable reference peak in the IR spectrum.

In order to account for most of the oxidation functionalities that may occur, the area under the IR peaks between 1530–1840 cm^{-1} was divided by the area under the reference peak between 1410–1530 cm^{-1}. The ratio of areas obtained for the untreated PP was then subtracted from the ratios obtained for the various treated samples. The error in determining these IR ratios is estimated to be 0.02–0.03 units.

3. RESULTS

The operating conditions for each process were chosen to yield the specific ESCA ΔO/C atomic ratios at a 38° electron take-off angle of 0.12 for PP and 0.13 for PET. These target levels of surface oxidation correspond to the typical O/C ratios obtained with high levels of corona treatment [8]. Corona treatment was chosen as the standard for comparison because it is the most widely used surface oxidation process. Although we targeted specific atomic ratios, in practice the multi-sample-average O/C ratios reported in this study vary somewhat from this target.

Tables 1 and 2 summarize the contact-angle measurements and ESCA atomic ratios at a 38° electron take-off angle for the various surface-treated PP and PET films, respectively. It is immediately apparent that the corona, flame, and remote-plasma processes can rapidly oxidize polymer surfaces. The target O/C atomic ratios are easily attained at exposure times of 0.5 s or less. By contrast, the various UV/ozone treatments require orders of magnitude more exposure time to reach the same levels of surface oxidation.

As we have discussed in other published papers [6–8, 17, 36], all of these surface-oxidation treatments can generate a water-soluble surface consisting of low-molecular-weight oxidized material (LMWOM). These water-soluble components complicate the interpretation of wettability measurements made on treated samples. When mesuring the contact angle of water in air on any of the surface-oxidized PP or PET samples, dissolution of LMWOM is likely to alter the localized surface tension of the water drop. In addition, the surface energy of the LMWOM itself may be different from the insoluble underlying material. The combination of these two factors causes difficulties in interpreting the contact-angle data. Despite these problems, the contact-angle measurements do yield at least a *semi-quantitative* measure of the wettability of the surface-treated films.

Table 1.
Surface properties[a] of treated polypropylene films

Treatment	Exposure time (s)	ESCA O/C atomic ratio[b]	θ_a (°)	θ_r (°)
None	—	0.0	117	95
Corona (1.7 J/cm^2)	0.5	0.12	71	52
Corona (0.17 J/cm^2)	0.05	0.07	74	50
Flame	0.04	0.12	73	24
Remote air plasma	0.1	0.12	83	33
Ozone only	1800	0.13	85	63
UV/air	600	0.085	83	51
UV/air plus O_3	600	0.14	75	53

[a] θ_a and θ_r are the advancing and receding water contact angles, respectively, in air. Typical standard deviations are 2–3° for the contact-angle measurements and ±0.02 O/C for the treated PP films.

[b] At a 38° electron take-off angle with respect to the surface.

Table 2.
Surface properties[a] of treated PET films

Treatment	Exposure time (s)	ESCA ΔO/C atomic ratio[b]	θ_a (°)	θ_r (°)
None	—	0.0	70	53
Corona (1.7 J/cm^2)	0.5	0.13	42	< 5
Corona (0.17 J/cm^2)	0.05	0.10	48	< 5
Flame	0.04	0.12	29	< 5
Remote air plasma	0.1	0.11	44	< 5
Ozone only	3000	0.07	51	20
UV/air	300	0.15	48	15
UV/air plus O_3	300	0.17	41	14

[a] θ_a and θ_r are the advancing and receding water contact angles, respectively, in air. Typical standard deviations are 2–3° for the contact-angle measurements and ±0.03 O/C for the treated PET films.

[b] At a 38° electron take-off angle with respect to the surface.

3.1. Corona, flame, and plasma

Corona, flame, and remote-plasma treatments all attain approximately the same ESCA O/C ratios at a 37–38° electron take-off angle at comparable exposure times. Despite this, the wettabilities of the various surface-treated films are somewhat different. For PP, the flame-treated and corona-treated materials have the lowest advancing water contact angles, while the flame-treated films clearly have the lowest receding angle. For PET, the flame-treated films have both the lowest advancing and receding contact angles. A general observation is that flame treatment tends to generate the most wettable films.

As shown in Tables 1 and 2, most of the surface-modified films have approximately the same O/C atomic ratio in the outermost 5.5 nm, which corresponds to the approximate effective sampling depth of ESCA using an Al K_α photon source at a 37–38° electron take-off angle [37, 38]. Thus, the *overall* extent of surface oxidation in the outermost 5–6 nm of the polymer must be about the same for all of the treated films. Yet the wettabilities of the various samples can be significantly different. This apparent discrepancy may be explicable if we consider the different sampling depths of ESCA and contact-angle measurements. While ESCA at a 37–38° take-off angle effectively samples 5–6 nm deep into the polymer, contact-angle measurements are affected by only the outermost 0.5 nm of a polymer surface [35]. Thus, the surface chemistry probed by contact-angle measurements can be considerably different than the deeper-sampling ESCA technique.

Some idea of the trends in surface oxidation as a function of distance from the outer surface can be obtained by performing ESCA at a variety of electron take-off angles. The effective sampling depth of ESCA is proportional to sin θ, where θ is the electron take-off angle with respect to the surface. The use of 18°, 38°, and 68° take-off angles enables sampling depths of approximately 3, 5.5, and 8 nm, respectively, assuming that the carbon 1*s* electron escape depth is *ca.* 3 nm [38]. While the 3 nm sampling depth at an 18° take-off angle is still a factor of six greater than the region probed by contact-angle measurements, the use of still smaller take-off angles is generally not recommended [39].

The ESCA O/C atomic ratios of the surface-modified PP and PET films as a function of the electron take-off angle are shown in Table 3. The flame-treated polymers clearly have a larger gradient in oxygen concentration than any other surface-oxidized film. As compared with the corona-treated materials, the flame-treated samples have considerably more surface oxidation in the outermost 2–3 nm, even though the total amount of incorporated oxygen is about the same in the outermost 5–6 nm. Our observation agrees with Briggs *et al.* [9] who found that the depth of oxidation for flame treatments is generally shallower than for corona treatments. This greater extent of oxidation near the outermost surface of the treated films is the most likely reason for the improved wettability of the flame-treated materials.

Another possible explanation of the better wetting properties of flamed polymers could be a different mix of chemical functional groups in the region probed by contact-angle measurements. Different oxidized functionalities will make varying contributions to the wettability of a surface-oxidized polymer. The oxidized groups on flamed PP and PET may be a more-wettable 'mix' than those present on corona-treated or

Table 3.
ESCA ΔO/C atomic ratios as a function of the electron take-off angle with respect to the surface for surface-treated PP and PET

Treatment	ESCA electron take-off angle 18°	38°	68°
PP			
None	0	0	0
Corona (1.7 J/cm^2)	0.14	0.12	0.09
Corona (0.17 J/cm^2)	0.09	0.07	0.06
Flame	0.20	0.12	0.08
Remote air plasma	0.14	0.12	0.10
Ozone only	0.10	0.12	0.08
UV/air	0.10	0.09	0.06
UV/air plus O_3	0.12	0.13	0.10
PET			
None	0.00	0.00	0.00
Corona (1.7 J/cm^2)	0.17	0.13	0.10
Flame	0.22	0.12	0.06
Remote air plasma	0.15	0.13	0.07
Ozone only	0.06	0.06	0.04
UV/air	0.09	0.12	0.07
UV/air plus O_3	0.10	0.17	0.11

plasma-treated films. Unfortunately, the high-resolution carbon 1*s* ESCA spectra of the various surface-oxidized materials are not easily resolved into component peaks nor are they different enough between the treatments to enable us to draw any conclusions about the relative concentrations of the various oxidized groups on the treated films. Nevertheless, small differences in the mix of chemical species on the surface are not likely to be that important in determining surface wettability when compared with the large differences in the O/C atomic ratio of the outer 2–3 nm of the treated films.

A final factor that can influence surface wettability is the water-soluble LMWOM present on these treated surfaces. As discussed above, because of the presence of LMWOM, 3–5° differences in the contact angles should not be considered too significant.

Another technique that can provide some information on the relative depth of surface oxidation is FTIR-ATR. For the surface-modified PP films, Fig. 4 shows the ratio of the area of IR peaks associated with oxidized functional groups to the area of IR peaks associated with unmodified carbon–hydrogen bonds. As described earlier, our IR analysis probes to depths of *ca.* 1.4 μm with PP. Thus, as expected, IR examines depths much greater than even ESCA at high electron take-off angles. As shown in Fig. 4, the amount of oxidation detected by IR on PP treated by flame, low-energy corona, or plasma is quite small. This indicates that all three of these processes oxidize only a fairly thin surface region, more of the order of the ESCA sampling depth (10 nm or so).

3.2. UV/ozone treatments

In contrast to corona, flame, and plasma, the various combinations of UV and ozone exposure are relatively slow methods of oxidizing polymer surfaces, requiring many minutes of exposure to reach our target levels of oxidation. For PP films, the O/C atomic ratio generated by the UV/air treatment is somewhat lower than the target value of 0.12. Even at exposure times of 30 minutes, PP samples treated in UV/air do not have O/C ratios greater than *ca.* 0.09. By contrast, both the UV/air plus ozone and the ozone-only treatments generated a PP surface with an O/C ratio of 0.12 for exposure times of 10 and 30 minutes, respectively. The reduced O/C atomic ratios on the PP treated with UV/air were not accompanied by any noticeable differences in either the high-resolution carbon 1*s* ESCA spectra or in the IR spectra taken of these samples. As seen in Fig. 4, the IR ratio for the UV/air sample falls between that for the UV/air-plus-ozone sample and the ozone-only sample.

For PET films, 5 minute exposures to UV/air or to UV/air plus ozone were sufficient to attain our target values for the ESCA O/C atomic ratio. For identical exposure times, the use of supplemental ozone appears to give somewhat more oxygen incorporation than the simple UV/air treatment. For PET, we were unable to obtain ΔO/C ratios above *ca.* 0.10 with the ozone-only treatment, even when the exposure times were lengthened to 90 minutes.

From this information, we can conclude that exposure to ozone without UV is a more effective method for treating PP than for treating PET. Conversely, treatment of the PP

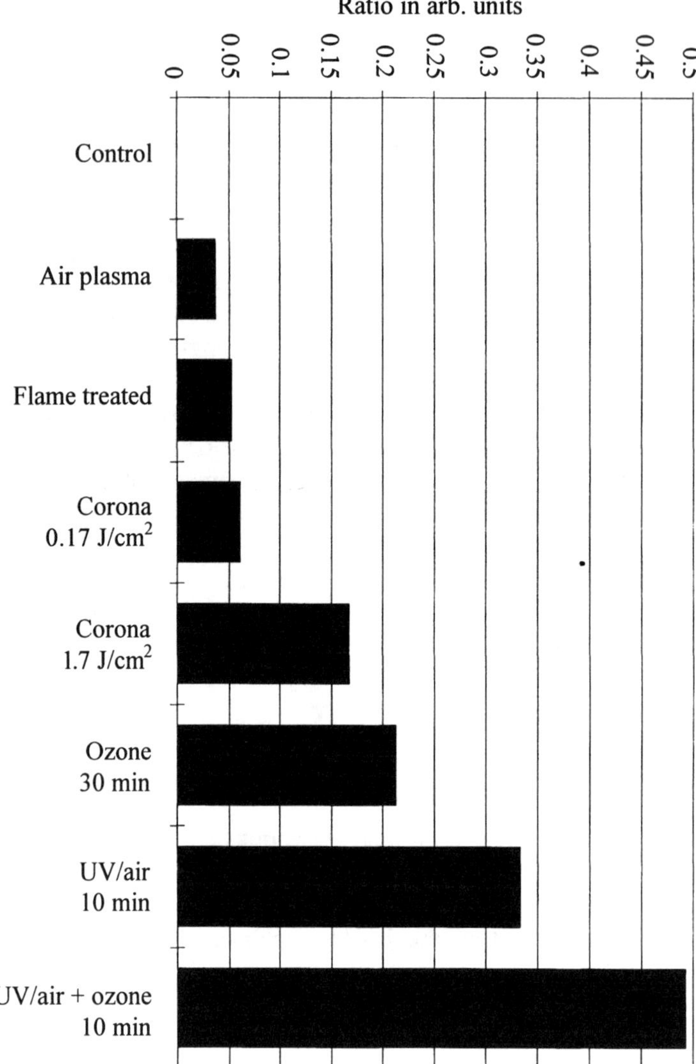

Figure 4. The IR ratios of the surface-oxidized PP films. The IR ratio is the ratio of the area of the IR peaks between 1530–1840 cm^{-1} associated with oxidation of the PP to the area of the reference carbon–hydrogen stretching, bending, and scissoring peaks between 1410–1530 cm^{-1}.

in the UV/air condition, where the concentration of ozone is relatively low, is fairly ineffective, while for PET the UV/air and the UV/air plus ozone conditions lead to extensive oxidation of the PET surface. This behavior is not surprising if we consider the chemical nature of the two polymers. PET strongly absorbs UV radiation while PP is largely transparent to UV at many wavelengths [40, 41]. Thus, the conditions involving UV exposure are more likely to induce surface-chemical changes on PET than on PP. In general, polyolefins appear to be more readily oxidized by active oxygen than PET [42–44] so we might expect PP to be more reactive toward ozone than PET.

The high-resolution carbon 1*s* spectra of the various treated PET films were quite similar, with no significant formation of functional groups having binding energies higher than the ester/carboxylic acid group. The primary changes in the carbon 1*s* spectra that occurred after treatment were the appearance of a peak associated with carbonyl-type (C=O) functionalities and the growth in the relative areas of the C–O and C–OOR/C–OOH peaks.

Again, it is useful to examine the IR data (Fig. 4) and the ESCA depth profiles (Table 3) for the UV/ozone samples. The ESCA depth profiles for the various UV/ozone treatments do not show the strong decrease in the extent of oxidation as a function of depth from the outer surface that we observed for the corona-, flame-, and plasma-treated materials. In fact, for the various UV/ozone treatments, there is little indication of enhanced oxidation at the outermost surface of the treated films. In most instances, the ESCA O/C atomic ratios at the 18° electron take-off angle are actually smaller than the O/C ratios at the 38° take-off angle. This may indicate a region of enhanced oxidation lying just beneath the surface of these UV/ozone-treated films. We do not believe that hydrocarbon contamination is the cause of the lowered O/C atomic ratios at the 18° take-off angle. Extensive previous studies [7] of corona-treated films showed that hydrocarbon contamination of the outermost surface does not occur under the storage conditions used in this study. There is no evidence that adventitious hydrocarbon is present on any of the surface-treated films studied. Therefore, we conclude that the UV/ozone-treated films are not enriched with oxidized functionalities at or very near the outermost surface.

The wettabilities of the various UV/ozone-treated PP and PET films seem to be poorer than the films treated by corona, flame, or plasma. This poorer wettability of the UV/ozone-treated films could easily be attributed to the less-oxidized outermost surface seen on these films. Of course, as mentioned above, the presence of LMWOM and the relative distribution of the various types of oxidized functionalities can also influence the contact angles.

For the various UV/ozone treatments, the ESCA depth profiles shown in Table 3 imply little variation in the concentration of oxidized groups as a function of distance in the outermost 10 nm or so of the materials. This lack of a gradient in the ESCA O/C atomic ratio is often associated with a surface treatment that extends much farther into the bulk of the treated material. The IR data (Fig. 4) show that the PP exposed to the two UV/air treatments have the highest ratios of oxidized species to hydrocarbon species. This indicates that oxidized functionalities are present relatively deep into the treated polymers.

4. DISCUSSION

This study illustrates the need to exercise care in the comparison of ESCA compositional data with contact-angle measurements. Relative to the extreme surface sensitivity of contact-angle measurements, ESCA can be considered to be a 'bulk' analysis. To properly characterize a surface-modified polymer, some idea of the concentration gradients within the near-surface region is needed. The non-destructive depth profiling capability of ESCA performed at several electron take-off angles is ideally suited for obtaining this type of information. While even shallow-angle ESCA is not as surface-sensitive as contact-angle measurements, an ESCA depth profile can detect compositional gradients that can help to explain wettability results. IR examination of the surface is also useful if there are modifications to the surface that extend to depths of 100 nm or more.

Of the treatments studied, flame appears to be the 'shallowest'; that is, the oxygen incorporated by the treatment is most concentrated near the outer surface of the treated film. Much of the oxygen incorporated by flame treatment is concentrated in the outermost 2–4 nm. Corona and plasma treatments appear to affect somewhat deeper areas of the polymer, but still alter the surface to depths of only *ca.* 10 nm. At the other extreme, because the modifications can be detected by IR and because the ESCA depth profiles reveal little gradient in the O/C atomic ratios as a function of sampling depth, the UV/ozone treatments must modify farther into the bulk polymer, in the range of 10 nm to possibly 1 μm or so.

Because of the great potential for undesired thermal damage to the film, flame treatment must be a short-exposure-time process when used to treat commodity polymer films. The flame treatments used in this work, with an exposure time of the film to the flame of only 0.04 s, are clearly representative of flame processing as practiced in the commodity-film business. Longer-exposure-time flame treatments are not easily performed nor are they meaningful from an industrial perspective. By comparison, both plasma and corona treatments can easily span an extremely wide range of exposure times and normalized energies, ranging from hundredths of a second to many minutes and from 0.1 J/cm^2 (or less) to 100 J/cm^2 (or more). In this study, we specifically chose to emphasize exposure times of less than 1 s as these are the types of treatments that are most amenable to industrial use. As shown in this study, short-exposure air plasma and corona treatments lead to shallow depths of oxidation of the order of that achieved by gas flaming. Other published studies of corona and plasma treatments [6, 45] have shown that a much deeper oxidation can be achieved with longer exposures to the corona or the air plasma. In these cases, the depth of oxidation is more comparable to that obtained by the various UV/ozone treatments. The 1.7 J/cm^2 corona treatment is intermediate between the low-energy treatments commonly used in industry and the long-exposure-time corona and plasma treatments often discussed in the literature.

The results of this study clearly demonstrate the primary reason that corona and flame treatments are the most widely used methods for modifying the surfaces of industrial polymer films. Both corona and flame are effective, extremely rapid methods of surface oxidation that can be performed at atmospheric pressure.

While corona treatment and flame treatment are well known as efficient means of oxidizing polymer surfaces, the ability of plasma treatments to rapidly oxidize polymer surfaces is not as widely appreciated. Many of the published studies of plasma treatment used exposure times of the polymer to the plasma from 5 s up to several minutes. This can lead to the erroneous impression that air-plasma treatment is an inherently slower method of modifying polymer surfaces than corona or flame treatments. However, as shown in this study and in the work of Foerch *et al.* [17,

29–30], remote plasmas can oxidize polymer surfaces as rapidly as either corona or flame treatments. For the surface oxidation of commodity polymer films, plasma treatments cost considerably more than corona and flame treatments because of the need to use vacuum equipment. However, for applications such as metallization, in which the polymer is subjected to further vacuum processing, remote air plasmas offer an effective, rapid means of oxidizing polymer surfaces. In addition, unlike coronas, plasmas are easily amenable to the treatment of three-dimensional objects. While flame treatment is routinely used to modify three-dimensional objects, plasmas can be used to treat temperature-sensitive materials that cannot be subjected to the heat of a flame treatment.

Although the reaction times for the UV/ozone treatments are significantly longer than those needed to attain equivalent extents of oxidation by corona, flame, or plasma treatments, an advantage to the UV/ozone approach is its ability to modify the surface of oddly-shaped, three-dimensional objects at atmospheric pressure. The reaction times necessary for sufficient modification are not as long as previously reported [24–27] so that the UV/ozone process is well-suited to applications where corona or flame cannot be used or where plasma treatment is too costly.

5. CONCLUSIONS

Corona, flame, and remote-plasma processes can rapidly oxidize polymer surfaces, attaining XPS ΔO/C atomic ratios of greater than 0.10 in less than 0.5 s. In contrast, the various UV/ozone treatments require orders of magnitude more exposure time to reach the same levels of surface oxidation. While corona treatment and flame treatment are well known as efficient means of oxidizing polymer surfaces, the ability of plasma treatments to rapidly oxidize polymers is not so widely appreciated.

Of the techniques studied, flame treatment most readily yields a highly wettable surface. Flame treatment also appears to have the oxygen incorporated by the treatment most concentrated near the outer surface of the film. Corona and plasma treatments appear to affect somewhat deeper areas of the polymers. Both plasma and corona treatments can be used over a wide range of exposure times, thereby generating various depths of oxidation. At the other extreme, the UV/ozone treatments tend to modify considerably deeper into the bulk polymers.

Acknowledgements

We would like to thank John Thomas for his assistance in obtaining some of the ESCA results and Stewart McIntyre and Joan Strobel for their helpful insight.

REFERENCES

1. D. Briggs, in: *Surface Analysis and Pretreatment of Plastics and Metals*, D. M. Brewis (Ed.), Ch. 9, pp. 199–226. Macmillan, New York (1982).
2. D. Briggs, in: *Practical Surface Analysis by Auger and X-ray Photoelectron Spectroscopy*, D. Briggs and M. P. Seah (Eds), Ch. 9, pp. 383–391. John Wiley, Chichester (1983).

3. R. J. Ashley, D. Briggs, K. S. Ford and R. S. A. Kelly, in: *Industrial Adhesion Problems*, D. M. Brewis and D. Briggs (Eds), Ch. 8, pp. 213–218. John Wiley, New York (1985).
4. L. J. Gerenser, J. F. Elman, M. G. Mason and J. M. Pochan, *Polymer* **26**, 1162–1166 (1985).
5. J. M. Pochan, L. J. Gerenser and J. F. Elman, *Polymer* **27**, 1058–1062 (1986).
6. M. Strobel, C. Dunatov, J. M. Strobel, C. S. Lyons, S. J. Perron and M. C. Morgen, *J. Adhesion Sci. Technol.* **3**, 321–335 (1989).
7. J. M. Strobel, M. Strobel, C. S. Lyons, C. Dunatov and S. J. Perron, *J. Adhesion Sci. Technol.* **5**, 119–130 (1991).
8. M. Strobel, C. S. Lyons, J. M. Strobel and R. S. Kapaun, *J. Adhesion Sci. Technol.* **6**, 429–443 (1992).
9. D. Briggs, D. M. Brewis and M. B. Konieczko, *J. Mater. Sci.* **14**, 1344–1348 (1979).
10. F. Garbassi, E. Occhiello and F. Polato, *J. Mater. Sci.* **22**, 207–211 (1987).
11. F. Garbassi, E. Occhiello, F. Polato and A. Brown, *J. Mater. Sci.* **22**, 1450–1456 (1987).
12. I. Sutherland, D. M. Brewis, R. J. Heath and E. Sheng, *Surf. Interface Anal.* **17**, 507–510 (1991).
13. E. Sheng, I. Sutherland, D. M. Brewis and R. J. Heath, *Surf. Interface Anal.* **19**, 151–156 (1992).
14. E. Papirer, D. Y. Wu, G. Nanse and J. Schultz, in: *Chemically Modified Surfaces*, H. A. Mottola and J. R. Steinmetz (Eds), pp. 369–384. Elsevier, Amsterdam (1992).
15. E. Papirer, D. Y. Wu and J. Schultz, *J. Adhesion Sci. Technol.* **7**, 343–362 (1993).
16. E. M. Liston, L. Martinu and M. R. Wertheimer, *J. Adhesion Sci. Technol.* **7**, 1091–1127 (1993).
17. R. Foerch, G. Kill and M. J. Walzak, *J. Adhesion Sci. Technol.* **7**, 1077–1089 (1993).
18. S. L. Kaplan and P. W. Rose, in: *Plastics Finishing and Decoration*, D. Satas (Ed.), pp. 91–100. Van Nostrand Reinhold, New York (1991).
19. R. G. Bosisio, C. F. Weissfloch and M. R. Wertheimer, *J. Microwave Power* **7**, 327–346 (1972).
20. T. Sugano, *Applications of Plasma Processing to VLSI Technology*. Wiley, New York (1985).
21. M. Morra, E. Occhiello and F. Garbassi, *J. Adhesion Sci. Technol.* **7**, 1051–1063 (1993).
22. J. R. Vig, *J. Vac. Sci. Technol. A* **3**, 1027–1034 (1985).
23. N. S. McIntyre, R. D. Davidson, T. L. Walzak, R. Williston, M. Westcott and A. Pekarsky, *J. Vac. Sci. Technol. A* **9**, 1355–1359 (1991).
24. J. Peeling and D. T. Clark, *J. Polym. Sci.: Polym. Chem. Ed.* **21**, 2047–2055 (1983).
25. J. F. Rabek, J. Lucki, B. Rånby, Y. Watanabe and B. J. Qu, in: *Chemical Reactions on Polymers*, pp. 187–200. American Chemical Society, Washington, DC (1988).
26. J. Peeling, G. Courval and M. S. Jazzar, *J. Polym. Sci.: Polym. Chem. Ed.* **22**, 419–428 (1984).
27. T. Sasamoto, P. Desai and A. S. Abhiraman, *Polym. Mater. Sci. Eng.* **67**, 393–394 (1992).
28. M. J. Walzak, S. Flynn, R. Foerch, J. M. Hill, E. Karbashewski, A. Lin and M. Strobel, *J. Adhesion Sci. Technol.* (submitted).
29. R. Foerch, N. S. McIntyre and D. H. Hunter, *J. Polym. Sci.: Polym. Chem. Ed.* **28**, 193–204 (1990).
30. R. Foerch, J. Izawa and G. Spears, *J. Adhesion Sci. Technol.* **5**, 549–564 (1991).
31. N. Novis, M. Chtaib, J. Vohs, J. J. Pireaux, R. Caudano, P. Lutgen and G. Feyder, in: *Metallized Plastics 1: Fundamental and Applied Aspects*, K. L. Mittal and J. R. Susko (Eds), pp. 193–204. Plenum Press, New York (1989).
32. E. C. Onyiriuka, L. S. Hersh and W. Hertl, *J. Colloid Interface Sci.* **144**, 98–102 (1991).
33. C. E. Bryson III, *Surface Sci.* **189/190**, 50–59 (1987).
34. S. Wu, *Polymer Interface and Adhesion*. Ch. 1. Marcel Dekker, New York (1982).
35. M. Morra, E. Occhiello and F. Garbassi, *Adv. Colloid Interface Sci.* **32**, 79–116 (1990).
36. J. M. Hill, E. Karbashewski, A. Lin, M. Strobel and M. J. Walzak, *J. Adhesion Sci. Technol.* (submitted).
37. A. Dilks, in: *Characterization of Polymer Molecular Structure by Photon, Electron, and Ion Probes*, T. J. Fabish, H. R. Thomas and D. Dwight (Eds). Ch. 18. American Chemical Society, Washington, DC (1981).
38. R. F. Roberts, D. L. Allara, C. A. Pryde, D. N. E. Buchanan and N. D. Hobbins, *Surf. Interface Anal.* **2**, 5–10 (1980).
39. C. Fadley, *Prog. Solid State Chem.* **11**, 265–343 (1976).

40. B. G. Rånby and J. F. Rabek, *Photodegradation, Photooxidation, and Photostabilization of Polymers: Principles and Applications*. Ch. 4. Wiley, London (1975).
41. N. S. Allan and M. Edge, *Fundamentals of Polymer Degradation and Stabilization*, p. 78. Elsevier, Barking, UK (1992).
42. F. Clouet and M. K. Shi, *J. Appl. Polym. Sci.* **46**, 1955–1966 (1992).
43. M. K. Shi, J. Christoud, Y. Holl and F. Clouet, *Pure Appl. Chem.* **A30**, 219–239 (1993).
44. S. J. Moss, A. M. Jolly and B. J. Tighe, *Plasma Chem. Plasma Process.* **6**, 401–416 (1986).
45. L. J. Gerenser, *J. Adhesion Sci. Technol.* **7**, 1019–1040 (1993).

Polymer Surface Modification: Relevance to Adhesion, pp. 253–272
K. L. Mittal (Ed.)

UV and ozone treatment of polypropylene and poly(ethylene terephthalate)

MARY JANE WALZAK,[1,*] SUSAN FLYNN,[2] RENATE FOERCH,[1,†] JOSEPHINE M. HILL,[1] ELIZABETH KARBASHEWSKI,[2,‡] AMY LIN[1] and MARK STROBEL[3]

[1]*Surface Science Western, Western Science Centre, The University of Western Ontario, London, Ontario, N6A 5B7, Canada*
[2]*3M Canada, PO Box 5757, London, Ontario, N6A 4T1, Canada*
[3]*3M Company, 3M Center, Building 208-1, St. Paul, MN 55144, USA*

Revised version received 13 March 1995

Abstract—The effects of exposure to ultraviolet (UV) light and ozone, separately and in combination, were investigated with respect to polypropylene (PP) and poly(ethylene terephthalate) (PET) surfaces. Three combinations of UV light and ozone were studied: ozone only, UV light in air (producing ozone), and UV light in air (producing ozone) supplemented by additional ozone in the incoming air. The effect of the exposure time of the PP and PET to each treatment was studied. The samples were analyzed by X-ray photoelectron spectroscopy (XPS) to determine the surface composition, and by dynamic contact angle to determine the water wettability. The results showed that the effect of the treament was dependent on the properties of the exposed polymer, with PET being more sensitive to the UV light and PP being more sensitive to the reactive species in the gas. The exposure times studied ranged from 1 to 90 min. By monitoring the oxygen uptake levels, we were able to determine that surface modification occurred within minutes. The possible reactive species and mechanisms are discussed.

Keywords: Polypropylene; poly(ethylene terephthalate); ultraviolet light; ozone; atomic oxygen; XPS; contact angle; surface modification.

1. INTRODUCTION

Printing and adhesion problems are often encountered in the commercial use of many polymers. This is a result of the inherently low surface energy of polymers. Gas-phase surface modification processes such as corona discharge, plasma treatment, and

---

*To whom correspondence should be addressed.

†Present address: Institut für Microtechnik Mainz, Physical Technology Division, Carl Zeisstr. 18–20, 55129 Mainz, Germany.

‡Present address: Novacor Chemicals Ltd., Technical Centre, 3620 32nd Street N.E., Calgary, Alberta, T1Y 6G7, Canada.

flame treatment can oxidize polymer surfaces, making them more wettable. There are, however, drawbacks to each of these surface modification methods.

Corona-discharge treatment is the method of choice for modifying commodity polymer film because it is an effective process for high-speed, in-line applications. Corona treatment, however, is difficult to use on three-dimensional objects. Plasma treatment can be very effective in treating non-flat objects but this technique requires a high-vacuum environment and, as a result, has higher capital costs than corona or flame treatments. Flame treatment is a relatively low-cost process but it can be uneven when used on three-dimensional objects and may not be suitable for thermally sensitive materials. Awareness of these limitations has highlighted the need for additional surface modification techniques.

Ozone-generating UV light has been used for many years to clean organic contaminants from various surfaces [1]. The individual and combined effects of UV light and ozone on polyethylene films were studied by Peeling and Clark [2], using longer-exposure-time experiments than will be discussed in this paper. Treatment times were of the order of hours in an oxygen atmosphere, resulting in fully oxidized surfaces within the depth sampled by X-ray photoelectron spectroscopy (XPS). Peeling and Clark state that '...the time scale for ozonation and photo-oxidation to produce fully oxidized surfaces is of the order of several hours.' A fully photo-oxidized surface of polyethylene was defined as having an XPS O:C atomic ratio of 0.17. Peeling *et al.* [3] studied the photo-oxidation of poly(ethylene terephthalate) in an oxygen atmosphere at treatment times of up to 30 min.

Rabek *et al.* [4] also studied the effects of ozone and atomic oxygen on polypropylene in the presence of UV light. Samples were irradiated for 1–8 h in order to characterize the mechanisms of oxidation in weathering situaions. Extensive oxidation in these experiments led to embrittlement of the polymer, indicating some effect on the bulk polymer. Dasgupta [5] investigated the oxidation of polypropylene and polyethylene in water through which ozone was passed. While this proved to be an effective method for modification of the polymer for inclusion in wood-pulp blends, the technique has limited applicability.

The mechanisms for ozone formation and destruction in the presence of UV light have been extensively studied and reported [6]. The low-pressure mercury vapour lamp used in this study generates two main wavelengths of interest, one at 184.9 nm and one at 253.7 nm. The intensity ratio of the light emitted at 184.9 nm relative to that at 253.7 nm is generally between 0.1 and 0.3.

Molecular oxygen, $O_2(^3\Sigma_g^-)$, absorbs the 184.9 nm light to form an excited-state oxygen molecule, $O_2^*(^3\Sigma_u^-)$:

$$O_2(^3\Sigma_g^-) + h\nu \text{ (184.9 nm)} \rightarrow O_2^*(^3\Sigma_u^-). \tag{1}$$

This excited state, $O_2^*(^3\Sigma_u^-)$, overlaps with the repulsive $O_2^*(^3\Pi_u)$ electronic state. This overlap allows a transition from the higher-energy electronic state to the lower-energy repulsive state:

$$O_2^*(^3\Sigma_u^-) \rightarrow O_2^*(^3\Pi_u). \tag{2}$$

This repulsive excited state can then dissociate to form two ground-state oxygen atoms (^3P):

$$O_2^*(^3\Pi_u) \rightarrow 2O(^3P). \tag{3}$$

The ground-state oxygen atom then reacts with molecular oxygen to form ozone:

$$O(^3P) + O_2(^3\Sigma_g^-) \rightarrow O_3(^1A). \tag{4}$$

The overall quantum yield for a reaction is a measure of its efficiency. The quantum yield for ozone formation is approximately 2, meaning that for every photon absorbed, two ozone molecules are formed.

Ozone has a strong absorption maximum at 255 nm and two weak ones at 305 and 440 nm. Ozone absorption of the 253.7 nm light emitted by the mercury vapour lamp is responsible for its photodecomposition to $O(^1D)$ and molecular oxygen:

$$O_3(^1A) + h\nu \text{ (253.7 nm)} \rightarrow O(^1D) + O_2(^1\Delta_g \text{ or } ^1\Sigma_g^+). \tag{5}$$

The quantum yield for ozone photolysis at 253.7 nm is approximately 0.9 ± 0.2; therefore, there is one oxygen atom formed for each ozone molecule destroyed. These are the main pathways of ozone formation and decomposition. The overall quantum yield of ozone, taking into account both the formation and the decomposition reactions, is only 0.5, meaning that it takes two photons of light to generate one ozone molecule.

Our preliminary studies showed that polymer surfaces could be effectively modified in terms of minutes rather than hours. The objective of this work was to investigate the efficacy of various combinations of UV light and ozone in modifying the surface chemistry of polypropylene and poly(ethylene terephthalate) films. As part of this, the effects of distance from the UV source, air flow rate, and ozone concentration were studied.

2. EXPERIMENTAL

2.1. Materials

Thermally extruded, biaxially oriented polypropylene (PP) and poly(ethylene terephthalate) (PET) films were used in this study. The 0.03 mm thick PP film was produced from a homopolymer resin ($M_w = 1.9 \times 10^5$, polydispersity = 6.0). The base resin contains 500–1000 ppm each of an inorganic acid scavenger and a high-molecular-weight phenolic anti-oxidant. The PP was produced on a tenter-frame film line and quenched at 45°C prior to orientation. The machine-direction (MD) draw ratio was 5.2:1 and the transverse-direction (TD) draw ratio was 9:1.

The 0.1 mm thick PET film was fabricated from a homopolymer resin with a degree of polymerization of 100 ($M_w = 25\,000$). The PET base resin contains approximately 300 ppm of residual metal catalyst and 50 ppm of organometallic thermal stabilizer. Biaxially oriented film was produced on a tenter-frame film line (MD draw ratio

3.4:1, TD draw ratio 4.6:1). The finished PET film had a glass transition temperature of 83°C.

As discussed in a previous paper [7], the additives contained in these films are not surface-active as they were not detectable by X-ray photoelectron spectroscopy (XPS) or static secondary ion mass spectroscopy (SSIMS) on treated or untreated polymers.

2.2. Reactor design

The UV light source was an ozone-generating mercury vapour grid lamp manufactured by BHK Inc. The radiation intensity at a wavelength of 254 nm was 12–15 mW/cm^2 at a distance of 2.54 cm from the lamp. A Pillar Technologies ozone generator, using a silent discharge, was used to generate supplemental ozone. A schematic diagram of the UV/ozone system is shown in Fig. 1. Film samples, 12×13 cm in size, were treated on a shelf mounted 1.5 cm from the UV lamp. The treatment took place in a sealed, 15 cm diameter stainless steel reactor. A cooling coil immersed in an ice/brine bath chilled the incoming air prior to entry into the ozone generator (whether or not the ozone generator was in use). The gas used was Liquid Carbonic extra-dry air (dew point −65°C). Air flow rates were controlled by an MKS Inc. mass flow controller. The air inlet was situated so that the air flowed between the sample surface and the UV lamp. A thermocouple was used to monitor the reactor air temperature between the sample and the lamp. The ozone concentration generated by the ozone generator was approximately 0.536% by volume at a flow rate of 2000 sccm as measured by a Resonance Ltd. Model UVTRANS2 ozone transmissiometer. The accuracy of the ozone concentration measurement was ±10%.

2.3. Treatment regimes

The reaction conditions for each of the treatments are listed below. The reactor temperature was controlled to less than 35°C during the reactions.

UV/air. The ozone present was only that generated by the 184.9 nm radiation from the mercury vapour lamps. The dry-air flow rate was 1000 sccm.

UV/air + ozone. The gas coming into the reactor was pretreated in the ozone generator so that it had significant quantities of ozone. The mercury vapour lamps were also on. The dry-air flow rate was 1000 sccm.

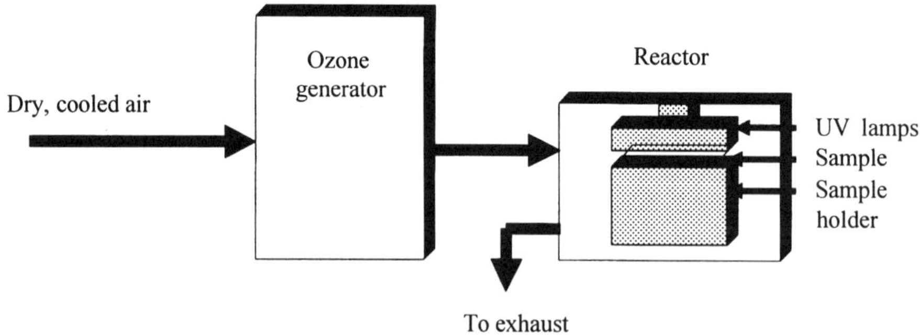

Figure 1. Schematic diagram of the reactor.

Ozone only. The air coming into the reactor at a flow rate of 2000 sccm was pretreated in the ozone generator. This was the only source of ozone because the mercury vapour lamps were off.

To determine the rate of modification, the samples were treated for 1–90 min under the above conditions. They were then analyzed for oxygen uptake using XPS and for wettability using contact-angle measurements.

To determine whether there were any effects attributable to unstable conditions in the reactor before equilibrium was reached, the reactor was modified to allow conditions in the reactor to stabilize for 5 min before exposing the polymer surface. This was achieved by employing a shutter system that allowed us to keep the polymer surface covered with a thick glass plate while the reactor was coming to steady state.

2.4. Distance study

The effect of distance from the UV source on oxygen uptake and water wettability was investigated by treating PET and PP films at distances of 1.5, 3, 6, and 8.5 cm under *UV/air* and *UV/air + ozone* conditions. The PET was treated for 5 min with *UV/air* and 4 min with *UV/air + ozone*. The PP was treated for 15 min with *UV/air* and 10 min with *UV/air + ozone*. Each run was done in duplicate for a total of 32 experiments. Within each group, the order of the experiments was random. The treatment times corresponded to the optimum times chosen for the subsequent washing and ageing studies [8].

2.5. Analysis

The surface-treated PP and PET samples were analyzed within 4 h of treatment by XPS and dynamic contact-angle measurements. XPS spectra were obtained on a Surface Science Laboratories SSX-100 spectrometer using a monochromatic Al K_α photon source. Spectra were taken at an electron take-off angle relative to the surface of 37°. Broad scans were obtained using a 600 μm X-ray spot size with a pass energy of 150 eV, while high-resolution scans were obtained using a spot size of 150 μm and a pass energy of 50 eV. Sample charging was controlled using the flood gun/screen technique [9]. Spectra were referenced with respect to the 285.0 eV carbon 1s level observed for hydrocarbons. For each treatment time and regime, between 2 and 11 replicates were generated and analyzed. The typical standard deviation for the O:C ratio of the treated PET is $\pm$0.04, and for the treated PP $\pm$0.03. These standard deviations are a combination of the reproducibility of the treatment and the precision of the XPS analysis. Because PET is an oxygen-containing polymer, the oxygen uptake from the surface treatment is expressed as 'ΔO:C', the difference between the XPS O:C atomic ratio of a treated sample and the O:C ratio of untreated PET. This method of presenting the XPS data for PET enables easier comparisons with the PP experiments. The ΔO:C values are simply the increase in the amount of oxygen as detected by XPS.

Advancing and receding contact angles in air of deionized, filtered water (surface tension 72.6 mN/m) were determined using the Wilhelmy plate technique on a Cahn

Dynamic Contact Angle Analyzer Model 322. A three-layer laminate was prepared using double-coated tape (3M Scotch Brand #410 Tape), mounting the treated sides of the film outwards. To prevent contamination during the preparation of this laminate, the treated surfaces only contacted untreated films. The laminate was cut to a width of 25.4 mm and a length of approximately 25 mm. The platform speed was set at 100 μm/s with a travel distance of 15 mm. The advancing and receding contact angles were calculated by a routine supplied with the Cahn instrument that uses linear regression for the buoyancy correction. The data were not smoothed prior to analysis. The typical error in the advancing and receding contact angles measurement is $\pm 3°$.

Contact angles are affected by changes in surface chemistry and changes in surface topography [10]. Although the treated films showed some topographical features not present in the controls, these were estimated to be 70–90 nm in size as observed using scanning electron microscopy (SEM) ($\times$40 000 magnification) and atomic force microscopy (AFM) ($\times$500 magnification). Surface topographical features having dimensions of less than 100 nm should not significantly affect contact-angle measurements [10]. As a result, all changes in contact angle are attributed to changes in the surface chemistry of the treated polymers.

The advancing water contact angle is most sensitive to the low-energy (unmodified) components of the surface. The receding contact angle tends to be more sensitive to the high-energy, oxidized groups introduced by the surface treatment [10]. Therefore, the receding contact angle is most characteristic of the modified component of the surfaces. In many industrial coating processes, liquids are physically forced to wet a film surface. Whether or not the coating continues to uniformly wet the film during the drying or curing of the coating can depend on the wetting characteristics of the substrate that are probed by the receding contact-angle measurement. Therefore it is important to examine both the advancing and the receding contact angles on all surface-modified materials.

As we have discussed in other published papers [7, 8, 11, 12], surface-oxidation treatments can generate a water-soluble surface consisting of low-molecular-weight oxidized material (LMWOM). These water-soluble components complicate the interpretation of wettability measurements made on treated samples. When measuring the contact angle of water in air on any of the surface-oxidized PP or PET samples, dissolution of LMWOM is likely to alter the localized surface tension of the water. In addition, the surface energy of the LMWOM itself may be different from the insoluble underlying material. The combination of these two factors causes difficulties in interpreting the contact-angle data. Despite these problems, the contact-angle measurements do yield at least a semi-quantitative measure of the wettability of the surface-treated films.

3. RESULTS

3.1. Ozone concentration

The ozone concentration was measured using an ozone detector mounted at the gas outlet of the reactor. The lamps were at room temperature at time 0. In Fig. 2 the

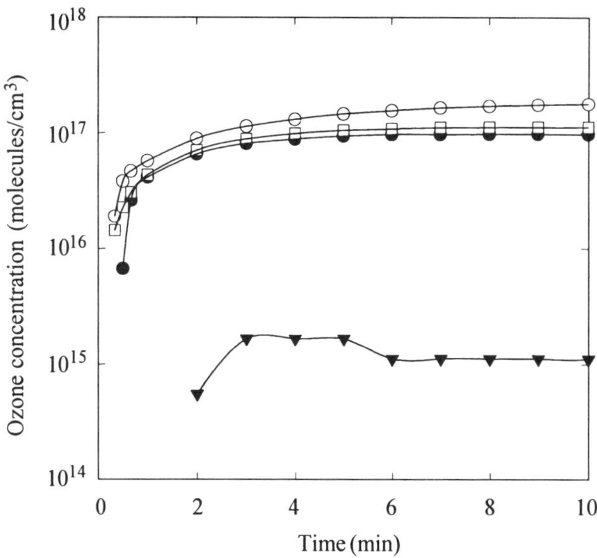

Figure 2. Change in ozone concentration as a function of time (ozone detector at the reactor gas outlet). ○, Ozone only (1000 sccm); □, Ozone only (2000 sccm); ●, UV/air + ozone (1000 sccm); ▼, UV/air (1000 sccm).

ozone concentration for each of the three treatment regimes is plotted as a function of time. Under these conditions, the concentration is highest for the *ozone-only* regime, slightly less for the *UV/air + ozone* regime, and approximately 100 times less for the *UV/air* regime. At comparable flow rates, the concentration of ozone in the *UV/air + ozone* regime is half that of the *ozone-only* regime. When the air flow rate is increased to 2000 sccm for the *ozone-only* treatment, the ozone concentration at the reactor outlet is similar to that seen in the *UV/air + ozone* regime at 1000 sccm. The ozone concentration at the reactor outlet approaches a maximum within 2–3 min of turning on the UV lights and/or the ozone generator.

To ensure that this 2–3 min lag time before a steady state is reached was not adversely affecting the interpretation of the treatment-time data, the following experiment was devised. The polymer film was covered with a thick glass plate and loaded into the reactor as described in Section 2. For each treatment regime, the reactor was allowed to come to a steady state for 5 min prior to removing the glass plate and exposing the polymer to the treatment conditions. PP and PET were exposed to each of the three treatment regimes for times less than that shown to give maximum uptake (Table 1). The O:C ratios of the polymers treated after the reactor had reached a steady state did not differ significantly from the ratios obtained under the normal, non-steady-state conditions. The ozone concentration is therefore not the limiting factor in determining the extent of reaction oxidation.

3.2. Time study: polypropylene

Figure 3 is a plot of the O:C ratio for PP as a function of the exposure time for the three treatment regimes. It is evident from this plot that the PP was oxidized when

Table 1.
Comparison of O:C ratios of PP and PET on exposure to normal treatment conditions (non-steady-state conditions) and on exposure for the same treatment times after the reactor had achieved steady-state conditions

Treatment regime	ΔO:C		
	Exposure with glass plate in place	Exposure after glass plate is removed	Normal treatment[a]
PET UV/air, 2 min, 1000 sccm	0	0.10	0.08
PET UV/air + ozone, 2 min, 1000 sccm	0	0.11	0.12
PET ozone only, 25 min, 2000 sccm	0	0.04	0.04

	O:C		
	Exposure with glass plate in place	Exposure after glass plate is removed	Normal treatment[a]
PP UV/air, 5 min, 1000 sccm	0	0.05	0.05
PP UV/air + ozone, 5 min, 1000 sccm	0	0.13	0.10
PP ozone only, 15 min, 2000 sccm	0.01	0.09	0.09

[a] Values obtained from Figs 3 and 5. These represent the normal, non-steady-state conditions.

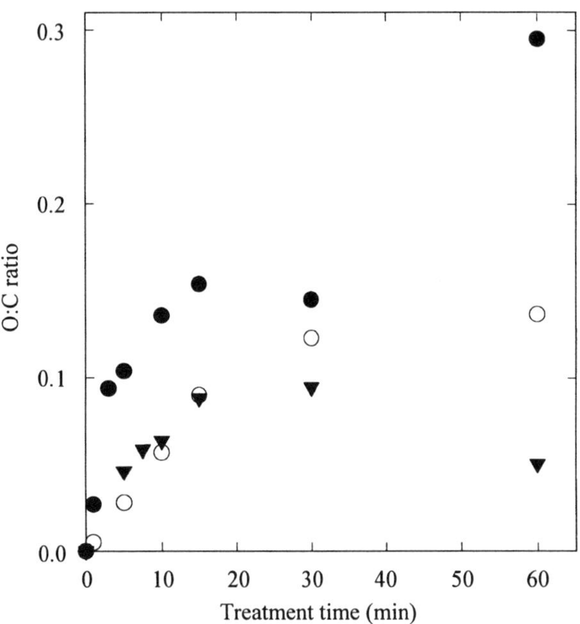

Figure 3. Effect of treatment time on the oxygen uptake on PP for different treatment regimes. ○, Ozone only (2000 sccm); ●, UV/air + ozone (1000 sccm); ▼, UV/air (1000 sccm).

treated with *UV/air*, *UV/air + ozone*, and *ozone only*, relative to the control PP, which contained no detectable oxygen.

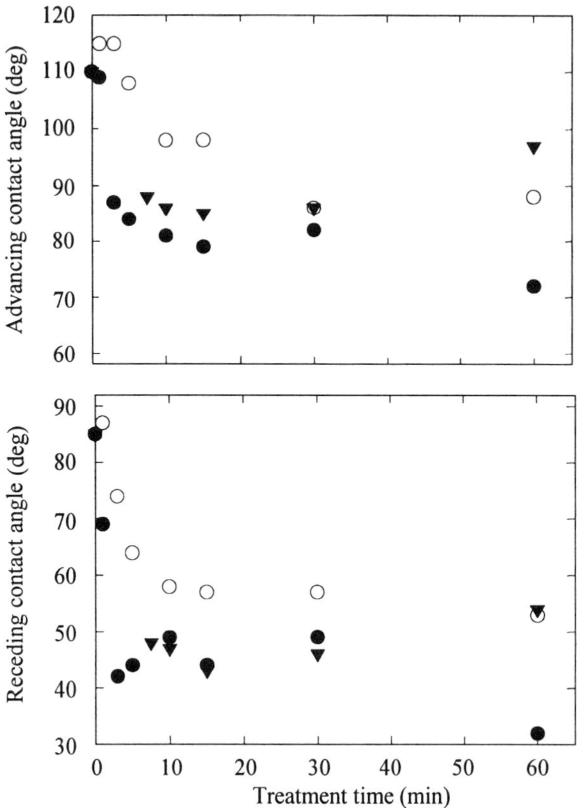

Figure 4. Effect of treatment time on the contact angles of water on PP. ○, Ozone only (2000 sccm); ●, UV/air + ozone (1000 sccm); ▼, UV/air (1000 sccm).

3.2.1. UV/air treatment. Treatment of the PP with *UV/air* led to an O:C ratio of 0.09 after 15 min of treatment, at which point the uptake levelled off and then decreased to 0.05 on treatment for 60 min. The contact-angle data plotted in Fig. 4 show a decrease in the contact angle with increasing oxygen uptake.

3.2.2. UV/air + ozone treatment. Treatment with *UV/air + ozone* resulted in an increase in the O:C ratio to approximately 0.17 after 15 min of treatment followed by a further increase in the O:C ratio to 0.3 after 60 min of treatment (Fig. 3). The advancing and receding contact angles decreased by approximately 30° and 40°, respectively, after approximately 10 min of treatment and further decreased by 10–15° after a 60 min treatment (Fig. 4). *UV/air + ozone* treament resulted in the fastest initial reaction rate and also the highest final O:C ratios.

3.2.3. Ozone-only treament. The PP samples treated with *ozone only* exhibited steadily increasing O:C ratios as a function of time and appeared to reach a plateau of 0.12 after 30 min of treatment. The advancing contact angle also reached a plateau after 30 min of treatment, while the receding contact angle leveled off after 10 min of treatment. Although the surface concentration of oxygen in the *ozone-only* treatment can be made similar to that achieved with *UV/air*, the wettabilities of the surfaces are not necessarily the same. The *ozone-only* treatment resulted in a polymer surface with consistently higher advancing and receding contact angles.

3.3. *Time study: poly(ethylene terephthalate)*

3.3.1. UV/air treatment. During *UV/air* treatment of the PET, the ΔO:C ratio reached a maximum of 0.24 after 15 min (Fig. 5). The receding contact angles were close to 0° after only 3 min of treatment (Fig. 6). The advancing contact angle decreased to a minimum of 53° after 15 min of treatment and then increased slightly to 60° for treatment times of 30 and 60 min. The behaviour of the advancing contact-angle data corresponds to the changes in the O:C ratio.

3.3.2. UV/air + ozone treatment. The *UV/air + ozone* treatment reached a similar plateau of 0.24 in terms of the ΔO:C ratio after 10 min, while the receding contact angle of the film decreased to near 0° within 3 min. The advancing contact angle decreased to approximately 50° after 10 min.

3.3.3. Ozone-only treatment. Detectable changes to the ΔO:C ratio at a take-off angle of 37° occurred only after treatment times of 30 min (Fig. 5). In contrast, the advancing contact angle decreased by 10° within the first 15 min of treatment and the receding contact angle decreased to near 0° within 3 min (Fig. 6).

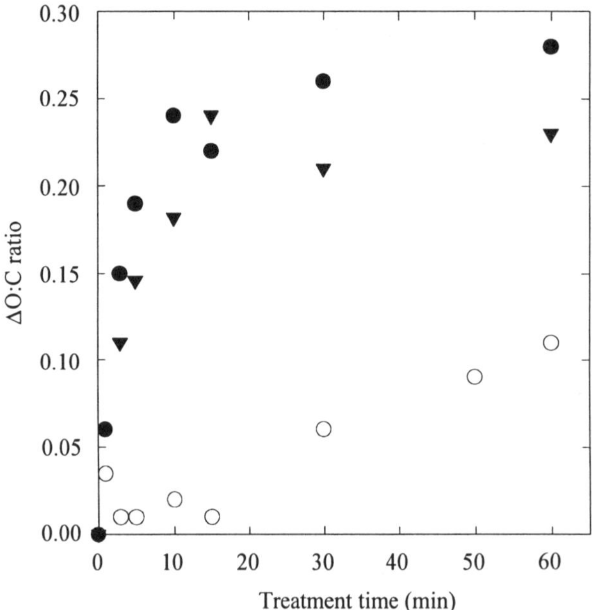

Figure 5. Effect of treatment time on the oxygen uptake on PET for different treatment regimes. ○, Ozone only (2000 sccm); ●, UV/air + ozone (1000 sccm); ▼, UV/air (1000 sccm).

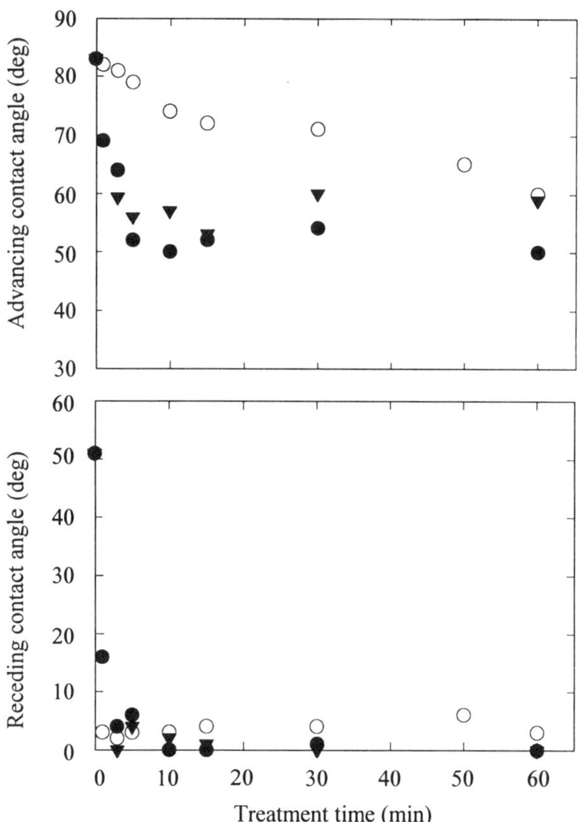

Figure 6. Effect of treatment time on the contact angles of water on PET. ○, Ozone only (2000 sccm); ●, UV/air + ozone (1000 sccm); ▼, UV/air (1000 sccm).

3.4. Distance study: polypropylene and poly(ethylene terephthalate)

Figure 7 is a plot of the ΔO:C ratios for PET as a function of the distance from the lamp to the polymer film for two treatment regimes (*UV/air* and *UV/air + ozone*). It is evident that the distance from the lamps has an effect on oxygen uptake because the oxygen content decreases significantly with increasing distance. The *UV/air + ozone* condition appears to be slightly more sensitive to distance as the change in oxygen uptake is greater than that for the *UV/air* treatment. The contact angles as plotted in Fig. 8 show similar trends of less water wettability as the distance from the lamps increases. The receding contact angles for the *UV/air + ozone* treatment initially increase more quickly than the *UV/air* treatment but are very similar at a distance of 8.5 cm from the lamps.

With PP, the changes are very similar. The O:C ratio tends to drop off as a function of the distance from the lamps and again the *UV/air + ozone* treatment seems to be more sensitive to this change. The drop in O:C ratio is quite sharp between 1.5 and 3 cm (Fig. 9). The contact angles reflect the O:C ratios in that they increase

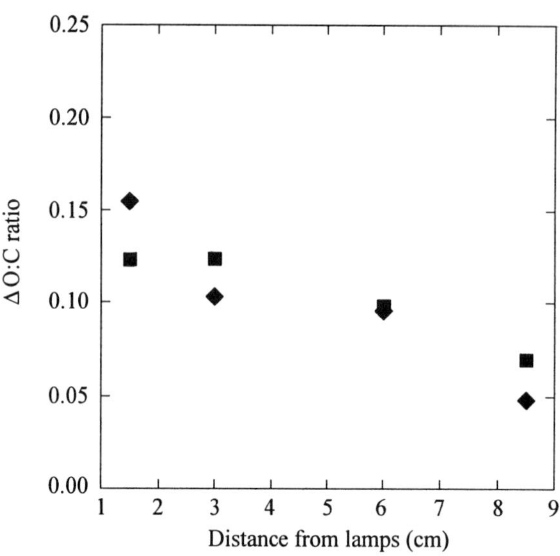

Figure 7. Effect of the distance from the lamps on the oxygen to carbon ratio for PET. ♦, UV/air + ozone (1000 sccm); ■, UV/air (1000 sccm).

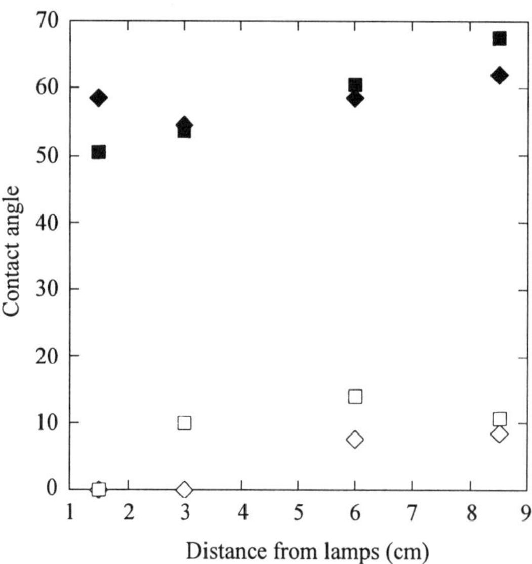

Figure 8. Effect of the distance from the lamps on the contact angles of water on PET. ♦, UV/air, advancing angle; ◇, UV/air, receding angle; ■, UV/air + ozone, advancing angle; □, UV/air + ozone, receding angle.

with distance from the lamp and the receding contact angles for the *UV/air + ozone* treatment increase dramatically when the treatment distance is increased from 1.5 to 3 cm (Fig. 10).

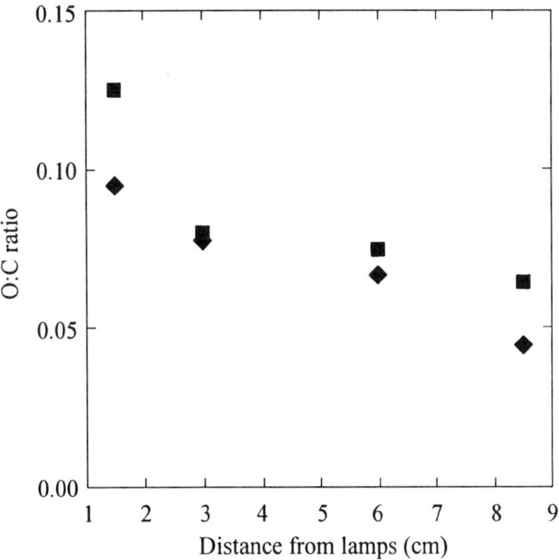

Figure 9. Effect of distance from the lamps on the oxygen to carbon ratio for PP. ◆, UV/air; ■, UV/air + ozone.

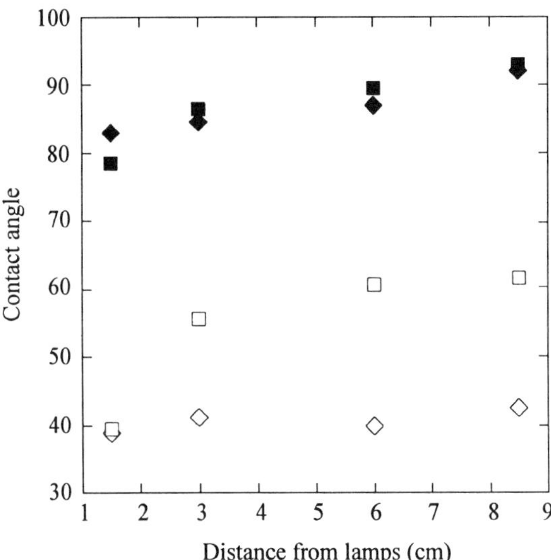

Figure 10. Effect of the distance from the lamps on the contact angles of water on PP. ◆, UV/air, advancing angle; ◇, UV/air, receding angle; ■, UV/air + ozone, advancing angle; □, UV/air + ozone, receding angle.

4. DISCUSSION

4.1. Discussion: polypropylene

During the photo-oxidative treatment of polymers, the major reactive gaseous species are ozone, atomic oxygen, and singlet molecular oxygen [6]. All of these species are present in a large excess of ground-state triplet molecular oxygen. Ozone, atomic oxygen, and singlet molecular oxygen are constantly formed and destroyed in the reactor by the UV light.

A comparison of the concentration of ozone in the *ozone-only* and *UV/air + ozone* regimes indicates the effect of the UV light on the incoming ozone. In the *UV/air + ozone* case, the ozone concentration should be similar to the *ozone-only* case at the same flow rate. The concentration is, instead, half of that value. According to equation (5), the ozone photodecomposes to atomic oxygen $O(^1D)$ with a quantum efficiency of 0.9 ± 0.2. Therefore the decrease in the ozone concentration in the reactor in the presence of UV light should be approximately equal to the concentration of atomic oxygen formed in the reactor.

Using the three different treatment regimes of *ozone only*, *UV/air*, and *UV/air + ozone* allows us to evaluate the relative importance of ozone, UV light, and atomic oxygen, respectively, in modifying the surface of PP and PET. With the *ozone-only* treatment, the concentration of ozone is very high: 1.4×10^{17} molecules/cm^3 at a flow rate of 2000 sccm in the absence of UV light. These conditions of high ozone concentration and the absence of UV light allow us to evaluate the effects of ozone without the complication of significant quantities of atomic oxygen formed by the absorption of UV light by the ozone or absorption of UV light by the polymer itself.

In the *UV/air* treatment, we can study the effects of UV light in the presence of low concentrations of atomic oxygen and ozone. The formation of ozone under these conditions gives a quantum yield of 0.5 [13] because of the competing formation and destruction reactions. In this treatment, the concentration of ozone is approximately 1×10^{15} molecules/cm^3, 100 times less than *ozone only* or *UV/air + ozone*. Using this concentration of ozone and the path length in the reactor in the Beer–Lambert law [6] gives the amount of 254 nm light transmitted to the surface of the polymer. At this concentration of ozone, 98% of the 254 nm light is transmitted. If one calculates the amount of 185 nm light transmitted under these same conditions, one finds that 93% of the 185 nm light is transmitted to the surface of the polymer. Therefore, with the *UV/air* conditions, there are low concentrations of ozone and atomic oxygen and high transmission of both the 185 and the 254 nm lines of the UV light.

The *UV/air + ozone* regime allows us to differentiate the effects of high levels of atomic oxygen from the effects of high levels of ozone alone. Under the conditions of this regime, approximately 2×10^{17} molecules of ozone per cm^3 enter the reactor and 50% of these are photodecomposed to atomic oxygen. Using the Beer–Lambert law and this concentration of ozone, we find that only 3% of the 254 nm light is transmitted to the polymer surface while 93% of the 185 nm light is transmitted. Therefore, in this regime we have high concentrations of ozone and atomic oxygen, but the transmission of the 185 nm light is the same as the UV/air regime while the transmission of 254 nm light is significantly reduced.

Initially, the treatments of PP with *ozone only* and *UV/air* appear to be very similar with respect to their oxygen uptake rates (Fig. 3) but differences soon become apparent as the treatment time is increased. The contact-angle measurements for the PP treated with *ozone only* are much higher than those for the *UV/air* regime in the first 15 min when the oxygen uptakes were the same. At longer treatment times, the oxygen uptake on the PP treated with *ozone only* remained at a fairly high level while the concentration of oxygen on the PP treated with *UV/air* started to decrease. The significant difference between the two treatments is the presence of UV light with low levels of atomic oxygen and ozone in the *UV/air* treatment as opposed to the high levels of ozone and no UV light in the *ozone-only* treatment. The data indicate that the treatment with *ozone only* yields a surface which has less of a high-energy component than the *UV/air*, at least in the initial stages of treatment.

We believe that this difference is a direct result of the different reaction conditions to which the polymer is subjected. The *ozone-only* treatment is causing a fast initial reaction of ozone with PP to form polymer peroxy radicals that subsequently react more slowly to form carbonyls, carboxylic acids, and hydroperoxides. This type of reaction does not result in the significant amounts of chain scission necessary to form LMWOM. We would therefore expect the ozone reactions with PP not to result in significant amounts of chain scission. This interpretation is consistent with the mechanism discussed by Rabek *et al.* [4]:

$$
-CH_2-\underset{\displaystyle CH_3}{\underset{|}{CH}}- \; + \; O_3 \longrightarrow
\begin{cases}
\longrightarrow -\overset{\displaystyle OO^{\bullet}}{\overset{|}{CH}}-\underset{\displaystyle CH_3}{\underset{|}{CH}}- \; + \; {}^{\bullet}OH \\[2ex]
\longrightarrow -CH_2-\overset{\displaystyle OO^{\bullet}}{\overset{|}{\underset{\displaystyle CH_3}{\underset{|}{C}}}}- \; + \; {}^{\bullet}OH
\end{cases}
$$

where the initial fast reaction of ozone with PP results in a polymer peroxy radical ($POO^{\bullet}$). This radical undergoes subsequent slow reactions to form carbonyls, carboxylic acids, and hydroperoxides:

$$
POO^{\bullet} \longrightarrow
\begin{cases}
\longrightarrow -\overset{\displaystyle O}{\overset{\|}{C}}-CH_2- \\[2ex]
\longrightarrow -\overset{\displaystyle O}{\overset{\|}{C}}-OH \\[2ex]
\longrightarrow -\overset{\displaystyle OOH}{\overset{|}{CH}}-CH_2-
\end{cases}
$$

Only one of these slow reactions results in overall chain scission, while the remaining reactions result in addition to the chain. One therefore would not expect changes in the O:C ratio or the contact angle on water washing since there is no significant formation of LMWOM. Previous studies have shown that water washing of corona-treated PP samples having LMWOM at the surface usually results in a loss of oxygen and an increase in the contact angle [11, 12]. In fact, with *ozone-only* treated PP we observed no effect of water washing on the O:C ratio or on the contact angle [8], a slow decrease in the contact angle with treatment time, and a slight gradient in the O:C ratio of the angle-resolved XPS data [14] of the sample treated for 30 min. These results support the reaction mechanisms of chain addition rather than scission. The effects of ageing and water washing on PP modified by each of these three treatments were studied and the results are published separately [8].

The *UV/air* treatment, with the high transmission of UV light and low levels of reactive species, should result in much more chain scission and crosslinking of the polymer as well as oxidation at the surface [15]. We saw embrittlement of the polymer under these conditions after only 30 min. The mechanisms of reactions to form LMWOM and crosslinked chains are a mixture of reactions involving UV light, ozone, atomic oxygen, and molecular oxygen, and have been discussed elsewhere [4, 15]. Water washing of the *UV/air* sample treated for 15 min results in a 40% loss of oxygen at the surface and essentially no change in the contact-angle measurement [8]. The evident loss of oxygen coupled with the lack of change in the contact angle indicates that the treatment is deeper than the 10 nm probed by the XPS [14], that a significant amount of oxygen at the surface is tied up in the LMWOM, and that the material beneath the LMWOM is functionalized to a significant degree. This functionalized material below the LMWOM can be made up of higher-molecular-weight oxidized chains which would not be water-soluble. In addition to being functionalized, there is evidence that the material below the LMWOM is also crosslinked because of the embrittlement seen in the polymer after exposure to *UV/air*.

Rabek *et al.* [4] propose a mechanism involving atomic oxygen in the ground state because they believe that any $O(^1D)$ formed in the photodecompostion of ozone will be rapidly deactivated by collision with excess ozone molecules. From their Experimental section it appears that the mercury lamps were 30 cm from the samples. In our reactor, the samples were 1.5 cm from the lamps and the concentration of ozone was 10 to 100 times smaller than that used by Rabek *et al.* We believe that the $O(^1D)$ is deactivated by indiscriminate reaction with the polymer surface and there are subsequent reactions with ozone and molecular oxygen at the surface that lead to the formation of LMWOM. Where the polymer radicals are formed farther into the polymer bulk, they react with each other to form crosslinked chains. This theory is supported by the increased dependence of the *UV/air + ozone* treatment on the distance from the lamps and can be explained in terms of the shorter lifetime of the atomic oxygen (Figs 9 and 10). When the polymer is closer to the lamp, the atomic oxygen has an increased chance of being formed near the polymer surface and reacting with it. As the polymer is moved farther from the lamps, there is less UV light available near the surface to form atomic oxygen. The atomic oxygen can recombine or react with other gas molecules before it reaches the polymer surface.

We reacted PP with *ozone only* at the same low concentration levels as those seen in the *UV/air* treatment, to rule out extensive reaction of ozone at these low concentration levels as being the main reactant in the surface modification under *UV/air* conditions. The PP took up only small amounts of oxygen ($O:C = 0.008$) when treated with low concentrations of ozone alone. This shows that the reactions that we see in the presence of UV light are not arising solely from the reaction with ozone. Some reaction might be expected with the UV light but PP is a non-absorber with a cut-off of less than 180 nm [16]. Although PP has been known to photodegrade in UV light, the degradation takes place on a much longer time scale than that studied here [15]. Therefore in this study we would not expect UV light to play a major role in the modification. This leaves atomic oxygen as the likely reactant to initiate the modification.

PP treated with *UV/air + ozone* is modified more quickly and to a greater extent than PP exposed to *UV/air* or *ozone only*. When analyzed by AFM, the resulting surface, after 10 min of treatment, showed distinct mounds that were removed on water washing. These mounds are very similar to those seen by Strobel *et al.* [11] on corona-treated PP using SEM. These mounds demonstrate the presence of LMWOM at the surface of the polymer after treatment. The FTIR-ATR spectra of the samples treated for 10 min show a high concentration of carbonyl functionalities in the top 1–2 μm [14]. Angle-resolved XPS analysis of these samples did not indicate a gradient of oxygen concentration in the top 10 nm [14]. Washing studies of the treated PP did not show significant changes in the O:C ratio or in the contact angle. This indicates that the treatment results in chain scission, leading to the formation of LMWOM, and that penetration of the oxygen is of the order of micrometres within 10 min. As mentioned previously, the Beer–Lambert law shows that only 3% of the 254 nm light and 93% of the 185 nm light are transmitted under these conditions. This is much less of the 254 nm light than is transmitted under *UV/air* conditions. It has also been shown that high levels of ozone do not account for this extent of modification (*ozone-only* treatment). However, exposure of the ozone to UV light results in a decrease in the ozone concentration to 50%. According to equation (5), ozone decomposes to atomic oxygen $O(^1D)$ and molecular oxygen on exposure to light of wavelength 253.7 nm. This reaction, with a quantum yield of 0.9 ± 0.2, leads to approximately equal concentrations of ozone and atomic oxygen in the reactor. It must be the atomic oxygen that modifies the surface of the PP. The high concentration of reactants in the gas phase results in a more highly modified surface than that seen with *UV/air*, as indicated by the higher O:C ratio. This modified surface, rich in LMWOM, is one through which the reactive gas species can easily pass because these reactants are then able to modify layers that are farther from the surface. In the *UV/air + ozone* treatment, the limiting factor is the amount of UV light that can be absorbed by the ozone to form atomic oxygen. For the *UV/air* treatment, the limiting factor is the small amounts of ozone and atomic oxygen that are formed in equilibrium with each other in the presence of the UV light.

4.2. *Discussion: poly(ethylene terephthalate)*

Polymers such as PET absorb light because of the presence of chromophoric functionalities in the polymer backbone (cut-off of 310 nm) [16]. The changes that can occur in PET on absorption of UV light include the formation of carboxylic acid end-groups, terminal vinyl groups, phenols, and the evolution of CO and CO_2. These changes take place by direct photocleavage via Norrish I and Norrish II scission processes involving the ester group [15, 16]. Support for the formation of these groups is provided in part by the work of Peeling *et al.* on the UV/oxygen treatment of PET at low oxygen flow rates [3]. The reactions that take place at the surface of the PET are known to be complex and give rise to a variety of functional groups; however, the groups mentioned above are those most commonly encountered at the modified surface [15]. In our work, the changes in the XPS C1*s* spectra are consistent with the formation of phenols/alcohols and carboxylic acid groups.

With respect to the ΔO:C ratio and the contact angle, PET behaved much the same under the *UV/air* and *UV/air + ozone* conditions. The ΔO:C ratio increased very quickly to levels above 0.20, surpassing a ΔO:C ratio of 0.12, which, in corona treatment, is generally considered to be a heavily modified surface, within approximately 3–5 min [14]. The graphs of contact angle as a function of the treatment time were very similar for the two treatments. The similarity of the changes in ΔO:C and contact angles for these two treatments suggests that it is the influence of the UV light rather than the concentration of ozone or atomic oxygen that is the driving force in the surface modification reaction because the concentration of ozone is 100 times greater in the *UV/air + ozone* treatment than in the *UV/air* treatment. This dependence on UV light exposure is further substantiated by the similar effects seen in the *UV/air* and *UV/air + ozone* treated PET when the distance from the lamps was changed (Figs 7 and 8). The evidence of chain scission is further supported by the washing behaviour seen in samples treated in *UV/air* for 5 min and in *UV/air + ozone* for 4 min. These samples, when water-washed, showed a decrease in the O:C ratio of approximately 65% and 70%, respectively [8]. There was a corresponding increase in the advancing contact-angle measurements.

The changes in ΔO:C ratio are similar to those seen by Peeling *et al.* [3] in their UV/oxygen treatment of PET surfaces at low flow rates. Because they saw similar changes in ΔO:C and the contact angles with respect to the oxidation time, Peeling *et al.* proposed that the oxidation of the surface is occurring uniformly throughout the outermost 5 nm of the PET. No further changes were seen on extended treatment to 30 min, except for the appearance of a peak attributed to carbonyl functionalities in the high-resolution C1*s* XPS spectrum.

The *ozone-only* treatment does not modify the surface of the PET at a rate comparable to or to the same extent as the treatments involving UV light. Because there is no UV light available to initiate the formation of functionalized groups, the rate of modification is much slower. The ΔO:C ratio remains near zero for the first 20 min of treatment. Although we did not see significant changes in the ΔO:C ratio, the advancing contact angle decreased by approximately 10° in this time and the receding contact angle decreased to near 0° levels. This type of behaviour is consistent with small amounts of oxygen being added to the outermost surface of the PET or with chain scission of the PET that results in the generation of different oxidized functional groups but without any overall increase in the oxygen content of the surface. In an attempt to resolve which mechanism was dominant at the surface on exposing PET for short reaction times, we analyzed the surface using variable-angle XPS. A sample of PET was exposed to *ozone only* for 10 min and subsequently analyzed by XPS using a take-off angle of 18°. At this take-off angle, XPS is much more surface-sensitive than it is at 37°. The ΔO:C ratio at 18° was 0.02, which is the same, within experimental error, as the values taken at 37°. This indicates that it is the chain scission mechanism which is dominant. When PET was exposed to *ozone only* for 50 min and then water-washed, the decrease in the ΔO:C ratio was only 30%, indicating that there is much less chain scission than that seen with modification involving UV light [8].

The kinetics of the reaction in the ozone-only treatment may be dependent on the ability of oxygen and ozone to penetrate the surface of PET. The kinetics of penetration of these gases may be dependent on the oxygen transmission rate of PET, which is between 4.8 and 9 $cm^3/(100\ in.^2)(24\ h)(atm/mil)$ or between 19 and 39 $\times 10^{-10}$ $m^3/(m^2)(24\ h)(atm/m)$. This is a low transmission rate when compared with the value of 150 $cm^3/(100\ in.^2)(24\ h)(atm/mil)$ or 5.9×10^{-8} $m^3/(m^2)(24\ h)(atm/m)$ for PP. We believe that the oxygen and ozone are restricted to the surface of PET until sufficient modification of the surface has taken place to allow them to penetrate to lower molecular layers. This theory is consistent with the changes in the advancing contact angle. The rate of change in the advancing contact angle decreases after 15 min of treatment, just as the ΔO:C ratio starts to increase. This indicates that the modification of PET is now taking place at the deeper molecular levels that are probed by the XPS but not by the contact-angle measurements. The receding contact angles levelled off at between 5° and 7° after 3 min of treatment and this substantiates the theory that the top monolayer is modified within the first few minutes of the treatment.

5. CONCLUSIONS

Surface modification of PP and PET films can be accomplished by exposure to UV light and ozone or to ozone alone. The length of time needed for modification is between 3 and 10 min when the reaction is performed in the presence of UV light. After approximately 3 min of treatment, the receding contact angle has decreased to near 0° and at 10 min, the ΔO:C ratio has increased to a level which would indicate a heavily modified surface if the polymer were corona-treated. Although the times needed for significant oxidation using UV light are similar for these two polymers, the mechanisms of reaction are very different. When PP is exposed to UV light, the modification is governed by the atomic oxygen formed from the ozone exposed to UV light. If the concentration of reactive oxygen species in the gas phase is low, then the radicals generated at the surface of the PP react with radicals on other polymer chains to form a crosslinked polymer. PET is directly affected by the UV light itself, causing chain scission and modification at the surface regardless of the concentration of the reactive oxygen species in the gas phase.

Reactions of PET with *ozone only* typically need longer times to reach the same contact angles and ΔO:C ratios as those obtained by the *UV/air* and *UV/air + ozone* treatments. In the case of PET, the ΔO:C ratio does not reach the same levels within 90 min but the receding contact angles do reach close to 0° within 3–10 minutes.

This apparent contradiction has been explained in terms of the relative probe depths of XPS and contact-angle measurements.

UV light in the presence of ozone can be an effective method for the treatment of polymer surfaces, whether the polymer is modified by the UV light itself or by the atomic oxygen formed by the decomposition of ozone. Ozone treatment of polymer surfaces in the absence of UV light follows a different reaction pathway, resulting in a slower treament with less chain scission.

Acknowledgements

We thank Chris Lyons, Duncan Hunter, and Stewart McIntyre for their many useful suggestions and discussions. We also thank the Natural Sciences and Engineering Research Council of Canada for their financial assistance.

REFERENCES

1. J. R. Vig, *J. Vac. Sci. Technol. A* **3**, 1027–1034 (1985).
2. J. Peeling and D. T. Clark, *J. Polym. Sci.* **21**, 2047–2055 (1983).
3. J. Peeling, G. Courval and M. S. Jazzar, *J. Polym. Sci. Polym. Chem. Ed.* **22**, 419–428 (1984).
4. J. F. Rabek, J. Lucki, B. Rånby, Y. Watanabe and B. J. Qu, in: *Chemical Reactions on Polymers*, J. L. Benham and J. F. Kinstle (Eds), pp. 187–200. American Chemical Society, Washington, DC (1988).
5. S. Dasgupta, *J. Appl. Polym. Sci.* **41**, 233–248 (1990).
6. J. G. Calvert and J. N. Pitts, Jr, *Photochemistry*, pp. 205–209. John Wiley, New York (1966).
7. M. Strobel, C. S. Lyons, J. M. Strobel and R. S. Kapaun, *J. Adhesion Sci. Technol.* **6**, 429–443 (1992).
8. J. Hill, E. Karbashewski, A. Lin, M. Strobel and M. J. Walzak, *J. Adhesion Sci. Technol.* (accepted).
9. C. E. Bryson III, *Surface Sci.* **189/190**, 501–559 (1987).
10. M. Morra, E. Occhiello and F. Garbassi, *Adv. Colloid Interface Sci.* **32**, 79–116 (1990).
11. M. Strobel, C. S. Dunatov, J. M. Strobel, C. S. Lyons, S. J. Perron and M. C. Morgen, *J. Adhesion Sci. Technol.* **3**, 321–335 (1989).
12. R. Foerch, G. Kill and M. J. Walzak, *J. Adhesion Sci. Technol.* **7**, 1077–1089 (1993).
13. B. DuRon, in: *Handbook of Ozone Technology and Applications*, R. G. Rice and A. Netzer (Eds), Vol. 1, pp. 77–84. Ann Arbor Science, Ann Arbor, MI (1982).
14. M. Strobel, M. J. Walzak, J. M. Hill, A. Lin, E. Karbashewski and C. S. Lyons, *J. Adhesion Sci. Technol.* **9**, 365–383 (1995).
15. B. Rånby and J. F. Rabek, *Photodegradation, Photo-oxidation and Photostabilization of Polymers; Principles and Applications*, pp. 134, 231–236. John Wiley, New York (1975).
16. D. J. Carlsson and D. M. Wiles, in: *Encyclopedia of Polymer Science and Engineering*, J. I. Kroschwitz (Ed.), Vol. 4, pp. 630–696. John Wiley, New York (1985).

Polymer Surface Modification: Relevance to Adhesion, pp. 273–289
K. L. Mittal (Ed.)

Effects of aging and washing on UV and ozone-treated poly(ethylene terephthalate) and polypropylene

JOSEPHINE M. HILL,[1] ELIZABETH KARBASHEWSKI,[2,†] AMY LIN,[1] MARK STROBEL[3] and MARY JANE WALZAK[1,*]

[1]*Surface Science Western, Western Science Centre, The University of Western Ontario, London, Ontario, N6A 5B7, Canada*
[2]*3M Canada, PO Box 5757, London, Ontario N6A 4T1, Canada*
[3]*3M Company, 3M Center, Building 208-1, St Paul, MN 55144, USA*

Revised version received 14 June 1995

Abstract—In this study we investigated the stability of poly(ethylene terephthalate) (PET) and polypropylene (PP) surfaces modified using three combinations of UV light and ozone: ozone only, UV light in air (producing ozone), and UV light in air supplemented by additional ozone in the incoming air. Analysis was done using X-ray photoelectron spectroscopy and dynamic contact angle measurements. Our results showed that PET film is oxidized using these treatment conditions and it changes significantly within the first week of aging and after washing with water. These changes are reflected in the decrease in the Δ(O : C) ratio and the increase in the contact angle. Conversely, PP changes very little on aging or washing. Low-molecular-weight oxidized material (LMWOM), produced on the polymer surfaces treated with UV/air or UV/air + ozone, is easily removed with water washing. On aging PET, a number of the oxidized groups at the surface disappear, seeming to migrate into the bulk. The PP, however, does not favour migration as a path to reduce the overall free energy of the system, so the oxidized groups remain at the surface. Treatment with ozone only, in the absence of UV light, is a much different modification process in terms of the mechanism and the functional groups formed on the surface. This is reflected in the aging and washing behaviour of both the PET and the PP treated with ozone only.

Keywords: Polypropylene; poly(ethylene terephthalate); UV treatment; ozone treatment; aging; washing; surface modification; X-ray photoelectron spectroscopy; ESCA; contact angle.

1. INTRODUCTION

In spite of their desirable bulk properties, the use of many organic polymers is hindered by their low surface energy; the resultant lack of wettability results in printing and adhesion problems. These problems can be overcome by using surface modification

*To whom correspondence should be addressed.

†Present address: Novacor Chemicals Ltd., Technical Centre, 3620 32nd Street NE, Calgary, Alberta T1Y 6G7, Canada.

techniques such as corona discharge treatment [1–5], plasma treatment [6–10], or ultraviolet light (UV)/ozone treatment [11–14].

The UV/ozone treatment has been shown to be very successful in modifying poly(ethylene terephthalate) (PET) and polypropylene (PP) surfaces [13]. Traditionally, UV/ozone has been, and still is, used as a means of removing organic contaminants from various surfaces [15, 16]. More recently, however, UV/ozone has been used to increase the wettability of PET [13, 14], polyethylene [11], polystyrene [12] and PP [13, 17].

A mercury vapour lamp is typically used as the UV source for the modification of polymers. The two wavelengths of interest, generated by the mercury lamps, are at 184.9 and 253.7 nm [18]. Ozone is produced when oxygen absorbs light at 184.9 nm. On the other hand, ozone absorbs light at 253.7 nm and decays to atomic oxygen and singlet oxygen, both of which are very reactive with most polymers [13, 17].

The ozone produced by the UV light can be supplemented by using an ozone generator. The ozone generator is usually a 'silent' discharge that produces ozone by applying a high voltage across a gap through which air or oxygen is flowing. Our previous study [13] showed that ozone alone (ozone only), UV light in air (UV/air), and a combination of UV light in air supplemented by ozone in the incoming air (UV/air + ozone) each have different effects on polymer surfaces and that these effects are polymer-dependent. In terms of the rate of oxygen uptake and water wettability, UV/air + ozone was the most rapid treatment, modifying PP and PET surfaces in minutes. Exposure to UV/air required slightly longer treatment times (one to five additional minutes of treatment). Treatment with ozone only, however, required 30–60 min to attain a level of surface modification comparable to that achieved with UV/air and UV/air + ozone within 4–15 min.

Unfortunately, although polymer surfaces may be sufficiently modified for use immediately after treatment, the surfaces can change when aged or washed [1, 3–8, 14, 19–24]. Aging behaviour is a function of both the polymer and the treatment regime. From the available literature, however, some generalizations can be made:

- PP does not age significantly after corona- or oxygen-plasma treatment [21, 23, 24];
- PET films age rapidly (within days) after corona, UV/ozone, or plasma treatment [1, 6, 24];
- surface rearrangement on aging is enhanced by increasing the storage temperature and is hindered by storing the samples in liquid nitrogen [8, 21–23];
- the receding contact angles of water on oxygen-plasma [7]- or air-corona [23]-treated PP and on air-corona-treated PET [24] samples that have been aged tend to remain near the initial values.

The aging behaviour of corona-treated or photo-oxidized PET has been attributed to the migration of low-molecular-weight oxidized material (LMWOM) into the bulk and to the reorientation of oxidized functionalities that were formed on relatively immobile polymer chains [1, 14, 24]. The loss of wettability that occurs after treated PET is washed is mainly due to the removal of the LMWOM. In fact, Strobel *et al.* [24]

found that 75% of the oxidized groups added to PET by corona treatment were removed by water washing.

The aging and washing behaviours of PP are different from those of PET. As PP is non-polar, it is more difficult for oxidized material to migrate into, or interact with, the bulk PP region [2, 21–23]. The changes on corona- or oxygen-plasma-treated PP that occur during aging, therefore, are primarily attributed to rearrangements in the surface region [21, 23]. On water washing, some oxidized groups are removed [2, 14, 24]; the amount of material removed is dependent on the type of surface oxidation and the extent of modification.

From these previous studies, it is evident that treated polymer surfaces do change on aging and when washed with water. Treatment stability is an important industrial consideration because polymers are often treated and then stored before use in coating processes that may involve water or organic-based solutions. This paper studies the effect of ambient aging and water washing on PET and PP films treated with UV light in the presence of air (UV/air), UV light plus ozone from a generator (UV/air + ozone), and ozone from a generator in the absence of UV light (ozone only).

2. EXPERIMENTAL

2.1. Materials

Thermally extruded, biaxially oriented polypropylene (PP) and poly(ethylene tereph-thalate) (PET) films were used in this study. The 0.03 mm thick PP film was produced from a homopolymer resin ($M_w = 1.9 \times 10^5$, polydispersity = 6.0). The base resin contains 500–1000 ppm each of an inorganic acid scavenger and a high-molecular-weight (HMW) phenolic anti-oxidant. The PP was produced on a tenter-frame film line and quenched at 45°C prior to orientation. The machine-direction (MD) draw ratio was 5.2 : 1 and the transverse-direction (TD) draw ratio was 9 : 1.

The 0.1 mm thick PET film was fabricated from a homopolymer resin with a degree of polymerization of 100 ($M_w = 25\,000$). The PET base resin contains approximately 300 ppm of residual metal catalyst and 50 ppm of organometallic thermal stabilizer. Biaxially oriented film was produced on a tenter-frame film line (MD draw ratio 3.4 : 1, TD draw ratio 4.6 : 1). The finished PET film had a glass transition temperature of 83°C.

As discussed in a previous paper [24], the additives contained in these films are not surface-active because they were not detectable by X-ray photoelectron spectroscopy (XPS) or static secondary ion mass spectroscopy (SSIMS) on either treated or un-treated polymers.

2.2. Reactor design

A sealed, 15 cm diameter, stainless steel reactor containing an ozone-generating mer-cury vapour grid lamp, manufactured by BHK Inc. (Pomona, California, USA), was used for the experiments (see Fig. 1). The radiation intensity of the lamp was

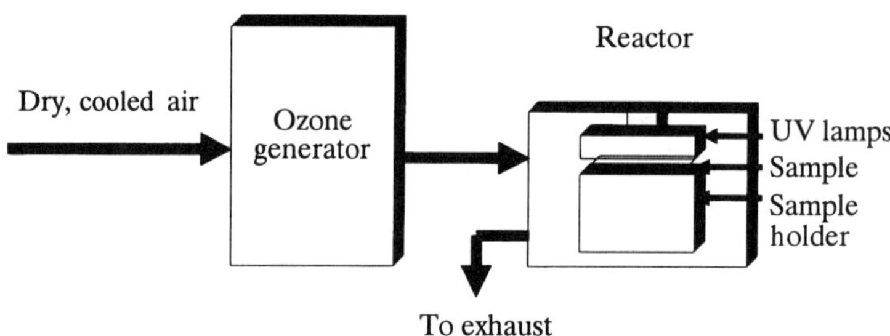

Figure 1. Schematic diagram of the reactor.

12–15 mW/cm^2 at a wavelength of 254 nm and a distance of 2.54 cm from the lamp. The supplemental ozone was provided by a Pillar Technologies Inc. ozone generator at a concentration of approximately 0.5 mmol/l at a flow rate of 1000 sccm and 0.3 mmol/l at 2000 sccm. A cooling coil immersed in an ice/brine bath chilled the incoming air prior to entry into the ozone generator (whether or not the ozone generator was in use). In all experiments, Liquid Carbonic extra-dry air was used (dew point −65°C). The air flow rates were controlled by an MKS Inc. mass flow controller. A thermocouple was used to monitor the reactor air temperature between the sample and the lamp. The polymer film samples (12 × 13 cm) were treated on a shelf mounted 1.5 cm from the UV lamp.

2.3. Treatment regimes

The reaction conditions for each of the treatments are listed below. The reactor temperature was controlled to < 35°C during the reactions.

UV/air. The ozone present was only that generated by the 184.9 nm radiation from the mercury vapour lamps. The dry-air flow rate was 1000 sccm. The ozone concentration was approximately 1×10^{15} molecules/cm^3.

UV/air + ozone. The gas coming into the reactor was pretreated in the ozone generator so that it had significant quantities of ozone. The mercury vapour lamps were also on. The dry-air flow rate was 1000 sccm. The ozone concentration was approximately 1×10^{17} molecules/cm^3.

Ozone only. The air coming into the reactor at a flow rate of 2000 sccm was pretreated in the ozone generator. This was the only source of ozone because the mercury vapour lamps were off. The ozone concentration was approximately 2×10^{17} molecules/cm^3.

The optimal treatment times for each of these regimes were determined using data from a previous study [13], so that treatment caused an increase in the O : C atomic ratio of approximately 0.12 as measured by XPS analysis (see Table 1). This extent of oxidation was chosen to provide sufficient oxygen uptake for accurate XPS

Table 1.
The treatment times necessary with each treatment regime to achieve an increase in the XPS O : C atomic ratio of *ca.* 0.12

	Treatment time (min)	
Treatment regime	PP	PET
UV/air	15	5
UV/air + ozone	10	4
Ozone only	30	50

analysis [24]. In all cases, the samples treated for shorter times showed aging and washing behaviour similar to that of the samples treated for the optimum time.

During the 28-day aging period, the samples were stored in aluminium foil at ambient conditions (20–25°C, approximately 50% relative humidity). Immediately after treatment, half of the samples were immersed in ultra-filtered water (Millipore filtering system, resistivity = 15–18 $M\Omega$ cm, surface tension = 72.6 mN/m) for 5 min with gentle intermittent agitation. After washing, the samples were allowed to dry until no water droplets were visible. The washed samples were then aged in the same manner as the unwashed samples.

Untreated (control) samples were also aged and washed along with the treated samples. The controls verified that any changes, on aging and/or washing, were caused solely by the surface treatments. Extensive previous studies [24] of corona-treated films showed that hydrocarbon contamination of the outermost surface does not occur under the storage conditions used in this study. There is no evidence that adventitious hydrocarbon is present on any of the surface-treated polymers studied.

2.4. Analysis

Oxygen content and contact angles were determined immediately after treatment and after 3, 7, 14, and 28–31 days, using X-ray photoelectron spectroscopy (XPS) and dynamic contact angle (DCA) analysis, respectively. The samples were initially analysed within 4 h of treatment. XPS spectra were obtained on a Surface Science Laboratories SSX-100 spectrometer using a monochromatic Al K_α photon source. An electron take-off angle of 37° relative to the surface was used. Broad scans were obtained, to determine elemental composition, using a 600 μm X-ray spot size and a pass energy of 150 eV, while high-resolution scans were obtained using a spot size of 150 μm and a pass energy of 50 eV. Sample charging was controlled using the flood gun/screen technique [25].

Spectra were referenced with respect to the 285.0 eV carbon 1*s* level observed for hydrocarbons. The typical standard deviation for the O : C ratio of the treated PET is ±0.04, and for the treated PP ±0.03. These standard deviations are a combination of the reproducibility of the treatment and the precision of the XPS analysis. Because PET is an oxygen-containing polymer, the oxygen uptake from the surface treatment is expressed as 'Δ(O : C)', the difference between the XPS O : C atomic ratio of a treated sample and the O : C ratio of untreated PET. This method of presenting the XPS data for PET enables easier comparisons with the PP experiments. The Δ(O : C) values are simply the increase in the amount of oxygen as detected by XPS.

Advancing and receding contact angles of deionized, filtered water (surface tension = 72.6 mN/m) on the treated and untreated films were determined using the Wilhelmy plate technique on a Cahn DCA Analyzer Model 322. A three-layer laminate was prepared using double-coated tape (3M Scotch Brand #410 Tape), to mount the treated sides of the film outwards. To prevent contamination during preparation of the laminate, the treated surfaces contacted only untreated films. Each laminate was cut to a width of 12 mm and a length of 25 mm. Samples were run in duplicate, using ultra-filtered water as the probe liquid. The platform speed was set at 100 μm/s with a travel distance of 15 mm. The advancing and receding contact angles were calculated using a routine supplied with the Cahn Instrument; the routine uses linear regression for the buoyancy correction. The data were not smoothed prior to analysis. The typical standard deviation in the advancing and receding contact angles measurement was $\pm 3°$. Note that receding contact angles below 10° are very difficult to measure accurately.

Contact angles are affected by changes in both surface chemistry and surface topography [26, 27]. Although the treated films showed some topographical features not present in the controls, these were estimated to be 70–90 nm in size as observed using scanning electron microscopy (SEM) ($\times$40 000) and atomic force microscopy (AFM) ($\times$500). Surface topographical features having dimensions of less than 100 nm should not significantly affect contact angle measurements [27]. As a result, all changes in the contact angles are attributed to changes in the surface chemistry of the treated polymers.

The advancing water contact angle is most sensitive to the low-energy (unmodified) components of the surface. The receding contact angle tends to be more sensitive to the high-energy, oxidized groups introduced by the surface treatment [27]. Therefore, the receding contact angle is most characteristic of the modified component of the surface. In many industrial coating processes, liquids are physically forced to wet a film surface. Whether or not the coating continues to wet the film uniformly during the drying or curing of the coating depends on the wetting characteristics of the substrate that are probed by the receding contact angle measurement. Therefore, it is important to examine both the advancing and the receding contact angles on all surface-modified materials.

The interpretation of contact angles can be complicated by the presence of low-molecular-weight oxidized material (LMWOM). When measuring the contact angle of water in air on a treated sample, dissolution of LMWOM is likely to alter the localized surface tension of the water. In addition, the surface energy of the LMWOM itself may be different from that of the insoluble underlying material. The combination of these two factors causes difficulties in interpreting the contact angle data. Despite these problems, the contact angle measurements do yield at least a semi-quantitative measure of the wettability of the surface-treated films.

3. RESULTS

3.1. Poly(ethylene terephthalate)

3.1.1. UV/air treatment. The results for the UV/air treatment (see Fig. 2) indicate that both aging and washing have distinct effects on the modified surface. Within the first 3 days of aging, the Δ(O:C) ratio of the unwashed samples decreased by one-half. After this initial loss, the oxygen content appeared to be stable. The contact angles changed relatively little (<10°) over the 28-day period. This demonstrates the difference between the depths probed by the two techniques. Even though there may be a great deal of restructuring, or even loss of material in the surface region, the contact angle provides information only on the outermost layer. The energy of the outermost surface, therefore, appears to remain relatively constant, while the composition of the region 5–10 nm deep has changed significantly.

After 4 weeks of aging, the receding contact angle was still over 40° lower than that of the control, even though the Δ(O:C) ratios indicated a significant loss of oxygen over this aging period. This demonstrates the sensitivity of the receding contact angle

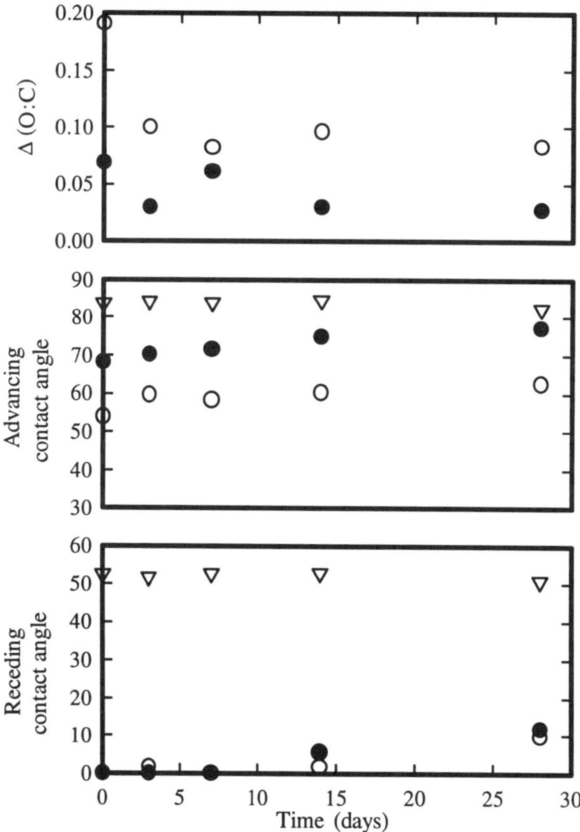

Figure 2. Effect of aging and washing on the Δ(O:C) ratio and the contact angle of water on PET treated for 5 min with UV/air. (○) Unwashed; (●) washed; (▽) control.

to the oxidized groups added to the surface [7]. The receding contact angle of the sample decreases significantly with small increases in the surface oxidation.

After washing, the oxygen content of the treated PET surface decreased by approximately 60%; this is likely caused by the dissolution of the LMWOM. Correspondingly, the advancing contact angle was approximately 10° higher for the washed samples. After aging, little change in the Δ(O : C) ratios of the washed samples was seen. Over the 28-day period, however, the contact angles (both advancing and receding) did increase by approximately 10°. This suggests that rearrangement is occurring within the XPS sampling region of the washed samples.

3.1.2. UV/air + ozone treatment. The aging behaviour of the samples treated with UV/air + ozone was very similar to that of the samples treated with UV/air (see Fig. 3). The Δ(O : C) ratio decreased sharply in the first 3 days of aging and then continued to decrease to a value slightly lower than that for the UV/air-treated sample after 28 days of aging.

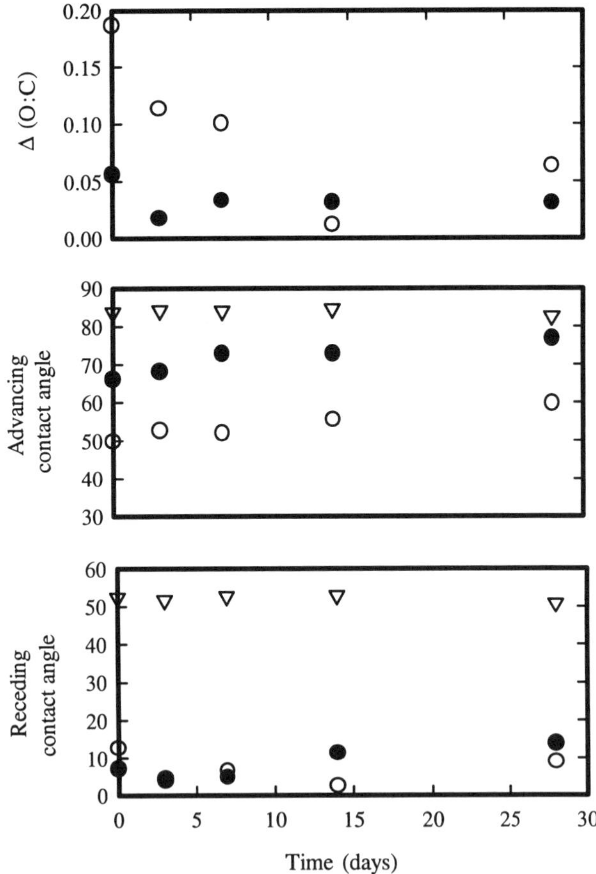

Figure 3. Effect of aging and washing on the Δ(O : C) ratio and the contact angle of water on PET treated for 4 min with UV/air + ozone. (○) Unwashed; (●) washed; (▽) control.

Slightly more oxygen was removed on washing (approximately 70%) than for the UV/air-treated sample. Again, the advancing contact angle increased by approximately 10° after washing but the receding contact angle remained near 0°.

3.1.3. Ozone-only treatment. PET is difficult to treat in an ozone-only regime because UV light is not present [13]. The terephthalate group in PET absorbs UV light, causing chain scission [13]; therefore, in the absence of UV light, the surface-oxidation reactions are much slower. After 50 min of treatment, the Δ(O : C) ratio was only 0.07 (see Fig. 4). The advancing and receding contact angles decreased significantly, but to values approximately 10° and 15° higher than for the UV/air and UV/air + ozone treatments, respectively.

Washing decreased the Δ(O : C) ratio by approximately 30% and this decrease is smaller than that seen for samples treated with UV/air or UV/air + ozone. The smaller effect of washing on the Δ(O : C) ratio of PET treated with ozone only is consistent with less LMWOM being produced in the absence of UV light due to less chain scission [13]. On aging, the washed and unwashed samples reached very similar

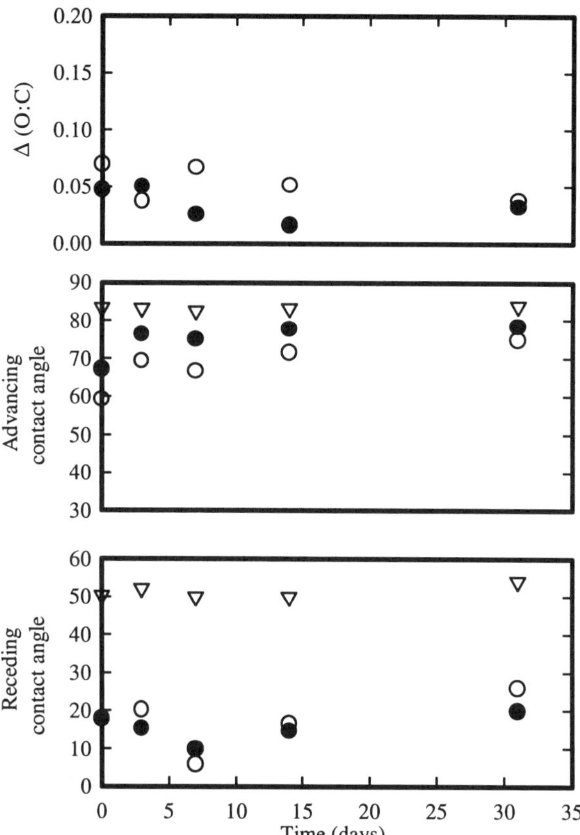

Figure 4. Effect of aging and washing on the Δ(O : C) ratio and the contact angle of water on PET treated for 50 min with ozone only. (○) Unwashed; (●) washed; (▽) control.

values for the Δ(O : C) ratio and both contact angles. For the other treatments, only the receding angles had similar values at the end of the aging period.

3.2. Polypropylene

In general, both aging and washing had a much smaller effect on the surface of the treated PP than they did on the surface of the PET.

3.2.1. UV/air. Washing did decrease the surface oxygen content of the UV/air-treated PP but had very little effect on the contact angles (see Fig. 5). The treated surfaces, either washed or unwashed, were relatively unaffected over time. The receding contact angle of the treated PP remained more than 30° lower than the control following aging for 28 days.

3.2.2. UV/air + ozone. The results for the UV/air + ozone-treated samples were essentially the same as those for the UV/air treated samples (see Fig. 6). The one

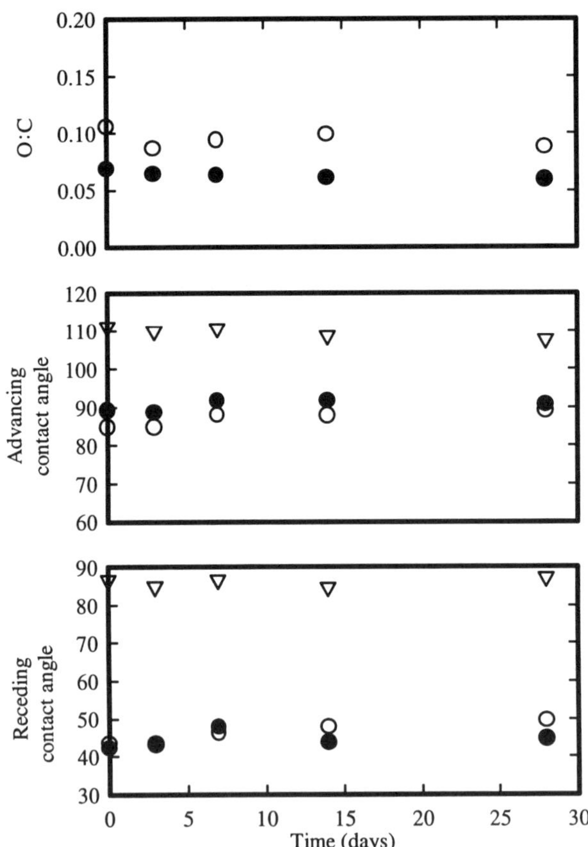

Figure 5. Effect of aging and washing on the O : C ratio and the contact angle of water on PP treated for 15 min with UV/air. (○) Unwashed; (●) washed; (▽) control.

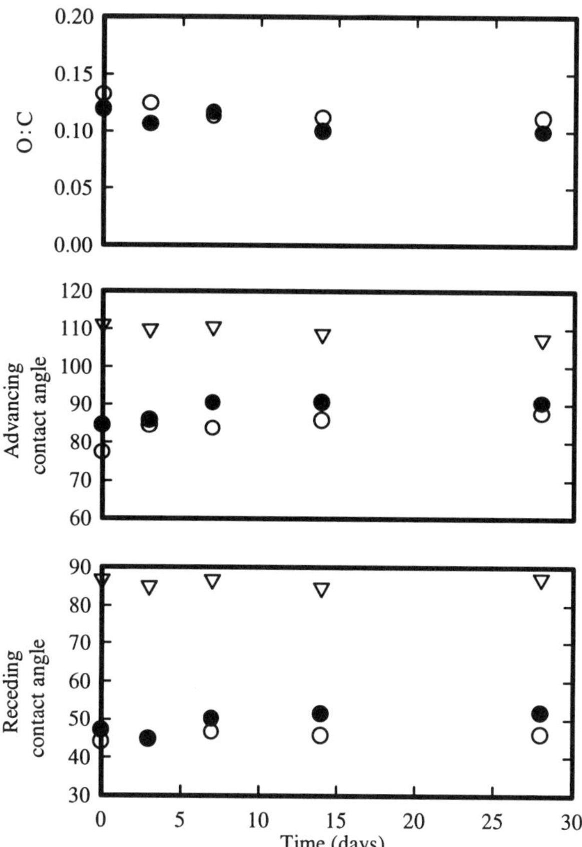

Figure 6. Effect of aging and washing on the O : C ratio and the contact angle of water on PP treated for 10 min with UV/air + ozone. (○) Unwashed; (●) washed; (▽) control.

difference was that washing had less of an effect on the O : C ratio of the UV/air + ozone-treated samples. Preliminary studies using AFM have identified the presence of LMWOM on the treated surface in the form of 'mounds' similar to those seen on corona-treated PP [2]. After washing, the mounds disappear and yet the O : C ratio has not changed. According to previous FTIR analysis [28], the modification is of the order of 1 μm deep.

3.2.3. Ozone only. The results for the ozone only-treated samples were different from those of the UV/ozone treatments. On aging the treated sample, the surface oxygen content increased (see Fig. 7). Washing, however, did not change the O : C ratio, nor did aging of the washed sample. The contact angles were the same for the washed and unwashed samples, and did not change over the 28-day aging period. The advancing contact angle was similar for all three treatments, while the receding angle was over 10° higher for the ozone only-treated sample as compared with the UV/ozone-treated sample.

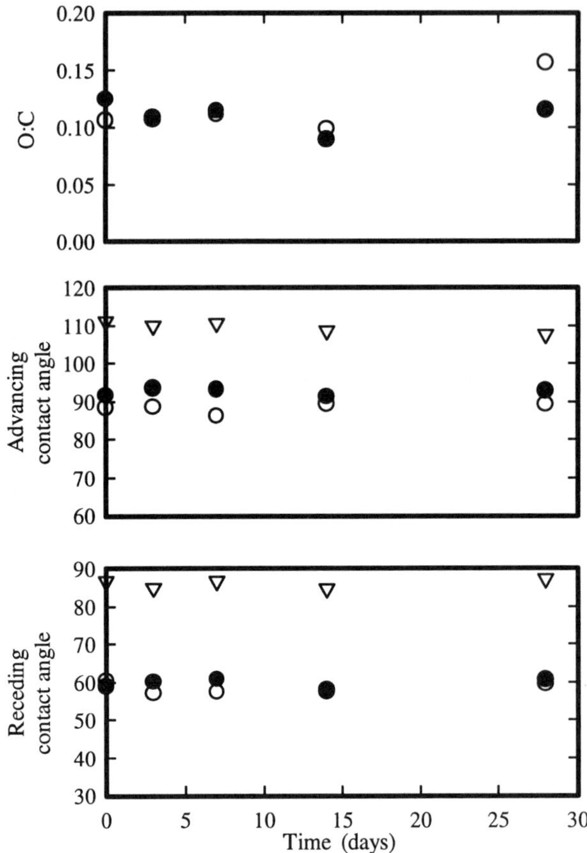

Figure 7. Effect of aging and washing on the O : C ratio and the contact angle of water on PP treated for 30 min with ozone only. (○) Unwashed; (●) washed; (▽) control.

4. DISCUSSION

4.1. Discussion: poly(ethylene terephthalate)

In this discussion UV/ozone refers to both the UV/air and the UV/air + ozone treatments. Because the results for the ozone-only treatment are distinctly different from the other treatments, the ozone-only results will be discussed separately. In general, the results of this study are consistent with the results of other aging studies of corona- and plasma-treated PET [1, 6, 19, 24].

Most likely the aging and washing behaviours seen for the UV/ozone treatments in this study are due to the same phenomena as those described by Briggs *et al.* [1] and Strobel *et al.* [24] for corona-treated PET. That is, although the polymer is well oxidized by the UV/ozone treatments, the resulting surface is mainly in the form of LMWOM. The LMWOM is formed by direct photocleavage via Norrish I and Norrish II reactions of the ester group. The LMWOM can be removed quickly by washing or slowly by migration away from the surface region into the bulk during aging. Because the PET bulk contains ester functionalities that can interact with oxidized materials, the potential for migration of the LMWOM is enhanced. Briggs *et al.* [1] also note that the likelihood of reorientation of the higher MW oxidized species is increased because of the chain mobility. They also state that the increased mobility, which could be aided by the plasticizing effect of the LMWOM, allows the surface free energy to be minimized at room temperature.

Strobel *et al.* [24] studied the aging behaviour of air-corona-treated PET. After 28 days of aging, the corona-treated samples had higher contact angles and higher $\Delta(O:C)$ ratios than the samples washed immediately after treatment. Washing removes the LMWOM, whereas aging involves a combination of migration of LMWOM, migration of some water-insoluble oxidized material, and a reorientation of higher MW oxidized species.

Similarly, Peeling *et al.* [14] found that with photo-oxidized PET, ambient aging and water washing immediately after treatment are not equivalent processes. After 2 weeks of aging, the 'static' contact angle of water on a sample treated for 30 min was 12° higher than for a sample washed immediately after treatment. They also state that 'the changes in the ESCA spectra on aging are greater than those observed when washing a photo-oxidized sample', although no O : C ratios are explicitly given. Their explanation of the aging and washing behaviour was the same as that given by Strobel *et al.* [24]. After either washing or 2 weeks of aging, the surface was still highly oxidized.

In the present study, the rate of recovery of the surfaces is somewhat different. After 28 days of aging, the advancing contact angles of the UV/ozone-treated samples were still slightly lower than for the immediately washed sample, while the receding angles were similar. The $\Delta(O:C)$ ratios for the aged samples, however, reached values similar to those of the samples washed immediately after treatment. This behaviour indicates that the distribution of oxidized species within the surface layer (5–10 nm) is different for the UV/ozone-treated films than it is for the films previously examined by other investigators. In the previous experiments on corona-treated PET [24] and photo-oxidized PET [14], aging (for 28 or 22 days, respectively) affected the contact angles more than washing immediately after treatment. The reverse is true in this study.

Based on the $\Delta(O:C)$ ratios, it appears that slightly less LMWOM is lost on washing when the PET is treated with UV/ozone than when it is treated with corona — 64% for UV/air and 70% for UV/air + ozone compared with approximately 75% for corona-treated PET [24] — but these differences are not statistically significant considering the accuracy of the XPS analysis. We do believe, however, that slightly less oxygen is lost in the UV/ozone treatments as compared with the corona treatment and that the UV/ozone treatments may be producing more higher MW oxidized material, providing less opportunity for the LMWOM to migrate from the surface region. Evidence for the continued presence of the LMWOM at the surface is provided by the advancing contact angle, which remains lower after 1 month of aging than after washing immediately after treatment. As with washed corona-treated PET, washed UV/ozone-treated PET displayed an increase in the contact angle and a decrease in the $\Delta(O:C)$ ratio on aging. This aging behaviour likely arises from the reorientation of higher MW oxidized material.

Ozone only-treated PET displays a completely different behaviour on aging or water washing. In the absence of UV light, less chain scission occurs; as a result, less LMWOM is formed. This was verified by the fact that on washing only 30% of the oxidized material was lost. The aging results for the ozone-only-treated samples also suggest that less LMWOM was present initially; the $\Delta(O:C)$ ratios and contact angles of water on unwashed and washed samples became similar much more quickly than those on the UV/ozone-treated samples, indicating that the washed and unwashed surfaces were more similar initially. Changes with ozone only treatment must involve primarily HMW materials, with little contribution from the water-soluble LMWOM.

4.2. Discussion: polypropylene

Lane and Hourston [29] suggest that 'the surface oxidation rapidly reaches an equilibrium at a relatively shallow depth of oxidation' for corona-treated PP. Plasma, flame, and pressing against aluminium treatments all produce oxidation to a depth of only a few monolayers (4–9 nm). In the case of UV/ozone and ozone-only treatments, we believe that the surface modification extends quite deeply, well beyond a few monolayers. In fact, according to attenuated total reflectance-Fourier transform infrared (ATR-FTIR) analyses [28], the treatment depth is of the order of 1 μm for all of the treatment regimes. The depth of penetration is considerably less for corona- or plasma-treated samples, and is very shallow for flame treatment. Little or no gradient was seen in the O:C ratios obtained from variable angle XPS analysis of the UV/ozone or ozone-only-treated samples, while a slight gradient was seen in the corona- or plasma-treated samples and a steep gradient was seen in the flame-treated sample [28].

Garbassi and co-workers [7, 21, 22] aged oxygen-plasma-treated PP and found that within 10 days the advancing contact angle had returned to the value of untreated PP. The receding angle, however, remained well below the value for the control after 30 days of aging. Their results were explained using a model of an ideal heterogeneous surface composed of high- and low-energy components. During aging, the O:C ratio essentially did not change. They therefore concluded that their results were due to macromolecular rearrangements in the top 5 nm (the XPS sampling depth) that redirected the polar groups away from the uppermost surface which was analysed by contact angle measurements. SSIMS analysis showed that the increases in the advancing angles were indeed directly related to an increase in the $CH^-/^{18}O^-$ ratio, supporting the theory of macromolecular rearrangement. However, the receding angle, which is far more sensitive to the presence of trace numbers of oxidized groups on the surface, did not return to the value of the untreated control. When the treated samples were aged in water, a highly polar solvent, the rearrangement seen on aging in air did not occur. In the experiments with UV/ozone and ozone only, there was very little change in the O:C ratio or in the contact angles on aging. Thus, any macromolecular motions did not significantly affect the number of oxygen-containing groups at the surface. Because the treated layer is fairly thick, rearrangement of the functional groups to form a hydrophobic surface is not favoured. This is what Van der Mei *et al.* [8] observed after aging polyethylene which had been treated repeatedly in an oxygen-glow-discharge. The contact angles and O : C ratios remained lower after multiple treatments, thus giving a 'more stable hydrophilic surface even after "infinite" recovery'.

Strobel *et al.* have extensively studied the LMWOM produced on air-corona-treated PP [2] and the aging behaviour of these films [23, 24]. After water washing, the variable-angle XPS analysis showed that the amount and depth of oxidation on the samples decreased significantly; this was attributed to the loss of LMWOM. In contrast, both the advancing and the receding contact angles increased only slightly and this is consistent with our present results. In our experiments, the O : C ratio of only the UV/air-treated sample decreased significantly (by 35%), although no change occurred in the contact angles. The loss in O : C ratio on washing indicates that treatment with UV/air produces some LMWOM through chain scission. The depth of treatment is such that even after washing the remaining surface is sufficiently oxidized so that the contact angles remain the same as those for the unwashed samples.

Water washing had a much more significant effect on the corona-treated samples than aging did. As with our treated films, the corona-treated samples, washed or unwashed, did not change in terms of O : C ratio on aging [23]. Strobel *et al.* [24] did see a slight decrease in wettability on aging (only a 4° increase in the advancing and receding contact angles).

Strobel *et al.* attributed the stability of PP on aging to the high MW of the base PP resin and the lack of interaction between the corona-treated surface and the unmodified bulk [23]. As untreated PP is relatively inert, interaction between the highly oxidized surface layer and the bulk is inhibited. Surface migration and reorientation are also hindered by the HMW nature of the PP bulk. As a result, the slight changes in the contact angles on aging were likely due to a reorientation within the XPS sampling depth, as was the case for plasma-treated PP [7].

It is interesting to note that evidence of LMWOM on the UV/air + ozone-treated samples was seen in AFM images of the surface. Before washing, the treated surface had 'mounds' similar to those seen on corona-treated PP using an SEM [2]. These mounds were approximately 500–700 nm wide and 50–80 nm high. After washing, the mounds disappeared but there was no significant change in the O : C ratio. There was, however, a small increase of 8° in the advancing contact angle. This increase may be due to the absence of the LMWOM, which can interfere with the measurement of the advancing contact angle on the unwashed sample, or it could be the result of the change in the topography of the surface. The treatment obtained with UV/air + ozone is much deeper than with corona, which would explain why the removal of the LMWOM did not decrease the O : C ratio. Atomic oxygen and ozone must be able to penetrate the layer of LMWOM present at the surface to react with polymer chains which are of the order of 1 μm from the surface. This ability of the reactive gas species to penetrate into the polymer is shown by the FTIR-ATR spectrum [28], which indicates significant modification of the surface in the top 1–2 μm. The modification taking place farther from the surface does not result in LMWOM that can be removed by water washing, but rather results in higher MW material which is not water-soluble.

On the ozone-only-treated samples, there was no evidence in the AFM images of the mounds formed from LMWOM; there was no change in the O : C ratio on washing; and there was none of the embrittlement which was observed after long UV/air treatments, [13] resulting from extensive crosslinking. Since there is no evidence of scission or extensive crosslinking, one proposed mechanism for treatment is the formation of multiple hydroperoxide groups [13]. The reactive species appear to be able to penetrate the polymer without breaking significant numbers of bonds. The absence of LMWOM clearly distinguishes ozone-only treatment from UV/ozone, air plasma, corona, and flame treatments of PP. In each of these other processes, chain scission through the reaction of atomic oxygen causes the formation of LMWOM. The reaction with ozone only leads to the formation of polymer peroxy radicals that do not necessarily result in chain scission but rather undergo slow reactions to form carbonyls, carboxylic acids, and hydroperoxides [17].

5. CONCLUSIONS

The aging and washing behaviours of PET and PP, each treated with UV/air, UV/air + ozone, and ozone only, have been described. The observed behaviours were consistent with the previously proposed [13] mechanisms of oxidation by the three treatment regimes.

Treatment of PET with UV/ozone resulted in considerable LMWOM on the surface, which was readily removed by water washing. Some of the LMWOM was lost from the surface in the first 3 days after UV/ozone treatment. The loss of LMWOM resulted in a decrease in the Δ(O : C) ratio and an increase in the advancing contact angle. The receding contact angle remained very low for the UV/ozone-treated samples after a month of aging. Ozone-only treatment did not result in as much chain scission as the UV/ozone treatment, so washing and aging did not have as significant an effect. The treatments penetrated so deeply that even after washing there was still sufficient modification remaining at the surface to influence the Δ(O : C) ratio and the contact angle.

The treated PP is very stable in terms of both O : C ratio and contact angle for all of the treatments. When using UV/ozone treatments, significant amounts of LMWOM were formed and could be washed away. The underlying material of higher MW was also significantly oxidized. There was no evidence of LMWOM formation on the ozone-only-treated samples.

Acknowledgements

We wish to thank the Natural Sciences and Engineering Research Council of Canada for their financial support during this work. We would also like to thank N. S. McIntyre, D. H. Hunter, and Chris Lyons for their helpful comments and insightful discussions.

REFERENCES

1. D. Briggs, D. G. Rance, C. R. Kendall and A. R. Blythe, *Polymer* **21**, 895–900 (1980).
2. M. Strobel, C. Dunatov, J. M. Strobel, C. S. Lyons, S. J. Perron and M. C. Morgen, *J. Adhesion Sci. Technol.* **3**, 321–335 (1989).
3. J. Adelsky, *Tappi* **72**, 181–184 (1989).
4. H. L. Spell and C. P. Christenson, *Tappi* **62**, 77–81 (1979).
5. L. J. Gerenser, J. F. Elman, M. G. Mason and J. M. Pochan, *Polymer* **26**, 1162–1166 (1985).
6. Y.-L. Hsieh and E. Y. Chen, *Ind. Eng. Chem. Prod. Res. Dev.* **24**, 246–252 (1985).
7. M. Morra, E. Occhiello and F. Garbassi, *J. Colloid Interface Sci.* **132**, 504–508 (1989).
8. H. C. Van der Mei, I. Stokroos, J. M. Schakenraad and H. J. Busscher, *J. Adhesion Sci. Technol.* **5**, 757–769 (1991).
9. F. Normand, J. Marec, Ph. Leprince and A. Granier, *Mater. Sci. Eng.* **A139**, 103–109 (1991).
10. M. Strobel, C. S. Lyons and K. L. Mittal (Eds), *Plasma Surface Modification of Polymers: Relevance to Adhesion*. VSP, Zeist, The Netherlands (1994).
11. J. Peeling and D. T. Clark, *J. Polym. Sci., Polym. Chem. Ed.* **21**, 2047–2055 (1983).
12. J. Peeling, M. S. Jazzar and D. T. Clark, *J. Polym. Sci., Polym. Chem. Ed.* **20**, 1797–1805 (1982).
13. M. J. Walzak, S. Flynn, R. Foerch, J. M. Hill, E. Karbashewski, A. Lin and M. Strobel, *J. Adhesion Sci. Technol.* (in press).
14. J. Peeling, G. Courval and M. S. Jazzar, *J. Polym. Sci., Polym. Chem. Ed.* **22**, 419–428 (1984).
15. J. R. Vig, *J. Vac. Sci. Technol.* **A3**, 1027–1034 (1985).
16. N. S. McIntyre, R. D. Davidson, T. L. Walzak, R. Williston, M. Westcott and A. Pekarsky, *J. Vac. Sci. Technol.* **A9**, 1355–1359 (1991).
17. J. F. Rabek, J. Lucki, B. Ranby, Y. Watanabe and B. J. Qu, in: *Chemical Reactions on Polymers*, J. L. Benham and J. F. Kinstle (Eds), pp. 187–200. ACS Symposium Series No. 364. American Chemical Society, Washington, DC (1988).
18. J. G. Calvert and J. N. Pitts, *Photochemistry*. Wiley, New York (1966).
19. J. M. Pochan, L. J. Gerenser and J. F. Elman, *Polymer* **27**, 1058–1062 (1986).
20. S. K. Bhateja, *J. Appl. Polym. Sci.* **28**, 861–872 (1983).
21. F. Garbassi, M. Morra, E. Occhiello, L. Barino and R. Scordamaglia, *Surface Interface Anal.* **14**, 585–589 (1989).
22. E. Occhiello, M. Morra, G. Morini, F. Garbassi and P. Humphrey, *J. Appl. Polym. Sci.* **42**, 551–559 (1991).
23. J. M. Strobel, M. Strobel, C. S. Lyons, C. Dunatov and S. J. Perron, *J. Adhesion Sci. Technol.* **5**, 119–130 (1991).
24. M. Strobel, C. S. Lyons, J. M. Strobel and R. S. Kapaun, *J. Adhesion Sci. Technol.* **6**, 429–443 (1992).
25. C. Bryson, *Surface Sci.* **189/190**, 50–59 (1987).
26. R. E. Johnson, Jr and R. H. Dettre, *J. Phys. Chem.* **68**, 1744–1750 (1964).
27. M. Morra, E. Occhiello and F. Garbassi, *Adv. Colloid Interface Sci.* **32**, 79–116 (1990).
28. M. Strobel, M. J. Walzak, J. M. Hill, A. Lin, E. Karbashewski and C. S. Lyons, *J. Adhesion Sci. Technol.* **9**, 365–383 (1995).
29. J. M. Lane and D. J. Hourston, *Prog. Org. Coat.* **21**, 269–284 (1993).

Polymer Surface Modification: Relevance to Adhesion, pp. 291–301
K. L. Mittal (Ed.)

Surface pretreatment of plastics for adhesive bonding

A. KRUSE, G. KRÜGER, A. BAALMANN* and O.-D. HENNEMANN
Fraunhofer-Institut für Angewandte Materialforschung, Bereich Klebtechnik und Polymere, Neuer Steindamm 2, D-28719 Bremen, Germany

Revised version received 17 August 1995

Abstract—Many plastics have a poor tendency to bond to other materials because of their inherent inert chemical structure and thus require a pretreatment. Wet chemical methods are expensive because of the disposal of the waste liquids. In this study, the corona treatment (Ional process), the low-pressure plasma process, and the fluorination process were tested and compared with each other. The following plastics were tested: PP (polypropylene), PBT (polybutyleneterephthalate), PBT blends, and a high-temperature thermoplastic, PEEK (polyetheretherketone). In particular, the low-temperature plasma process results in excellent adhesion strength. In addition, we have shown that the stability of freshly plasma-treated surfaces could be maintained for time periods of at least several days.

Keywords: Surface pretreatment; plastics; corona; fluorination; plasma; adhesion; adhesive bonding.

1. INTRODUCTION

In many application areas, adhesive bonding technology is the only approach that is useful for the joining of materials. This is particularly true for the joining of plastics. Since welding is limited to thermoplastics, for the joining of thermosetting plastics and high-temperature plastics, adhesive bonding technology is the preferred solution for bonding. On the other hand, mechanical joining processes result in an inhomogeneous stress distribution, which is undesirable.

As a prerequisite for a high adhesive bond strength, an effective pretreatment which includes cleaning and activation of the polymer surface is needed. Polymers such as polyethylene, polypropylene (PP), and polyetheretherketone (PEEK) have quite low surface energies due to their chemical structure, and, therefore, a chemical modification at the surface is necessary.

Plastics such as PP and polybutyleneterephthalate (PBT) can be easily processed and recycled, and can be used at elevated temperatures (>80°C). The group of high-temperature plastics comprises plastics such as polyphenylenesulfide (PPS), polysulfone (PSU), and PEEK. Because of their good resistance to high temperatures and their

*To whom correspondence should be addressed.

high mechanical strength, they are used in the automotive, aircraft, and electronics industries for different applications.

Several methods of surface treatment for plastics, e.g. corona, chemical etching, flame treatment, and plasma, have been used [1–3] but a great deal of work is still required to develop the technical potential of these methods and to understand the physical and chemical mechanisms involved. In this paper we discuss some promising methods, namely the Ional process (which is a special type of corona treatment), the low-temperature plasma treatment, and gas phase fluorination.

2. EFFECTS OF THE COMPONENTS OF POLYMERIC MATERIALS ON THEIR ADHESION BEHAVIOUR

Compared with metals, it is more difficult to obtain good adhesion to polymers. There are several reasons for this:

Table 1.
Components of industrial polymeric materials

Component	Examples	Typical behaviour
Antioxidants	Alkyl phenols	
	Alkylidene bisphenols	
	Hydroxybenzyl compounds	Volatile
	Aromatic amines	
	Phosphites	
	Organotin mercaptides	Unstable to hydrolysis
Hydrolysis stabilizers	Organic silicone compounds	
UV absorbers	Sterically hindered amines	
Lubricants	Metal soaps	Hinder adhesion
	Paraffin waxes	Create weak boundary layers
	Fatty alcohols	
	Silicones	
Antistatic additives	Distearin	Hinders adhesion
	Polyglycols	Migrate to surface
	Polyalcohols	
Metal deactivators	Hydrazines	
	Carboxylic acid amides	
Flame retardants	Chlorine and bromine-containing compounds	Toxic
	Antimony trioxide	
	Phosphoric acid	
Plasticizers	Dioctylphthalates	Dubious physiological effects
	Tricresylphthalates	
	Fatty acid esters	Some hinder adhesion

(i) the number of different types of plastics available is very high and they differ widely in their behaviour and characteristics;
(ii) the temperature dependence of the mechanical properties of plastics and adhesives is much higher than it is for metals;
(iii) all plastics have a low surface energy. This necessitates some kind of pretreatment to make plastics wettable by adhesives;
(iv) plastics contain numerous components and these can vary considerably in a single group of plastics and some of them, especially lubricants and plasticizers, hinder adhesion severely.

Table 1 summarizes the typical components of industrial polymeric materials and how these affect adhesion. The effect often depends on the migration of the components from the bulk to the surface and also on the temperature dependence of their mobility.

3. PRETREATMENT PROCESSES FOR PLASTICS

To obtain an optimum strength of adhesive bonds to plastics, it is necessary to increase the surface energy of the plastic substrate by specific pretreatment processes. These processes can be divided into three groups:

(1) Mechanical processes
- — sand-blasting
- — grinding, brushing

(2) Chemical processes
- — CSA (chromic-sulfuric acid) treatment
- — pickling
- — ozone treatment
- — gas phase fluorination*
- — coating with amphiphilic substances

(3) Physico-chemical processes
- — low-pressure plasma*
- — corona treatment, Ional process*
- — thermal treatment
- — flame treatment
- — ion etching
- — laser
- — UV light.

In studying the influence of different pretreatment processes on the bond strength of plastics, the dry chemical processes marked with an asterisk were performed on a series of plastic substrates and characterized by mechanical tests as well as by analytical methods.

3.1. Gas phase fluorination

To increase the surface energy and to decrease simultaneously the permeation of non-polar organic fluids through polyolefins, fluorination of the surface to a thickness of 0.1 mm was carried out. With this treatment in a partly fluorine-containing atmosphere [4, 5], a considerable increase in surface energy could be observed. To evaluate the influence of this pretreatment on the performance of the adhesive bond, we carried out many experiments with different parameter sets and different adhesives.

For the gas phase fluorination experiments, a chemical reactor was used. A schematic diagram of the experimental set-up is shown in Fig. 1. The fluorine–nitrogen mixture contains up to a maximum 10 vol% fluorine, in order to prevent overheating due to the exothermic fluorination reaction. It is important to minimize the content of oxygen or air in the reactor, because in the presence of oxygen the fluorination reaction on the substrate is affected and unwanted reactions are possible.

The materials tested were PBT: Pocan B1501 (Bayer AG); PBT + ABS blend: Pocan S1506 (Bayer AG); PPS-Gf30: Tedur (Bayer AG); PP: Hostalen PPN1060 (Hoechst AG); PP-Gf30: Hostacom G3N01 (Hoechst AG); and PE-UHMW: GUR 412 (Hoechst AG). These materials were supplied as sheets of 5 mm thickness.

The unreacted fluorine and the reaction product HF are converted into CaF_2 by chemical reaction, so there is no gas emission into the atmosphere.

In order to determine the thickness range of the polymer in which the fluorination reaction takes place, ellipsometric measurements were carried out. The results for PBT and ABS-modified PBT are listed in Table 2.

The lap shear strength data for PBT (modified and unmodified) are listed in Table 3.

The adhesive used for the specimens of Table 3 was Araldite AW 106/HY 953U (thickness 0.1 mm, CIBA AG). The failure mode in these single lap shear tests was

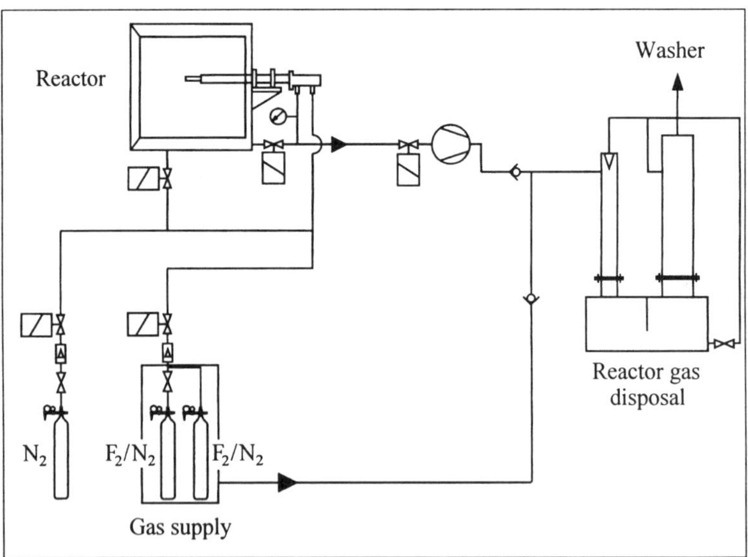

Figure 1. Experimental set-up of the fluorination equipment with a revolving reactor.

Table 2.
Ellipsometric measurements to determine the thickness of the fluorinated layer on PBT

Material	Fluorine treatment		Thickness of the fluorinated layer
	Concentration (vol%)	Time (min)	(nm)
PBT, Pocan B1501	3	1	5.8
(Bayer AG)	3	3	8.6
	3	5	10.3
PBT, Pocan S1506	3	1	28.6
(Bayer AG)	3	3	33.8
	3	5	33.8

Table 3.
Lap shear strength (DIN 53283) and failure mode of unmodified PBT before and after gas phase fluorination

Fluorination condition	Lap shear strength (N/mm^2)
Without fluorination	0.2
1 s, 0.5 vol% F	3.6 cohesive failure
10 s, 0.5 vol% F	6.1 cohesive failure
2 min, 0.5 vol% F	5.3 cohesive failure
15 min, 0.5 vol% F	4.7 cohesive failure
2 min, 1 vol% F	5.4 cohesive failure
2 min, 5 vol% F	3.3 cohesive failure

cohesive in the polymer itself. For materials such as the ones tested that are not perfectly rigid, in the single lap shear test the mechanical stress concentrates along a diagonal line across the adhesively bonded zone (overlap region) of the polymer sheets. According to this effect, in the tests performed here the propagation line of fracture follows this diagonal. The differences in the lap shear strength values for different fluorination treatments are considerable.

For industrial applications it is important to note that a high adhesive joint strength is obtained even after short times of fluorination.

Further results of fluorination treatments are presented in Figs 2–5. Also shown are the lap shear strength and the bending peel resistance values for some other plastics.

3.2. The Ional process — a special kind of corona treatment

A promising process for the pretreatment of polymers is treatment in a corona discharge, which contains highly energetic particles. In a corona-like process, named Ional (Softal GmbH, Hamburg, Germany), a special type of corona discharge is used

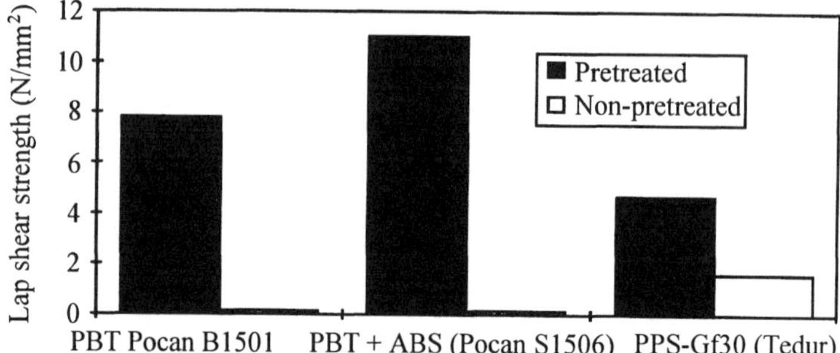

Figure 2. Lap shear strength (DIN 53283) of PBT, the PBT–ABS blend, and PPS with and without fluorination pretreatment. The fluorination treatment was 3 min in 3 vol% F.

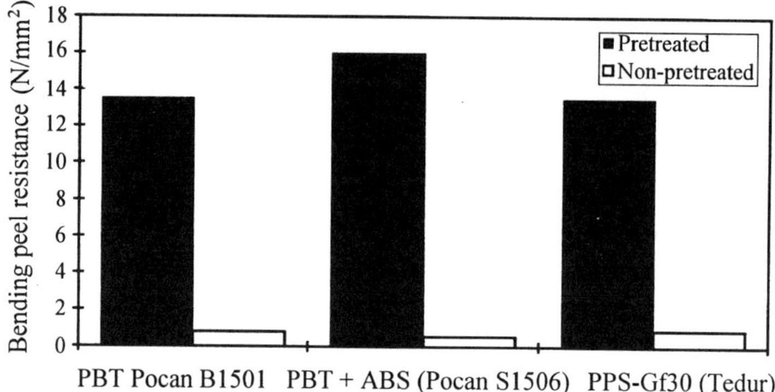

Figure 3. Bending peel resistance (DIN 54461) of PBT, the PBT–ABS blend, and PPS with and without fluorination pretreatment. The fluorination treatment was 3 min in 3 vol% F.

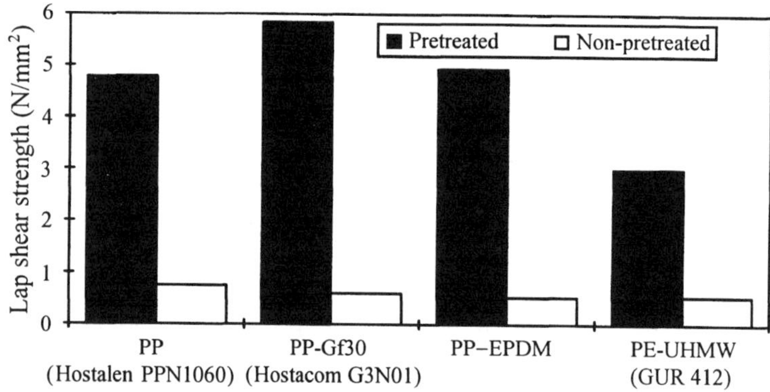

Figure 4. Lap shear strength of PP, PP-Gf30, the PP–EPDM blend, and PE-UHMW with and without fluorination pretreatment. The fluorination treatment was 3 min in 3 vol% F.

for the surface treatment of plastic substrates. In this process (see Fig. 6), the electrical discharge is generated between two parallel electrodes (electrically isolated) close to the surface of the substrate. By a gas feeding tube, a controlled gas flow directed to the substrate is generated in the discharge region. Compressed air, nitrogen, oxygen,

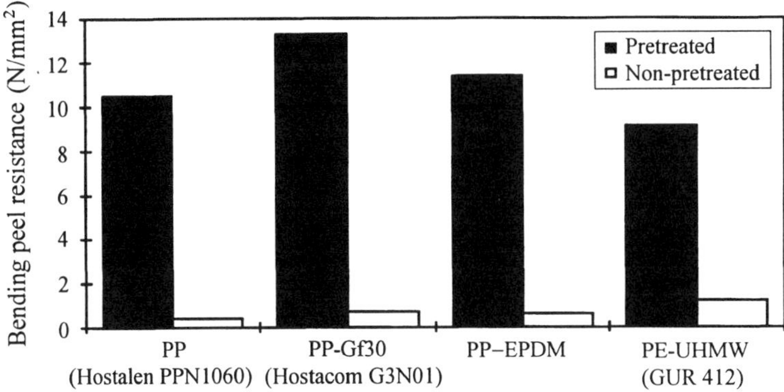

Figure 5. Bending peel resistance of PP, PP-Gf30, the PP–EPDM blend, and PE-UHMW with and without fluorination pretreatment. The fluorination treatment was 3 min in 3 vol% F.

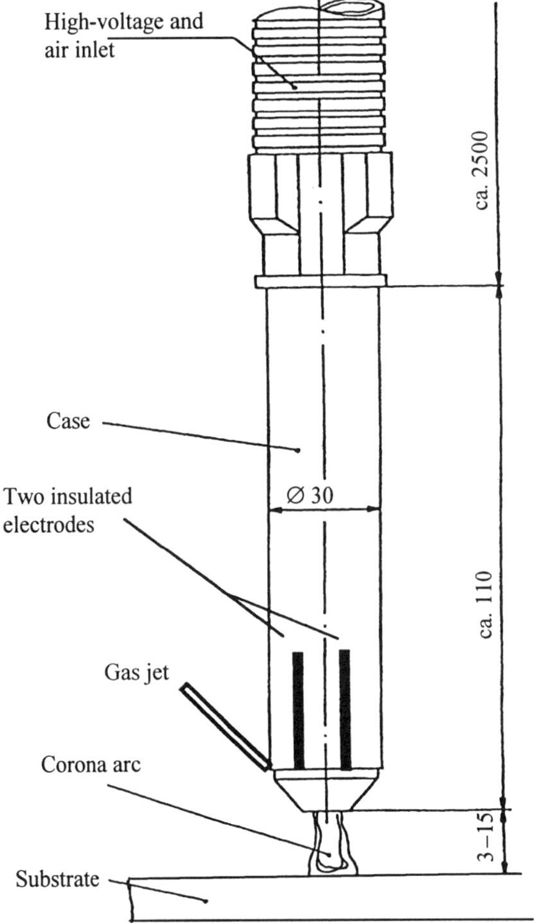

Figure 6. Ional process equipment. All dimensions given are in mm.

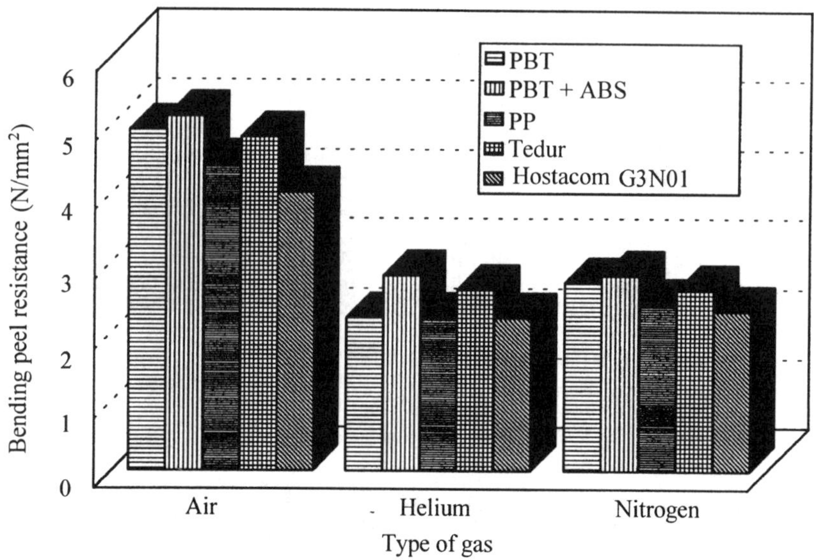

Figure 7. Effect of the type of gas in the Ional treatment on the bending peel resistance (DIN 54461) of various plastics.

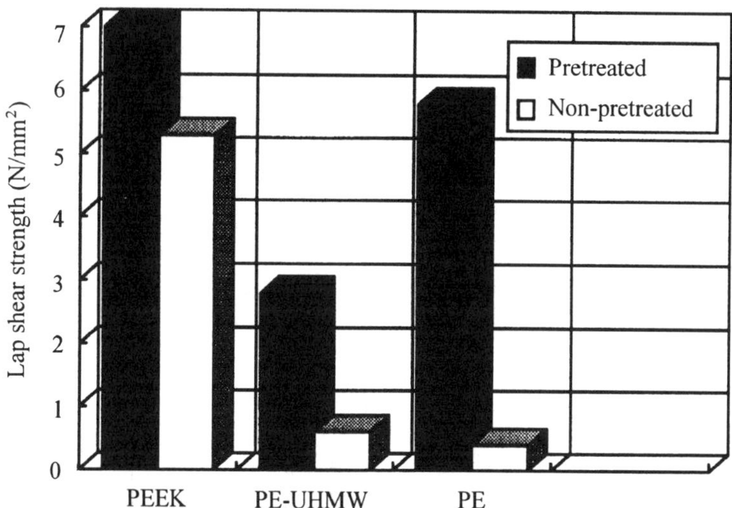

Figure 8. Lap shear strength of Ional-treated PEEK, PE-UHMW, and PE.

or rare gases are the typical gases used. In Fig. 7, some results of a bending peel test (DIN 54461) are given. For this test, an acrylate-based adhesive tape was used. Depending on the treatment, a significant increase in the peel strength is obtained. Further results are presented in Fig. 8 for an Ional treatment on PEEK, PE-UHMW, and PE.

3.3. Plasma treatment

PEEK represents one of the most interesting engineering materials. It shows processing flexibility, along with many advantages such as chemical stability and an excellent thermal and mechanical performance.

PEEK is used for high-temperature cable isolation; for injection moulding parts in the automotive, aircraft, and electronics industries; and for gears, bearings, and sealings in the mechanical industry. In these fields of application, the adhesive bonding of PEEK is a useful and attractive technique to make use of the advantages of PEEK, but a sufficient pretreatment is necessary [7, 8].

The PEEK composites (APC2/AS4, ICI) were exposed to a microwave plasma-discharge (2.45 GHz, 10 l reactor, power 30 W) in an argon or oxygen atmosphere. In addition to this treatment, some other pretreatments were tested:

(i) degreasing with MEK (methylethylketone) in an ultrasonic bath;

(ii) corona treatment;

(iii) CSA (chromic-sulfuric acid) etching;

(iv) SACO (silicate coating).

In the SACO coating process [6], the surface was grit-blasted with a chemically precoated grit, resulting in a silicate-like coating on the surface.

Following the pretreatment, the test specimens were adhesively bonded with AF 191, a well-known epoxy film adhesive (0.08 mm thick, 3M company) with applications in the aircraft industry.

The effect of this pretreatment was tested using a single lap shear strength test in accordance with DIN 53283. These tests were performed under different conditions (room temperature; room temperature after storage in water at 70°C for 42 days; 150°C; 150°C after storage in water at 70°C for 42 days). The results are shown in Figs 9 and 10. When tested at room temperature, both plasma treatments (argon and oxygen) resulted in lap shear strengths higher than 30 N/mm^2. Actually the argon plasma gave the best performance, resulting in a shear strength higher than

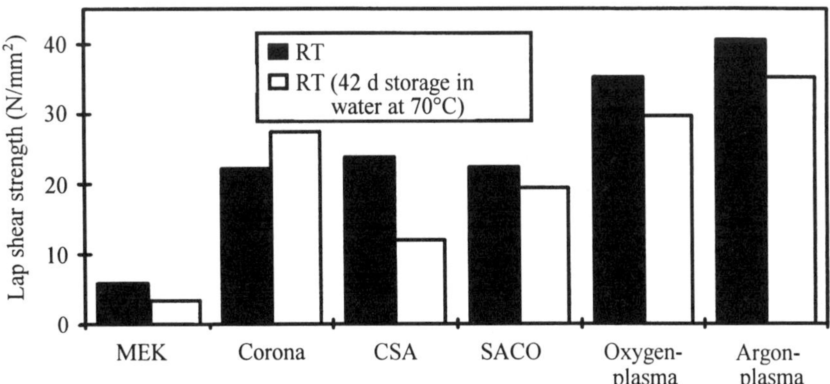

Figure 9. Lap shear strength of PEEK at room temperature after various pretreatments.

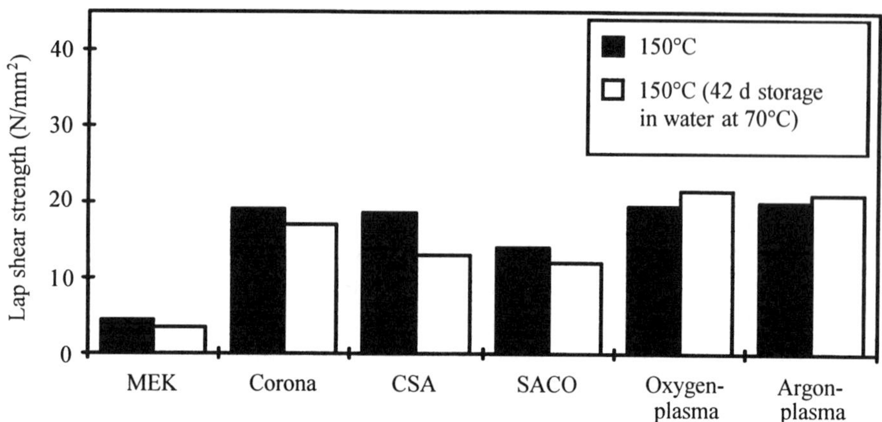

Figure 10. Lap shear strength of PEEK at 150°C after various pretreatments.

40 N/mm^2 (Fig. 9). Compared with other surface treatments such as corona discharge and chromic sulfuric acid etching, the plasma treatment showed a significantly better shear strength. All of the plasma-treated samples failed interfacially (determined by microscopic inspection) at room temperature, whereas the failure at 150°C was mainly cohesive (in the adhesive).

It is well known that some plasma-treated surfaces quite often change their surface properties when they are exposed to air. To study this effect for the oxygen-plasma treatment, freshly treated specimens were stored in ambient atmosphere for 1–4 weeks prior to the adhesive bonding. Even after a storage of 4 weeks a reasonable shear strength of 29 N/mm^2 could be obtained.

4. CONCLUSIONS

Polymers have a low surface energy and/or contain some components that can lower the surface energy. So an effective pretreatment is necessary to obtain good adhesion to polymer materials. Physical methods such as the Ional process (corona-like) or the low-pressure plasma treatment are suitable for increasing the adhesion strength remarkably. For an oxygen plasma treatment the increase could also be sustained even after a 4-week storage of the treated samples prior to bonding. Another useful method is the fluorination of the polymer surface in the gas phase. Considerable improvements in adhesion were recorded as a result of such treatment especially on polyolefins, PBT, and PEEK.

REFERENCES

1. E. M. Liston, *J. Adhesion* **30**, 199 (1989).
2. T. Hjertberg, B. A. Sultan and E. M. Sörvik, *J. Appl. Polym. Sci.* **37**, 1183 (1989).
3. D. Briggs and C. R. Kendall, *Int. J. Adhesion Adhesives* **2**, 13 (1982).
4. O. -D. Hennemann and G. Krüger, *Kunststst.-Berat.* **37**, 25 (1992).

5. D. M. Brewis, *J. Adhesion* **37**, 97 (1992).
6. R. Guggenberger and P. Koran, *Adhesion* **34**, 40 (1990).
7. G. K. A. Kodokian and A. J. Kinloch, *J. Mater. Sci. Lett.* **7**, 625 (1988).
8. C. Jama, O. Dessaux, P. Goudmand, L. Gengembre and J. Grimblot, *Surface Interface Anal.* **18**, 751 (1992).

Polymer Surface Modification: Relevance to Adhesion, pp. 303–317
K. L. Mittal (Ed.)

'Surface photografting' onto polymers — a new method for adhesion control

BENGT RÅNBY

Department of Polymer Technology, The Royal Institute of Technology, S-100 44 Stockholm, Sweden

Revised version received 5 August 1994

Abstract—Surface photografting onto polymers is a new method for surface modification of sheet, film, filament and yarn. It was invented in our laboratories in 1985 and was developed as two processes: *A batchwise process* where photoinitiator and monomer are transferred to the substrate through vapor phase and the monomer is grafted to the surface of the substrate by UV irradiation; and *a continuous process* where photoinitiator and monomer in solution are transferred to the substrate in liquid phase by 'presoaking' a strip of film, a filament or a yarn in the solution and the monomer is grafted to the substrate surface by UV irradiation on line.

The batch process has been applied to film, sheet and molded plates of various commercial polymers which have been surface photografted with acrylic acid (AA), acrylamide (AM), 4-vinyl pyridine (4-VP) and glycidyl acrylate (GA) using benzophenone (BP) as initiator. The grafted layers are analyzed by electron spectroscopy for chemical analysis (ESCA) and reflection IR spectroscopy, contact angle measurements with water, and dye adsorption measurements, and analyzed by microtitration to be maximum 10 nm thick. The epoxide groups of GA-grafted polymer surfaces were reacted with stabilizers, amines and proteins to secondary grafted layers.

The effect of surface photografting on adhesion was studied for molded low density polyethylene (LDPE) plates which were surface photografted with AA, AM, 4-VP and GA using the batch process. The adhesion of Scotch tape to the LDPE plates increased 5 to 8 times after grafting of the plates with the four monomers, measured by 90° peel tests.

Strips of polypropylene film and filaments and yarns of polyethylene, polypropylene and polyester (all commercial samples) were surface photografted with the four monomers mentioned using the continuous process. The measured adhesion of the substrates to epoxy resin, measured by pull-out tests from the cured resin, increased by a factor of 3 for AA-grafting and a factor of 6 to 7 for AM-grafting. The bulk tensile properties of high strength PE fibers are not affected by the grafting.

Keywords: Surface modification; fiber and film; polyethylene; polypropylene; poly(ethyleneterephthalate); acrylic monomers; vinylpyridine; benzophenone; photoinitiated grafting; adhesion; peel test; shear test; pullout test.

1. INTRODUCTION

Several methods for modification of a polymer surface have been developed in order to increase the adsorption of dyes and the printability, improve the wetting with water and increase the adhesion to other materials. Of the chemical methods described, the following are commonly applied.

Surface oxidation by flame treatment, corona discharge or etching with reactive agents [1]. These treatments cause degradation, radical formation and subsequent reactions with oxygen at the surface of the substrate.

Plasma treatment with electric discharge through gas or vapor at low pressure using a high voltage DC or a radiofrequency AC electric field [2]. The plasma may contain ionized gas molecules or radicals of vapor molecules which react at the surface of the substrate and form a crosslinked layer.

High energy radiation with gamma rays or accelerated electron beam in the presence of a monomer is a widely studied and applied method for polymer modification of both surface and bulk properties [3]. The radicals formed at the surface of the substrate react with the monomer by grafting. The radiation also penetrates the substrate and affects the bulk properties.

Surfactants added to the polymer may migrate to the surface. Surface active monomers are polymerized at the surface by UV-irradiation and form entangled coated layers [4]. A surface active block copolymer can also be used.

Ultraviolet radiation of long wavelength (300–400 nm) is not absorbed by most common polymers but selectively excites UV-initiators. With initiator and monomer present in the surrounding of the polymer substrate, UV light initiates grafting with low efficiency [5, 6]. Our research on 'surface photografting' started in 1983 and was developed as two processes in 1985 [7, 8]. In the *batch process* the substrate is irradiated with UV light in an atmosphere of initiator and monomer vapor. In the *continuous process* film strips and fibers are presoaked in a solution of initiator and monomer which gives a thin liquid layer on the surface of the substrate. The irradiation excites the initiator which abstracts hydrogen from the surface of the substrate. The radicals formed at the surface add monomer and form grafted chains on the substrate with high efficiency compared to earlier methods. The photografted films and fibers give increased adsorption of dyes and increased adhesion to resins as first published in 1988 [9]. A recent review of the surface photografting methods describes these developments with references to the original papers [10].

2. EXPERIMENTAL

2.1. The batch process for surface photografting

For the batch process a 'UV Cure' irradiator was used (Fig. 1). In principle, the substrate sample (S) is enclosed in a cell (C) with quartz window (Q) together with an open vessel (V) containing a solution of initiator and monomer both of which are volatile. The cell is temperature controlled by a thermostat and oxygen is removed by a flow of pure nitrogen through the cell (not shown). At a preset temperature close to the boiling point of the solvent, the sample is irradiated with a high pressure UV lamp, e.g. HPM 15 from Philips, which has high intensity at 300–400 nm.

With benzophenone (BP) as initiator and an acrylic monomer (M) for grafting, the photoreaction proceeds as follows:

$$\underset{\text{(BP)}}{\mathrm{Ph_2C{=}O}} \xrightarrow{h\nu} \underset{(\mathrm{BP}^{*\mathrm{S}})}{\mathrm{Ph_2C{=}\overline{\underline{O}}|^{*}_{S}}} \xrightarrow{\text{ISC}} \underset{(\mathrm{BP}^{*\mathrm{T}})}{\mathrm{Ph_2\dot{C}{-}\dot{\overline{\underline{O}}}|^{*}_{T}}} + \mathrm{PH} \longrightarrow \underset{\text{(K)}}{\mathrm{Ph_2\dot{C}{-}OH}} + \mathrm{P}\cdot \qquad (1)$$

BP molecules are excited to singlet state, revert rapidly to excited triplet state by intersystem crossing (ISC) and abstract hydrogen from the polymer substrate (PH) and give a polymer radical (P·) (equation (1)). The ketyl radical formed (K) is rather inactive and may dimerize to benzpinacol or combine with another radical. The polymer radicals (P·) at the surface of the substrate add monomer (M) and form grafted polymer chains (equation (2)).

$$\mathrm{P}\cdot + \mathrm{M} \longrightarrow \mathrm{PM}\cdot + n{\times}\mathrm{M} \longrightarrow \mathrm{PM}_{n+1}\cdot \qquad (2)$$

The grafted chains are rather short due to termination by radical combination of two chain end radicals or with other radicals present, e.g. ketyl radicals. The grafted chains form a thin surface layer (2 to 8 nm) which is studied by measurements of contact angle with water (Fig. 2), adsorption of dyes and adhesion to other materials [11]. The grafted layer is most conveniently analyzed by the ESCA technique, e.g. for a polystyrene film grafted with acrylic acid (Fig. 3). The intensity ratio of the O_{1s} and C_{1s} ESCA peaks gives a relative measure of the extent of grafting, i.e. in this case the ratio of oxygen and carbon in the surface layer as analyzed by the ESCA spectra. The grafted layers are too thin to be analyzed by weight increase (< 0.1%) [11]. An absolute measure of the amount of grafted polymer is analyzed by microtitration, e.g. of the carboxyl groups in grafted AA polymer, which can be

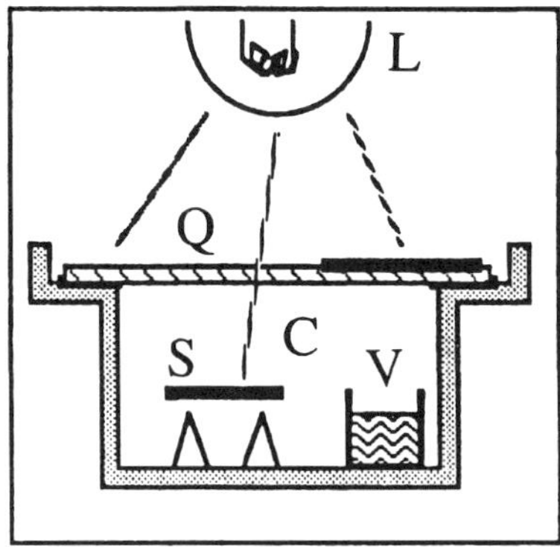

Figure 1. Reactor used for surface photografting with the batch process of a sample S irradiated with the UV lamp through the quartz window Q in a cell C containing vapor of initiator and monomer from a solution V.

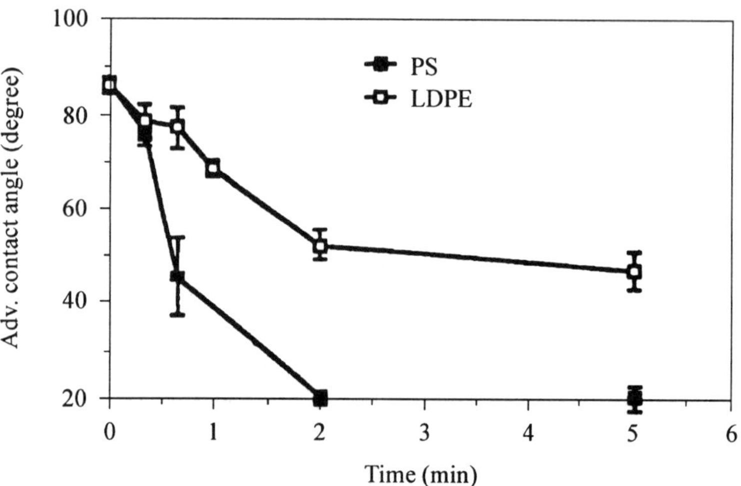

Figure 2. The advancing contact angle of water against surfaces of low density polyethylene (LDPE) and polystyrene (PS) grafted with acrylic acid for different irradiation times.

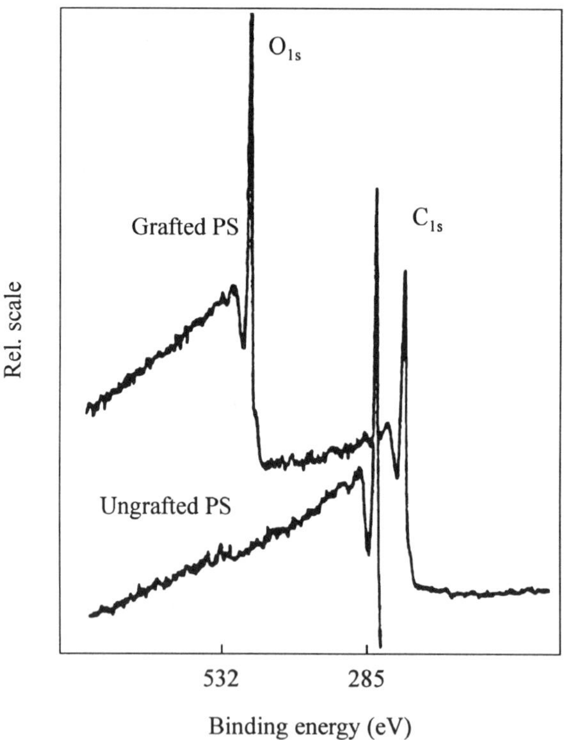

Figure 3. ESCA spectra of ungrafted and grafted polystyrene surfaces. Monomer: acrylic acid. Irradiation time: 5 min. The ESCA spectrum of grafted PS is shifted 45 eV.

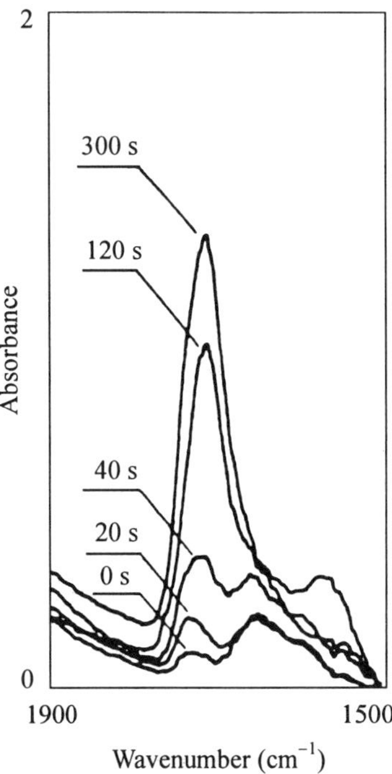

Figure 4. Carbonyl group absorption at 1715 cm^{-1} measured with ATR-IR. Low density polyethylene surface photografted with acrylic acid for different irradiation times.

applied only to substrates of very large surface area, e.g. textile yarn. Infrared reflection spectroscopy (ATR-IR) is also used for analysis of surface grafted polymer, e.g. from recording the carbonyl absorption at 1715 cm^{-1} for LD polyethylene grafted with acrylic acid during different irradiation times (Fig. 4). Polymer surfaces grafted with glycidyl acrylate (GA) are further modified by secondary reactions of the epoxy groups with stabilizers containing amine groups [11]. The very thin grafted layer of stabilizers gives an efficient protection against degradation.

3. RESULTS AND DISCUSSION

3.1. Adhesion measurements of solid surfaces grafted with the batch process

Using the batch process a series of injection-molded LD polyethylene plates were prepared (10 cm diameter, 3 mm thick) and surface photografted together with strips of commercial polyester film (PET, $60 \times 7 \times 0.124$ mm) with AA, AM, 4-VP and GA monomers. Grafted layers of a thickness 5 to 10 nm were formed as measured by ESCA technique. The grafted PET film strips were pressed onto the grafted LDPE plates according to Table 1. The grafted PET strips showed no reproducible increased adhesion to the ungrafted LDPE plates. We had expected grafted AA (acid) to adhere to grafted 4-VP (base) and grafted AM (amide) to react with grafted GA (epoxide). The negative result is probably due to the uneven surface structure of the molded

Table 1.
Grafted PET film strips glued to ungrafted and grafted LDPE plates under a 10 mm thick glass plate at 60°

LDPE plates	PET strips grafted with
Blank (ungrafted)	AA, AM, 4-VP, GA
Grafted with	
AA	4-VP
AM	GA
4-VP	AA
GA	AM

Table 2.
180° shear test results for Scotch tape glued to PET film samples grafted with four different monomers

Grafted monomer	Shear strength (N/m)
Blank	3600
AA	4543
GA	3838
AM	4219
4-VP	4257

PET sample size: 60 × 7 × 0.124 mm.
Scotch tape, commercially available.
Crosshead speed: 20 mm/min; full scale: 200 N.

LDPE plate material. The very thin grafted layers on the stiff PET films touched the grafted layers on the LDPE plates only at a few separate small areas.

Adhesion experiments with grafted surfaces using Scotch tape were more successful. In the first series PET film strips (7 mm wide) were grafted with AA, GA, AM and 4-VP monomers and bonded to clear Scotch tape (19 mm wide) by gentle pressing at room temperature. The tape/film assembly was mounted in an Instron (Fig. 5) and tested at a crosshead speed of 20 mm/min (full scale is 200 N). The shear strength recorded shows moderate increases after grafting with the four monomers (Table 2 and Fig. 6). The soft adhesive layer of the Scotch tape used is about 0.1 mm thick which is more than 10 000 times thicker than the grafted layers. Therefore, the shear test described (Figs 5 and 6) measures mainly the shear modulus of the tape adhesive. The increased shear strength due to grafting would prevent slipping at the interface of tape and PET film.

Peel tests of the grafted surfaces are more conclusive. One end of Scotch tape is bonded to the grafted surfaces of LDPE plates by gentle pressing at room temperature. The assembly is mounted in an Instron as shown in Fig. 7 and the 90° peel strength measured at a crosshead speed of 20 mm/min (full scale is 200 N). Typical

180° shear test

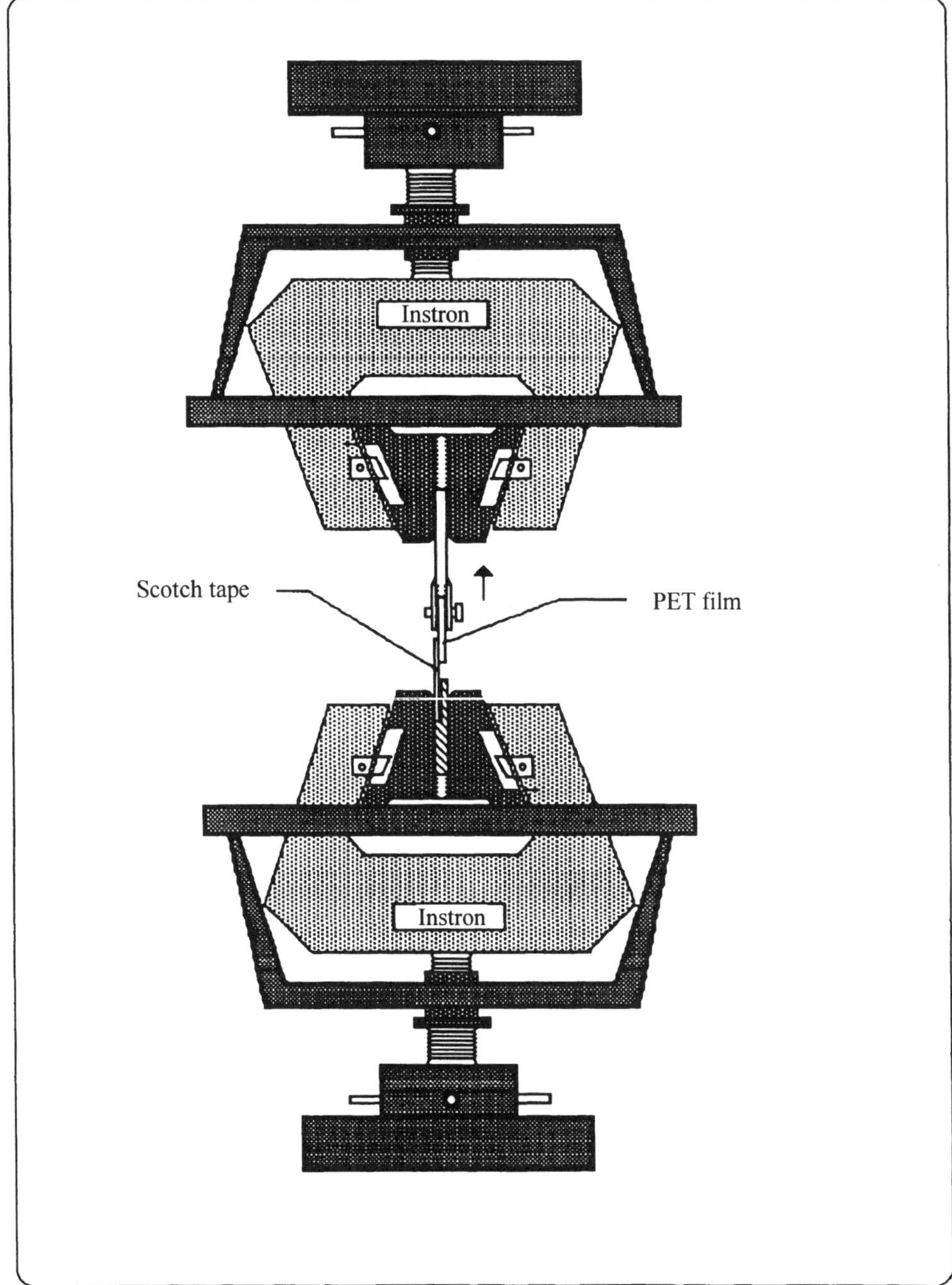

Figure 5. Instron measurements of adhesion of Scotch tape to strips of poly(ethylene terephthalate) (PET) film as 180° shear test.

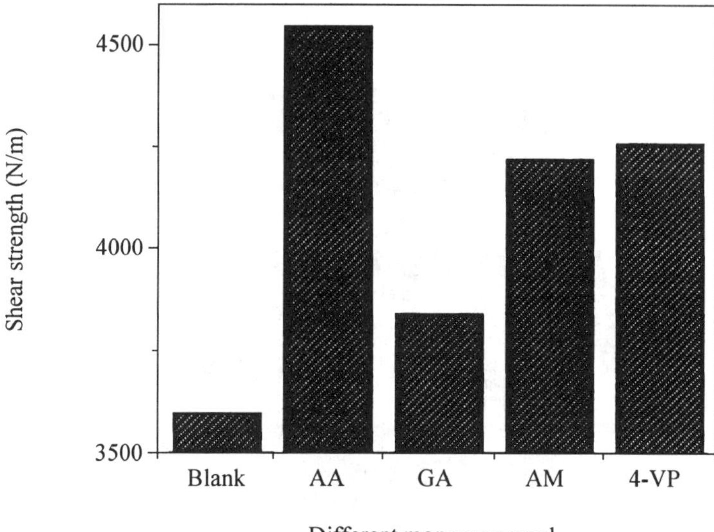

Figure 6. Shear strength of the assembly of ungrafted ('blank') and grafted polyester film samples (PET) adhering to Scotch tape, measured as 180° shear tests shown in Fig. 5. The PET film strips are grafted with four different monomers: acrylic acid (AA), glycidyl acrylate (GA), acrylamide (AM), and 4-vinyl pyridine (4-VP).

Table 3.
90° peel tests of Scotch tape (width 19 mm) glued onto LDPE plates before ('blank') and after grafting with four different monomers

Grafted monomer	Peel strength (N/m)	Relative increase (ratio)
Blank	14.4	
GA	69.5	4.8
AM	90.5	6.3
AA	96.8	6.7
4-VP	115.8	8.0

recordings for measurements with blank LDPE plate and AA grafted LDPE plate are given in Fig. 8. The peel strength was recorded by the Instron instrument and the values at 90° peel angle are given in Table 3. The 90° peel strength increased 5 to 8 times by grafting with GA and 4-VP, respectively. After peeling-off the Scotch tape the LDPE surfaces showed no visible trace of adhesive residue. This indicates that 90° peel test measures the adhesion of the tape to the LDPE surface at the interface. The surface photografting onto LDPE is quite effective for adhesion control.

For tape bonded to grafted PET film, reproducible 90° peel strength could not be measured. The PET strips are not rigid enough for mounting like the LDPE plates (Fig. 7).

90° peel test

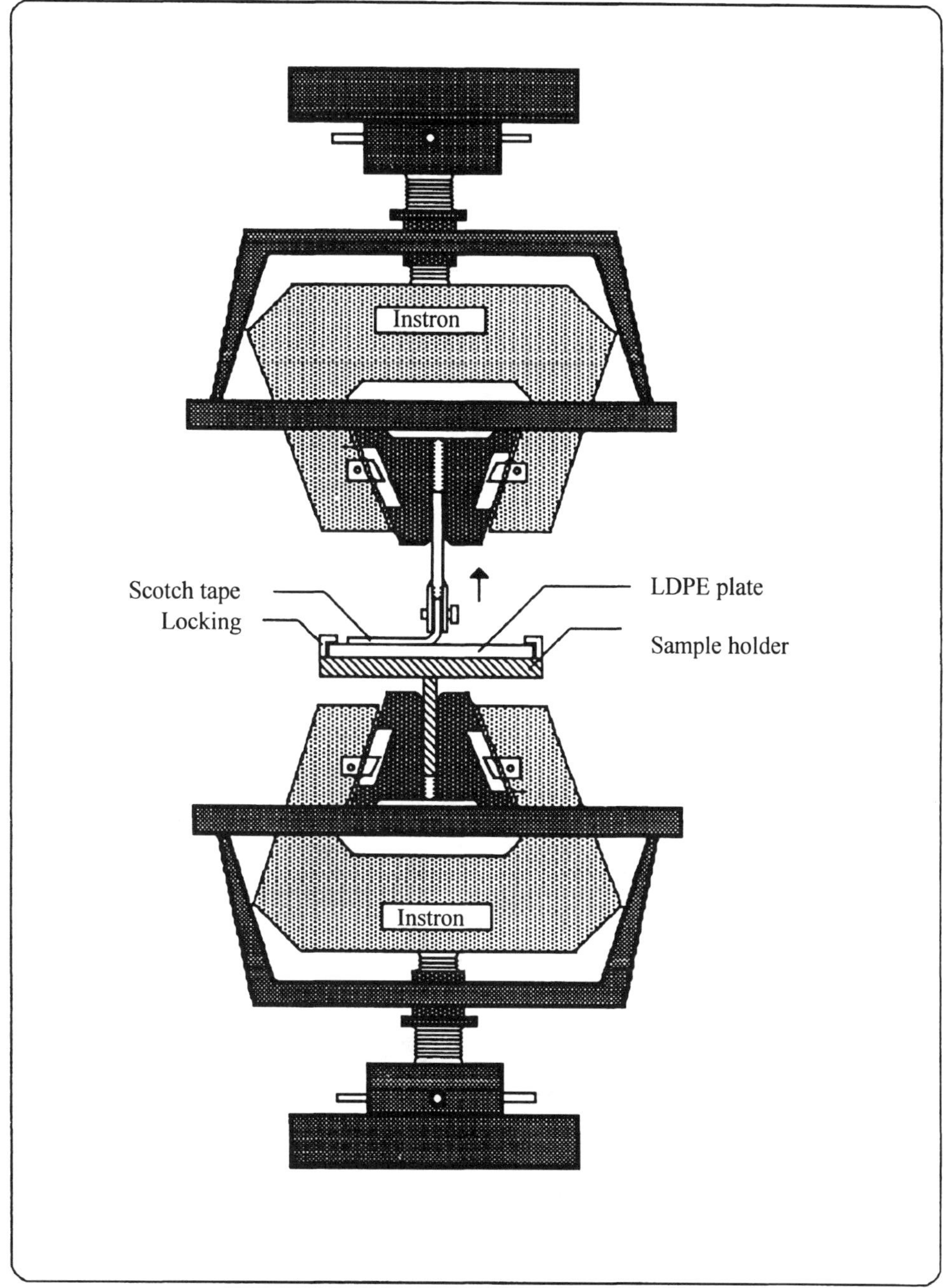

Figure 7. Instron measurements of adhesion of Scotch tape to plates of low density polyethylene (LDPE) as 90° peel tests.

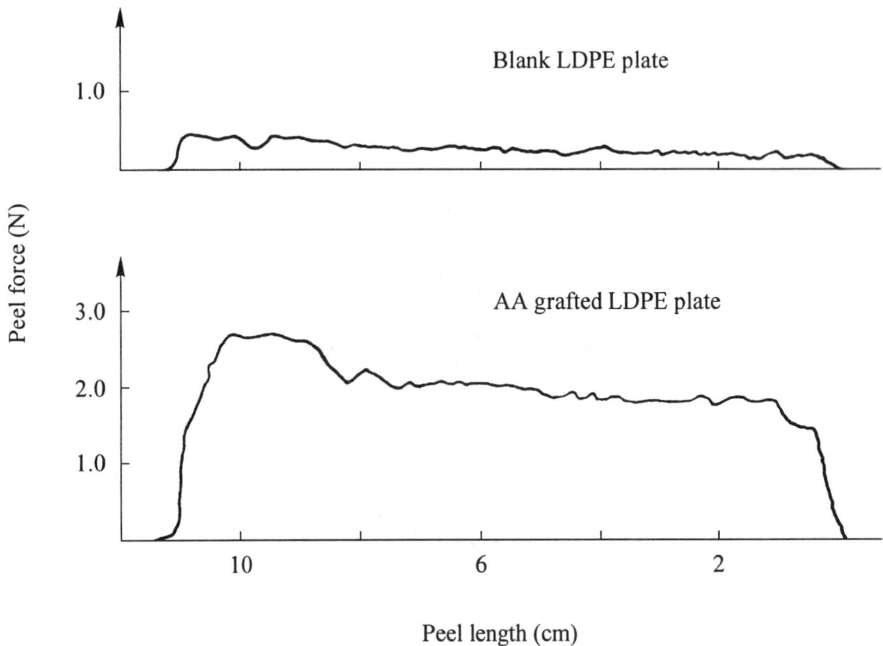

Figure 8. Recorded peel force curves of Scotch tape adhering to LDPE plates before and after grafting with acrylic acid, measured as 90° peel test shown in Fig. 7.

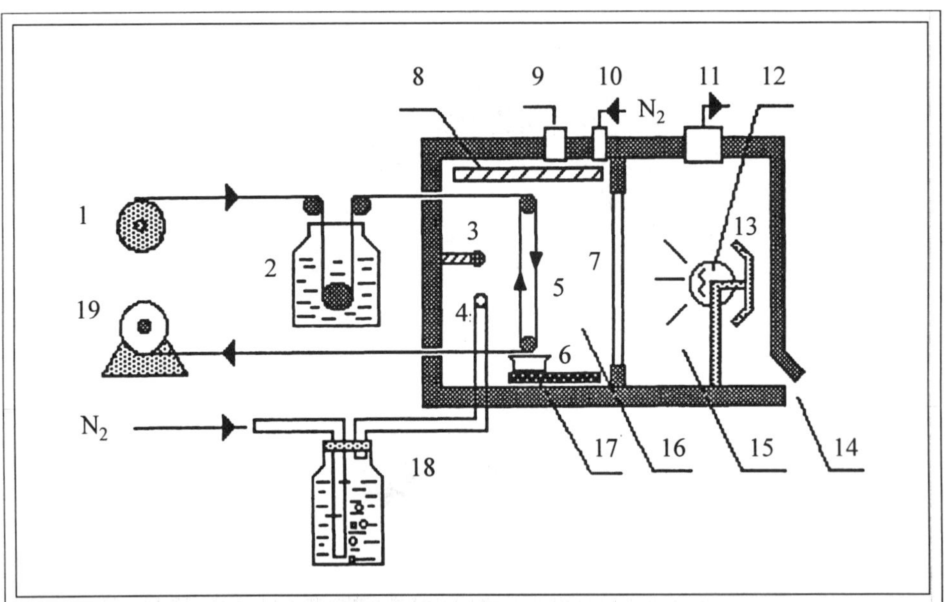

Figure 9. Device for surface photografting of film strips, filaments and yarns using the continuous process with presoaking and UV irradiation on line. The various parts are identified and the operation of the device is explained in the text.

3.2. Adhesion measurements for film and fiber 'surface photografted' with the continuous process

For the continuous surface photografting process a device shown in Fig. 9 was constructed and built in our laboratory [12]. The substrate, a strip of film, a filament or a yarn on a roll (1) is 'presoaked' in a solution of initiator and monomer (2) and pulled into a reactor (5) on line where it is moved on rollers while it is irradiated from the lamp house (15) with a high pressure UV lamp (12) with reflector (13) through the quartz window (7). The sample is pulled by a motor (19). The atmosphere in the reactor (16) is N_2 which is bubbled through the solution of initiator and monomer (18). An extra N_2 inlet (10) is used to remove air through the outlet (9). A vessel (17) can be used for solution of initiator and monomer. The temperature is regulated by a thermostat (3) with heater (6) and water cooling tubes (8). The lamp house is aircooled by ventilation with fan (11) with air inlet (14). In this device, the surface photografting reaction takes place in a thin layer of solution on the surface of the substrate. The reaction is the same as in the batch process (equations (1) and (2)). The initiator and monomer are transferred in the liquid phase to the substrate surface by presoaking which is more efficient than the vapor phase transfer in the batch process. Therefore, the reaction times in the continuous process are much shorter (5–10 s) than in the batch process (1–3 min). The grafting efficiency in the continuous process is 70 to 80%, i.e. only 20 to 30% of the polymer formed is homopolymer which is removed by washing in a proper solvent. These values are based on ESCA spectra of grafted films before and after washing.

The adhesion of polypropylene tape film to epoxy resin was measured before and after grafting with acrylamide (AM) (Fig. 9). The relative degree of grafting is given by ESCA spectra as the relative intensity ratio RI of the two lines N_{1s}/C_{1s}. The adhesion to Araldite epoxy resin is measured by pull-out tests as Newton (N) per mm^2 of film or fiber surface (Fig. 10). The results for PP tape film show an increase from 0.3 to 1.7 N/mm^2 with increased degree of grafting to RI ratio = 0.2 (Fig. 11).

Polypropylene (PP) fibers photografted as yarn with acrylamide (AM) show increased adhesion to epoxy resin measured on single filaments by a factor of about 3 for grafting to RI (N_{1s}/C_{1s}) = 0.085 (Fig. 12). In this case the mechanical properties of the PP filaments are slightly affected [13]. The tensile strength decreases by about 3% and Young's modulus increases by about 15% (Table 4). PP is known to be sensitive to UV degradation. In addition the tensile properties of the PP fibers may be affected by the acetone used as solvent. These changes in tensile properties are acceptable for the PP yarn as textile material.

Of particular interest is the surface modification of high strength polyethylene fibers (HSPE). A commercial HSPE yarn Spectra 900® from Allied Signal Inc., USA, was surface photografted with acrylic acid (AA) and acrylamide (AM) to RI-values of 0.10 to 0.15 [13]. In this case the surface photografting had no effect on the bulk properties measured on single filaments [14]. The tensile strength is 2.65 ± 0.15 GPa for all samples before and after grafting and Young's modulus 55 ± 3 GPa (Table 5). The adhesion to epoxy resin showed very significant increases as a result of surface grafting (Fig. 13). For grafting with AA the adhesion increased by a factor of about 3

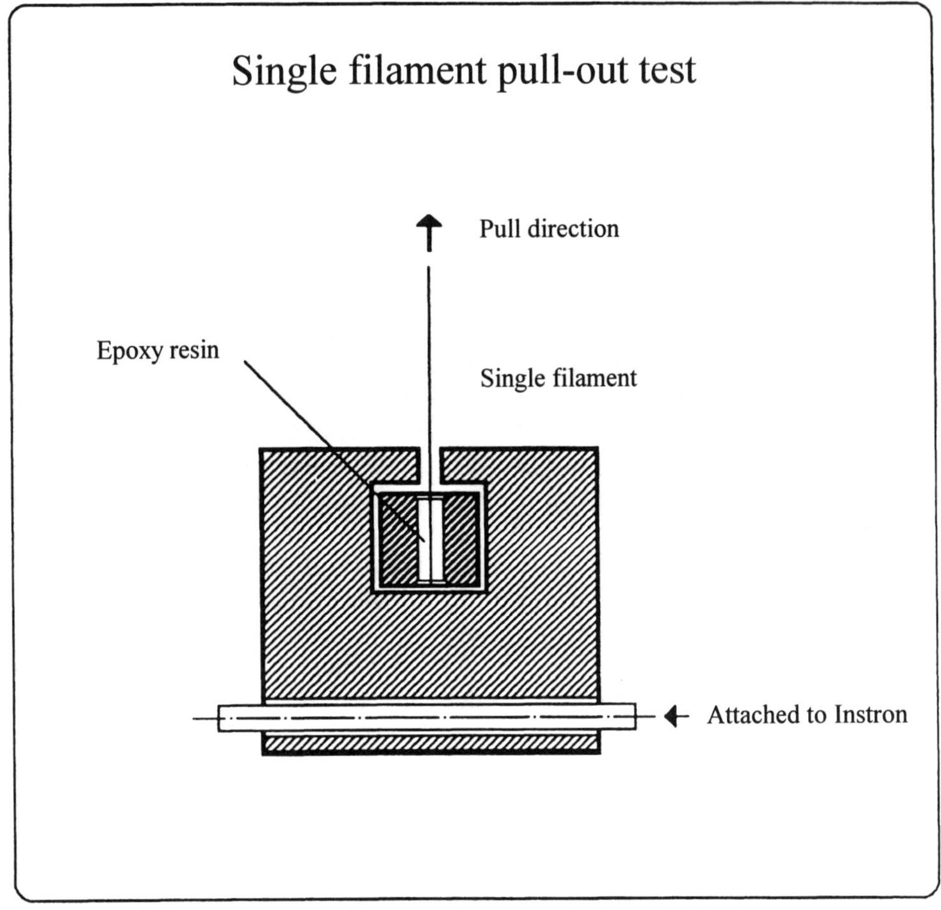

Figure 10. Holder for pull-out test for a filament or a strip of film after embedding and curing in epoxy resin. The filament or film strip is pulled out in an Instron testing instrument.

Table 4.
Bulk tensile properties of PP fibers before ('blank') and after surface photografting with acrylamide (AM) as monomer, benzophenone as photoinitiator and acetone as solvent

RI of N_{1s}/C_{1s} (by ESCA)	Tensile strength (GPa)	Young's modulus (GPa)
Blank	0.404	5.1
0.018	0.399	5.7
0.043	0.395	6.0
0.061	0.398	5.8
0.075	0.392	5.8

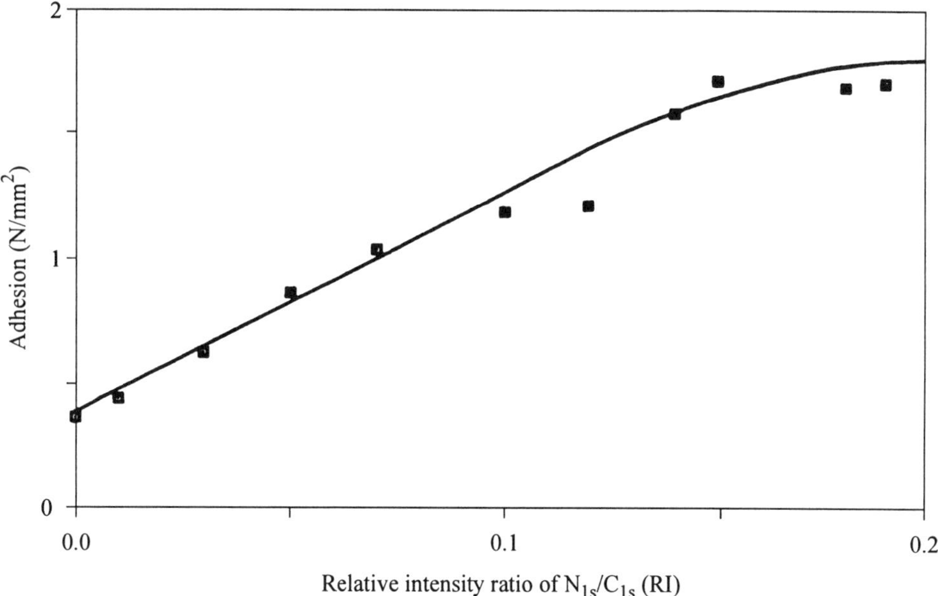

Figure 11. Adhesion of acrylamide grafted polypropylene tape film strips to epoxy resin measured by pull-out tests (Fig. 10) for samples with increased degree of grafting given as relative intensity ratio, RI = N_{1s}/C_{1s} in ESCA spectra.

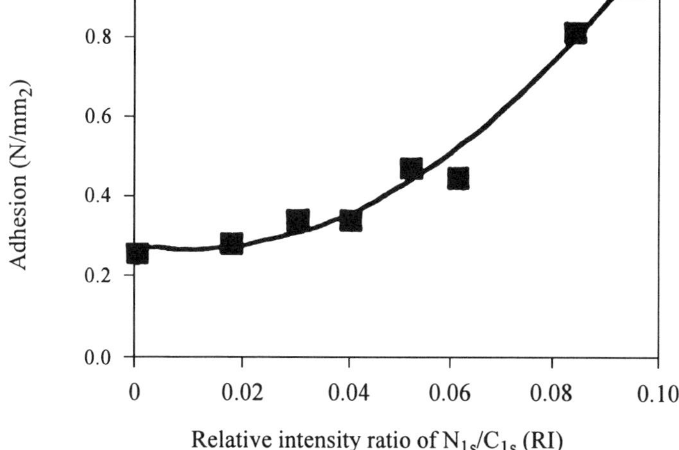

Figure 12. Adhesion of polypropylene fibers to epoxy resin measured by pull-out tests. The fibers are surface photo-grafted with acrylamide to increasing degree of grafting given as relative intensity ratio RI = N_{1s}/C_{1s} in ESCA spectra.

while grafting with AM increased the adhesion by a factor of 5 to 6. These results may be interpreted on a molecular level. The amide groups in the grafted AM are known to react with epoxy groups and form valence bonds (equation (3)). This would give a strong increase in adhesion.

Table 5.
Bulk tensile properties of HSPE fibers before ('blank') and after grafting with acrylamide (AM) and acrylic acid (AA) to increasing degrees of grafting measured by ESCA as relative intensity ratio (RI) of N_{1s}/C_{1s} and O_{1s}/C_{1s}, respectively

Grafted monomer	Irradiation time (s)	RI (ESCA)	Tensile strength (GPa)	Elongation at break %	Young's modulus (GPa)
Blank	0	—	2.6	5.0	52
AM	156	0.040	2.5	4.5	57
AM	18	0.053	2.8	4.7	59
AM	17	0.059	2.7	4.5	59
AM	84	0.081	2.7	5.3	52
AM	30	0.114	2.8	5.4	52
AM	126	0.115	2.8	5.0	56
AA	14	0.168	2.7	4.9	55

The bulk tensile properties are measured for single filaments at 20°C. All data in this table are given as three-sample averages.

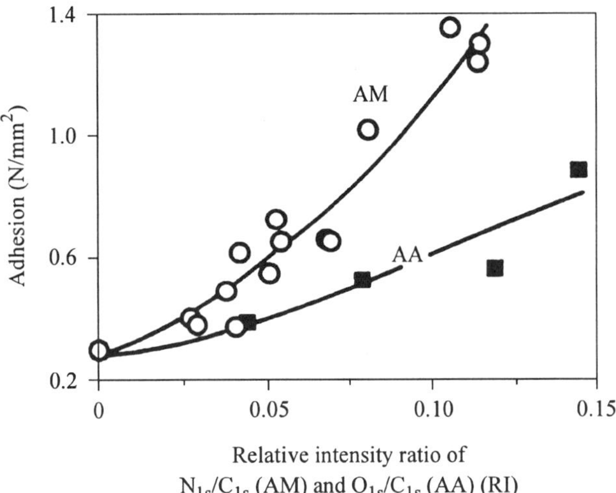

Figure 13. Adhesion to epoxy resin of high strength polyethylene fibers grafted with acrylic acid (AA) and acrylamide (AM) to increasing RI-values of O_{1s}/C_{1s} and N_{1s}/C_{1s}, respectively, in ESCA spectra. The adhesion is measured by pull-out tests (Fig. 10).

$$-C(=O)NH_2 \; + \; -CH\overset{O}{\frown}CH_2 \longrightarrow -\overset{O}{\overset{\|}{C}}-NH-CH_2-\overset{OH}{\overset{|}{C}H}- \qquad (3)$$

Under these conditions, the carboxyl groups of grafted AA are not expected to react with epoxy groups. Therefore, the smaller increase in adhesion after AA grafting is interpreted as due to increased hydrogen bonding with no new valence bonds formed.

4. CONCLUSIONS

1. The 'surface photografting' method developed as a batchwise process has been applied to plates of low density polyethylene. Grafting with functional acrylic monomers increases the adhesion to Scotch tape by a factor of 5 to 8 measured as 90° peel strength.
2. The 'surface photografting' method developed as a continuous process has been applied to strips of film, filaments and yarns of polyethylene (PE), and polypropylene (PP) grafted with acrylic acid (AA) and acrylamide (AM). The adhesion to cured epoxy resin, measured by pull-out tests, increases about 3 times after AA-grafting and up to 6 times after AM-grafting. The bulk tensile strength of high-strength PE fibers is not affected in the grafting process while that of PP fibers is decreased by a few percent.
3. The two photografting processes are promising for adhesion control of polymer surfaces in industrial applications.

Acknowledgements

The research work reported in this paper has been supported by grants from the National Swedish Board for Technical Development (STU) and the Carl Trygger Foundation for Scientific Research (CTS) which is gratefully acknowledged.

REFERENCES

1. D. Briggs, in: *Surface Pretreatments of Plastics and Metals*, D. M. Brewis (Ed.), pp. 199–226. Applied Science Publishers, London (1982).
2. H. Yasuda, *Plasma Polymerization*. Academic Press, New York (1985).
3. F. Yamamoto, S. Yamakawa and Y. Kato, *J. Polym. Sci., Polym. Chem. Ed.* **16**, 1883, 1897 (1978).
4. M. Torstensson, B. Rånby and A. Hult, *Macromolecules* **23**, 126 (1990).
5. G. J. Howard, S. R. Kim and R. H. Peters, *J. Society of Dyers Colourists* **85**, 468 (1969).
6. S. Tazuke, T. Matoba, H. Kimuri and T. Okada, *ACS Symp. Ser.* **121**, 217 (1980).
7. B. Rånby, Z. M. Gao, A. Hult and P. Y. Zhang, *Polymer Preprints (ACS)* **27** (2), 38 (1986).
8. B. Rånby, Z. M. Gao, A. Hult and P. Y. Zhang, *ACS Symp. Ser.* **364**, 168 (1988).
9. P. Y. Zhang, Surface Modification by Continuous Graft Copolymerization, Inaugural Dissertation, The Royal Institute of Technology, Stockholm, Sweden, October 14, 1988.
10. B. Rånby, *Makromol. Chem., Macromol. Symp.* **63**, 55 (1992).
11. K. Allmér, A. Hult and B. Rånby, *J. Polym. Sci., Polym. Chem. Ed.* **26**, 2099 (1988); **27**, 1641, 3405 (1989).
12. P. Y. Zhang and B. Rånby, *J. Appl. Polym. Sci.* **40**, 1647 (1990); **41**, 1459, 1469 (1990); **43**, 621 (1991).
13. Z. G. Feng and B. Rånby, *Angew. Makromol. Chem.* **195**, 17 (1992); **196**, 113 (1992); **199**, 33 (1992).

Polymer Surface Modification: Relevance to Adhesion, pp. 319–325
K. L. Mittal (Ed.)

Photochemical surface modification of polymers for improved adhesion

MELVIN J. SWANSON* and GARY W. OPPERMAN

BSI Corporation, 9924 West 74th Street, Eden Prairie, MN 55344, USA

Revised version received 6 August 1994

Abstract—A method using photoactivatable reagents is described to modify organic polymer surfaces without changing the bulk properties of the material. The reagents contain a benzophenone or other photoactivatable group which, when exposed to light of appropriate wavelength, generates highly reactive intermediates that covalently bond with nearly any organic material. Some general surface characteristics that can be achieved by this approach, on a wide range of materials, are good wettability, good lubricity, passivation, and priming for either adhesion or immobilization of other molecules. This technology provides tremendous flexibility for tailoring surface characteristics for a broad range of applications. Some materials that have desirable bulk properties for specific applications, however, have surface characteristics that make bonding them to other materials difficult. By photocoupling water-soluble polymers onto the surfaces of such materials, the surface properties can be modified to achieve greatly increased bond strengths with conventional adhesives. For example, using such techniques, the strength of bonding two pieces of high-density polyethylene to each other using a cyanoacrylate adhesive was increased by about 17-fold. Similarly, in preliminary experiments, the bond strengths of silicone rubber to polyvinyl chloride, using cyanoacrylate adhesive, were increased by more than 18-fold. This technology offers great potential for surface modification for improved adhesion.

Keywords: Photochemistry; surface modification; adhesion; hydrophilic polymers.

1. INTRODUCTION

Adhesives are a critical part of many manufacturing processes. A large number of adhesive formulations are available to cover a broad array of applications. However, there are limitations of adhesives that are being addressed by photochemical surface modification. In order to reduce damage to the environment as well as health risks for production personnel, there is a need to decrease or eliminate volatile organic solvents (VOCs) in adhesive formulations. There has been much progress in waterborne adhesives technology [1]. However, there is a need to improve waterborne adhesives

* To whom correspondence should be addressed.

for use with low surface energy materials such as polyolefins. Greater resistance to moisture is also needed for waterborne adhesives. In addition to moisture causing bond failure, contaminants in plastics, such as polyolefins, can contribute to failure of the bonds [2].

Various methods have been used for surface modifications to improve adhesion, including plasma treatments [3] and graft polymerizations [4]. Plasma treatments are often variable. Graft polymerizations are usually relatively complex procedures. BSI has developed a unique photochemical coating process (PhotoLink^{TM}) that is simple and amenable to most manufacturing processes [5]. In contrast to graft polymerizations, which often use photoinitiators for achieving the polymerization [4, 6, 7], this method does not involve polymerization but rather photochemical coupling of preformed polymers onto surfaces. This coating method involves applying photoactivatable reagents to the surface and exposing them to light to achieve covalent coupling to the surface. The photoreagents are typically polymers or biomolecules to which photoactivatable groups have been attached. However, they can also be heterobifunctional coupling reagents having both photogroups and thermochemically reactive groups.

Aromatic ketones, such as benzophenone derivatives, have commonly been used [5, 8] as the photoactivatable species for such photoreagents, although aryl azides [9] have also been used. Benzophenone derivatives (Fig. 1) are attached to the polymers or other molecules to be immobilized onto the surface. Upon illumination with ultraviolet (UV) light, benzophenones undergo excitation to an excited single state followed by intersystem crossing (ISC) to the triplet state which is capable of inserting into carbon–hydrogen bonds. This results in abstraction of a hydrogen atom from the surface, followed by collapse of the resulting radical pair to form a new carbon–carbon bond. The ability of the activated intermediate to return to the

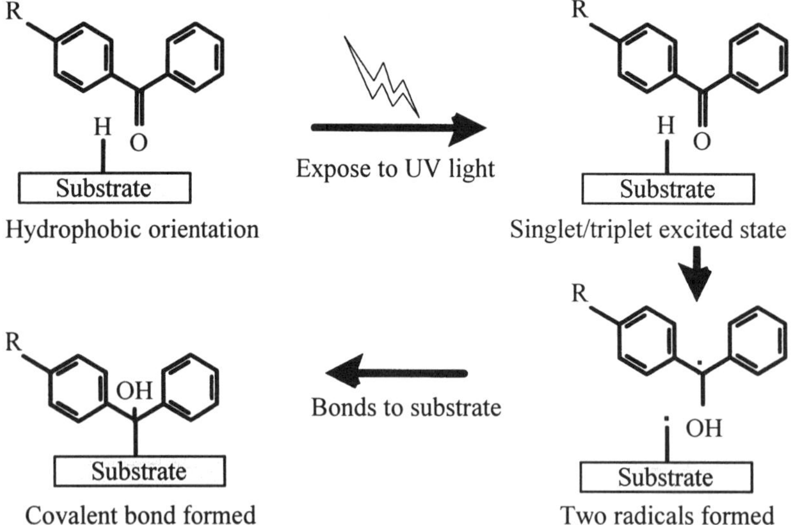

Figure 1. Mechanism of photogroup coupling to the surface. R refers to the polymer or other molecule to be immobilized onto the surface.

ground state if no coupling occurs and the tendency of the benzophenone derivatives to associate with hydrophobic surfaces contribute to the efficiency of the photocoupling. Photochemical covalent coupling of benzophenone derivatives to two different substrates has been demonstrated by coupling radiolabeled 4-benzoylbenzoic acid to thin films of polystyrene and polyvinylidenedifluoride, then dissolving the polymers and passing them through gel permeation HPLC columns [10]. In both cases, the presence of radioactivity in the isolated polymer peak confirmed the covalent coupling of the photogroup to the solid polymer surface during the UV illumination.

This general method has been used to achieve a variety of surface characteristics on a broad range of materials, including most organic materials as well as some inorganics, such as glass and metals, when used with an organosilane precoat. Some surface characteristics achieved by this approach are good wettability [11], good lubricity, reduced protein fouling, reduced microbial adherence [8], enhanced cell attachment, reduced thrombogenicity, improved biocompatibility [12], immobilization of biomolecules, barrier coatings, and increased adhesion. Some of the materials on which useful coatings have been achieved by this coating technology are polystyrene, polyvinylchloride, polyethylene, polypropylene, polycarbonate, nylon, Dacron, polymethylmethacrylate, polysulfone, polyurethane, polyvinylidenedifluoride, silicone rubber, and latex rubber.

2. METHODS

The general method for coating with these reagents is to apply the photoreagent by dipping, spraying, or brushing (depending on the application) photopolymers onto the surface and then to illuminate with light of appropriate wavelength for a few seconds to a few minutes (depending on the material and light intensity). The surface may then be washed to remove any uncoupled reagent. For some substrate materials and coatings, pretreatment of the surface is required in order to clean the surface or to increase the interaction of the photoreagent with the surface. Pretreatments can be with gas plasmas or solvents.

Photoactivatable hydrophilic polymers of acrylamide, vinylpyrrolidone, and other vinyl monomers are synthesized at BSI by including in the polymerization reaction low levels of comonomers having functional groups to which benzophenone derivatives can be coupled. As one pertinent example, photoactivatable polyvinylpyrrolidone [poly(vinylpyrrolidone/benzoylbenzoyl-aminopropylmethacrylamide)] is synthesized by copolymerization of *N*-vinylpyrrolidone with *N*-(3-aminopropyl)methacrylamide at 60 °C using azobisisobutyronitrile (AIBN) as an initiator. The polymer is purified by extensive dialysis against deionized water. Benzoylbenzoyl chloride (BBA-Cl), which is a photoactivatable benzophenone derivative, is then coupled to the amines using Schotten–Baumann conditions [13]. Using excess BBA-Cl over primary amines, a chloroform–water reaction mixture is shaken vigorously to generate an emulsion with triethylamine as an acid scavenger. The emulsion is centrifuged to separate the phases, after which the aqueous phase is extensively dialyzed against deionized water. Photogroup incorporation is measured by UV absorbance at 265 nm.

The photopolymers were applied to the surfaces from largely aqueous solutions (most commonly 1% photopolymer), either after plasma pretreatment or after prewetting with polar organic solvents. The photopolymers were applied by dipping the surface to be modified into an aqueous solution of photoactivatable hydrophilic polymer, drying, and illuminating with a Dymax PC-2 light welder (activation wavelength is about 335 nm) for 1–3 min at about 18 cm from the source at a temperature of about 50 °C. After photoactivation, the surfaces were thoroughly rinsed with deionized water and dried. The dry thickness of the photopolymer layer is typically approximately 0.5 μm but it can be varied significantly.

Adhesion experiments were performed by bonding the coated surfaces (and uncoated controls) with a commercial cyanoacrylate adhesive (Devcon Super Glue). Lap shear pull tests were performed with a tensile tester having a computer interface for data storage and processing.

3. RESULTS AND DISCUSSION

An example of the ability to modify the surfaces of a variety of materials using this technique is demonstrated by making the materials wettable as shown in Fig. 2. The materials were all coated with the same photopolymer formulation and have similar contact angles after modification. Theoretically, if the surface is completely coated with the polymer, the surfaces should have essentially the same wetting characteristics. The results indicate that this is the case.

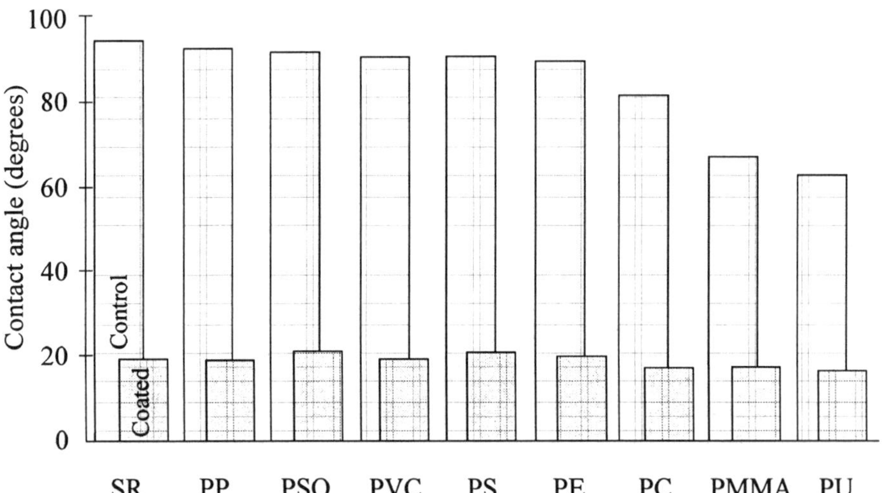

Figure 2. Wettability of surfaces coated with a mixture of photoactivatable polymers containing polyacrylamide, polyvinylpyrrolidone, and polyethylene glycol. The surfaces, silicone rubber (SR), polypropylene (PP), polysulsulfone (PSO), polyvinylchloride (PVC), polystyrene (PS), polycarbonate (PC), polymethylmethacrylate (PMMA), and polyurethane (PU), were first wiped with isopropanol (IPA) and then dipped into the polymer solution in 30% isopropanol in water. After photoactivation, the surfaces were rinsed and dried and then tested for water contact angles with a goniometer.

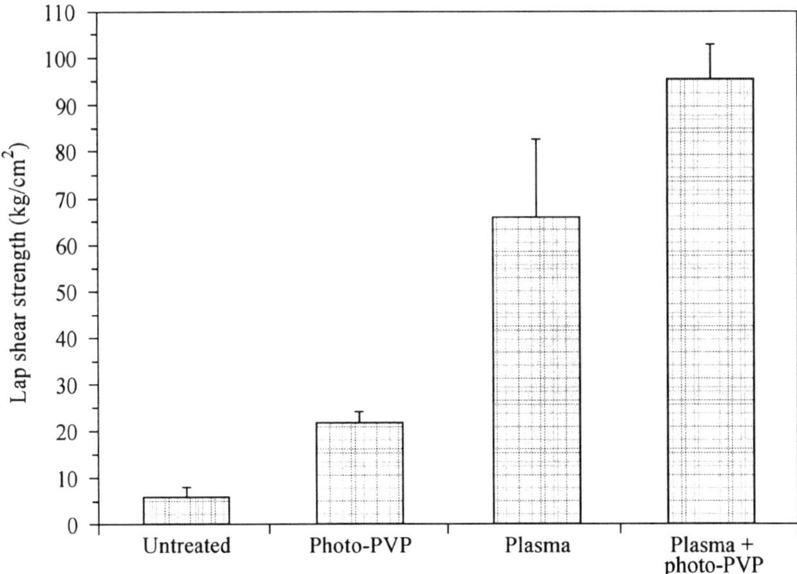

Figure 3. Lap shear bond strengths of high-density polyethylene (HDPE) with cyanoacrylate adhesive. The polyethylene was pretreated with an oxygen plasma for 2 min at 200 W. Photo-PVP was applied by dipping cleaned pieces of HDPE in a 10 mg/ml solution of photopolymer in 25% isopropanol in water and illuminated with a Dymax PC-2 light welder for 3 min. Pieces were bonded with commercial cyanoacrylate (Devcon Super Glue) and allowed to cure overnight before testing.

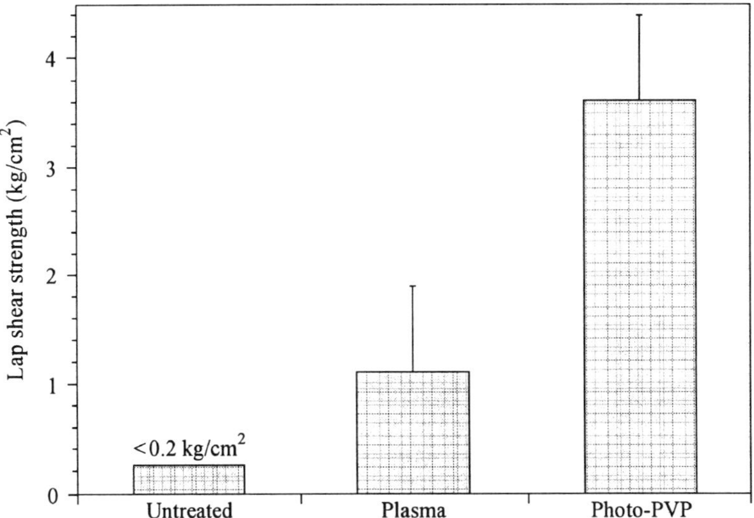

Figure 4. Lap shear strengths of silicone rubber bonded to polyvinylchloride (PVC) with cyanoacrylate adhesive. Medical grade silicone rubber was coated with photo-PVP using the same conditions as those described in Fig. 3. Untreated and treated silicone rubber pieces were bonded to PVC with cyanoacrylate adhesive and allowed to cure overnight before testing. Untreated samples fell apart before they could be fastened into the testing apparatus.

As one example of the usefulness of this technique to promote adhesion, a photoactivatable derivative of polyvinylpyrrolidone (photo-PVP) was coated onto high-density polyethylene surfaces. Pieces of coated and uncoated polyethylene were bonded together with a cyanoacrylate adhesive. Shear strength measurements were made of single lap joints of polyethylene onto polyethylene. The results are shown in Fig. 3. The failure appeared to be between the photopolymer and the polyethylene, although with plasma + photo-PVP some deformation of the polyethylene occurred before bond failure.

Photoactivatable polymers were also used to increase adhesion strength on silicone rubber. Silicone rubber was coated with photo-PVP, using a solvent pretreatment. Coated and uncoated silicone rubber pieces were bonded to polyvinylchloride using a cyanoacrylate adhesive. Figure 4 shows the shear strength results from this type of experiment. Bond failure appeared to be between the photopolymer and the silicone.

4. CONCLUSION

Although this surface modification technology has been under development and in use for several years, its application to improving adhesive bonding was begun fairly recently. This approach offers promise for providing a useful method for improving the adhesion of difficult-to-bond surfaces. In addition to making hydrophobic surfaces wettable, this general method can also be used to improve bonding by making the surfaces of two dissimilar materials the same.

The development of this application is continuing and is expected to result in significant improvements in surface modification for adhesives applications. Goals for continuing development include elimination of the need for plasma pretreatment and the use of solvents (except as required for surface cleaning for certain applications), achieving covalent bonds between the photopolymer and the adhesive in addition to between the substrates and the photopolymer, and further increases in the strengths of adhesive bonds.

REFERENCES

1. D. L. Prentice, E. J. McConnell, D. G. Pierson, W. R. Webster, G. Alexander, R. L. Cline and L. J. Maiorano, *Adhesives Age* **35**, 18 (Feb. 1992).
2. J. J. Bickerman, *J. Appl. Chem.* **11**, 80 (1961).
3. M. Strobel, C. S. Lyons and K. L. Mittal (Eds), *Plasma Surface Modification of Polymers: Relevance to Adhesion*. VSP, Zeist, The Netherlands (1994).
4. K. Yamada, H. Tsutaya, S. Tatekawa and M. Hirata, *J. Appl. Polym. Sci.* **46**, 1065 (1992).
5. P. E. Guire, S. G. Dunkirk, M. W. Josephson and M. J. Swanson, US Patent No. 5,002,582 (1991).
6. K. Yamada, T. Kimura, H. Tsutaya and M. Hirata, *J. Appl. Polym. Sci.* **44**, 993 (1992).
7. S. Tazuke and H. Kimura, *Makromol. Chem.* **179**, 2603 (1978).
8. S. G. Dunkirk, S. L. Gregg, L. W. Duran, J. D. Monfils, J. E. Haapala, J. A. Marcy, D. L. Clapper, R. A. Amos and P. E. Guire, *J. Biomater. Applications* **6**, 131 (1991).
9. P. E. Guire, *Methods Enzymol.* **44**, 280 (1976).

10. R. A. Amos, A. B. Anderson, D. L. Clapper, P. H. Duquette, L. W. Duran, S. G. Hohle, D. J. Sogard, M. J. Swanson and P. E. Guire, in: *Handbook of Biomaterials and Applications*. Marcel Dekker, Inc. (accepted for publication).
11. M. J. Swanson, S. G. Dunkirk, J. A. Pietig and P. E. Guire, *Chemtech* **22**, 624–626 (1992).
12. S. M. Kirkham and M. E. Dangel, *Ophthalmic Surg.* **22**, 455–461 (1991).
13. R. T. Morrison and R. N. Boyd, *Organic Chemistry*, p. 476. Allyn and Bacon, Boston (1959).

Polymer Surface Modification: Relevance to Adhesion, pp. 327–336
K. L. Mittal (Ed.)

Surface properties of polyethylene grafted with triallyl cyanurate in the presence of an electron beam

J. KONAR[1] and ANIL K. BHOWMICK[2,*]

[1]*Department of Chemistry and* [2]*Rubber Technology Centre, Indian Institute of Technology, Kharagpur 721302, India*

Revised version received 2 June 1994

Abstract—The surface energy and thermodynamic work of adhesion of polyethylene grafted with triallyl cyanurate in the presence of an electron beam have been determined. The surface energy increased with the grafting level and with the irradiation dose up to 10 Mrad. A similar trend was observed with the work of adhesion. The changes in these surface properties were correlated with the concentration of the polar groups as measured by IR (infrared) spectroscopy and ESCA (electron spectroscopy for chemical analysis) studies.

Keywords: Polyethylene; grafting; surface energy; work of adhesion; electron beam; ESCA.

1. INTRODUCTION

Electron beam-initiated grafting of monomers onto polymers is a relatively new technique with certain advantages over conventional grafting processes. Absence of catalyst residue, complete control of the temperature, a solvent-free system, and a source of an enormous amount of radicals and ions are some of the reasons why this technique has gained commercial importance in recent years. The modification of polyethylene (PE) for heat-shrinkable products using this technique has been reported by us [1, 2]. Such modification is expected to alter the surface properties of PE and lead to improved adhesion and dyeability. In this paper, our work on the characterization of PE grafted with triallyl cyanurate in the presence of an electron beam, by contact angle measurements and ESCA studies is reported. We have used these techniques earlier to study the surface properties of PE functionalized by dibutyl maleate and vinyl trimethoxysilane [3].

Measurement of the contact angle at a solid–liquid interface is a widely used method for the determination of the surface energy of solid polymers. Fowkes [4] first proposed that the surface energy of a pure phase, γ_a, could be represented by the sum of

*To whom correspondence should be addressed.

the contributions from different types of force components, especially the dispersion and the polar components, such that

$$\gamma_a = \gamma_a^d + \gamma_a^p, \tag{1}$$

where γ_a^d is the dispersion force component and γ_a^p is the polar force component. Fowkes [5] then suggested that the geometric mean of the dispersion force components was a reliable prediction of the interaction energies at the interface caused by dispersion forces. Thus, the interfacial free energy, γ_{ab}, between phases a and b may be given by

$$\gamma_{ab} = \gamma_a + \gamma_b - 2(\gamma_a^d \gamma_b^d)^{1/2}, \tag{2}$$

where γ_a and γ_b are the surface free energies of phases a and b, respectively. Owens and Wendt [6] and Kaelble and Uy [7], by analogy with the work by Fowkes, proposed another energy term, $2(\gamma_a^p \gamma_b^p)^{1/2}$, to account for the effect of the non-dispersion forces (γ_a^p and γ_b^p are the polar components of the surface free energy of phases a and b, respectively). Thus,

$$\gamma_{ab} = \gamma_a + \gamma_b - 2(\gamma_a^d \gamma_b^d)^{1/2} - 2(\gamma_a^p \gamma_b^p)^{1/2}. \tag{3}$$

Considering a solid–liquid system, this relationship may be combined with the well-known Young's equation to eliminate the interfacial free energy. Hence,

$$\cos\theta = -1 + \frac{2(\gamma_s^d \gamma_l^d)^{1/2}}{\gamma_l} + \frac{2(\gamma_s^p \gamma_l^p)^{1/2}}{\gamma_l}, \tag{4}$$

assuming the spreading pressure to be negligible and s and l to represent solid and liquid, respectively. The thermodynamic work of adhesion could also be obtained from the equilibrium contact angle θ and the surface tension of the liquid as follows:

$$W_A = \gamma_l(1 + \cos\theta). \tag{5}$$

Combining equations (4) and (5),

$$W_A = 2(\gamma_s^d \gamma_l^d)^{1/2} + 2(\gamma_s^p \gamma_l^p)^{1/2}. \tag{6}$$

Various workers including us have used these equations extensively to understand the surface energetics of polymeric solids.

The change in the surface properties of modified polymers is due to the generation of surface groups, which can be quantified by using electron spectroscopy for chemical analysis (ESCA). A detailed account of the use of ESCA for the investigation of polymers has been published [8]. A number of other publications on this topic have also appeared in the past [9, 10]. We used ESCA earlier to characterize functionalized PE [3].

2. EXPERIMENTAL

2.1. *Materials*

Polyethylene (PE) (Indotherm 16 MA 400) of density 916 kg/m^3 and melt flow index (MFI) 40 g/10 min was obtained from IPCL, Baroda, India.

Triallyl cyanurate, TAC (Nauryset 300),

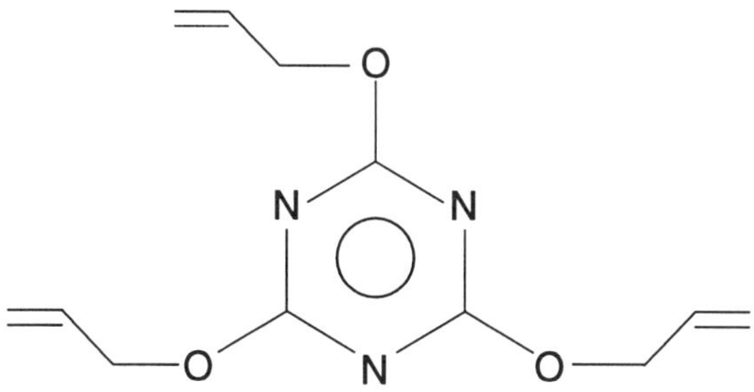

TAC

of density 1115 kg/m^3, flash point $> 142^\circ$C, and thermal decomposition temperature $> 150^\circ$C was obtained from Akzo Chemie Rubber Chemicals, The Netherlands. TAC is a polyfunctional unsaturated monomer, which is very efficient in producing high yields of radicals during irradiation.

2.2. *Grafting of TAC onto PE*

PE samples were mixed with TAC at various doses in a Brabender Plasticorder, PLE-330, at 60 rpm at 120°C for 4 min. The samples were then compression-molded at 150°C for 3 min in the form of rectangular sheets (150 $\times$ 150 mm) of 2.5 mm thickness, which were then subjected to electron beam irradiation in air at 25°C at Bhabha Atomic Research Centre, Bombay, India. Irradiation doses of 2, 5, 10, 15, and 20 Mrad were used. The formulations of the samples are given in Table 1 and the specifications of the electron beam accelerator in Table 2. All the samples were extracted with xylene to remove ungrafted TAC from the treated PE sheets before IR/ESCA/contact angle measurements. The samples were stored in a refrigerator.

2.3. *Degree of grafting*

The degree of grafting was determined by IR spectroscopy (Perkin Elmer, model 837) of thin films. From the absorbance at the 1140 cm^{-1} peak (corresponding to the $-O-CH_2-$ stretching of TAC) with respect to that at 1470 cm^{-1} (due to the scissor vibration of $>CH_2$ groups), the relative grafting level was calculated with the help of a calibration curve drawn by using physical mixtures of PE and TAC at different ratios.

Table 1.
Formulations of the samples

PE (parts)	TAC level (parts)	Irradiation dose (Mrad)	Sample designation
100	0	0	$T_{0/0}$
100	1	2	$T_{1/2}$
100	1	5	$T_{1/5}$
100	1	10	$T_{1/10}$
100	1	15	$T_{1/15}$
100	1	20	$T_{1/20}$
100	0.5	15	$T_{0.5/15}$
100	1.5	15	$T_{1.5/15}$
100	2	15	$T_{2/15}$
100	3	15	$T_{3/15}$
100	—	2	$T_{0/2}$
100	—	5	$T_{0/5}$
100	—	10	$T_{0/10}$
100	—	15	$T_{0/15}$
100	—	20	$T_{0/20}$

Table 2.
Specifications of the electron beam accelerator

Energy range	0.5–2.0 MeV
Beam power over the whole energy range	20 kW
Beam energy spread	±10%
Average current (E – 1.5 MeV)	15 mA
Adjusting limits for current	0–30 mA
Accelerating voltage frequency	100–120 MHz
Duration	400–700 μs
Repetition rate	2–50 Hz
Pulse curent: maximum	900 mA
minimum	400 mA
Power supply (PS) voltage	3 × 380/220 V
PS voltage frequency	50 Hz
Consumption of power (total)	150 kW

2.4. Contact angle measurements

The contact angles were determined using a contact angle meter (Kernco, model GII). The sessile drop method, using water and formamide as probe liquids, was used for contact angle measurements. Each contact angle quoted is the mean of at least five measurements with a maximum error in θ of $\pm 2^\circ$. All investigations were carried out using a polymer plate of dimensions 10 mm × 10 mm × 2 mm in vapor-saturated air at $(20 \pm 2)^\circ$C in a closed sample box.

2.5. *Electron spectroscopy for chemical analysis (ESCA)*

ESCA data were obtained using a VG Scientific ESCA LAB MK II spectrometer employing exciting radiation of 1487 eV (Al K_α). The samples examined were in the form of rectangles (20 mm × 8 mm), cut from untouched sheet and mounted with double-sided adhesive tape onto the probe tip. The working pressure in the spectrometer was 4.3×10^{-9} Torr. All spectra were referenced to the C 1*s* peak for neutral carbon, which was assigned a value of 284.5 eV, and were recorded at an electron take-off angle of 90° with respect to the polymer surface.

3. RESULTS AND DISCUSSION

3.1. *Surface energy of grafted PE*

The equilibrium contact angles for PE and grafted PE with water and formamide are presented in Table 3. The thermodynamic work of adhesion, W_A, calculated by

Table 3.
Contact angle θ and thermodynamic work of adhesion (W_A) for PE and grafted PE

Sample	θ_{H_2O} (degrees)	$W_A(H_2O)$ (mJ/m^2)	θ_{HCONH_2} (degrees)	$W_A(HCONH_2)$ (mJ/m^2)
$T_{0/0}$	92	70	75	73
$T_{0.5/15}$	88	75	72	76
$T_{1.5/15}$	82	83	67	81
$T_{2/15}$	79	87	64	84
$T_{3/15}$	75	92	61	86
$T_{1/2}$	72	95	60	87
$T_{1/5}$	69	99	58	89
$T_{1/10}$	65	104	53	93
$T_{1/15}$	70	98	60	87
$T_{1/20}$	72	95	62	85
$T_{0/2}$	88	75	73	75
$T_{0/5}$	82	83	72	76
$T_{0/10}$	78	88	68	80
$T_{0/15}$	80	85	69	79
$T_{0/20}$	84	80	69	79

equation (5) is also included in the table. The polar and dispersion force contributions to the surface free energy of PE were calculated using equation (4) and the following values of γ_l^d and γ_l^p for the probe liquids:

Water : $\gamma_l = 72.8\ \text{mJ/m}^2$; $\gamma_l^d = 21.8\ \text{mJ/m}^2$; $\gamma_l^p = 51.0\ \text{mJ/m}^2$;

Formamide : $\gamma_l = 58.2\ \text{mJ/m}^2$; $\gamma_l^d = 39.5\ \text{mJ/m}^2$; $\gamma_l^p = 18.7\ \text{mJ/m}^2$.

Table 3 reveals that the grafting of TAC onto PE decreases the equilibrium contact angles of water and formamide from 92° to 65° and from 75° to 53°, respectively. This decrease is a function of the monomer level and the irradiation dose. At a fixed irradiation dose of 15 Mrad, variation of the TAC level from 0.5 to 3 parts causes a reduction in the contact angles of water by 13° (from 88° to 75°) and of formamide by 11° (from 72° to 61°). This is due to the fact that the concentration of polar groups on the surface increases as a result of grafting of the polyfunctional unsaturated monomer. These polar groups are evident from the IR and ESCA studies described later. With increasing irradiation dose at a constant TAC level of 1 part, the contact angles of water and formamide are lowered by 7° up to an irradiation dose of 10 Mrad, and then increased for up to 20 Mrad irradiation dose. The results can be explained in terms of the level of surface modification described later. The work of adhesion calculated using equation (5) follows a similar but opposite trend to that of the contact angles. For example, there is an increase in the work of adhesion from 70 to 92 mJ/m^2 with the increase of TAC up to 3 parts at 15 Mrad itrradiation dose. Similarly, the highest value of W_A using water as the probe liquid for the sample with 1 part TAC irradiated by a 10 Mrad dose is 104 mJ/m^2. The effect of irradiation of samples in the absence of TAC on the contact angles is also reported in Table 3. Modification of the surface takes place, as is evident from the decrease in the contact angles of water and formamide. The change, which is maximum at an irradiation dose of 10 Mrad, is due to the generation of polar functionalities on the surface. This is also corroborated from the IR/ESCA studies described later. The contact angles are lowered further when TAC is incorporated in the system (cf. $T_{0/5}$ with $T_{1/5}$, $T_{0/15}$ with $T_{1/15}$, etc.).

The surface energy of grafted PEs was calculated using equation (1). The results are reported in Table 4. It is observed that the dispersion component of the surface energy, γ_s^d, does not increase with the incorporation of TAC, though there is an approximately two- to three-fold increase in the polar component of the surface energy, resulting in an overall increase of the surface energy value. A change in the total surface energy of the samples without TAC due to irradiation (i.e. $T_{0/2}$, $T_{0/5}$, $T_{0/10}$, etc.) is also evident.

With the increase in irradiation dose, however, there is an optimum value of the surface energy at 10 Mrad irradiation. In order to explain these results, the grafting level was calculated from the IR spectra [1] and is given in Table 4. It is clear that the changes in surface energy values are generaly in accord with the changes in the grafting level. As the grafting level increases, the surface energy values increase. The variation of grafting level with the variation of irradiation dose or TAC level

Table 4.
Surface energy of grafted PEs

Sample	Grafting level (mmol)	γ_s^d (mJ/m^2)	γ_s^p (mJ/m^2)	γ_s (mJ/m^2)
$T_{0/0}$	—	20	4	24
$T_{0.5/15}$	4.8	20	6	26
$T_{1.5/15}$	5.1	20	8	28
$T_{2/15}$	15.0	21	10	31
$T_{3/15}$	40.0	20	12	32
$T_{1/2}$	5.7	19	15	34
$T_{1/5}$	6.7	18	17	35
$T_{1/10}$	7.4	20	19	39
$T_{1/15}$	4.6	17	17	34
$T_{1/20}$	2.6	17	16	33
$T_{0/2}$	—	18	6	24
$T_{0/5}$	—	15	10	25
$T_{0/10}$	—	16	12	28
$T_{0/15}$	—	16	11	27
$T_{0/20}$	—	19	7	26

was explained in our earlier work in terms of the mechanism of the grafting reaction [1].

IR studies of irradiated samples without TAC indicate absorbance peaks at 1140 and 1732 cm^{-1} due to the generation of $-O-CH_2-$ and $>C=O$ functionalities. The ratio of the peak at 1732 cm^{-1} with respect to that at 1470 cm^{-1} was calculated for all samples. On introduction of TAC, the ratio increases by four- to eight-fold depending on the amount of TAC and the irradiation dose.

3.2. X-ray photoelectron spectroscopy (ESCA)

The chemical compositions of modified surfaces are difficult to analyse by internal reflection IR spectroscopy because of its deep penetration ($\sim$ 10 μm); concomitantly, chemical changes in a thin surface layer are missed. On the other hand, ESCA with its shallow penetration ($\sim$ 1 nm) greatly enhances the sensitivity of the surface analysis. Figure 1 displays the high-resolution spectra for PE in the O 1*s*, N 1*s*, and C 1*s* regions, and illustrates several points of interest. Three representative samples, i.e. control PE and grafted PEs at 15 Mrad irradiation dose using 2 parts TAC ($T_{2/15}$) and at 10 Mrad irradiation dose with 1 part TAC ($T_{1/10}$), are shown in the figure. The C 1*s* spectrum for the control PE film surface shows a peak at 284.5 eV indicative of the carbon bonds in PE with a small shoulder, probably related to the presence of an O 1*s* peak. This control sample shows a small percentage of oxygen but no nitrogen on the surface before grafting. After grafting, several percent oxygen and a small percentage of nitrogen are incorporated into the PE surface, and the C 1*s* spectrum shows new peaks towards higher binding energy, indicating the formation of carbon–oxygen and carbon–nitrogen functionalities. A simple deconvolution of the spectrum for C 1*s* in Fig. 1c is shown. The high binding energy region of the C 1*s* spectrum can be fitted with three peaks corresponding to carbon atoms with a single bond to oxygen or nitrogen at 287.0 eV, carbon atoms with two bonds to oxygen at 288.0 eV, and carbon atoms with three bonds to oxygen at 290.0 eV.

The O 1*s* spectrum of the control sample shows a peak at about 532.2 eV, while the O 1*s* peak for grafted PEs shifts slightly towards higher binding energies (532.9 eV) (Table 5). This shift confirms that the relative concentration of carboxyl and carbonyl groups increases in grafted PE. It is known that most oxygen functional groups present O 1*s* binding energies at 532.0 eV, while the ester oxygen in carboxyl groups shows a binding energy at 533.0 eV [10]. Also, the O 1*s* peak area increases three times on modification. The grafted samples show a N 1*s* peak at 400.7 eV and the incorporation of a nitrogen peak suggests a grafting reaction. It must be pointed out here that the samples even without TAC show a surface concentration of oxygen on irradiation in line with the IR studies. The O/C ratio, however, increases significantly for the TAC grafted PE.

However, the $T_{1/10}$ sample shows a higher concentration of TAC on the surface as reflected by the O 1*s* and N 1*s* peak areas. Comparison of the grafting level and surface energy values of the $T_{2/15}$ and $T_{3/15}$ samples also indicates that although the surface energies of the two samples are almost the same, there is a three-fold increase in the grafting level as calculated from IR spectroscopy. These results indicate that the concentration of TAC on the surface is not the same as that of the bulk. Also, at higher irradiation doses, there is the possibility of degradation of TAC grafted PE and the loss of small molecules, resulting in a reduction in the concentration of oxygen and nitrogen, as well as the surface energy.

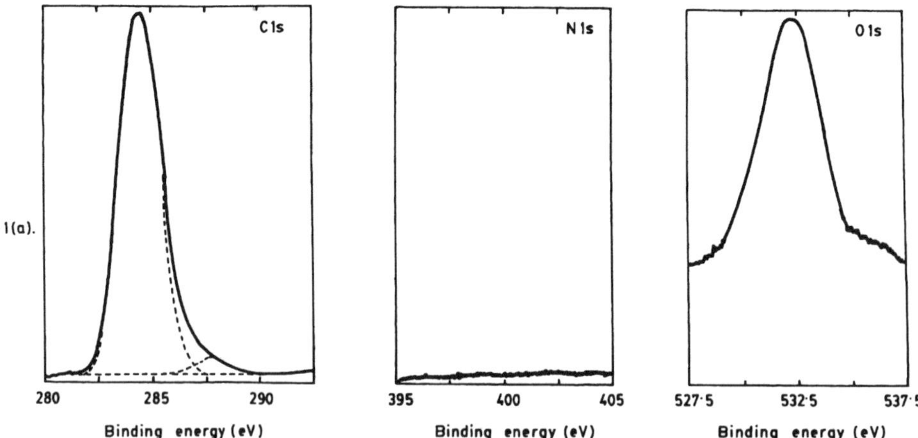

Figure 1a. Core level spectra of polyethylene ($T_{0/0}$).

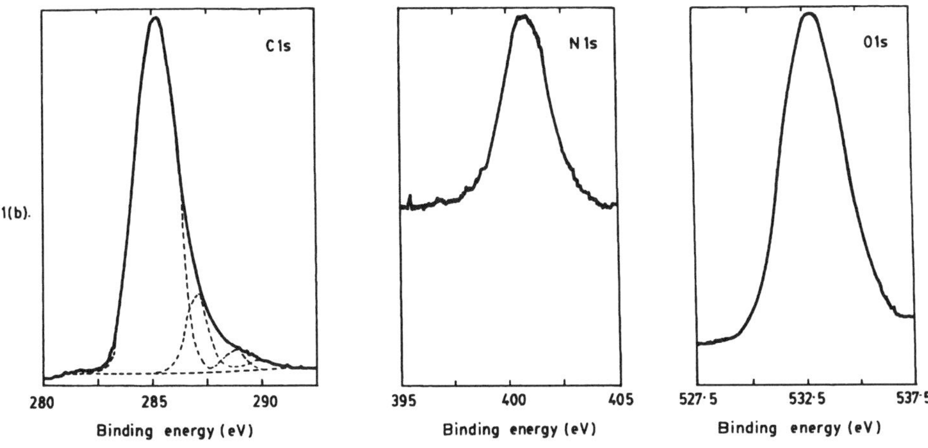

Figure 1b. Core level spectra of polyethylene grafted with 2 parts TAC at an irradiation dose of 15 Mrad ($T_{2/15}$).

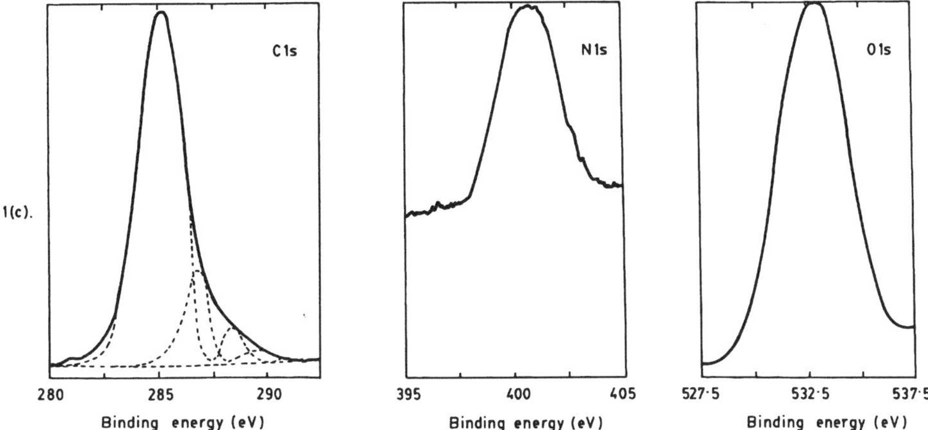

Figure 1c. C 1*s*, N 1*s*, and O 1*s* regions in the ESCA spectrum of polyethylene grafted with 1 part TAC at an irradiation dose of 10 Mrad ($T_{1/10}$).

Table 5.
XPS details of C 1*s*, N 1*s*, and O 1*s* spectra from grafted PE

Sample	C 1*s*		N 1*s*		O 1*s*	
	Peak centre (eV)	Area (eV/ms)	Peak centre (eV)	Area (eV/ms)	Peak centre (eV)	Area (eV/ms)
$T_{0/0}$	284.5	20.34	—	—	532.2	2.02
$T_{2/15}$	285.3	20.29	400.7	0.89	532.9	7.30
$T_{1/10}$	285.5	20.31	400.9	0.91	533.0	7.41
$T_{0/10}$	284.9	20.35	—	—	532.4	2.23
$T_{0/15}$	285.0	20.28	—	—	532.6	2.64

4. CONCLUSIONS

With the increase in the extent of modification of PE by TAC at a constant irradiation dose, the contact angles measured using water and formamide are lowered. The surface energy and the work of adhesion increase accordingly. IR and ESCA studies indicate a higher concentration of polar groups on the surface. With the increase in radiation dose at a constant TAC level, the surface energy and the work of adhesion pass through a maximum in line with the level of grafting. Careful control of the irradiation dose and TAC level can produce modified polyethylenes of higher surface energies.

REFERENCES

1. T. K. Chaki, S. Roy, A. B. Majali, R. S. Despande, V. K. Tikku and A. K. Bhowmick, *J. Appl. Polym. Sci.* **53**, 141 (1994).
2. T. K. Chaki, A. B. Majali, R. S. Despande, V. K. Tikku and A. K. Bhowmick, *Angew. Makromol. Chem.* **217**, 61 (1994).
3. J. Konar, A. K. Sen and A. K. Bhowmick, *J. Appl. Polym. Sci.* **48**, 1579 (1993).
4. F. M. Fowkes, *J. Phys. Chem.* **67**, 2538 (1963).
5. F. M. Fowkes, in: *Treatise on Adhesion and Adhesives*, R. L. Patrick (Ed.), Vol. 1, p. 344. Marcel Dekker, New York (1967).
6. D. K. Owens and R. C. Wendt, *J. Appl. Polym. Sci.* **13**, 1740 (1969).
7. D. H. Kaelble and K. C. Uy, *J. Adhesion* **2**, 50 (1970).
8. A. Dilks, in: *Electron Spectroscopy*, C. R. Brundle and A. D. Baker (Eds), Vol. 4, p. 415. Academic Press, London (1980).
9. D. T. Clark, W. T. Feast, D. Kilcast and W. K. R. Musgrave, *J. Polym. Sci., Polym. Chem.* **11**, 389 (1973).
10. D. Briggs, V. J. I. Zichy, D. M. Brewis, J. Comyn, R. H. Dahm, H. A. Green and M. B. Konieczko, *Surface Interface Anal.* **2**, 107 (1980).

Polymer Surface Modification: Relevance to Adhesion, pp. 337–348
K. L. Mittal (Ed.)

Surface modification of PVDF membranes by grafting of a vinylphosphonium compound

DER'E JAN, JOONG S. JEON and SRINI RAGHAVAN*

Department of Materials Science and Engineering, University of Arizona, Tucson, AZ 85721, USA

Revised version received 29 April 1994

Abstract—A technique to incorporate positively charged groups onto the surface of microporous polyvinylidene fluoride membrane filters has been developed. In this method, a water-soluble vinyltriphenylphosphonium bromide compound was grafted onto polyvinylidene fluoride membranes using ^{60}Co γ-irradiation. The electrical characteristics of prepared membranes were measured by streaming potential and anionic dye challenge tests. The compatibility of these charge-modified membranes with ultrapure water was also investigated. Results show that these charge-modified membranes are characterized by a positive zeta potential in the pH range from 4 to 9.3. From the dye challenge tests the density of positively charged sites was calculated to be approximately five times larger than that of unmodified polyvinylidene fluoride. The modified membranes released less than 1 ppb of total organic carbon (TOC) into ultrapure water and thus appear to have potential for use in DI water systems.

Keywords: PVDF membrane; vinyltriphenylphosphonium bromide; charge modification; zeta potential; filtration; surface modification.

1. INTRODUCTION

De-ionized (DI) water and liquid chemicals, including acids, solvents, photoresists, etchants, developers, etc. are used in large scale during semiconductor manufacturing. To minimize particulate contamination on semiconductor devices, these liquids need to be processed so that contamination is eliminated in the chemical distribution systems. Such a removal is commonly attempted but only partially accomplished using the technique of filtration. Selection of filter membranes for purification of these liquid chemicals and DI water is critical to contamination-free manufacturing. Charge-modified nylon 66 membranes have found widespread use in the production of ultrapure water (18 MΩ-cm resistivity) because of their high strength, flexibility, narrowly controlled pore size, minimal release of particulates, and high wettability in water [1], but they exhibit poor resistance to aggressive chemicals used in the semiconductor industry.

Polyvinylidene fluoride (PVDF) based membrane filters, because of their inertness, have been finding increased applications in the filtration of DI water and other

*To whom correspondence should be addressed.

chemicals. Since PVDF is hydrophobic, removal of macromolecular contaminants in ultrapure water, such as bacterial lipopolysaccharides and endotoxins which contain hydrophobic groups, is possible by hydrophobic adsorption [2]. Because hydrophobic adsorption does not involve charged groups, the removal mechanism is rather insensitive to the pH of the medium.

Incorporation of functional groups that can develop a charge in liquids may enhance the capacity of such membranes to remove charged contaminants by electrostatic adsorption. Colloidal contaminant removal by electrostatic adsorption mechanism relies on the nature of charge developed by the filter material and the contaminants in a given liquid medium. Most common colloidal contaminants, such as silica and lipopolysaccharides, are negatively charged in DI water and in basic chemicals. Hence, the development of positively charge modified microporous membranes has gained much attention over the last decade [3].

One technique that has been successfully used to incorporate charge modifying groups onto the surface of microporous membranes is γ-irradiation. Jan and Raghavan [4] have reported a method to synthesize positively charge modified polypropylene membranes by graft polymerizing 4-vinylpyridine using γ-rays in a methanolic medium and then quaternizing the pyridine groups in nitromethane. The modified membranes were characterized by a positive zeta potential in aqueous KCl solutions in the pH range of 4 to 9.3. One of the shortcomings of this method was the need to use nitromethane which required extensive post-washing with solvents of increasing polarity. One way to overcome this is to use a water-soluble monomer which can be grafted by γ-irradiation. Cationic vinylphosphonium salts are attractive in this respect.

Polymerization of vinylphosphonium salts into high molecular weight polymers has not been studied actively in the last decade. Pellon and Valan [5] and Rabinowitz and Marcus [6] successfully polymerized tributylvinylphosphonium bromide dissolved in aqueous solutions into high molecular weight polymers using X-rays. Other phosphonium salts such triethylvinylphosphonium bromide, dimethylphenylvinylphosphonium bromide and tricyclohexylvinylphosphonium bromide exhibited poor polymerization characteristics [6].

In this paper, a method for the charge modification of PVDF membranes by grafting of a water-soluble vinyltriphenylphosphonium bromide monomer using γ-rays is reported. The electrokinetic potentials of these charge-modified PVDF membranes have been characterized using a streaming potential method. The capacities of unmodified and charge-modified PVDF membranes to remove an anionic dye from DI water have been compared. The compatibility of synthesized charge-modified membranes with ultrapure water has also been investigated.

2. EXPERIMENTAL MATERIALS AND METHODS

2.1. Materials

Two 0.254 m $\times$ 0.203 m sheets of 0.45 μm PVDF membranes were obtained from Pall Corporation. Vinyltriphenylphosphonium bromide monomer (97%) was purchased from Aldrich Chemical Company. The anionic dye, metanil yellow, was obtained from Fluka Chemicals. Electronic grade water from a Millipore RO-DI system (Milli-Q 4-bowl) was used to make the required electrolyte solutions. All electrolytes used for the regulation of ionic strength and pH were of analytical grade.

2.2. Methods

An aqueous solution containing 2 wt% of vinyltriphenylphosphonium bromide monomer and 5 strips of PVDF films (0.15 mm × 60 mm × 50 mm) was degassed in a specially designed glass tube and then sealed. The tube was then irradiated in a ^{60}Co γ-ray facility at the Department of Nuclear and Energy Engineering, University of Arizona. The total amount of radiation was varied in the range of 0.61 to 1.22 Mrads at a dose rate 400 rads/min, as measured by Fricke's dosimeter. After irradiation, the membranes were thoroughly rinsed with DI water and then air dried. The reaction scheme is shown in Fig. 1.

The ATR-FTIR spectra of membranes were obtained with a Perkin-Elmer 1800 Fourier-transform spectrometer, in the region 400 to 4000 cm^{-1}. During the measurement, the sample chamber was purged with nitrogen gas to reduce moisture content. Twenty scans at a resolution of 4 cm^{-1} were signal averaged.

The zeta potential measurements were carried out using a streaming potential method with the experimental set-up depicted in Fig. 2. It consisted of a 3 liter PTFE reservoir, a valveless metering pump with a PTFE head (Fluid Metering Inc., Model QD1), a flow cell, a pressure transducer (Validyne DP 15-40) with a carrier

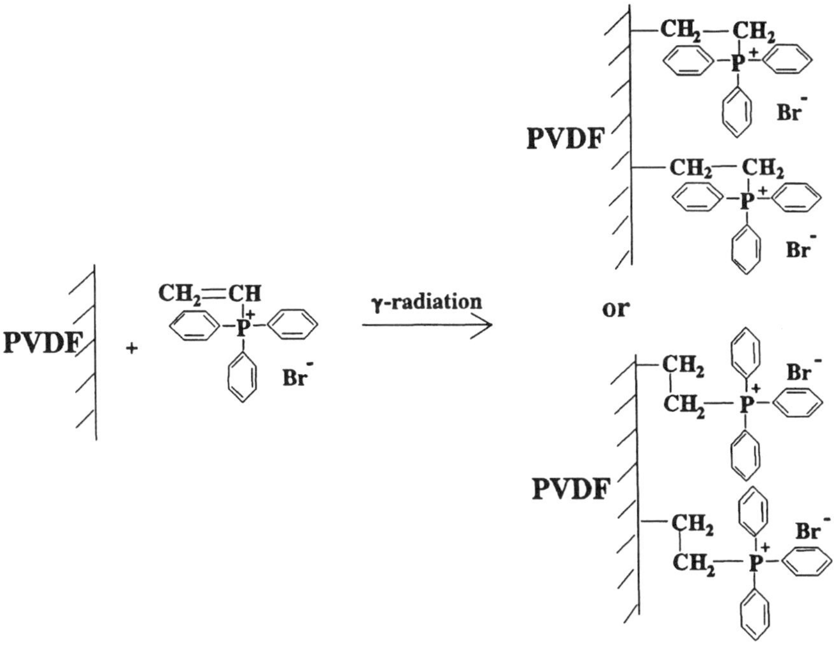

Figure 1. Reaction scheme for grafting vinyltriphenylphosphonium bromide onto PVDF membrane.

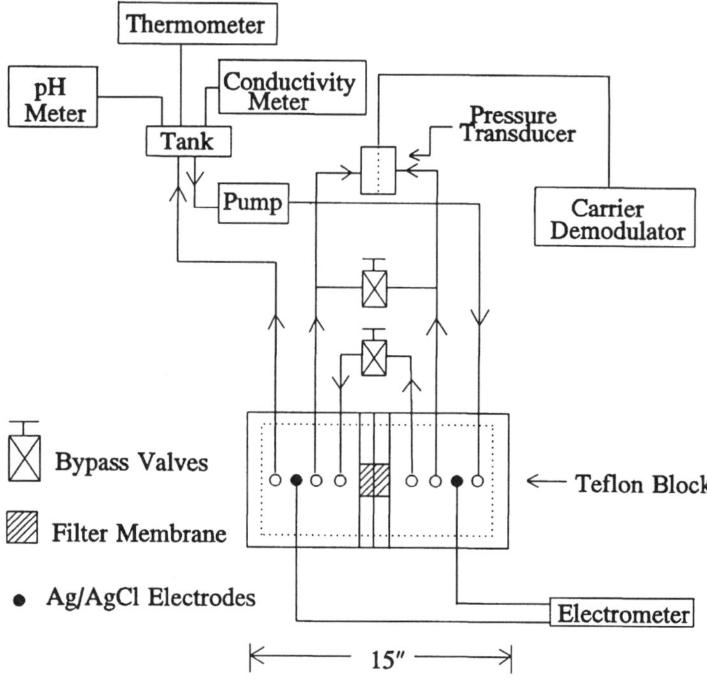

Figure 2. Experimental setup for streaming potential measurements.

demodulator, a high impedance electrometer (Keithley Model 614: input impedance 5×10^{13} ohms) and PTFE valves and lines. The flow cell consisted of two PTFE chambers with a plate in between to hold 47 mm diameter membrane samples. Silver billet electrodes (Ingold) positioned on either side of the membrane were used to measure the streaming potential. These electrodes were chloridized in 1 M HCl at a current of 0.0044 A for 40 min [7]. Additional details of this set-up have been published elsewhere [8]. The streaming potential was measured at pressure drop (P) values ranging from 351 to 3515 kg/m^2 and the slope of the streaming potential (E_s) versus pressure drop plot was used to calculate the zeta potential, ζ, according to the following equation:

$$\zeta = 4\pi\eta\frac{k}{\varepsilon} \times \frac{\Delta E_s}{\Delta P}. \tag{1}$$

In the above equation, η, ε, k, are the viscosity, dielectric constant and conductivity of the solution, respectively.

Dye challenge tests were also carried out in the streaming potential cell using an anionic dye, metanil yellow. In these tests, a dye solution of a given concentration was pumped through the membrane and the dye breakthrough was followed by measuring the concentration of the dye in the effuent using a UV-VIS spectrophotometer (Milton Roy 1201) at a wavelength of 426.6 nm.

It is critical that the charge modifying agents do not leach into the water and increase the level of total organic carbon (TOC). To characterize the compatibility

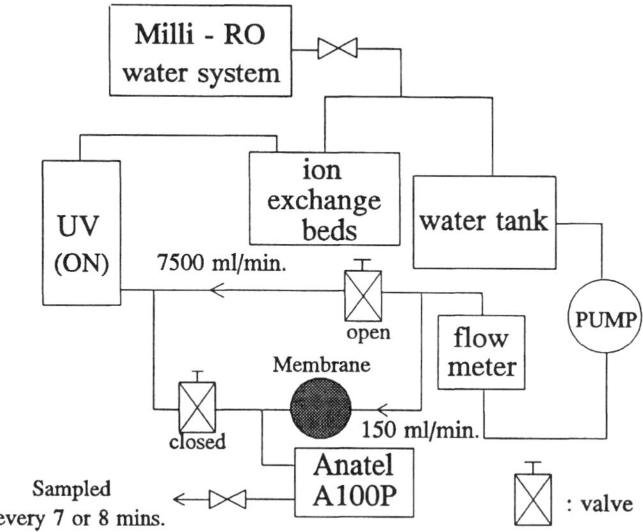

Figure 3. Test loop for filter TOC release tests.

of the modified membranes with DI water, TOC release tests were performed in the Sematech Ultrapure Water Pilot Plant located in the Department of Chemical Engineering at the University of Arizona. The filter with a diameter of 47 mm was tested in a special bypass line emanating from the main artery of the polishing loop of the water system (Fig. 3). The flow rate in the main artery was 7500 ml/min (2 gpm), while the flow rate in the bypass line was 150 ml/min. The membranes were wetted with IPA and thoroughly rinsed with DI water prior to loading in the bypass loop. The tests for PVDF and charge-modified PVDF lasted 200 min with TOC data points taken approximately every 7 to 8 min with an Anatel A-100P TOC analyzer in the outlet position.

In order to ascertain that the charge modification process did not cause pore blocking, the bubble point of the membranes was measured using isopropyl alcohol (IPA) as the displaced liquid phase. The membranes were first soaked in IPA and then clean air was forced through the membranes. The lowest pressure (ΔP) at which the air bubbles were generated in the IPA liquid was taken to be the bubble point. Assuming a hemispherical meniscus and complete wetting by IPA, the bubble point can be related to the radius of largest pore ($r_{\max}$) of the filter by the Laplace equation [9]:

$$\Delta P = \frac{2\gamma_{\mathrm{IPA}}}{r_{\max}}, \tag{2}$$

where γ is the surface tension (21.7 mN/m) of IPA.

The critical surface tension of the charge-modified and unmodified membranes was determined from the contact angles of IPA/water solutions measured using a Ramé-Hart contact angle goniometer. The surface tensions of IPA/water solutions were determined using a Cahn DCA 312 instrument.

The specific surface areas of unmodified and charge-modified membranes were measured by a Micromeritics ASAP 2000 Surface Area/Pore Volume Analyzer, using nitrogen as the adsorbate.

3. RESULTS AND DISCUSSION

3.1. Membrane characterization

The grafting of vinyltriphenylphosphonium bromide onto PVDF was confirmed by infrared analysis. Figure 4 shows the FTIR spectrum of charge-modified PVDF membrane obtained using an unmodified PVDF membrane as the reference. The charge-modified membrane has an absorption peak at 1100 cm^{-1} which is absent in the spectrum for unmodified PVDF membrane. This may be assigned to the stretching of the phenyl-phosphorus bond [10, 11]. The absorption peaks at 1050 and 1150 cm^{-1} are due to the vibration of phenyl groups [12].

As outlined in the experimental method section, the largest pore size in the modified and unmodified membranes was determined using the bubble point method. A bubble point at 7170 kg/m^2 which corresponds to a maximum pore size of 1.2 μm was measured for both unmodified and charge-modified PVDF membranes. This indicates that the largest pore size was not significantly changed during membrane modification. It may be noted that the bubble point may not correspond to the rated pore size (0.45 μm) of the membrane which is often determined by challenging the membrane with 0.6 μm *Serratia marcescenas* bacterium [13].

The specific surface areas of the unmodified and modified membranes were 15 m^2/g and 17 m^2/g respectively. A calculation of pore size distribution of pores less than 0.1 μm using the Barrett–Joyner–Halenda method [14] indicated that the mean pore

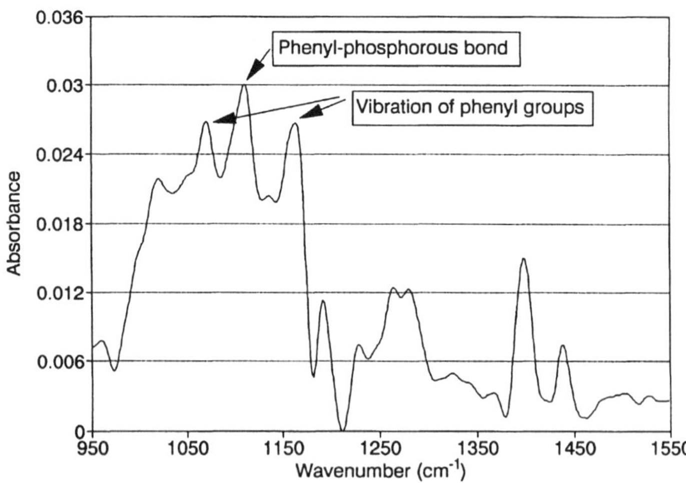

Figure 4. FTIR spectrum of charge-modified PVDF membrane (reference: unmodified PVDF).

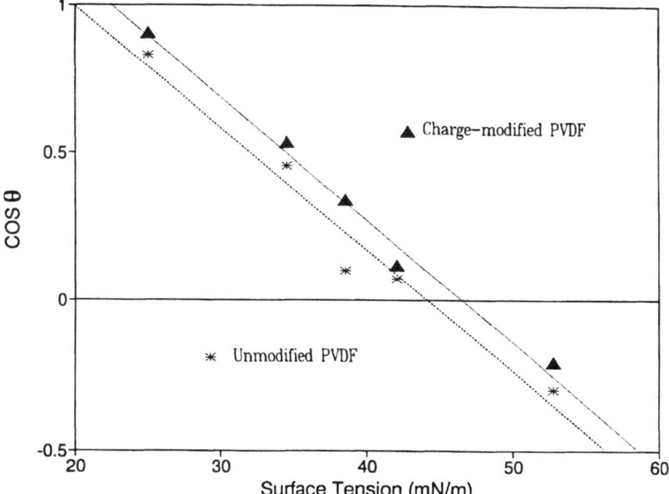

Figure 5. Cosine of contact angle as a function of surface tension of IPA/water solutions for unmodified and charge-modified PVDF membranes.

size in the modified membrane was slightly smaller than in the unmodified membrane (12 nm vs. 14 nm). In a membrane rated at 0.45 μm, this change in the mean pore size is not of much significance.

The results of the contact angle measurements on unmodified and charge-modified PVDF membranes are shown in Fig. 5 as Zisman plots. Isopropyl alcohol/water solutions of different compositions were used for these measurements. It can be seen that the critical surface tensions of unmodified and charge-modified PVDF membranes are 20 and 23 mN/m, respectively. This implies that the wettability of modified PVDF membrane is not significantly improved despite the incorporation of positive charge. This is most likely due to the presence of bulky phenyl groups around the phosphorus atom.

3.2. Zeta potential of membranes

The zeta potential values of unmodified and charge-modified PVDF membranes are plotted in Fig. 6 as a function of solution pH. These values were calculated from the measured streaming potential values using equation (1). Because of the poor wettability of the membranes in aqueous solutions, the membranes had to be prewetted with IPA prior to making streaming potential measurements. It may be seen that the unmodified PVDF membrane exhibits negative zeta potential values at pH values larger than 4, while the charge-modified PVDF membranes show positive zeta potential in the pH range from 4 to 9.3. The positive zeta potential of the charge-modified PVDF membranes is due to the grafting of phosphonium groups. Based on the zeta potential data, it appears that the extent of charge modification is independent of total γ-ray dosage in the range 0.61–1.22 Mrads.

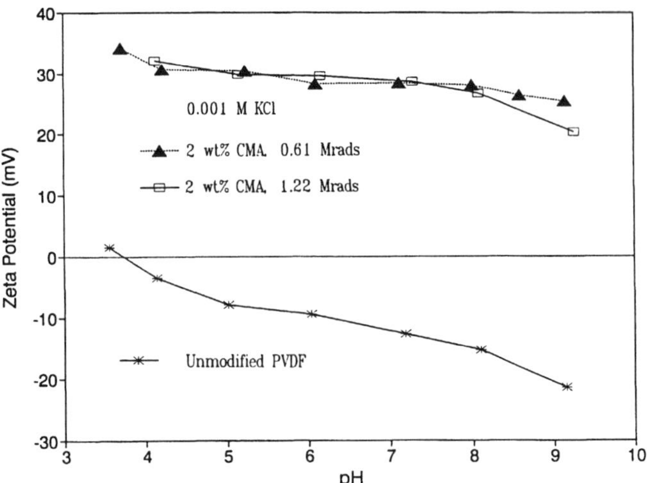

Figure 6. Zeta potential of unmodified and charge-modified PVDF membranes as a function of pH of 0.001 M KCl solutions (CMA refers to the charge modifying agent).

3.3. Dye challenge tests

Membrane manufacturers often use small anionic dye species to characterize the adsorptive capacity of positively charge-modified membranes [15]. In this work, metanil yellow, a water soluble anionic dye of molecular weight 375, was used to challenge the charge-modified and unmodified polypropylene membranes. From previous experience [16], a dye concentration of 9.3×10^{-6} M and a solution flow rate of 24.5 ml/min were chosen for these tests.

The results of the dye challenge tests carried out with charge-modified and unmodified PVDF (1.22 Mrads) membranes at a pH of 5.0 are shown in Fig. 7. This pH was the natural pH of the dye solution prepared with 18 MΩ-cm DI water and is most likely due to CO_2 abstracted from the atmosphere. This was difficult to prevent due to the high purity of the water. In the case of charge-modified PVDF membrane, the breakthrough of the dye occurred at about 10 min after the introduction of the dye in the holding tank. The streaming potential of the membrane was 150 mV in the absence of any dye in the solution, but it decreased in magnitude, changed sign and reached a value of −250 mV when the dye breakthrough was completed. The change in streaming potential is obviously due to the adsorption of negatively charged dye species onto the positively charged groups on the charge-modified PVDF membrane. It is interesting that the adsorption of the dye was able to reverse the sign of the streaming potential.

For the unmodified PVDF membrane, it took about 5 min after the addition of the dye solution to the holding tank for breakthrough to occur. The streaming potential of the membrane had a value of −135 mV before the introduction of the dye, but it changed to −160 mV when the uptake of the dye by the membrane was completed. Since PVDF is negatively charged at pH values greater than 4, the surface should mostly consist of negatively charged sites at a pH of 5.0. It is intriguing that the zeta potential becomes more negative as the dye is adsorbed onto the membrane. This

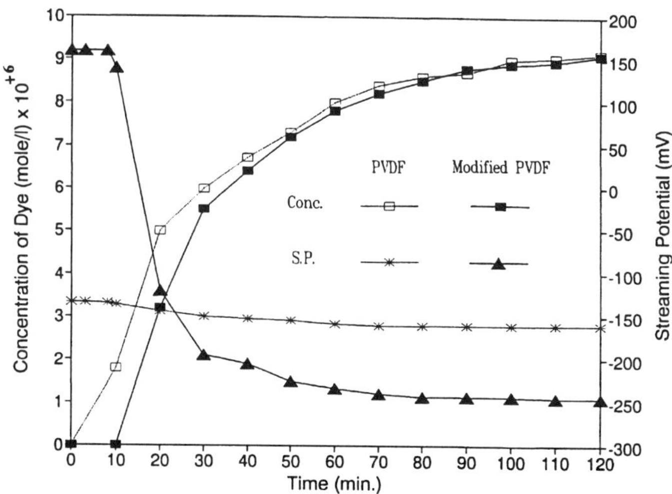

Figure 7. Change in streaming potential of unmodified and charge-modified PVDF membrane (1.22 Mrads) and dye breakthrough profiles during challenge testing with metanil yellow at a pH of 5.0 (flow rate = 24.5 ml/min, dye concentration = 9.3×10^{-6} M).

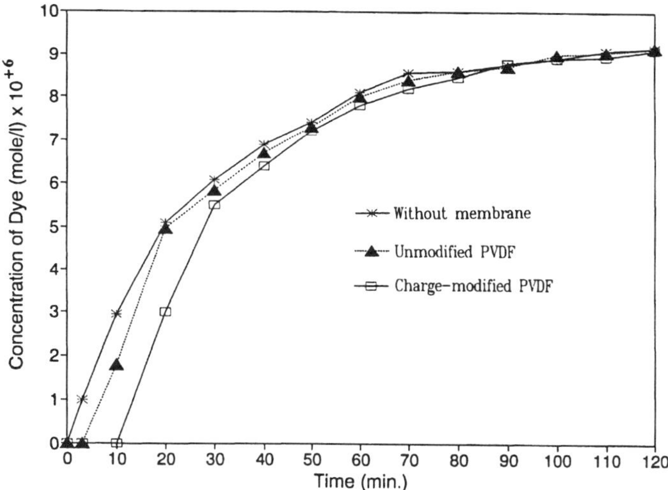

Figure 8. A comparison of dye breakthrough curves for charge-modified and unmodified PVDF membranes.

could be due to the neutralization of the small number of positively charged sites that may exist at a pH of 5.0 on PVDF or due to the specific adsorption of the dye.

The dye breakthrough curves obtained for both types of PVDF membranes are compared in Fig. 8. This figure also contains a curve obtained when no membrane was present in the experimental set up. By integrating the area between the curve for the no-membrane case and the curve for PVDF (or charge-modified PVDF), multiplying the area by the volumetric flow rate, and then dividing by the specific surface area, the extent of uptake of the dye (moles/cm^2) was calculated. At the dye concentration of 9.3×10^{-6} M, the uptake of the dye by charge-modified and unmodified PVDF membranes was calculated to be 1.5×10^{-10} moles/cm^2 and 0.31×10^{-10} moles/cm^2, respectively. It may be noted that a small amount of dye adsorbs onto the unmodified PVDF membrane in spite of a negative zeta potential exhibited by the membrane. A calculation of site density due to vinyltriphenylphosphonium groups on the charge-modified PVDF may be made by subtracting the dye adsorption density on PVDF membrane from the adsorption density for charge-modified membrane. Using this method, the positive surface site density due to vinyltriphenylphosphonium groups was calculated to be 1.19×10^{-10} moles/cm^2. The cross-sectional area of the vinyltriphenylphosphonium ion is roughly 1.5 nm^2. If adsorption of dye predominantly occurred on the vinyltriphenylphosphonium sites grafted onto the membrane, the net adsorption density of dye on charge-modified PVDF membranes will be approximately equivalent to 1.1 monolayer.

3.4. TOC (total organic carbon) release test

One of the requirements of a filter membrane to be used in an ultrapure water system is that it should not generate organic carbon compounds by leaching. The results of TOC measurements when PVDF and charge-modified PVDF membranes were installed in the bypass loop are shown in the Fig. 9. This figure also contains data for TOC background level of the ultrapure water system when no membrane was present in the experimental set-up. It should be mentioned that the first data point was collected seven minutes after loading but this was discarded due to lack of environmental control during handling. The results show that the initial TOC level was 4 ppb when an unmodified PVDF membrane was located in the system, but it decreased and reached a steady value of 3 ppb at about 70 min. On the other hand, the use of the charge-modified PVDF membrane resulted in an initial TOC level of

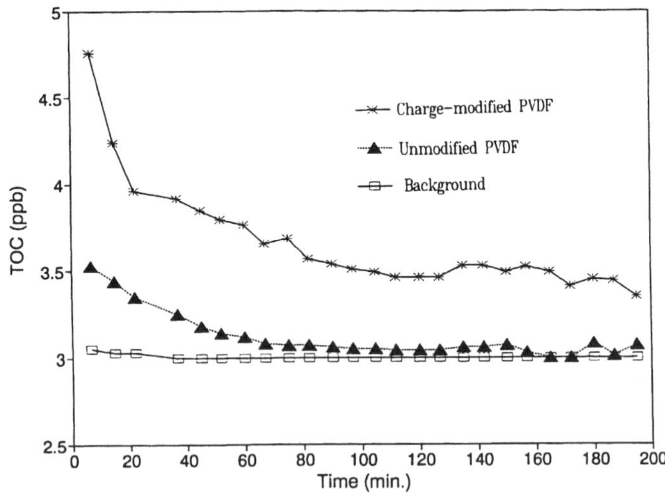

Figure 9. TOC release data during the testing of unmodified PVDF and charge-modified PVDF membranes in an ultrapure water system.

4.7 ppb, but the TOC level decreased to 3.5 ppb after 80 min. It thus appears that the charge modifying compound is rather strongly grafted onto the surface of PVDF and is not leached out to significant levels in DI water.

The conceived reaction scheme for the positive charge modification of PVDF membranes has improved the capacity of membranes to remove anionic compounds without compromising the stringent TOC requirements that should be met for use in ultrapure water systems. However, these modified membranes are still hydrophobic, and require prewetting with IPA prior to their use. To overcome this wettability problem, attempts are under way to cograft (or copolymerize) a hydrophilic acrylate monomer with the vinylphosphonium compound. The hydrophilic monomer may also function as a spacer to help in the polymerization of bulky vinyltriphenylphosphonium bromide. Preliminary results of this study have been very promising.

4. CONCLUSIONS

A method to incorporate positively charged groups onto the surface of PVDF membranes has been developed. This method involves grafting of a water-soluble monomer, vinyltriphenylphosphonium bromide, onto PVDF by γ-irradiation. Compared with unmodified PVDF membranes, the charge-modified membranes show positive zeta potential values in the pH range from 4 to 9.3 and possess a much higher capacity to remove metanil yellow dye. The developed charge modification scheme does not alter the bubble point of the membrane.

Acknowledgements

The authors wish to acknowledge the Sematech Center of Excellence at the University of Arizona for financial support to carry out this work. Sincere thanks are extended to Mr Harry Doane for help with the radiation experiments, Dr Roger Sperline for consultation in FTIR analysis and Mr Kon-Tsu Kin for his cooperation in carrying out water compatibility tests.

REFERENCES

1. Pall Corporation, *The Pall N66 Posidyne Membrane Filter Guide*, p. 16 (1986).
2. J. R. Robinson and C. Genovesi, in: *Depyrogenation, Technical Report*, No. 7, Chap. 6, pp. 54–69. Parenteral Drug Association, Betheseda, Maryland (1985).
3. E. A. Ostreicher, Paper presented at the Hershey Conference on Fibers, Filtration and Electrostatics. Hershey, Pennsylvania, August (1985).
4. D. Jan and S. Raghavan, *IEEE Trans. Semiconductor Manuf.* **6** (4), 367–372 (1993).
5. J. Pellon and K. J. Valan, *Chem. Ind. (London)* **32**, 1358 (1963).
6. R. Rabinowitz and R. Marcus, *J. Polym. Sci.* **A3**, 2063–2074 (1965).
7. G. D. Harford, Coupled Flow Phenomena in Clay-Water System, Ph.D. Dissertation, p. 155. University of California, Berkeley (1966).
8. I. Ali, Electrokinetic Characteristics of Particulate/Liquid Interfaces and Their Importance in Contamination from Semiconductor Process Liquids, Ph.D. Dissertation. University of Arizona (1990).
9. R. E. Kesting, *Synthetic Polymeric Membranes*. McGraw-Hill, New York (1971).

10. C. J. Pouchert, *The Aldrich Library of FT-IR Spectra*, Vol. 2, p. 545. Aldrich Chemical Co., Milwaukee, Wisconsin (1985).
11. L. W. Daaschand and D. C. Smith, *Anal. Chem.* **23** (6), 853–868 (1951).
12. G. Varsanyi, *Assignments for Vibrational Spectra of Seven Hundred Benzene Derivatives*, Vol. 1. John Wiley & Sons, New York (1974).
13. Personal Communication with Dr Barry Gotlinsky, Pall Corporation.
14. E. P. Barrett, L. G. Joyner and P. P. Halenda, *J. Amer. Chem. Soc.* **73**, 373–380 (1951).
15. R. A. Knight and E. A. Ostreicher, *Filtration and Separation*, 30–34 (Jan./Feb. 1981).
16. D. Jan and S. Raghavan, *Colloids and Surfaces* (in press).

Polymer Surface Modification: Relevance to Adhesion, pp. 349–362
K. L. Mittal (Ed.)

Surface modification of fluoropolymers

K. LUNKWITZ,* W. BÜRGER, U. LAPPAN, H.-J. BRINK and A. FERSE
Institute of Polymer Research, Hohe Str. 6, 01069 Dresden, Germany

Revised version received 15 June 1994

Abstract—COF groups are formed by electron irradiation of PTFE [poly(tetrafluoroethylene)] powders in air, especially at the surface and in near-surface regions which can be easily hydrolysed to carboxyl groups by air humidity. The application of special additives during irradiation leads to modified micropowders. Fourier transform infrared (FTIR) spectroscopy enables the detection of carboxyl and COF groups. γ-Irradiation of PTFE mainly causes degradation of the polymer; the concentration of carboxyl groups is much lower. Carboxylated micropowders created via radiation treatment retain the essential properties of PTFE. With increasing radiation dose, the increasing concentration of functional groups in the micropowders causes an increase in the surface free energy. This diminishes the strong water and oil repellency of PTFE in such a way that homogeneous incorporation into aqueous and organic liquids or other polymers is possible. So, the special properties of PTFE can be made effective in these media. Modified PTFE micropowders have been successfully tested in many application areas. The aim of our present work was to increase the concentration and vary the nature of functional groups by radiation-chemical methods or chemical conversion of COF groups (polymer-analogous reactions). A highly modified PTFE powder was used to reduce the repellent properties of PTFE diaphragms for application in brine electrolysis. The COF groups of the micropowders were modified by γ-aminopropyltriethoxysilane. The irradiation of FEP [poly(tetrafluoroethylene-co-hexafluoropropylene)] and PFA [poly(tetrafluoroethylene-co-perfluoroalkylvinylether)] yields products which contain a higher content of carboxyl groups than PTFE.

Keywords: PTFE modification; PTFE micropowders; high-energy irradiation; PTFE diaphragms; silane modification of PTFE; FEP modification; PFA modification.

1. INTRODUCTION

The formation of COF (carboxylic acid fluoride) groups during radiation degradation of polytetrafluoroethylene (PTFE) had been reported by several authors [1, 2] before we began with our investigations.

Our first investigations concerning the effect of electron radiation on PTFE already indicated that sufficient radiation doses (>10 000 kGy) caused complete degradation of the macromolecules to low-molecular-weight compounds, which could easily be identified. These experiments opened up new possibilities for interpreting reaction courses for the degradation process even at a relatively low radiation dose.

*To whom correspondence should be addressed.

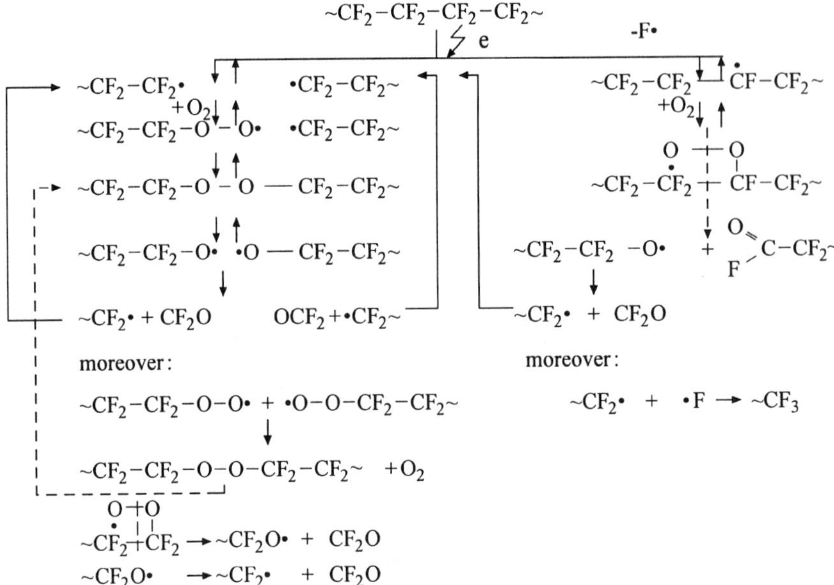

Figure 1. Probable course of radiation-induced degradation of PTFE in the presence of oxygen.

PTFE degradation in an inert atmosphere with a high radiation dose yielded a mixture of perfluoroparaffins and perfluoro-olefins with terminal or internal double bonds [3, 4]. The perfluoro-olefin content in the mixture was about 60 wt%. As for the frequently discussed problem of crosslinking reactions in PTFE degradation, our investigations have shown that they may be neglected, at least at temperatures above 70 °C. Straight-chain compounds with trifluoromethyl branches have been found in the degradation products.

We have also tried to obtain perfluorinated reaction products by adding reactants during radiation degradation. Degradation of PTFE at a high radiation dose in the presence of oxygen has been investigated in detail (Fig. 1) [4–6]. According to this mechanism, the yield was up to 65 wt% of carbonyl difluoride in addition to perfluorocarboxylic acid fluorides, which can be easily hydrolysed to perfluorocarboxylic acids.

The objective of this paper is to give a short report about our investigations in the field of incorporation of functional groups into fluoropolymers via high-energy radiation treatment at a relatively low dose ($<10\,000$ kGy) and about the application of the modified products obtained.

2. EXPERIMENTAL

2.1. Materials, irradiation, and modification experiments

The fluoropolymers used in the experiments were PTFE suspension grade (Hoechst AG; TF 1750, TF 1760), PTFE emulsion grade (Hoechst AG; TF 2025), FEP 532-8000 [poly(tetrafluoroethylene-co-hexafluoropropylene)] (Du Pont), and PFA 532-5011 [poly(tetrafluoroethylene-co-perfluoroalkylvinylether)] (Du Pont).

The irradiation experiments were performed with a universal electron-beam accelerator (ELT 1.5; produced at the Institute of Nuclear Physics, Novosibirsk) in the presence of air with an electron energy of 1.0 MeV and a beam current of 4.0 mA. The substances were irradiated in the dose range from 0.7 to 200 kGy in one step. Higher doses were applied on a step-by-step basis using 200 kGy each time.

After irradiation, the polymers were wetted with an acetone/water mixture and dried and heated (200°C, 60 min) to transform the near-surface COF groups into COOH groups.

For silanization of PTFE, TF 1760 was irradiated with 4000 kGy and dispersed in ethanol. Afterwards, an ethanolic solution of γ-(aminopropyl)triethoxysilane was added via a dropping funnel. After refluxing (4 h), the modified powder was separated by filtration, washed with ethanol and distilled water, and dried at room temperature.

2.2. Measurements

Infrared (FTIR) measurements were made on IFS 66 equipment (Bruker). The spectra were obtained using cold-pressed samples of PTFE and of copolymers.

The wettability of PTFE films was assessed using a modified contact angle meter 61 (Krüss GmbH). Contact angles using sessile drops were measured on cold-pressed material above the phase transition temperature of 19°C. n-Hexane, water, and methylene iodide were used as liquids for the contact angle measurements.

The content of near-surface COOH groups (ion-exchange capacity, IEC) was determined by potentiometric titration. The irradiated powders were dispersed in an aqueous solution of dimethylacetamide (DMA) and NaCl (200 g DMA, 150 g water, 20 g NaCl), stored for 24 h, and titrated with 0.05 N NaOH.

To obtain more information about the specific surface charge of the carboxylated PTFE powders, the method of charge-compensating polyelectrolyte titration by a PDC 02 particle charge detector (Mütek, Germany) was used. The consumption of a cationic titre [poly(diallyl-dimethylammonium chloride)] for a particle dispersion (acetone/water = 1:1, volume 50 ml for a solid content of 1 g) is a measure of the amount of negative surface charge at the solid/liquid interface.

3. RESULTS AND DISCUSSION

3.1. Modification of PTFE powders

Virgin PTFE is disintegrated by high-energy irradiation at a relatively low dose into a finely dispersed powder. A relatively fine PTFE powder can also be produced from moulded PTFE: PTFE chippings are an appropriate starting material. γ-Irradiation with an energy dose of about 300 kGy caused such an embrittlement of PTFE that it could be crushed down to a median particle size of 14 μm by special grinders.

Our investigations of electron-irradiated PTFE showed [7] that the content of functional groups influences decisively the incorporation of these micropowders into organic or aqueous media. COF groups formed by irradiation in air are subsequently hydrolysed to carboxyl groups by air humidity. The application of special additives during irradiation leads to differently modified micropowders.

IR spectroscopy enables the detection of carboxyl groups and COF groups (Fig. 2 and Table 1). Our FTIR spectra indicated that carboxylation, beginning at radiation doses above 50 kGy, is a significant reaction [8]. The characteristic carbonyl group stretching vibrations partially associated via hydrogen bonds in addition to bands of virgin PTFE can be seen in the spectrum of the irradiated sample. A higher radiation dose caused an increasing concentration of carbonyl groups. The band intensities of both the free and the hydrogen-bonded groups grew (Table 1: free 1810 cm^{-1}, hydrogen-bonded 1770–1780 cm^{-1}). At high doses (more than 800 kGy), hydrogen-bonded groups dominated. Furthermore, the association behaviour suggested that there are zones of both mainly near-surface associated carboxyl end-groups and isolated carboxyl end-groups in the bulk of the polymer.

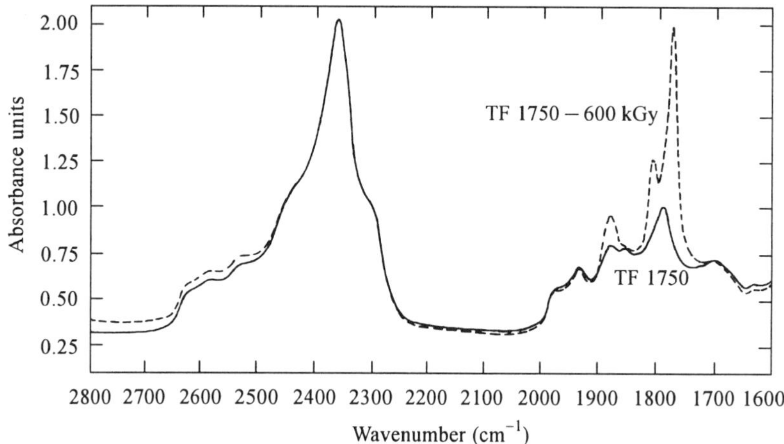

Figure 2. FTIR spectra of TF 1750 and electron-irradiated TF 1750 — 600 kGy.

Table 1.
Infrared absorption band assignments of electron-irradiated PTFE

Wavenumber (cm^{-1})	Assignment
3570	O–H, free
3100	O–H, hydrogen-bonded
2370	C–F overtone
1885	C=O, COF
1810	C=O of COOH, free
1790	C=C of $CF{=}CF_2$
1770–1780	C=O of COOH, hydrogen-bonded

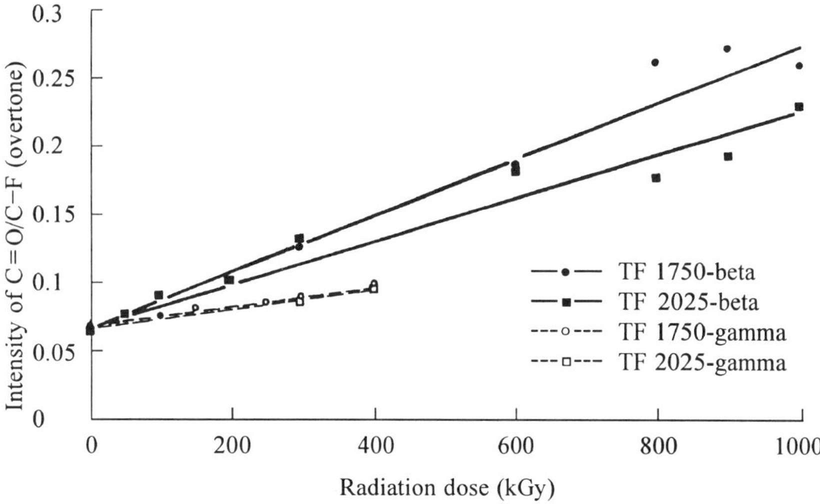

Figure 3. Intensity of FTIR bands of carboxyl groups of PTFE irradiated in air as a function of the radiation dose.

With increasing radiation dose (Fig. 3), the intensity ratio (a quotient of the absorbance of the C=O stretching vibration and a nonspecific C—F overtone) reflected the increase in carboxylation. Electron radiation was more efficient in the near-surface region of the PTFE particles than γ-radiation, i.e. the concentration of carboxyl groups was much higher. γ-Irradiation of PTFE mainly leads to degradation of the polymer.

In electron irradiation, the sufficiently high-energy flux density accelerates the degradation of PTFE to such a degree that oxygen cannot be provided fast enough in the inner parts of the polymer. Therefore carboxylation takes place mainly in the near-surface regions.

X-ray diffraction patterns combined with other methods have shown that at low radiation doses [9–11] internal radicals in crystalline regions are able to recombine because of their hindered mobility.

The change in surface energy was assessed by measurement of the sessile contact angles. Untreated PTFE showed a water contact angle of 109° (±3). In all cases of increasing doses, the contact angle was reduced (Fig. 4). With increasing radiation dose, the increasing concentration of carboxyl groups in the micropowder caused an increase in the surface free energy [4, 8]. But mainly the polar component of the surface free energy increased, while the dispersion component remained nearly the same. This means that radiation-chemically carboxylated micropowders retain the essential properties of PTFE.

The increase in the surface free energy of the PTFE micropowder with the radiation dose diminishes the strong water and oil repellency of PTFE so much that homogeneous incorporation into liquid systems or into other polymers is possible; consequently, the special properties of PTFE, such as its particular release and lubricating properties as well as its chemical resistance, can be favourably combined with desirable properties of other polymers.

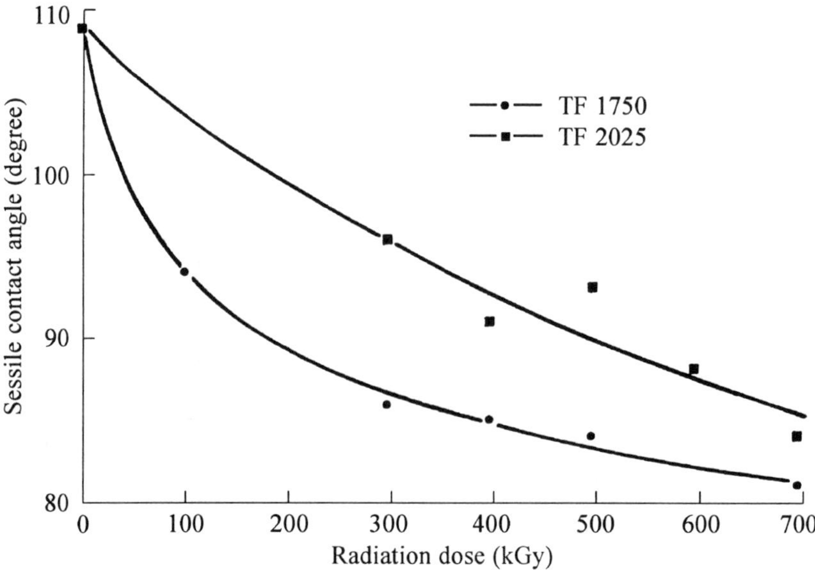

Figure 4. Sessile water contact angle on cold-pressed PTFE samples electron-irradiated in air as a function of the radiation dose.

We have successfully tested the PTFE micropowders modified with functional groups and reduced melt viscosity as:

- dry lubricant [7, 12];
- wire-drawing agent (unpublished results);
- mould-release agent [7];
- additive in lubricating greases and oils [7];
- additive in anti-friction varnishes [7];
- dispersed phase in PTFE dispersions [13, 14];
- chemically resistant binding component for disperse systems (shape-retaining asbestos diaphragms in brine electrolysis) [15, 16];
- maintenance-free slide bearing materials [17];
- additive in coatings (unpublished results);
- additive in glass-fibre-reinforced thermosets [18]; and
- hydrophilizing agent in fluoropolymers (PTFE membranes and diaphragms) [19–21].

A highly functionalized PTFE powder can be applied as a nearly polymer-identical dispersion agent for fluoropolymers in halogenated hydrocarbons [22].

The application range of PTFE micropowders can be extended if all possibilities of increasing the concentration and varying the nature of functional groups are employed. This is our present working field (Fig. 5) [21]. Radiation-chemical methods should be investigated in more detail, e.g. irradiation of PTFE in the presence of $SO_2 + O_2$, $SO_2 + Cl_2$, SO_2Cl_2, etc., or under inert conditions. The aim here is the

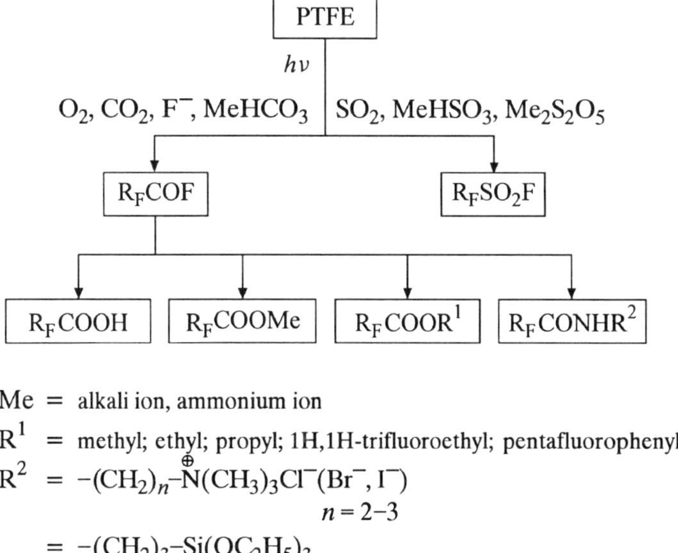

Figure 5. Surface modification of PTFE by means of accelerated electrons and end-group reactions of irradiated PTFE.

introduction of sulfonic groups into the polymer surface [23]. Chemical conversion (polymer-analogous reactions) of the radiation-chemically formed COF groups into amides or esters is also possible. The use of special reagents improves compatibility to such polymers which are to be modified with the micropowder. Esterification of COF groups and their reaction to amides [24] has been successfully performed. Further reactions of the amides lead to cationic PTFE powders which can be used in filters and membranes made of fluoropolymers [21].

The higher the radiation dose, the higher the concentration of COF groups, but it should be noted that the increase in the concentration of COF groups with increasing radiation dose leads to a distinct decrease of the average molecular weight of PTFE. It decreases from 10 000 to less than 1000 in the range of 2–10 MGy. (An average molecular weight could be estimated assuming that every perfluorinated chain possessed a carboxyl end-group.)

Figure 6 shows the dependence of the ion-exchange capacity on the radiation dose:

- The degree of carboxylation was greater for PTFE suspension grade (TF 1750) than for emulsion grade (TF 2025) if air was present and the radiation dose remained constant.
- Alkali carbonates and hydrogen sulphites give rise to synergistic effects during carboxylation.
- Addition of alkali fluorides to alkali bicarbonates also promoted the carboxylation.

Polyelectrolyte titrations showed (Fig. 7) that near-surface carboxyl groups are able to react with polycations, in this case with poly(diallyl-dimethylammonium chloride).

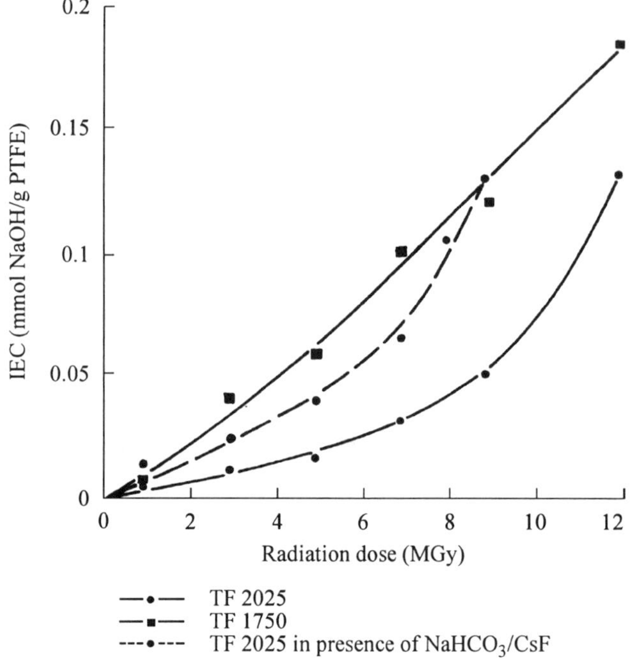

Figure 6. Ion-exchange capacity (IEC) of carboxylated PTFE (electron irradiation in the presence of air, titration with NaOH) as a function of the radiation dose.

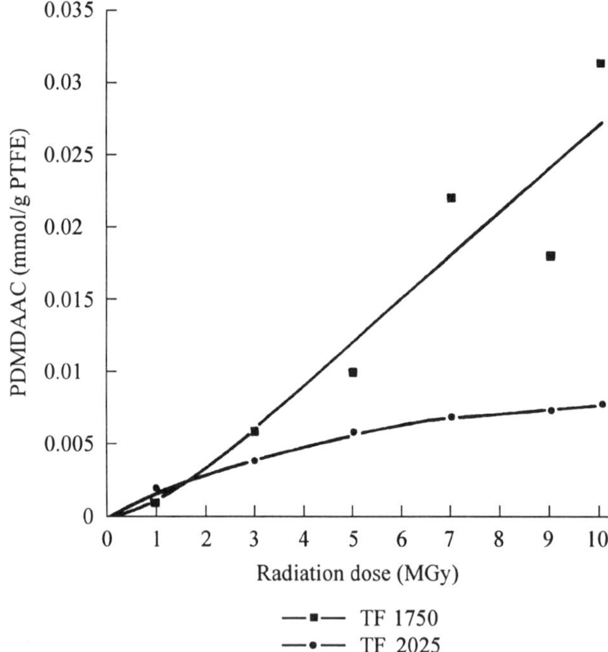

Figure 7. Polyelectrolyte titration of carboxylated PTFE with poly(dimethyl-diallylammonium chloride) (PDMDAAC) as a function of the radiation dose.

An increased number of carboxyl groups increased the interaction with poly(diallyldimethylammonium chloride).

3.2. Modification of PTFE membranes and PTFE diaphragms

A highly modified PTFE powder (irradiated in the presence of NH_4HSO_3 and air) was used to reduce the very high water- and oil-repellent properties of PTFE membranes and PTFE diaphragms for application in brine electrolysis [20]. The modified powders were incorporated into the diaphragms at a content of 1%. The usual ways for hydrophilizing PTFE membranes, summarized below, lead to insufficient or non-permanent effects in brine electrolysis:

- addition of hydrophilic inorganic substances, e.g. titanates, TiO_2, zirconates, ZrO_2;
- grafting (also radiation-chemically) of hydrophilic monomers, e.g. acrylic acid, methacrylic acid, acrylamide, styrene + sulphonation, maleic anhydride, *N*-vinylpyrrolidone;
- etching with naphthalene/Na in tetrahydrofuran;
- addition of functionalized fluorocarbons, e.g. fluorosurfactants;
- reaction with chlorosilanes and the formation of a silicate network.

The permeability of the diaphragm for brine solution increased with increasing content of functionalized PTFE (Fig. 8). The permeability also increased with increasing degree of functionalization at a constant percentage of modified PTFE. As expected, the negative value of the zeta potential increases with increasing content of functionalized PTFE.

Such diaphragms have been successfully tested in pilot cells in long-term investigations. They met all demands concerning the electrochemical and mechanical parameters for more than 3 years. Several further applications are possible.

Alternatively, we tried to modify PTFE diaphragms by grafting of hydrophilic monomers. Grafting was performed with γ-rays from the gaseous phase (Fig. 9). This method prevents the formation of homopolymers and the clogging of pores. The best results were obtained by grafting with styrene followed by sulphonation. Unfortunately, these grafted products are not stable in brine electrolysis. An increased permeability of diaphragms for aqueous liquids has also been achieved by grafting with inorganic compounds ($SO_2 + Cl_2$, SO_2Cl_2, $SO_2 + O_2$).

3.3. Silanization of PTFE

Because of the low surface energy of fluoropolymers, it is very difficult to bond PTFE to substrates such as glass or metal. It is necessary to utilize some type of adhesive compositions in order to obtain the required adhesion to the substrate. However, the adhesion is often not as high as desired for a particular application.

To overcome this disadvantage, we modified the COF groups of the micropowders by γ-aminopropyltriethoxysilane. Chain length variations of the fluorocompound and

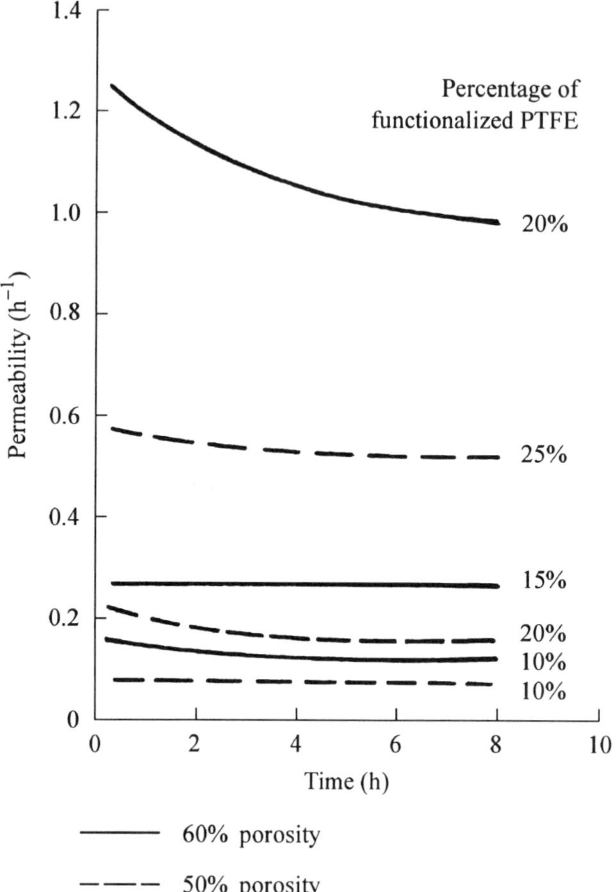

Figure 8. Influence of the percentage of functionalized PTFE in the PTFE diaphragm on the course of permeability for brine solution (PTFE emulsion grade irradiated in the presence of NH_4HSO_3 and O_2 with 3000 kGy).

the organo-silicon compound can be employed to produce suitable coupling agents to bond fluoropolymers to substrates such as glass or metal.

$$R_f{-}COF{+}H_2N{-}(CH_2)_3{-}Si{-}(OC_2H_5)_3 \rightleftharpoons R_f{-}CONH{-}(CH_2)_3{-}Si{-}(OC_2H_5)_3{+}HF.$$

We found that the attachment was effected by the formation of amide bonds between the acid fluoride end-groups present on the PTFE surface and the γ-aminopropyltriethoxysilane. Hydrolysis of bonded alkoxysilanes yielded a siloxane structure (Si—O—Si) at the surface.

An E-glass-fibre cloth was coated with an ethanolic suspension of silane-modified PTFE powder. The reaction scheme is shown below.

The coating was baked at 120°C. In this reaction, the concentration of OH groups is diminished and this leads to an increase of the zeta potential of the glass-fibre cloth (Fig. 10) [21].

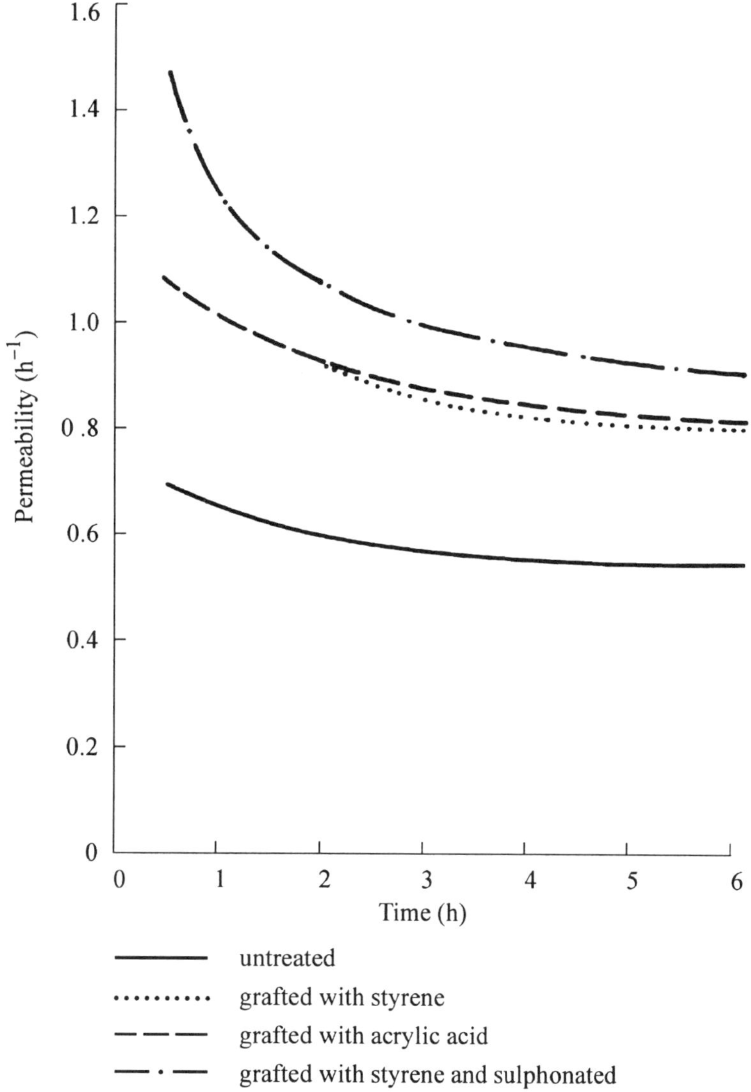

Figure 9. Courses of permeability for brine solution through PTFE diaphragms (55% porosity) grafted in the gaseous phase (grafting by influence of γ-rays).

3.4. Modification of perfluorinated copolymers

High-dose irradiation of perfluorinated copolymers yielded products with higher carboxyl group contents than PTFE (Fig. 11). FEP [poly(tetrafluoroethylene-co-hexafluoropropylene)] showed a higher degree of carboxylation than PFA [poly(tetrafluoroethylene-co-perfluoroalkylvinylether)]. In our opinion, the high degree of COF functionalization obtained in these copolymers is due to the lower crystallinities of FEP and PFA as compared with PTFE [21].

$$\text{glass}\begin{cases}-OH + H_5C_2-O-\underset{\underset{O-C_2H_5}{|}}{\overset{\overset{O-C_2H_5}{|}}{Si}}-(CH_2)_3-\underset{\underset{H}{|}}{N}-\overset{\overset{O}{\|}}{C}-R_f \\ -OH\end{cases} \xrightarrow[-C_2H_5OH]{}$$

$$\text{glass}\begin{cases}-O-\underset{\underset{O-C_2H_5}{|}}{\overset{\overset{O-C_2H_5}{|}}{Si}}-(CH_2)_3-\underset{\underset{H}{|}}{N}-\overset{\overset{O}{\|}}{C}-R_f \\ -OH\end{cases}$$

R_f → fluorinated chain (hydrophobicity)

OH → concentration is decreased → Zeta potential is increased

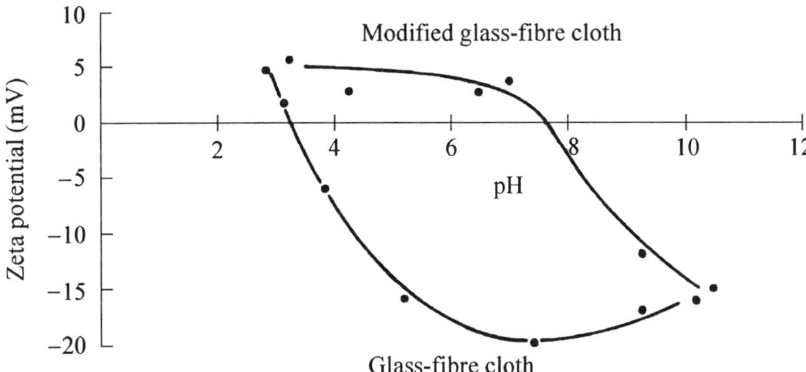

Figure 10. Zeta potential of E-glass modified by alkoxysilane-terminated PTFE micropowders versus the pH.

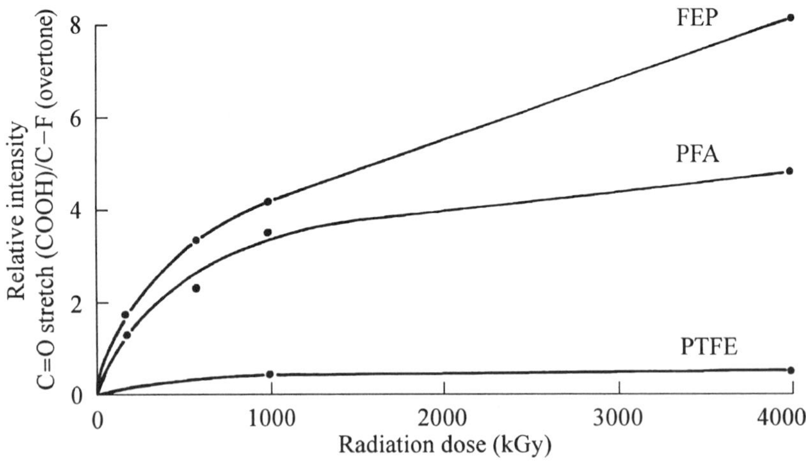

Figure 11. Dependence of the content of carboxyl groups in different fluoropolymers on the radiation dose.

4. CONCLUSIONS

Virgin PTFE is disintegrated by high-energy irradiation in air into a finely dispersed powder modified with COF groups which can subsequently be hydrolysed to carboxyl groups. A higher radiation dose causes an increasing concentration of carboxyl groups and an increased polar component of the surface free energy. Homogeneous incorporation of modified powders into organic or aqueous media and other polymers is possible. So, the special properties of PTFE, such as its particular release and lubricating properties as well as its chemical resistance, can be made effective. Changes in the nature of functional groups are possible, e.g. by chemical conversion.

Modified powders can be used for hydrophilizing fluoropolymer membranes and diaphragms. It should be possible to use such materials produced by the modification of radiation-chemically introduced COF groups with γ-aminopropyltriethoxysilane as coupling agents to bond fluoropolymers to substrates such as glass or metal.

FEP and PFA copolymers irradiated in air showed a higher degree of COF functionalization as compared with PTFE.

Acknowledgements

We are indebted to the Deutsche Forschungsgemeinschaft and the Fonds der Chemischen Industrie for financial support of this research.

REFERENCES

1. J. H. Golden, *J. Polym. Sci.* **45**, 534 (1960).
2. M. I. Bro, E. R. Lovejoy and G. R. McKay, *J. Appl. Polym. Sci.* **7**, 2121 (1963).
3. K. Lunkwitz, A. Ferse, P. Dietrich, G. Engler, U. Gross, D. Prescher and J. Schulze, *Plaste Kautsch.* **26**, 318 (1979).
4. K. Lunkwitz, H.-J. Brink, D. Handte and A. Ferse, *Radiat. Phys. Chem.* **33**, 523 (1989).
5. A. Ferse, K. Lunkwitz, H. Grimm, P. Dietrich, G. Engler, U. Gross, D. Prescher and J. Schulze, *Acta Polym.* **30**, 348 (1979).
6. U. Gross, P. Dietrich, G. Engler, D. Prescher, J. Schulze, K. Lunkwitz and A. Ferse, *J. Fluorine Chem.* **20**, 33 (1982).
7. A. Ferse, D. Handte and K. Lunkwitz, *Plaste Kautsch.* **29**, 458 (1982).
8. W. Bürger, K. Lunkwitz, G. Pompe, A. Petr and D. Jehnichen, *J. Appl. Polym. Sci.* **48**, 1973 (1993).
9. K. L. Hsu, D. E. Kline and J. N. Tomlinson, *J. Appl. Polym. Sci.* **9**, 3567 (1965).
10. D. M. Pinkerton and B. T. Sad, *Aust. J. Chem.* **23**, 1947 (1970).
11. H. N. Subrahmanyan and S. V. Subrahmanyan, *J. Polym. Sci.: Polym. Phys.* **25**, 1549 (1987).
12. P. Dietrich, G. Engler, A. Ferse, H. Grimm, U. Gross, D. Handte, K. Lunkwitz, D. Drescher, J. Schulze and W. Schenk, German Patent DD-WP 116 626 (1974).
13. A. Ferse, D. Handte, K. Lunkwitz, B. Poschardt and S. Uhlmann, German Patent DD-WP 138 986 (1978).
14. A. Ferse, K. Lunkwitz, G. Wiedemann and D. Handte, German Patent DD-WP 141 316 (1978).
15. K. Lunkwitz, A. Ferse, D. Handte, B. Klatt and M. Horx, *Angew. Makromol. Chem.* **148**, 137 (1987).
16. K. Lunkwitz, A. Ferse, D. Handte, B. Klatt and G. Grünzig, German Patent DD-WP 133 257 (1977).
17. K. Lunkwitz, D. Handte, A. Ferse, K. Zärl and D. Ludwig, German Patent DD-WP 146 715 (1979).
18. G. Wiedemann, H. Frenzel, A. Ferse, K. Lunkwitz, D. Handte, M. Matzke, H. Unganz, R. Radom and A. Haase, German Patent DD-WP 138 987 (1978).
19. H.-J. Brink, H. Kölling, K. Lunkwitz, B. Klatt, M. Horx, R. Herrmann and W. Bürger, Eur. Patent Announcement 91110744.9 (1991).

20. H.-J. Brink, H. Kölling, K. Lunkwitz, B. Klatt, M. Horx and R. Herrmann, German Patent Announcement 4117391 (1991).
21. U. Lappan, W. Bürger and K. Lunkwitz, Publication in preparation.
22. K. Lunkwitz, H.-J. Brink, W. Bürger and D. Handte, German Patent DD-WP 285 505 (1987).
23. W. Bürger, K. Lunkwitz, H.-J. Brink and D. Handte, German Patent DD-AP 285 890, 285 891 (1987).
24. W. Bürger, K. Lunkwitz and D. Handte, German Patent DD-AP 285 505 (1987).

Polymer Surface Modification: Relevance to Adhesion, pp. 363–378
K. L. Mittal (Ed.)

Modification of polyimide surface morphology: relationship between modification depth and adhesion strength

KANG-WOOK LEE
IBM T. J. Watson Research Center, PO Box 218, Yorktown Heights, NY 10598, USA

Revised version received 27 December 1993

Abstract—The morphology of a polyimide film surface is modified from a semicrystalline state to an amorphous state without altering the bulk properties. The outer layer (0.5–25 nm) of fully cured poly(pyromellitic anhydride-oxydianiline) (PMDA-ODA) and poly(biphenyl dianhydride-*p*-phenylene diamine) (BPDA-PDA) polyimides is chemically modified to polyamic acid, which is subsequently imidized at 230–250°C for 30 min to give a disordered polyimide surface. This disordered layer seems amorphous since it swells well in 1-methyl-2-pyrrolidinone (NMP) and the ions in the modified layer can be easily removed with a solvent such as water or ethanol. The modified surfaces are analyzed by surface-sensitive techniques such as contact angle measurement, X-ray photoelectron spectroscopy (XPS), ion scattering spectroscopy (ISS), secondary ion mass spectroscopy (SIMS), and external reflectance infrared spectroscopy (ERIR). The adhesion of polyimide layers onto the amorphous polyimides is greatly enhanced. Interdiffusion and subsequent mechanical interlocking are the major contributors to the polyimide–polyimide adhesion. The relationship between the depth of modification and the peel strength is studied. The deeper the modification depth, the greater is the peel strength.

Keywords: Adhesion; morphology; polyimide; surface modification; depth of modification; peel strength.

1. INTRODUCTION

Polyimides are used as dielectric layers in a variety of microelectronic applications, as they have good processability, a low dielectric constant, high thermal stability, low moisture absorption, and good mechanical properties [1]. Metal patterns are deposited onto polyimide surfaces to form electrically conducting lines on multichip module thin film packages such as the IBM ES/9000 microprocessor [2]. However, the adhesion of metals directly to polyimide is usually poor. Failure to activate a polymeric surface will normally cause subsequent coatings to be poorly adhered and easily cracked, blistered, or otherwise removed. Various surface treatments and adhesion-promoting layers have been used to enhance the metal-to-polyimide and polyimide-to-polyimide adhesion [3]. These methods can introduce foreign materials and/or undesirable modified layers into the interface, resulting in possible reliability failures. Introducing new functional groups may also cause reliability problems if these are unstable. It is challenging to improve the adhesion without introducing foreign materials, foreign functional groups, or adhesion promoters.

Polymers can be generally either semicrystalline or amorphous [4]. These categories are used to describe the degree of ordering of the polymer molecules. Amorphous polymers consist of randomly ordered tangled chains, i.e. amorphous polymers are highly disordered and intertwined with other molecules. Semicrystalline polymers consist of a mixture of amorphous regions and crystalline regions. The crystalline regions are more ordered and segments of the chains actually pack in crystalline lattices. Some crystalline regions may be more ordered than others. If crystalline regions are heated above the melting temperature of the polymer, the molecules become less ordered or more random. If cooled rapidly, this less-ordered feature is frozen in place and the resulting polymer is amorphous. If cooled slowly, these molecules can repack to form crystalline regions, and the polymer is crystalline.

One possible way to enhance the adhesion without introducing foreign materials into the interfaces is to change the morphology of the polyimide surface, coat the next layer (metal or polyimide), and then convert the modified morphology back to the original state. Semicrystalline polyimide consists of some amorphous poly-

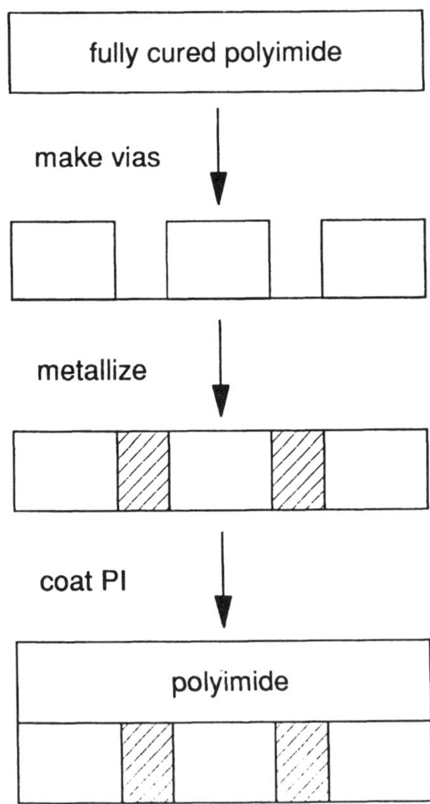

Scheme I. Thin film process.

imide and the amorphous polyimide contains a small amount of crystalline polyimide. However, the mechanical properties of the two materials are very different. The rate of water diffusion in amorphous polyimide is three times as fast as that in semicrystalline polyimide [5]. The adhesion property of a polymer is related to its surface properties. The morphology of a polyimide surface can be altered without changing the bulk morphology, and thus the bulk mechanical properties of the polymer remain unchanged. Saenger *et al.* [6] reported that polyimide-to-polyimide adhesion could be enhanced by swelling the adherend (substrate; bottom) layer of a cured poly(pyromellitic dianhydride-oxydianiline) (PMDA-ODA) in 1-methyl-2-pyrrolidinone (NMP). In contrast, we found that poly(biphenyl dianhydride-*p*-phenylene diamine) (BPDA-PDA) film did not swell in NMP at ambient temperature to give enhanced polyimide-to-polyimide adhesion. The swelling method, even though it improves adhesion, cannot be employed in a multilevel thin film package since the swelling destroys the electronic structure.

Brown *et al.* [7] studied the adhesion of a polyimide layer to a partially cured polyimide which is often a mixture of disordered polyimide and polyamic acid. However, in a real manufacturing process of a multilevel package the adherend layer should be fully cured at a high temperature (usually 350–400°C), followed by metallization and then coating with the polyimide for the next layer as illustrated in Scheme I. Thus, modification of the polyimide surface without changing the bulk properties is important to multilevel thin film packaging technology.

There are two major categories of surface modification techniques. One involves chemical reactions at the polymer surface with radiation, plasma, chemicals, etc.; consequently, the chemical structure of the polymer surface is altered. The other category consists of depositing a separate layer of another material onto the surface to obtain a desired property. A third category [8] is to change the morphology without altering the chemistry at all. The key to a successful surface modification technique is to just modify the surface without changing the bulk morphology. Here we report the modification of the polyimide surface morphology from a relatively more crystalline state to a relatively more amorphous state which provides improved polyimide-to-polyimide adhesion [8].

2. EXPERIMENTAL

2.1. Materials and methods

The polyimide precursors, PMDA-ODA and BPDA-PDA polyamic acids, were obtained from Du Pont. Tetramethylammonium hydroxide, KOH, HCl, acetic acid, 1-methyl-2-pyrrolidinone (NMP), and isopropanol were obtained from Aldrich. Polyamic acid in NMP was spin-coated onto a Si wafer coated with chromium and then cured to polyimide at 150°C for 30 min, 230°C for 30 min, 300°C for 30 min, and 400°C for 60 min. The purpose of the 100 nm thick layer of chromium was to enhance the wettability of the polyamic acid solution onto the Si wafer. Ten to one hundred nanometer thick layers of polyimides were employed to obtain the

ERIR spectra. The surface-modified samples for contact angle, SIMS, ISS, XPS, and ERIR investigations were dried under vacuum at ambient temperature for 12–24 h. Contact angle measurements were made using a Rame-Hart telescopic goniometer and Gilmont syringe with a 24-gauge flat-tipped needle. Distilled water was used as the probe liquid. Dynamic advancing and receding contact angles were determined by measuring the tangent at the intersection of the air/drop/surface while adding (advancing) and withdrawing (receding) water to and from the drop. External reflectance infrared (ERIR) spectra were obtained under nitrogen using a Nicolet-510P FTIR spectrometer with a Harrick reflectance attachment. ISS, XPS, and SIMS spectra were taken on a Perkin-Elmer Phi 5500 system. The thicknesses of the polyimide were measured with a Dek-Tak for thicknesses greater than 100 nm, and with a Waferscan ellipsometer equipped with a HeNe laser ($\lambda = 632.8$ nm) for thicknesses of 10–100 nm. The refractive indices employed for PMDA-ODA and BPDA-PDA were 1.730 and 1.848, respectively, at 632.8 nm [9].

2.2. Modification of the polyimide surface morphology

The PMDA-ODA polyimide samples were treated with 1 M KOH aqueous solution at 22°C for 1–20 min depending on the desired depth of surface modification, followed by washing with water and 2-propanol. The samples were dried under vacuum at ambient temperature for 12 h. The resulting modified surface was potassium polyamate, which was converted to polyamic acid by treating with 0.2 M HCl for 5 min and rinsing with water and 2-propanol. Curing the polyamic acid surface at 230°C for 30 min under nitrogen gave the amorphous polyimide surface.

The BPDA-PDA polyimide samples were initially modified with 1 M KOH at 55°C for 1–20 min to potassium polyamate which was transformed to polyamic acid. Curing the BPDA-PDA polyamic acid surface at 250°C for 30 min gave the amorphous polyimide surface. Some analyses of the modified surfaces have been described in our previous publications [10–12].

2.3. Adhesion measurement

For the polyimide-to-polyimide adhesion, the solution of the polyamic acid was spin-coated onto a chromium-coated Si water and cured at 150°C for 30 min, 230°C for 30 min, 300°C for 30 min, and 400°C for 60 min. The thickness of the polyimide layer was approximately 6 μm. A thin layer (20 nm) of gold was sputter-coated onto one side (20% of the total area) of the polyimide sample to initiate peel (polyimide has poor adhesion to gold). The surface of the polyimide was modified as previously described. Polyamic acid solution was spin-coated onto the surface-modified polyimide film and subsequently cured at 400°C under nitrogen. The thickness of the adherate layer (peel layer) after curing was approximately 20 μm and the width of the peel layer was 5 mm. The peel strengths were measured by a 90° peel test using an Instron at 1.5 mm/min peel rate. The reported values are the average of at least three measurements.

3. RESULTS AND DISCUSSION

3.1. Formation of the amorphous polyimide surface

Both PMDA-ODA and BPDA-PDA amic acids were cured at 150°C for 30 min, 230°C for 30 min, 300°C for 30 min, and 400°C for 60 min to the semicrystalline state polyimides (PIs). The PI surface layer (0.5–20 nm) was converted to the polyamic acid by a wet process in which the PI was reacted with aqueous KOH solution to yield potassium polyamate, which was subsequently acidified in a dilute acid [10–13]. The reaction sequences for the surface modification of BPDA-PDA polyimide are shown in Scheme II.

The contact angles are indicative of a surface polarity which has a close relationship to adhesion. The stability of a modified surface is important for practical applications and the contact angle is one of the surface-sensitive properties. The advancing/receding water contact angles (54°/9° on the BPDA-PDA polyamic acid surface and 76°/37° on the amorphous PI surface) in this work did not change over the 3-month period. Goldblatt *et al.* [14] reported that the advancing/receding water contact angles on a PI film surface treated with water vapor plasma increased from 7°/3° to 43°/28° on storage under ambient conditions for 1 week. Similar phenomena have been observed with other polymers [15, 16]. These phenomena have been explained by the fact that the polar functional groups, such as hydroxy (OH), aldehyde (CHO), and carboxylic acid (COOH), introduced to the end (at the surface) of the polymer chains reorient themselves with the polar groups turned toward the bulk. Thus, the water contact angles are increased and the surface energy is minimized [17]. However, the modified PI surface layer with the PI bulk may not provide enough mobility for the unmodified PI to diffuse to the surface through the polyamic acid layer. The molecules of the polyamic acid itself are expected to be very dynamic, but the reorientation of the polyamic acid chains would not change the number of carboxylic acid groups at the surface because if one carboxylic acid group is heading down, the other nearby will be heading up. This hypothesis explains why the contact angles of the modified PI surface modified by base-acid reactions do not change over a long period of time.

The X-ray photoelectron spectroscopy (XPS) survey spectrum of the BPDA-PDA polyamate surface displays new peaks at 378.5, 296.4, and 293.6 eV, which correspond to K_{2s}, $K_{2p1/2}$, and $K_{2p3/2}$, respectively. Figure 1 shows the XPS C_{1s} and K_{2p} regions of BPDA-PDA polyimide and the modified surfaces. Figure 1a displays the spectrum of the starting polyimide. The broad peak at 291.3 eV is due to the $\pi \rightarrow \pi^*$ transition of the phenyl groups. The sharp peak (FWHM = 0.82 eV) at 288.7 eV corresponds to the carbonyl carbons. There is only one carbonyl peak since the polyimide carbonyls have the similar nuclear environments. However, the spectrum (Fig. 1b) corresponding to the potassium polyamate surface exhibits a broad peak (FWHM = 1.08 eV) since the binding energies of carboxylate carbon and amide carbon are close, but they are not identical. Two carbonyl peaks are well separated in the case of polyamic acid. As shown in Fig. 1c, the XPS curve fitted very well when the carbonyl region was separated into two peaks (289.0 and 287.9 eV). The $K_{2p1/2}$ (296.4 eV) and $K_{2p3/2}$

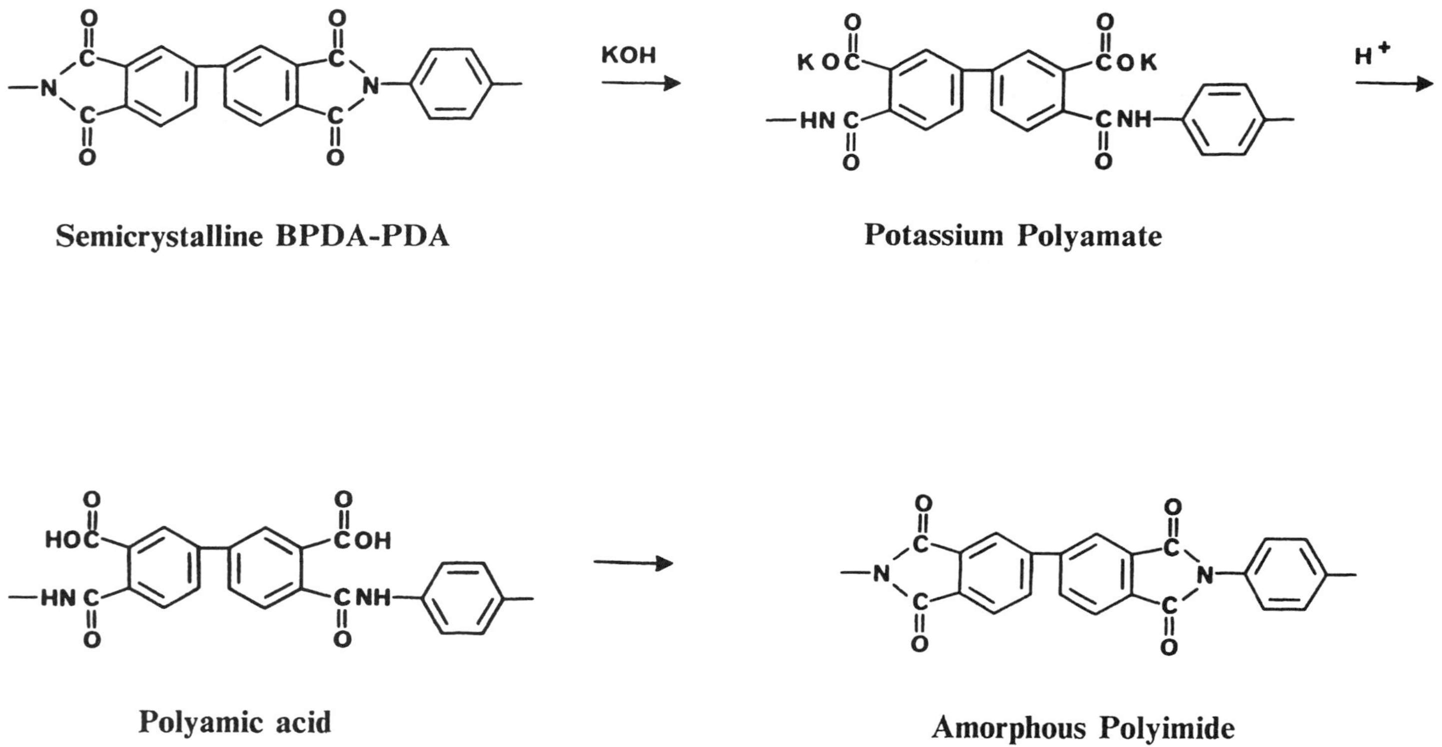

Semicrystalline BPDA-PDA

KOH

Potassium Polyamate

H^+

Polyamic acid

Amorphous Polyimide

Scheme II. Chemical reactions at the BPDA-PDA polyimide surface.

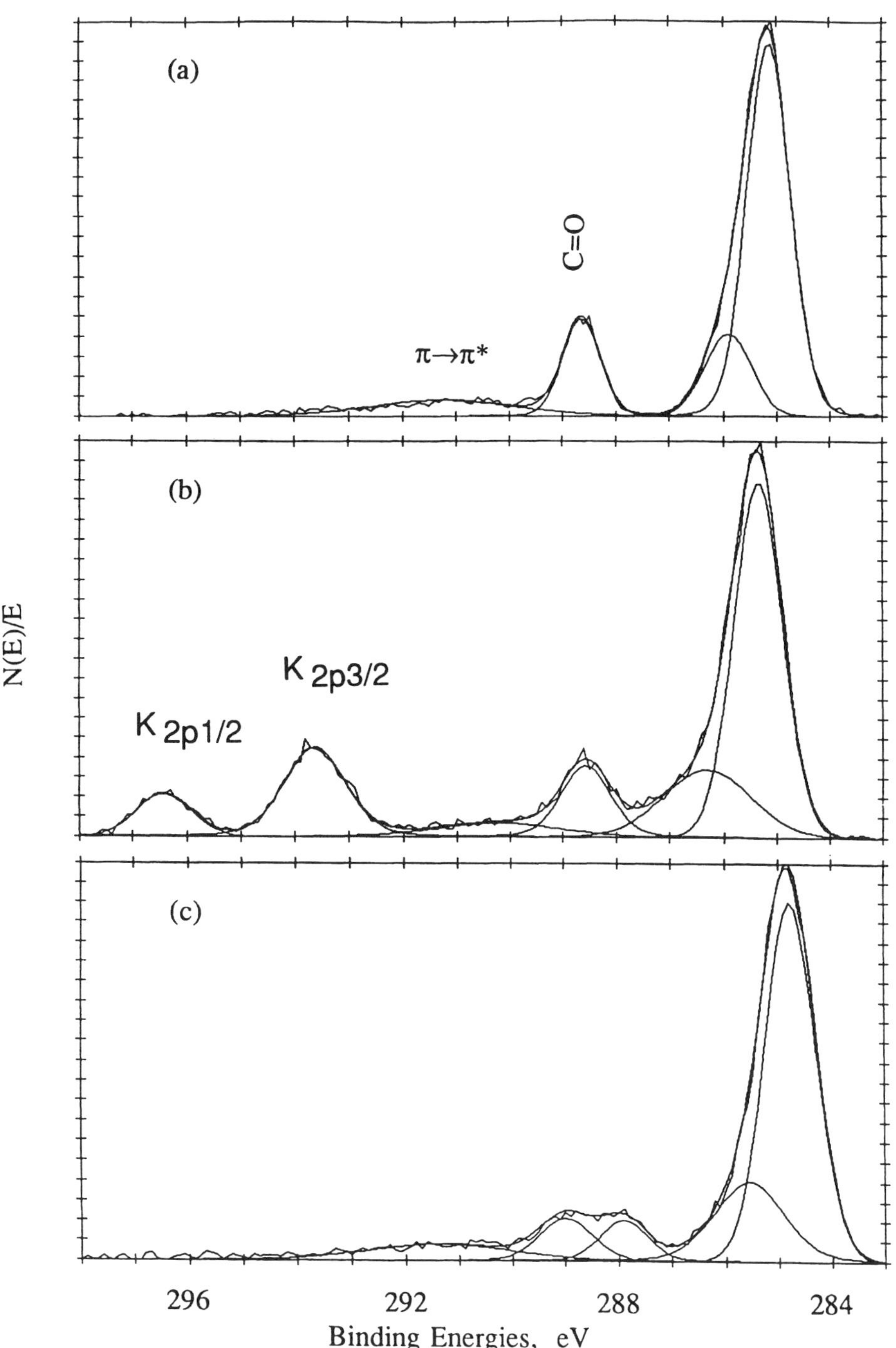

Figure 1. C_{1s} and K_{2p} XPS spectra of (a) virgin BPDA-PDA polyimide, (b) polyamate, and (c) polyamic acid. The electron take-off angle was 45°.

(293.6 eV) peaks appear in the spectrum of potassium polyamate and disappear in that of polyamic acid, as expected.

Figure 2 shows the O_{1s} XPS spectra of virgin BPDA-PDA and the modified surfaces. The BPDA-PDA polyimide contains only one kind of carbonyl, and thus only one XPS peak is observed (Fig. 2a). On the other hand, the polyamate and the polyamic acid have two different carbonyls: amide and carboxylate, amide and carboxylic acid, respectively. Thus, two peaks are observed (Figs 2b and 2c). These two peaks are fitted with two components: 533.4 and 531.7 eV for the polyamate, and 533.1 and 531.5 eV for the polyamic acid. The low binding energy peaks of the polyamate and the polyamic acid are assigned to carbonyl oxygen. As shown in the spectrum of the polyamate (Fig. 2b), a high binding energy peak appeared and this peak can be assigned to the carboxylate oxygen with a negative charge. The observed ratio of the two peaks (35/65) matches well with the calculated ratio ($2/4 = 33/67$) for the molecular structure of a repeat unit. The high binding energy peak of the polyamic acid corresponds to the OH [in C(O)OH] groups. The ratio of the two peaks in the case of polyamic acid (Fig. 2c) increased to 45/55. The calculated ratio for the polyamic acid is the same as that for the polyamate (the chemical structures are shown in Scheme II), but the observed ratio increased. The observed O/N ratios for the polyamate and the polyamic acid, obtained from the XPS atomic composition data, are identical ($3/1 = 75/25$), indicating that the oxygen concentration has not decreased. These results indicate that the carbonyl peak of the lower O_{1s} binding energy is shifted to the higher binding energy peak while the total amount of oxygen remains unchanged in comparison with the polyamate. The shifted portion probably corresponds to the carbonyls which form hydrogen bonds with the neighboring carboxylic acid. Thus, the binding energies of the electrons are changed and the XPS peak is shifted.

The incorporation of potassium and the subsequent removal of potassium by acidification can be followed by ion scattering spectroscopy (ISS), which is a good analytical technique for detecting a small amount of metal on a polymer surface. As shown in Fig. 3, a large potassium peak appears in the ISS spectrum of potassium polyamate and disappears as the surface is acidified to polyamic acid. The ISS technique is complementary to XPS. As shown in Fig. 1, the potassium incorporation can be followed by XPS, but the sensitivity is low. Even though the BPDA-PDA polyimide is initially modified at 55°C, the acidification is complete at 22°C in a few minutes. This result indicates that the aqueous acid solution easily penetrated into the modified layer and that the semicrystalline state of polyimide was converted to the amorphous state of potassium polyamate. The polyimide film surface can also be modified with other bases such as NaOH, NH_4OH, and $(CH_3)_4NOH$. The surface treated with tetramethylammonium hydroxide (TMAH) was analyzed by secondary ion mass spectroscopy (SIMS), and its spectrum is shown in Fig. 4. The peak at $m/e = 74$ corresponds to the tetramethylammonium cation. Acidification also converts the tetramethylammonium polyamate to polyamic acid. The polyamic acid surface was cured at 250°C for 30 min to disordered polyimide, which was analyzed by ERIR.

Lee *et al.* [10, 11] have explored the relationship between the modification depth of the PI surface and the adhesion strength. They also developed a method to non-destructively measure the average depth of modification using ellipsometry and

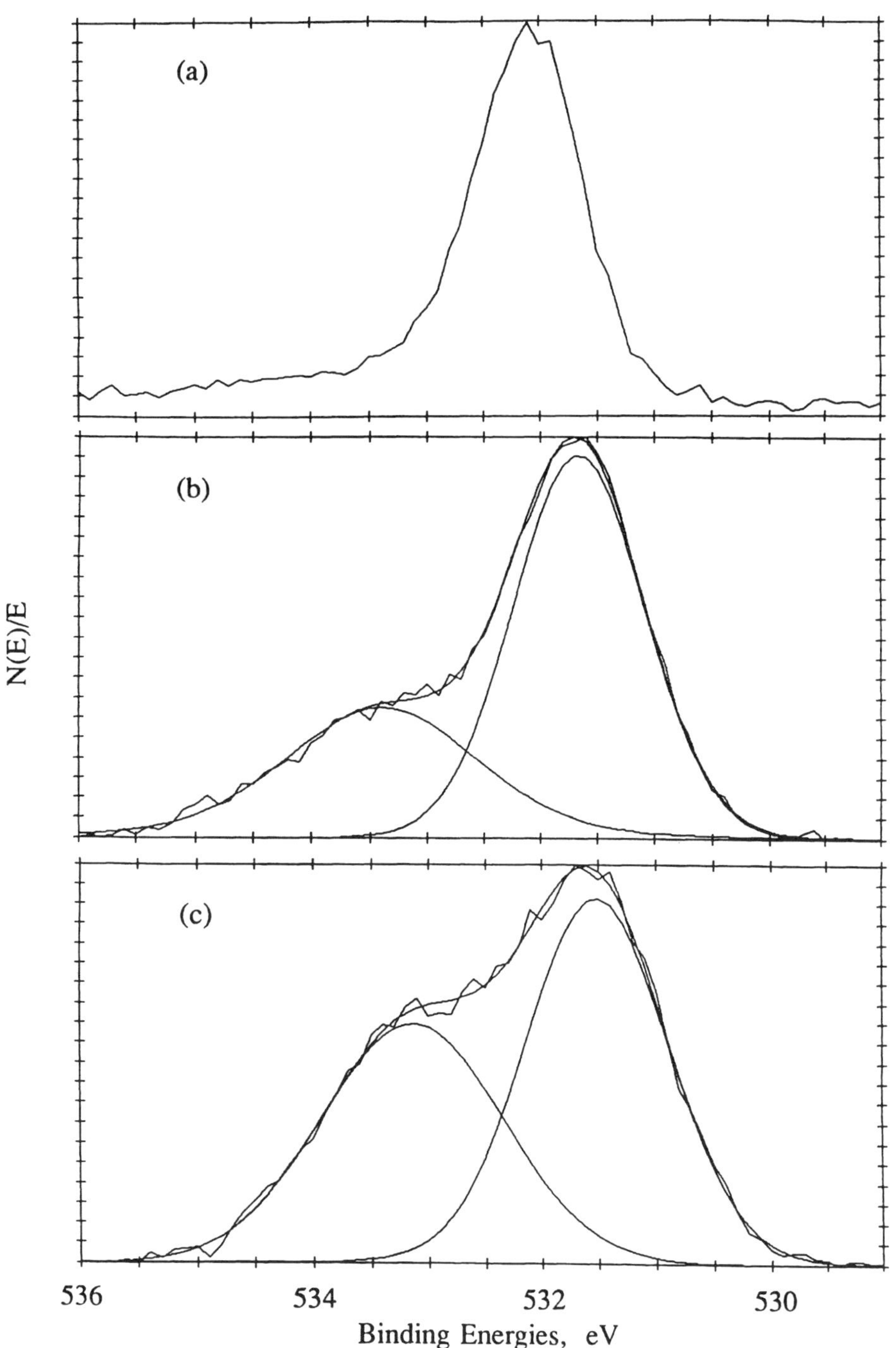

Figure 2. O_{1s} XPS spectra of (a) virgin BPDA-PDA polyimide, (b) polyamate, and (c) polyamic acid. The electron take-off angle was 45°.

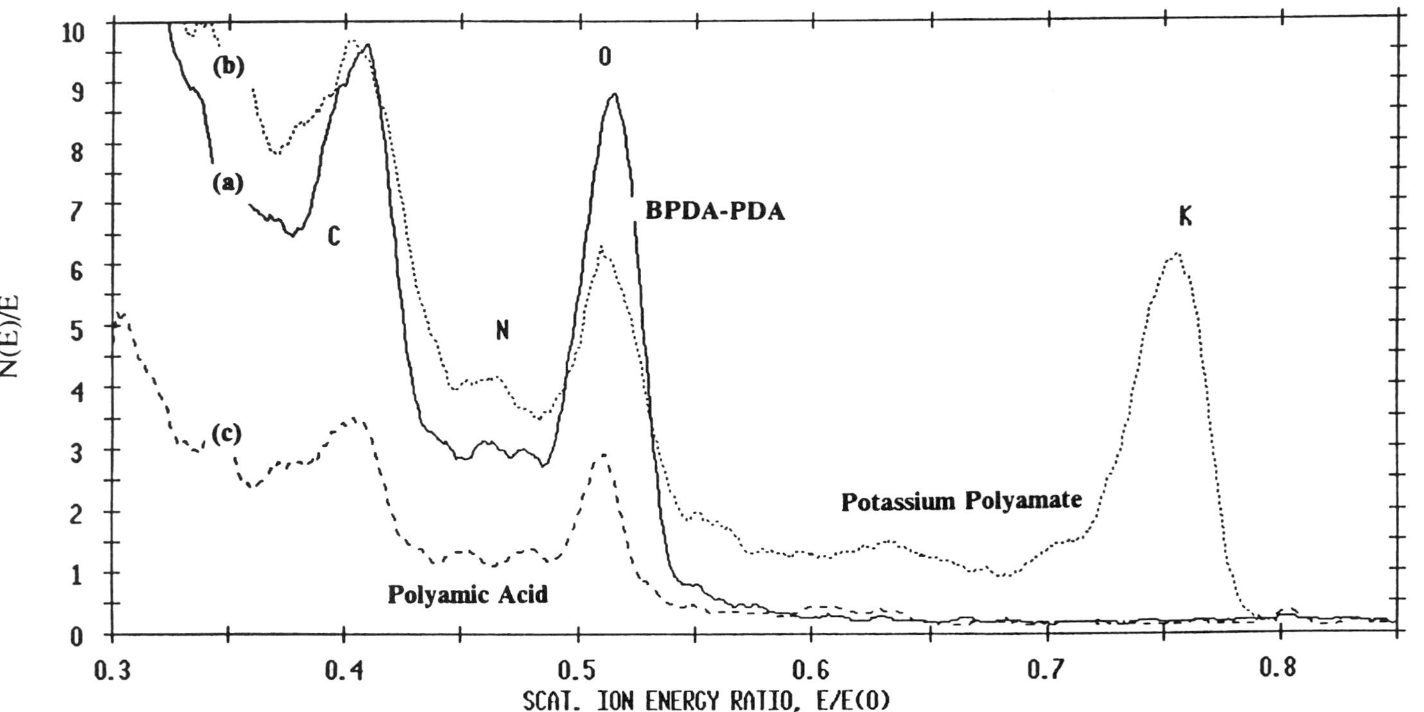

Figure 3. ISS spectra of BPDA-PDA polyimide, potassium polyamate, and polyamic acid.

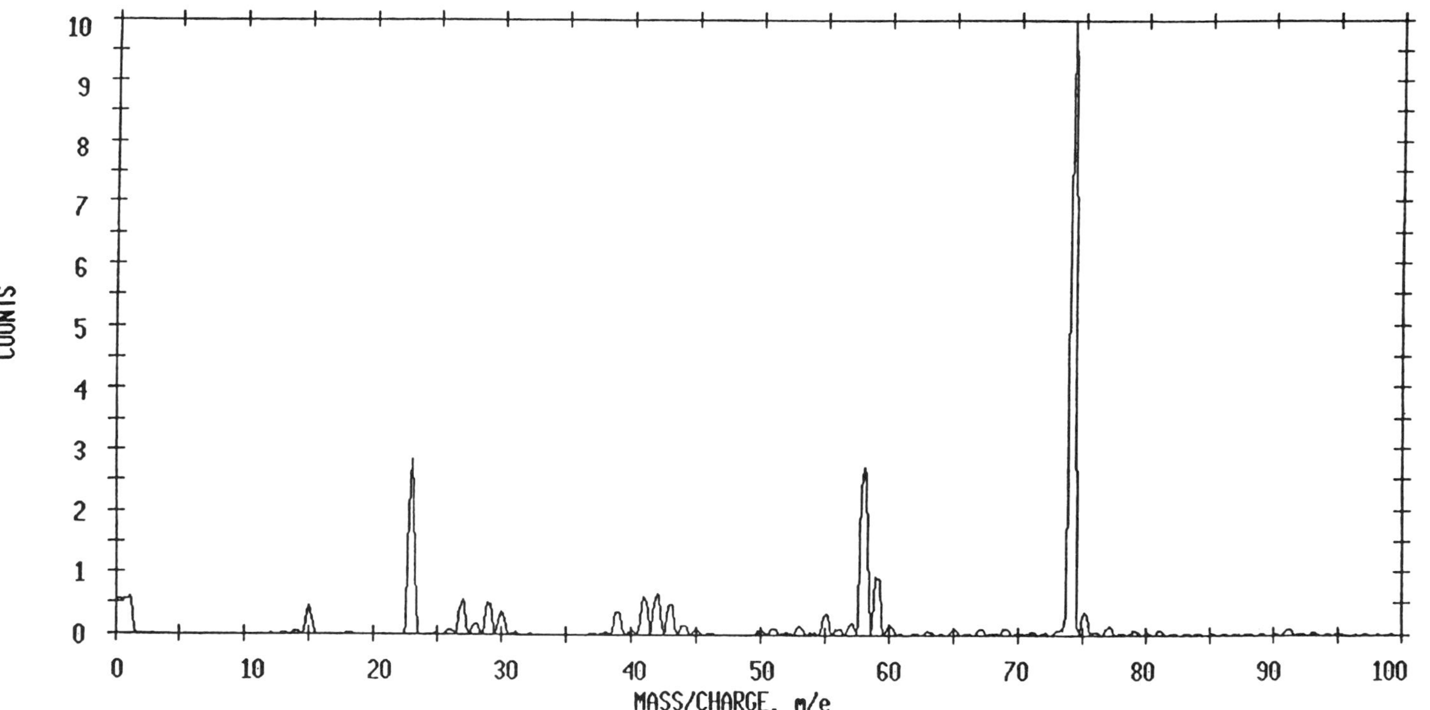

Figure 4. SIMS spectrum of BPDA-PDA tetramethylammonium polyamate. The peek at $m/e = 74$ corresponds to the tetramethylammonium cation.

ERIR techniques. There is a linear relationship between the film thicknesses (less than 100 nm) and the absorbances of the imide carbonyl stretching at 1739 cm^{-1}. Once the PI surface is modified under conditions under which the PI film is not etched, the thickness of the unmodified layer can be calculated by measuring the imide carbonyl absorbance of the modified (to polyamate) film. The average depth of modification can be obtained by subtracting the thickness of the unmodified PI film from the thickness of the starting polymer, which was adjusted to less than 100 nm for this study. In a real application, the polymer thickness is in the range of 1.0–25 μm, but it is reasonable to assume that the rate of reaction from the outer layer would not depend on the film thickness and thus the modification depth of a thick film should be same as that with a thin film. The polyamic acid surface can be imidized at 230–250°C for 30 min to a disordered PI. The chemical reactions are illustrated in Scheme II. To obtain vibrational spectra, a thin layer (around 50 nm) of BPDA-PDA polyamic acid was coated onto a Cr-coated Si wafer and then cured to polyimide. The whole layer was modified to polyamic acid by the wet process. The ERIR spectrum displays the typical peaks of polyamic acid: 1726 (s), 1670 (s), 1607 (m), 1547 (m), 1517 (s), 1408 (s), and 1322 (m, br) cm^{-1}. The sample was cured at 250°C for 30 min under nitrogen. The ERIR spectrum shows the same bands as those of fully cured polyimide at 1776 (w), 1739 (vs), 1709 (w, sh), 1621 (vw), 1517 (m), 1423 (w), and 1358 (vs) cm^{-1}, indicating that the modified PI surface is fully imidized.

It is known that PMDA-ODA polyimide cured at 230°C has a disordered structure and that upon treating at 300°C or greater it becomes more ordered to a relatively more crystalline state [18]. We carried out an experiment to determine whether this disordered polyimide was amorphous. A thin layer (around 50 nm) of PMDA-ODA polyimide was prepared on a silicon wafer and the whole layer was modified to polyamic acid, followed by curing at 230°C for 30 min to polyimide. As this sample was dipped into NMP solvent, NMP diffused into the modified layer and finally dissolved the whole layer. Another indication for an amorphous surface is that impurities such as ions can be easily rinsed out with a solvent. These results indicate that the disordered polyimide is relatively amorphous. The corresponding experiment with BPDA-PDA gave the same results.

3.2. Polyimide-to-polyimide adhesion

PMDA-ODA (or BPDA-PDA) polyamic acid was spin-coated onto the corresponding amorphous (modified) polyimide surface at 1200 rpm for 30 s. The sample was baked at 85°C for 30 min to evaporate the NMP solvent and then cured at 150°C for 30 min, 230°C for 30 min, 300°C for 30 min, and 400°C for 60 min. This procedure gave an approximately 20 μm thick PMDA-ODA (or BPDA-PDA) peel layer. The 90° peel strength was measured by peeling the 20 μm thick adherate polyimide layer. Table 1 shows the relationship between the surface modification depth and the peel strength.

These results show that the deeper the modification depth, the greater the peel strength. If the PMDA-ODA surface was modified to depths greater than 25 nm, the ultimate tensile strength of the PMDA-ODA peel layer was exceeded by the

Table 1.
Relationship between the surface modification depth and the peel strength

PMDA-ODA		BPDA-PDA	
Depth (nm)	Peel strength (J/m^2)	Depth (nm)	Peel strength (J/m^2)
0	40	0	30
0.5–2	400	1.7	160
10	850	16	680
20	1250	22	900
> 25	cnp[a]		

[a]cnp: cannot peel without tearing the peel layer.

interfacial adhesion strength. However, if the BPDA-PDA polyimide was employed as the peel (adherate) layer, the adhesion strength did not exceed the tensile strength of the BPDA-PDA polyimide, probably because the BPDA-PDA polyimide film has greater tensile strength at break (300 MPa) than PMDA-ODA (140 MPa).

The mechanism for polyimide–polyimide adhesion here seems quite different from the dry process, such as plasma treatment and ashing, in which the chemical reaction is the most important factor. However, the involvement of a chemical reaction in the adhesion to the amorphous polyimide in this work is excluded since the polyamic acid is not expected to react with fully imidized polyimide. Wettability and interdiffusion are the other possible mechanisms. But the water contact angles on the amorphous (less ordered) polyimide surface are similar to those on the semicrystalline structure (more ordered), while the peel strength corresponding to the amorphous polyimide adhesion in an optimized case is 30 times greater than that corresponding to the semicrystalline polyimide adhesion. This result indicates that in the case of polyimide-to-amorphous polyimide adhesion, as illustrated in Scheme III, the polyamic acid solution wets the polyimide surface well, but significant mechanical interlocking between the polyamic acid and the polyimide is involved only to the depths of the amorphous polyimide surface. Thus, the polyimide-to-semicrystalline (more ordered) polyimide adhesion is weak and the polyimide-to-amorphous (less ordered) polyimide adhesion is strong. Interdiffusion and subsequent mechanical interlocking are important factors in this system. The mechanical interlocking in a usual system involves the interpenetration of one material into the roughness (or pores) of the other, and thus surface roughening is a technique of surface modification [19]. However, in this work the adherate material diffuses into the outer layer of the adherend and subsequent curing induces mechanical interlocking of two phases, i.e. the incoming polyamic acid solution diffuses into the amorphous layer as NMP does. Upon curing, two polyimide phases are interlocked and the interlocking force will be greater for the more deeply modified surface, as shown in Table 1.

Thomas *et al.* [20] reported that the PMDA-ODA polyamic acid surface is converted to polyisoimide by treating with trifluoroacetic anhydride and pyridine. An approximately 20 nm outer layer of a PMDA-ODA polyimide film was modified to

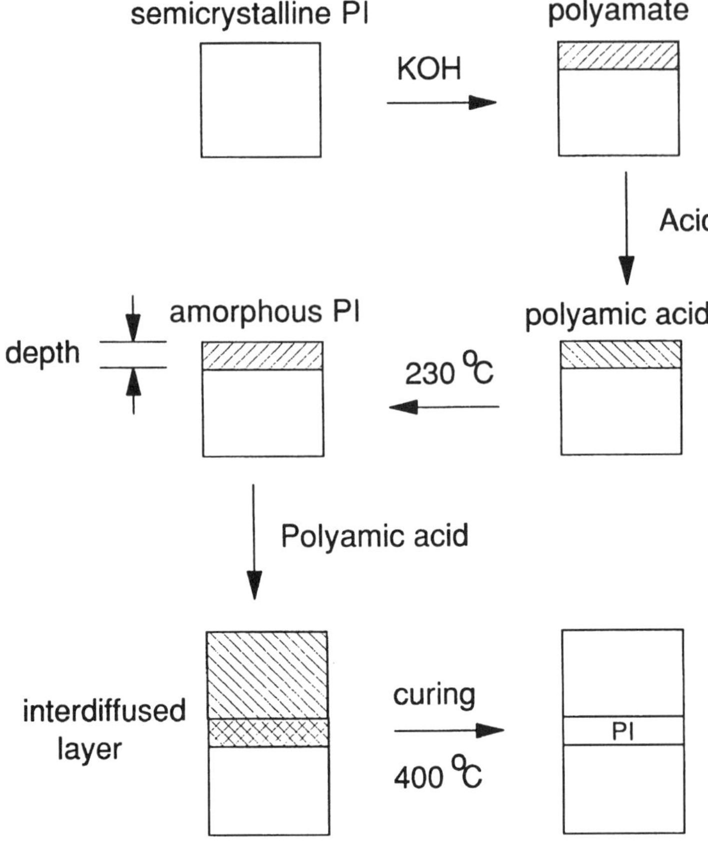

Scheme III. Polyimide surface modification and adhesion.

the polyamic acid by base hydrolysis and acidification. The sample was then treated with trifluoroacetic anhydride and pyridine to give polyisoimide. To evaluate the adhesion, a PMDA-ODA polyamic acid solution was coated onto the polyisoimide surface, followed by curing at 400°C. The 90° peel strength was 1200 J/m^2, which is comparable to the adhesion of polyimide to amorphous polyimide. The incoming polyamic acid solution may diffuse into the amorphous polyisoimide layer and upon curing, both the polyamic acid and the polyisoimide are cured to semicrystalline polyimide. Interdiffusion and mechanical interlocking could be major factors for the adhesion strength, as in the case of polyimide-to-amorphous polyimide adhesion. A possible reaction between polyamic acid and polyisoimide has a negligible effect on PI–PI adhesion.

4. CONCLUSION

The morphology of the polyimide surface was altered from a more crystalline state to a more amorphous state. Polyimide adhesion to the amorphous polyimide surface was greatly enhanced. The surface property can be controlled by tailoring the surface morphology. Morphology change can be one of the categories for polymer surface modification.

Acknowledgements

I would like to thank D. Hunt, E. Adamopoulos, and M. Norcott at IBM for technical help.

REFERENCES

1. (a) M. I. Bessonov, M. M. Koton, V. V. Kudryavtsev and L. A. Laius, *Polyimides: Thermally Stable Polymers*. Consultants Bureau, New York (1987).
   (b) D. Wilson, H. D. Stenzenberger and P. M. Hergenrother (Eds), *Polyimides*. Chapman & Hall (1990).
2. (a) C. Feger and C. Feger, in: *Multichip Module Technologies and Alternatives*, D. A. Doane and A. D. Franzon (Eds), pp. 311–348. Van Nostrand Reinhold, New York (1993).
   (b) R. R. Tummala and E. J. Rymaszewski (Eds), *Microelectronics Packaging Handbook*. Van Nostrand Reinhold, New York (1989).
   (c) R. R. Tummala, J. U. Knickerbocker, S. H. Knickerbocker, L. W. Herron, R. W. Nufer, R. N. Master, M. O. Neisser, B. M. Kellner, C. H. Perry, J. N. Humenik and T. F. Redmond, *IBM J. Res. Dev.* **36**, 889–904 (1992).
3. (a) L. P. Buchwalter and A. I. Baise, in: *Polyimides: Synthesis, Characterization, and Applications*, K. L. Mittal (Ed.), Vol. 1, p. 537. Plenum Press, New York (1984).
   (b) D. T. Clark and W. J. Feast (Eds), *Polymer Surfaces*. John Wiley, New York (1978).
   (c) W. J. Feast and H. S. Munro (Eds), *Polymer Surfaces and Interfaces*. John Wiley, New York (1987).
   (d) B. Chapman, *Glow Discharge Process*. John Wiley, New York (1980).
4. L. Mandelkern, in: *Physical Properties of Polymers*, J. E. Mark, A. Eisenberg, W. W. Grasessley, L. Mandelkern and J. L. Koenig (Eds), pp. 155–205. American Chemical Society, Washington, DC (1984).
5. B. S. Lim, A. S. Nowick, K.-W. Lee and A. Viehbeck, *J. Polym. Sci. Polym. Phys. Ed.* **31**, 545 (1993).
6. K. L. Saenger, H. M. Tong and R. D. Haynes, *J. Polym. Sci. Polym. Lett. Ed.* **27**, 235 (1989).
7. H. R. Brown, A. C. M. Yang, T. P. Russell, W. Volksen and E. J. Kramer, *Polymer* **29**, 1807 (1988).
8. A preliminary account of portions of this work has been reported: K.-W. Lee, *Mater. Res. Soc. Symp. Proc.* **227**, 347 (1991).
9. W. A. Pliskin, J. D. Chapple-Sokol and K.-W. Lee, Unpublished work (1990).
10. K.-W. Lee, S. P. Kowalczyk and J. M. Shaw, *Macromolecules* **23**, 2097 (1990).
11. K.-W. Lee, S. P. Kowalczyk and J. M. Shaw, *Langmuir* **7**, 2450 (1991).
12. K.-W. Lee, S. P. Kowalczyk, J. M. Shaw and E. Adamopoulos, *In Search of Excellence* **49**, 1108 (1991).
13. K.-W. Lee and S. P. Kowalczyk, in: *Metallization of Polymers*, E. Sacher, J.-J. Pireaux and S. P. Kowalczyk (Eds), Amer. Chem. Soc. Symp. Series, Vol. 440, Ch. 13. American Chemical Society, Washington, DC (1990).
14. R. D. Goldblatt, L. M. Ferreiro, S. L. Nunes, R. R. Thomas, N. J. Chou, L. P. Buchwalter, J. E. Heidenreich and T. H. Chao, *J. Appl. Polym. Sci.* **46**, 2189 (1992).
15. E. M. Cross and T. J. McCarthy, *Macromolecules* **23**, 3916 (1990).
16. S. R. Holmes-Farley, R. H. Reamey, R. Nuzzo, T. J. McCarthy and G. M. Whitesides, *Langmuir* **3**, 799 (1987).
17. J. D. Andrade, D. E. Gregonis and L. M. Smith, in: *Surface and Interfacial Aspects of Biomedical Polymers*, J. D. Andrade (Ed.), Ch. 2. Plenum Press, New York (1988).

18. N. Takahashi, D. Y. Yoon and W. Parrish, *Macromolecules* **17**, 2583 (1984).
19. (a) K. W. Allen, *J. Adhesion* **21**, 261–277 (1987).
(b) F. H. Chung, *J. Appl. Polym. Sci.* **42**, 1319–1331 (1991).
20. R. R. Thomas, S. L. Buchwalter, L. P. Buchwalter and T. H. Chao, *Macromolecules* **25**, 4559 (1992).

Polymer Surface Modification: Relevance to Adhesion, pp. 379–400
K. L. Mittal (Ed.)

Surface modification of synthetic vulcanized rubber

M. M. PASTOR-BLAS, M. S. SÁNCHEZ-ADSUAR
and J. M. MARTÍN-MARTÍNEZ*

Adhesion and Adhesives Laboratory, Department of Inorganic Chemistry, University of Alicante, 03080 Alicante, Spain

Revised version received 21 January 1994

Abstract—Surface modifications produced by treatments (mainly halogenation) of synthetic vulcanized styrene–butadiene rubber (SBR) leading to increased adhesion properties with polyurethane adhesives have been studied. T-peel tests, scanning electron microscopy (SEM), advancing contact angle measurements, infra-red (IR) spectroscopy, X-ray photoelectron spectroscopy (XPS) and differential scanning calorimetry (DSC) were used to analyze the nature of surface modifications produced in the rubber. Although some surface heterogeneities were created, physical treatments (ultrasonic cleaning, solvent wiping, abrasion) did not noticeably increase the adhesion strength because certain abhesive substances (e.g. zinc stearate, paraffin wax) cannot be removed from the rubber surface by such treatments. Chemical treatment (chlorination) was carried out using ethyl acetate solutions of trichloroisocyanuric acid (TCI) (1,3,5-trichloro-1,3,5-triazine-2,4,6-trione). Chlorination of SBR with trichloroisocyanuric acid produced a significant improvement in T-peel strength, due to the contribution of mechanical (surface roughness, microcracks), thermodynamical (increase of polar contribution to the surface energy) and chemical (removal of abhesive substances, creation of polar groups) rubber surface modifications. The strong adhesion between the chlorinated SBR surface and the polyurethane adhesive was due to the presence of oxidized species of >C=O, −C−OH and −COR type. Chlorination of SBR is a fast reaction which needs only a small concentration of chlorination agent (< 1 wt% TCI/ethyl acetate) to produce high adhesion levels. An increased amount of TCI facilitated the chlorination reaction progressing from the exterior to the internal rubber bulk; however, although a thicker layer of chlorinated rubber created no further increase in adhesion strength was obtained.

Keywords: Synthetic vulcanized rubber; surface treatments; halogenation; mechanisms of adhesion; polyurethane adhesives.

1. INTRODUCTION

Rubbers are fascinating materials widely used in industry due to their interesting properties: elasticity, chemical inertness and excellent mechanical properties (mainly abrasion and tear resistance) [1]. Their fields of application are very wide, including principally in tire manufacturing and other specific uses in the building industry

*To whom correspondence should be addressed.

(carpeting, noise insulator), automotive industry (bumpers, windshield wiper blades), packaging (tape manufacturing), aerospace (structural joints) and footwear industry (sole material). There are several types of synthetic rubbers which, for certain applications , need to be bonded to different substrates. Rubbers are most easily bonded to other materials during vulcanization, i.e. when they are crosslinked. Rubber/rubber and rubber/metal bonds are achieved by using one- or two-component adhesives [2]. In these joints the uncured rubber surface should be clean and free from contaminants (e.g. mould-release agents) and an adequate surface modification of the rubber substrate is necessary to obtain an optimum adhesion.

In some instances, especially when complicated rubber shapes need to be bonded, it may be necessary to bond rubbers after they have been vulcanized (post-vulcanization bonding). Generally these synthetic rubbers contain several components in their formulation to improve their processability and to adapt each elastomer to a given use. However, some additives for synthetic rubber may produce poor adhesion. There are several sources of adhesion problems which are due to components of the rubber itself (antioxidants, mould-release agents), adhesive-rubber incompatibility, or poor durability (migration of abhesive substances to the interface once the adhesive joint is formed). To overcome these problems one- or two-component primer/adhesive systems may be used (i.e. addition of reactive polyisocyanate to one-component polyurethane adhesives), but their effectiveness is not always high [3]. Therefore, various surface treatments of rubber have been proposed to take care of most of its adhesion problems and produce strong and long service life adhesive bonds.

The effectiveness of a surface treatment depends on the nature of the rubber [4]. Thus, natural rubber and synthetic styrene-butadiene rubber, for instance, are sensitive to the type of surface treatment, abrasion being the most effective. Polychloroprene is relatively insensitive. However, other cured rubbers such as butyl and ethylene-propylene-diene rubber are far more difficult to bond. Because of its industrial relevance [1], in this study only synthetic vulcanized styrene–butadiene rubber (SBR) is considered.

Depending on the kind of surface modification desired on the rubber surface, physical or chemical surface treatments of synthetic rubber can be utilized. Physical surface treatments do not modify the chemical nature of rubber, but abhesive compounds are removed by means of solvent wiping (which produces a certain degree of swelling on the surface) [5], ultrasonic cleaning [6] or surface roughening [7]. In general, these physical treatments of rubber enhance the mechanical interlocking with adhesives. Therefore, the rubber modified by means of physical treatments is adequate when the joint has to bear only a light load and/or abhesive components do not migrate to the surface, once the adhesive joint is formed [6]. Whenever a strong adhesion is needed or elimination of weak boundary layers is necessary, a chemical surface treatment is recommended.

Chemical treatments produce a chemical modification of the rubber surface by the creation of polar groups and formation of a particular surface topography, which results in an enhancement of the chemical and mechanical adhesion components. In general, high quality adhesive joints are formed, whose properties remain stable over long periods of time. Several chemical surface treatments for synthetic SBR rubber have been proposed in the literature [3, 6, 8–10].

i) Use of organic peroxides [11]. For such treatment to be effective infra-red radiation and two-component adhesives are needed.

ii) Cyclization [12]. This consists of the immersion of the rubber in concentrated sulfuric acid for 1–5 min, removing the excess acid by alkaline rinsing and thorough washing with tap water. This treatment yields a quite brittle layer of rubber at the surface, which develops external microcracks by flexing. It is very appropriate and effective for SBR, as microcracks are produced and oxidation occurs. It is not frequently applied, however, due to the difficulties associated with the manipulation of sulfuric acid and because the residual sulfuric acid that could be trapped in the cyclized layer may lead to an eventual deterioration of the adhesive bond [13].

iii) Use of reactive polyisocyanates [14]. Both the adhesion to polyurethane adhesives and the durability of adhesive joints are noticeably improved by this treatment. It must be applied prior to the formation of the adhesive bond. The parameters which determine the effectiveness of this treatment have not yet been well-studied, mainly because of the poor reproducibility of the experimental results [15].

iv) Plasma treatment [16]. The best results are obtained with helium, air and nitrogen plasma. Although plasma treatment is an effective method, its use is limited for economic reasons.

v) Halogenation [5–11]. This is the most common surface treatment for SBR rubber. Several agents for the halogenation of rubber have been proposed in the literature, and those precursors which liberate chlorine or bromine are the most common. Chlorine vapours [17], aqueous solutions of chlorine [7], and chlorine in CCl_4 [18] are some of the treating agents which have been used. The toxicity of these agents has restricted their use, and other less hazardous halogenation agents have been proposed. Immersion of rubber surfaces in an acidified sodium hypochlorite solution followed by rinsing in water has been extensively used [5, 8, 19] because of its high effectiveness; however, this causes a poor wettability of the rubber, water needs to be removed from the surface by heating and there is a release of chlorine into the atmosphere. The application of oxalic acid and chloramines mixtures to rubber surfaces [20–22] liberates chlorine gradually and these are effective halogenation agents for SBR rubbers. However, these halogenation solutions are unstable, the two components need to be mixed just prior to its application to the rubber surface, and the effectiveness depends on the nature of the rubber: the method is restricted to particular cases, e.g. mixtures with other halogenation agents, such as TCI [20]. TCI is the most common and useful chlorination agent for SBR rubber. In the solid compound chlorine is not free but it is liberated by reaction of TCI with butadiene groups of rubber. TCI is normally applied in an organic volatile solvent, so that penetration of the chlorination agent compound is favoured, a good wettability of the rubber is attained and there is a fast evaporation of the solvent. The effects produced on the chlorinated rubber with TCI remain for at least six months, although TCI solutions are only stable for two weeks.

The kind of reaction involved in the chlorination of rubber is not clear, although it has been suggested [23–26] that an electrophilic attack of C=C bonds by Cl_2 leads to substitution, addition, or cyclization reactions, depending on the nature of the rubber. In SBR the addition of chlorine to the C=C bond is the most important reaction [24, 25], although it has been suggested [27, 28] that the mechanism of the chlorination reaction of rubber with TCI involves the formation of charge-transfer complexes.

The detailed nature of the surface modifications induced by the treatment of the rubber which leads to its improved adhesion is not well understood. Physical treatments enhance the mechanical adhesion of rubber [2, 6], but in the case of chemical treatments the adhesion mechanism responsible for the increased adhesion of SBR has not yet been studied in detail. External physical changes in the rubber, migration of chlorine to the surface which could have some effect on the orientation of adhesive molecules, and other surface chemical modifications are some of the hypotheses proposed [5, 6, 8, 9]. Considering the lack of systematic studies dealing with the surface modifications of rubber by chemical treatments, the objective of this study was to analyze the modifications produced in the surface of a synthetic styrene-butadiene rubber. The mechanisms of adhesion of rubber favoured by chlorination (the most convenient and widely used chemical treatment of rubber) are also considered.

2. EXPERIMENTAL

2.1. Materials

A sulphur-vulcanized styrene-butadiene rubber (SBR) sheet, about 5 mm thick, was used in this study. Its formulation includes silica as filler, coumarone-indene resin, stearic acid, and zinc oxide. The oils and plasticisers content of this rubber is 12 wt% and the Shore A hardness is 77°.

Surface treatments applied to SBR were as follows:

i) *Ultrasonic cleaning*. SBR was cleaned in an ultrasonic bath containing doubly distilled water for 15 min and stored in vacuum for 60 min to remove any retained water.
ii) *Solvent wiping*. SBR was wiped with a tissue paper immersed in 2-butanone, or ethyl acetate solvent. Solvent was allowed to evaporate for one hour.
iii) *Abrasion*. A mechanical roughening of the SBR surface was carried out using sandpaper, until 0.5 mm of the external surface was removed.
iv) *Chlorination*. SBR was wiped with a tissue paper immersed in ethyl acetate and 15 minutes later, this wiped rubber was treated with a tissue paper previously immersed in TCI/ethyl acetate solutions (1–7 wt%). After one hour, a 25 wt% ethanol/water solution was applied to the chlorinated SBR surface. The influences of the time of chlorination and the concentration of chlorination agent were monitored.

A linear polyester–urethane polymer based on ε-polycaprolactone from Merquinsa S. A. (Barcelona, Spain) was used as the test adhesive. The adhesive solution was prepared by mixing 15 wt% of solid polymer with 2-butanone in a laboratory mixer, and stirring the mixture (400 rpm) for two hours at room temperature to obtain a homogeneous solution. Its Brookfield viscosity at 25°C was periodically measured and a value of 1.3 ± 0.1 Pa.s was obtained. This polyurethane adhesive was used to join the surface-treated rubber and to determine the T-peel strength.

2.2. *Methods*

1) *T-peel strength measurements.* Adhesive joints were made using SBR rubber strip test pieces (150×30 mm^2) which had been treated in the same way. The polyurethane adhesive (about 100 mg) was applied with a brush to each SBR sample and the samples were left to dry for 1 h. The adhesive film (0.01 mm thick) was heated under infra-red radiation to produce a temperature of 90°C, and then the two SBR test pieces were brought in contact and a pressure of 3 atm. (1 atm = 101325 Pa.) was immediately applied for 10 s to achieve a suitable joint. Afterwards, the test samples were kept at 23 ± 1°C and 50% relative humidity for 72 h (to obtain the highest cohesive strength of the polyurethane) before the T-peel strength was measured on an *Adamel-L'Homargy* (Paris, France) DY-32 instrument (peel rate = 0.1 m/min). The values obtained were the average of three replicates of the same test (deviations were less than 5%).

2) *Scanning Electron Microscopy (SEM).* SBR samples were analysed using a *JEOL* SEM JSM 840 microscope.

3) *Contact angle measurements.* Advancing contact angles were measured with a *Ramé-Hart* 100 goniometer. Single drops (2 μl) of different liquids (ethane diol (*Merck*, 99% minimum purity); *o*-tricresyl phosphate (*Probus*, 98% minimum purity); 1-bromonaphthalene (*Merck*, 98% minimum purity) and doubly distilled water, were placed on the SBR surface. Although the equilibrium was reached 10 minutes after the drops were placed on the surface of the rubber, the contact angle measurements were taken after 15 minutes as an experimental convenience. Average values of at least three drops on three different batches of the similarly treated rubber were taken, and the standard deviation was ± 2 degrees. The dispersion and polar components of the surface energy of the modified rubber were calculated following the Fowkes approximation [29]:

$$(1 + \cos\Theta)\gamma_L = 2\left(\gamma_s^d \gamma_L^d\right)^{1/2} + 2\left(\gamma_s^p \gamma_L^p\right)^{1/2},$$

where Θ is the contact angle, γ_L is the surface energy of pure liquid and γ_s is the surface energy of solid. The superscripts 'd' and 'p' represent dispersion and polar contributions, respectively.

4) *IR spectroscopy.* A Nicolet 205 FTIR spectrophotometer was used to analyze the modified SBR surfaces. Samples were placed on a thallium bromoiodide crystal (KRS-5) and a multiple attenuated transmission reflectance technique (ATR) was used. The incident angle was 45° and 80 scans per experiment were recorded.

5) *X-ray photoelectron spectroscopy* (*XPS*). XPS experiments on the modified rubber samples were carried out at VPI in Blacksburg, Virginia, USA. A PHI Perkin-Elmer model 5400 photoelectron spectrometer was used. Photoelectrons were generated by bombarding the sample with Mg K_α X-rays ($h\nu = 1253.6$ eV). The binding scale was calibrated by setting the C_{1s} hydrocarbon photoelectron peak at 285.0 eV. The atomic concentrations were calculated using peak areas, which were corrected for relative instrument response using an experimentally determined sensitivity factor. Although the approximate area of the sample introduced into the spectrometer was 20×20 mm^2, the photoelectron spectra were obtained on a 1×1 mm^2 area of the sample.

6) *Differential Scanning Calorimetry* (*DSC*). Glass transition temperatures ($T_g s$) of some modified-rubber samples were determined using a *Mettler* DSC-30 calorimeter.

3. RESULTS AND DISCUSSION

The SBR rubber selected in this study contains additives which are well-known abhesive substances [30]. The IR spectrum of SBR (Fig. 1) shows the presence of a paraffin wax (720, 1379, 1446, 2851, 2920 cm^{-1}) and zinc stearate (1539 cm^{-1}) on the rubber surface. Because of these components, there is no adhesion between SBR and solvent-based polyester–polyurethane adhesives, and therefore a surface treatment is required to achieve a good bond.

In our study three physical and one chemical surface treatments of SBR rubber have been investigated. As will be shown below, the chemical treatment produces the best results, although the nature of the surface modifications produced on the SBR by all the treatments has been considered.

Figure 1 shows the IR spectra of untreated and ultrasonic-cleaned SBR. Both IR spectra are very similar (silica band at 1089 cm^{-1} and C=C bands corresponding to 1,4-butadiene at 703, 760, 797, 910, 964 cm^{-1}), except for the presence of hydrogen bonds in the ultrasonic-treated SBR (wide band at 3100–3600 cm^{-1}) probably due to water remaining on the surface after the treatment. The external surfaces of untreated and ultrasonic-cleaned SBR were identical (SEM microphotographs not shown) and, as expected, the adhesion strength of SBR/polyurethane adhesive joints was not improved by this treatment.

Figure 2 shows the IR spectra of an untreated SBR surface, and of other surfaces wiped with ethyl acetate or 2-butanone. Ethyl acetate does not remove either zinc stearate or paraffin wax from the rubber surface although a certain aromaticity appears (3032, 3078 cm^{-1}). The wiping of the SBR surface with 2-butanone, however, reduces the concentration of zinc stearate and paraffin wax, and also some aromaticity is produced in the surface. On the other hand, the advancing contact angles for ethane diol at 25°C are very similar to those for untreated SBR (67°) and SBR wiped with ethyl acetate (64°) or 2-butanone (67°); thus, the removal of abhesive substances from the SBR surface wiped with 2-butanone does not cause an increase in its surface energy, probably because the residual amount of abhesive substances after the treatment

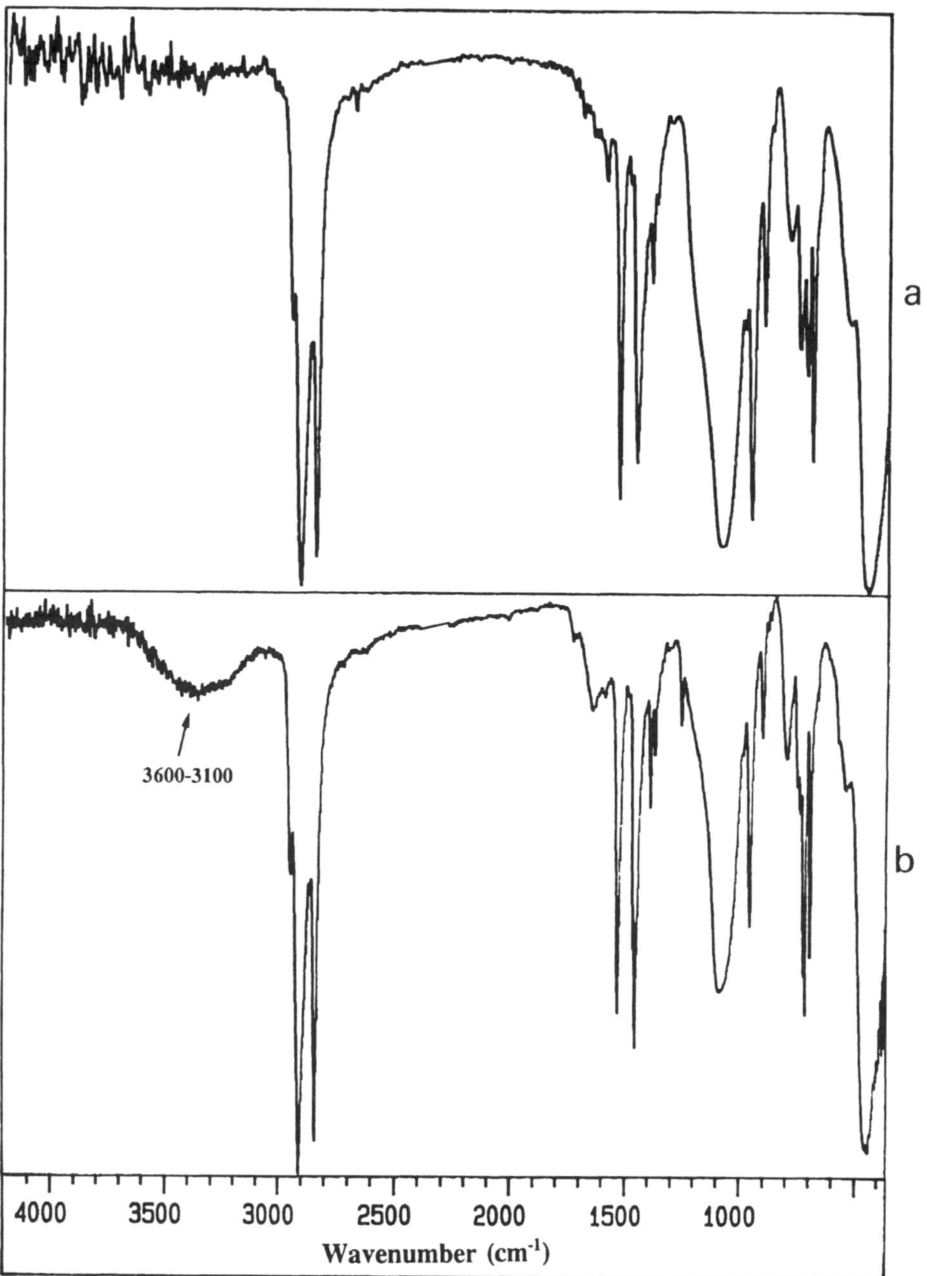

Figure 1. IR spectra of a) untreated and b) ultrasonic-cleaned SBR. Arrow shows the OH groups.

is relatively high. Therefore, no improvement in T-peel strength was found (values lower than 1 $kN\,m^{-1}$ were obtained) by solvent wiping of the SBR surface with ethyl acetate, or 2-butanone.

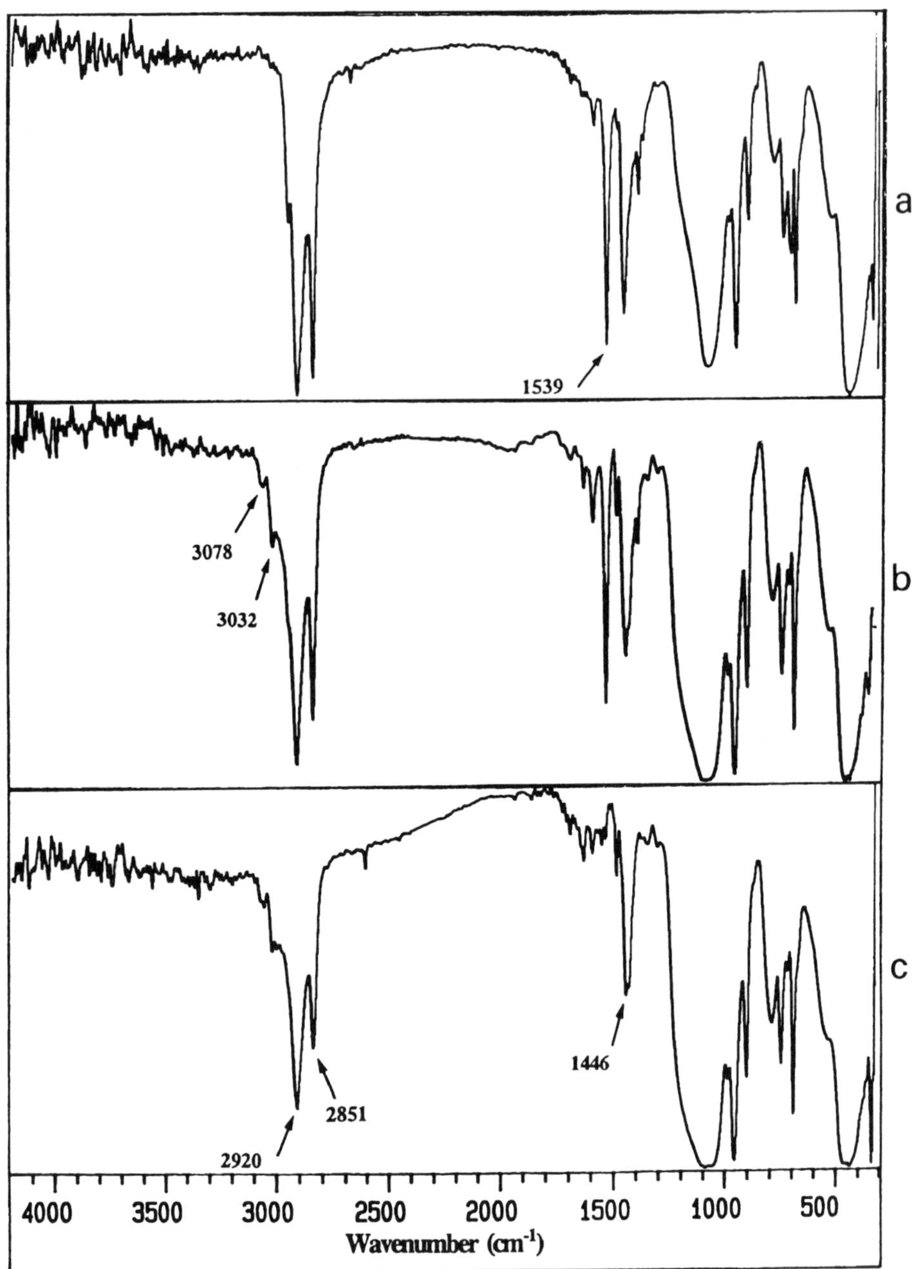

Figure 2. IR spectra of a) untreated, b) ethyl acetate, and c) 2-butanone wiped SBR. Arrows show the removal of the zinc stearate and the paraffin wax.

Figure 3 shows the IR spectra of untreated and roughened SBR surfaces. Roughening of SBR produces the mechanical elimination of abhesive substances (1539, 2851, 2920 cm^{-1}) from the surface and consequently a slight increase of T-peel strength

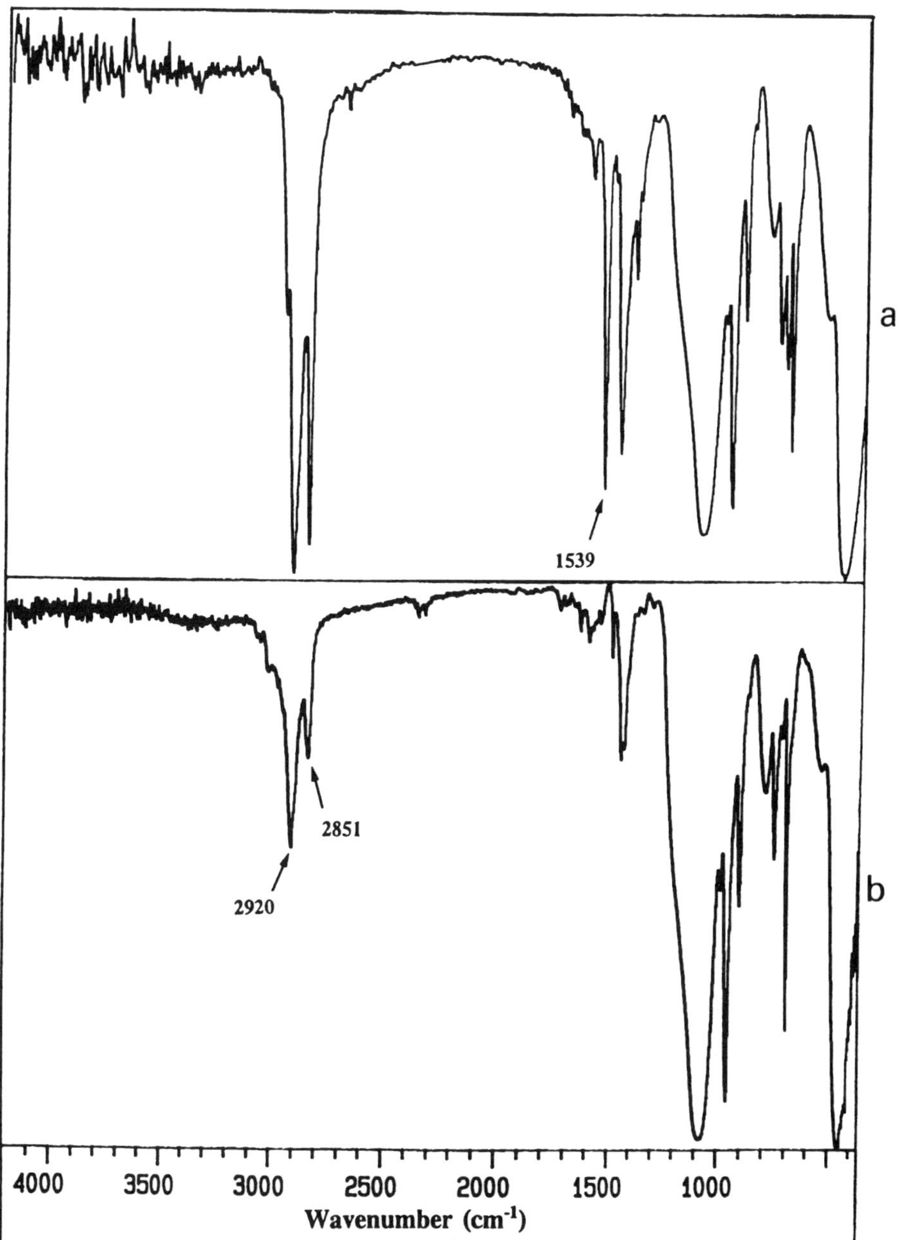

Figure 3. IR spectra of a) untreated and b) roughened SBR. Arrows show the removal of the zinc stearate and the paraffin wax.

(1.4 kN m^{-1}) is produced. At the same time, the increase of adhesion obtained by abrasion of the SBR surface produces a more heterogeneous surface which favours mechanical interlocking with the polyurethane adhesive and results in improved adhesion.

Chlorination treatment produced the most noticeable increase of adhesion in SBR (1.0 kN m^{-1} for SBR wiped with ethyl acetate versus 7.7 kN m^{-1} for chlorinated SBR with 2 wt% TCI; an adhesion failure was always found). This increase of adhesion is due to the modifications produced on the SBR surface by the treatment, which (as will be shown below) essentially produces three kinds of surface modifications: mechanical, thermodynamical, and chemical.

Mechanical surface modifications of SBR can be seen in Fig. 4. SEM micrographs of SBR wiped with ethyl acetate (0 wt% TCI) shows a quite homogeneous surface and

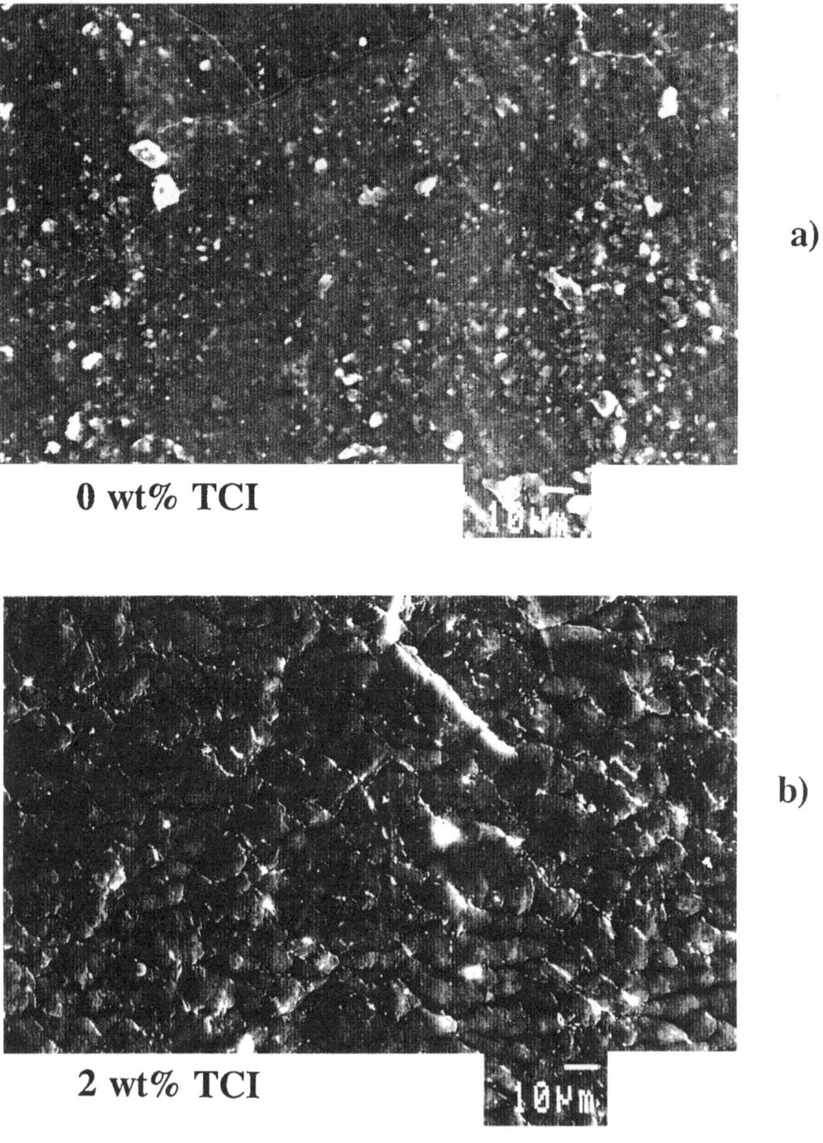

Figure 4. SEM micrographs of a) untreated (0 wt% TCI) and b) chlorinated (2 wt% TCI) SBR.

Table 1.
Surface energies ($mJ\,m^{-2}$) of untreated and chlorinated SBR

% TCI	Surface energy ($mJ\,m^{-2}$)			γ_s^p/γ_s
	γ_s^d	γ_s^p	γ_s	(%)
0	21.1	10.5	31.6	33
2	22.1	30.5	52.6	58

the presence of some white impurities (probably mould-release agents). The treatment of SBR with 2 wt% TCI produces several microcracks across the surface and gives a certain degree of surface roughness. The origin of these microcracks probably lies in a strong selective reaction of certain surface rubber components with TCI. Additionally, the chlorinated rubber shows significant surface degradation facilitated by the removal of rubber particles from the bulk with the tissue paper during the halogenation treatment. These microcracks and the surface roughness should favour the mechanical interlocking of the rubber with the polyurethane adhesive resulting in an increased T-peel strength. However, the increase found in T-peel strength (about 6-fold) of rubber-polyurethane joints cannot be justified only by an improvement in mechanical adhesion of SBR and therefore other surface modifications should also have a contribution.

Thermodynamical surface modifications of SBR can be analyzed by means of contact angle measurements. It has been shown previously [31] that a decrease of advancing contact angle is produced when styrene–butadiene is halogenated with TCI. In the chlorinated SBR a decrease of advancing contact angle is found, which corresponds to an increase of surface energy from 31.6 (0 wt% TCI) to 52.6 $mJ\,m^{-2}$ (2 wt% TCI) (Table 1). Surface energy of untreated SBR agrees well with the surface energy reported for several elastomers [32]. Chlorination of SBR, however, yields a noticeable increase of surface energy, which is mainly due to the improvement of the polar contribution (from 10.5 to 30.5 $mJ\,m^{-2}$, which means an increase from a 33 to a 58% in respect to the global surface energy), whereas there is no noticeable change in the dispersion contribution (from 21.1 to 22.1 $mJ\,m^{-2}$). The nature of the polar groups created on the SBR surface by chlorination has also been extensively studied by means of IR spectroscopy [6, 9, 31]. Figure 5 shows the IR spectra of 0 wt%, 2 wt% and 7 wt% TCI treated SBR. Chlorination modified the SBR surface as follows:

i) Zinc stearate (1539 cm^{-1}) is completely removed and the paraffin wax concentration (720, 2851, 2920 cm^{-1}) is drastically reduced.

ii) TCI is a powerful oxidation agent [26] and C=O groups (1710 cm^{-1}) are created. These C=O groups have been assessed as acid chloride species [8, 31].

iii) Chlorinated hydrocarbon chains with different degrees of substitution are created (534, 1237, 1387, 1420 cm^{-1}).

iv) There is weak band at 3296 cm^{-1} which corresponds to N–H groups of residual TCI.

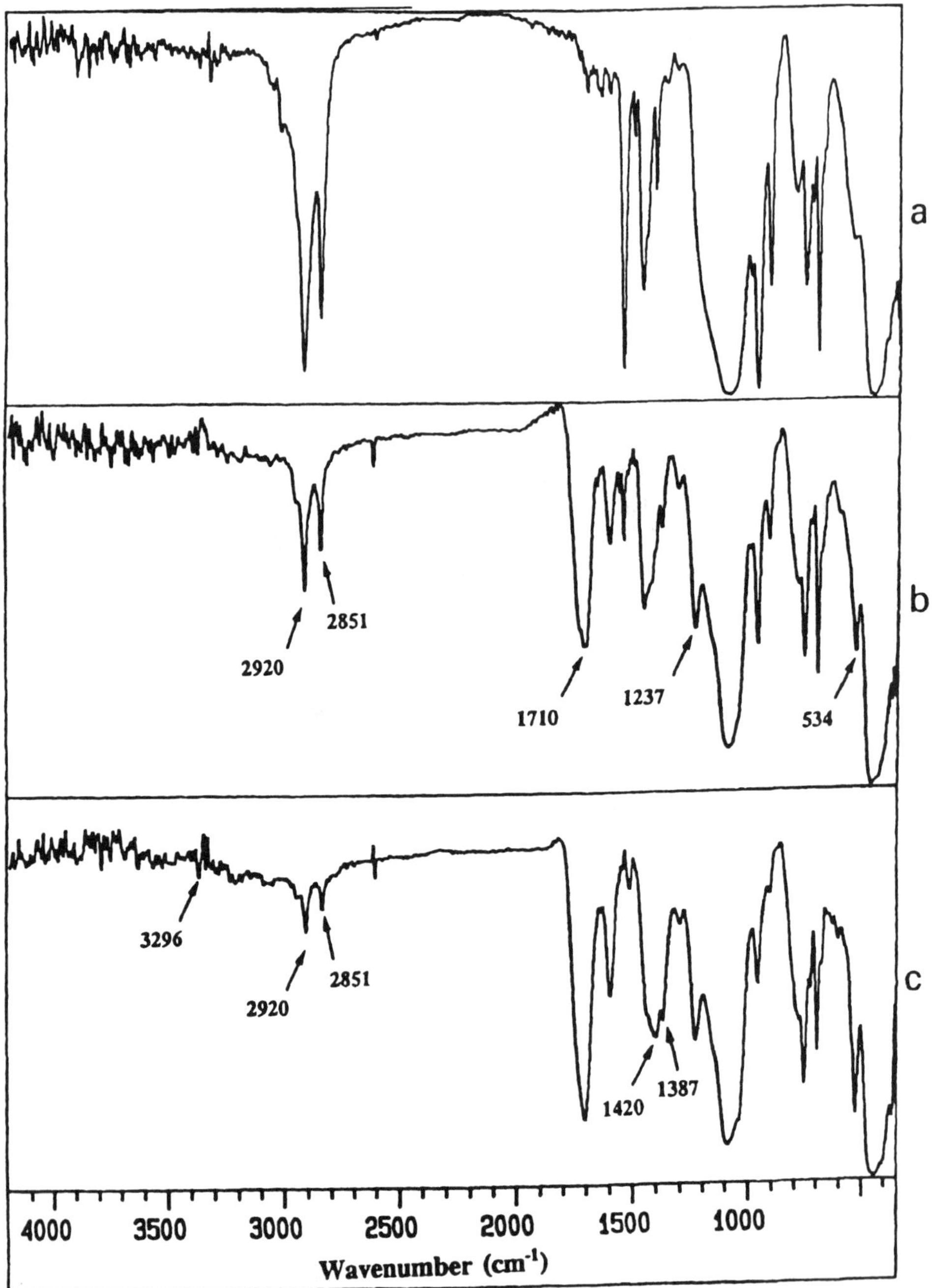

Figure 5. IR spectra of a) untreated (0 wt% TCI), b) chlorinated (2 wt% TCI) and c) chlorinated (7 wt% TCI) SBR. Arrows show the decrease in intensity of paraffin wax and the creation of C−Cl and C=O groups.

Therefore, according to the IR spectra of Fig. 5, acid chloride and chlorinated groups are the species which may contribute to the increase of the polar component of the surface energy of SBR. However, the IR spectra of this chlorinated rubber cannot be easily quantified and besides, this technique considers a surface depth of around 5 μm. For these reasons, XPS experiments on unchlorinated and chlorinated SBR surface were carried out (Fig. 6, Table 2). The key feature of this technique is that the information is obtained from about the outer 5 nm of the material, so an analysis of exclusively the surface chemistry of SBR material is obtained. The unchlorinated SBR is mainly composed of carbon, as hydrocarbon. A small concentration of oxygen is found (probably $-$OH type) and low concentrations of sulfur (sulfide), zinc (zinc II) and silicon are noted in the XPS spectrum (Fig. 6a, Table 2). In sample treated with 7 wt% TCI (Fig. 6b), the carbon chemistry included hydrocarbon, carbon singly-bonded to oxygen and/or chlorine, and carboxyl functionality. The shape of the carbon photoelectron peak in Fig. 6c is indicative of the presence of $>C{=}O$ and $-C{-}OH$ or $-COR$ functional groups. The binding energy for chlorine is indicative of organic chlorine (Fig. 6d). The apparent double peak (shoulder on the chlorine photoelectron peak) is due to the contribution from $Cl_{2p1/2}$ and $Cl_{2p3/2}$. The nitrogen binding energy is consistent with an amine-type nitrogen.

Quantitative analysis of XPS spectra (Table 2) shows that carbon is the main component of the untreated rubber, oxygen and silicon contributions being also relatively important. Chlorination of SBR reduces significantly the concentration of carbon and increases noticeably the amounts of chlorine and, mainly, oxygen and nitrogen at the surface. The chlorination of SBR causes a reaction of TCI with C=C bonds to produce mainly oxidized species, the chlorinated species being much less important. Therefore, the surface polarity of chlorinated SBR is mainly due to the presence of $>C{=}O$, $-C{-}OH$ and/or $-COR$ functional groups.

The chemical modifications of SBR surface after chlorination are expected to have a greater contribution to its improved adhesion to polyurethane adhesives than the contribution due to surface roughening, although the extent of each process is difficult to determine. This statement does not agree with previous publications [8, 13] where the

Table 2.
Elemental analysis (XPS spectra) of untreated and chlorinated SBR surfaces

Element	Elemental composition (%)		
	at 0% TCI	at 2% TCI	at 7% TCI
C	96.4	85.4	76.0
O	2.6	9.0	12.0
S	0.2	0.4	0.7
Zn	0.2	0.1	<0.1
Si	0.6	0.4	0.3
Cl	<0.1	2.5	3.6
N	<0.1	2.3	7.4

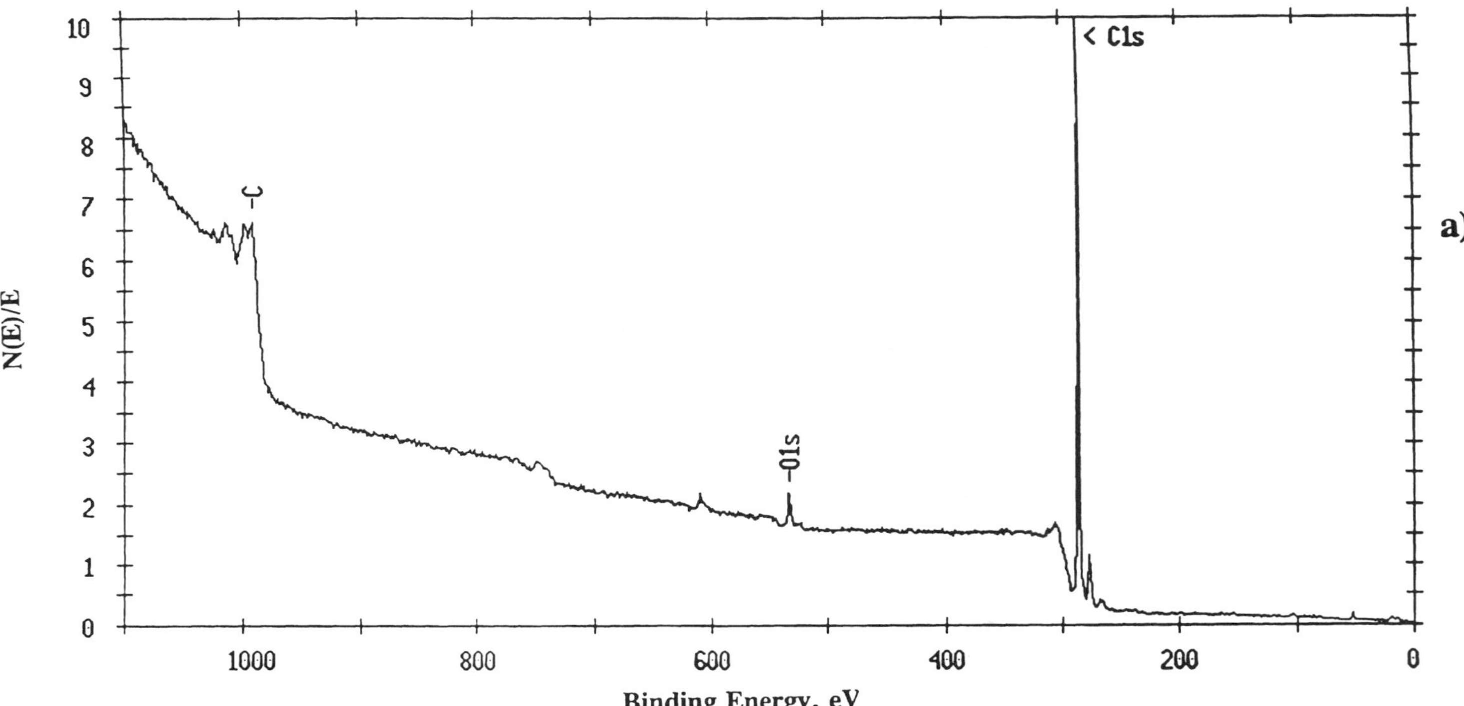

Figure 6a. XPS spectrum of untreated SBR (0 wt% TCI).

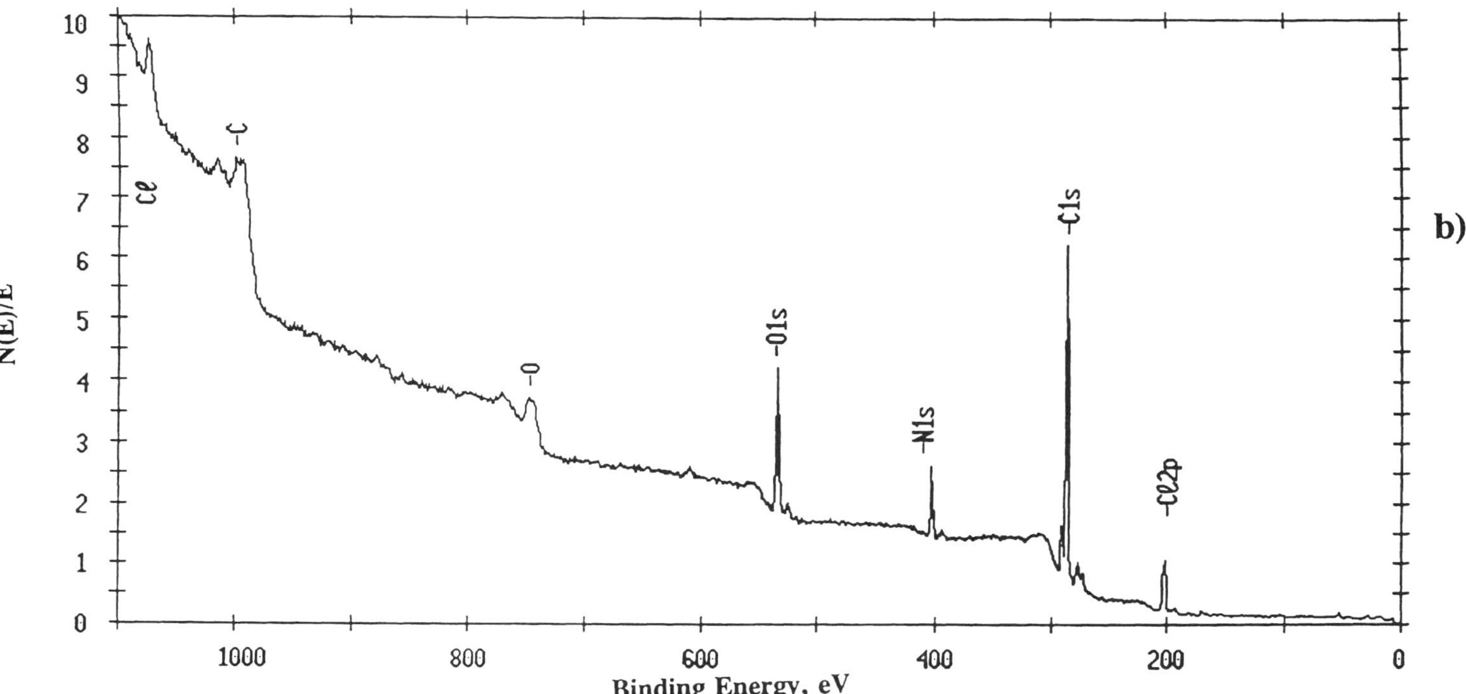

Figure 6b. XPS spectrum of chlorinated rubber (7 wt% TCl). General survey.

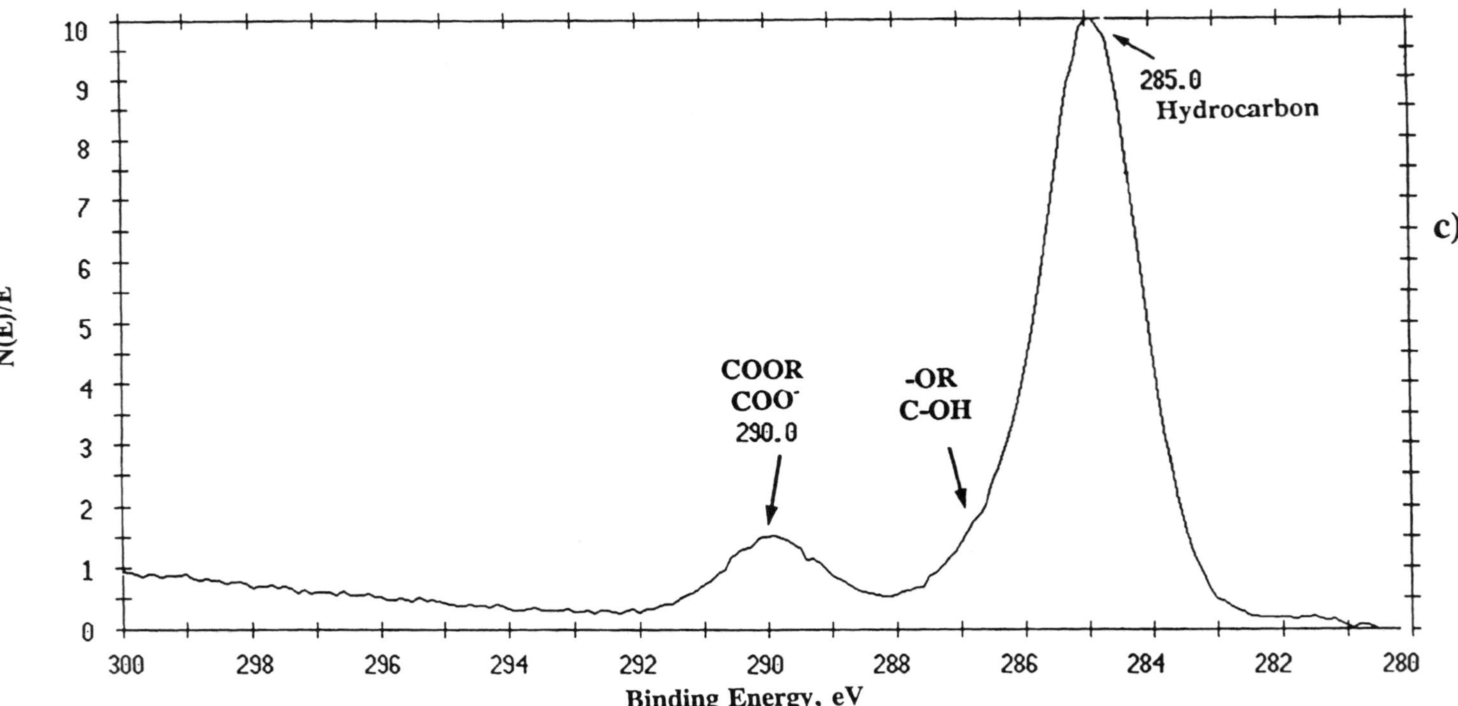

Figure 6c. XPS spectrum of chlorinated SBR (7 wt% TCI). C_{1s} spectrum.

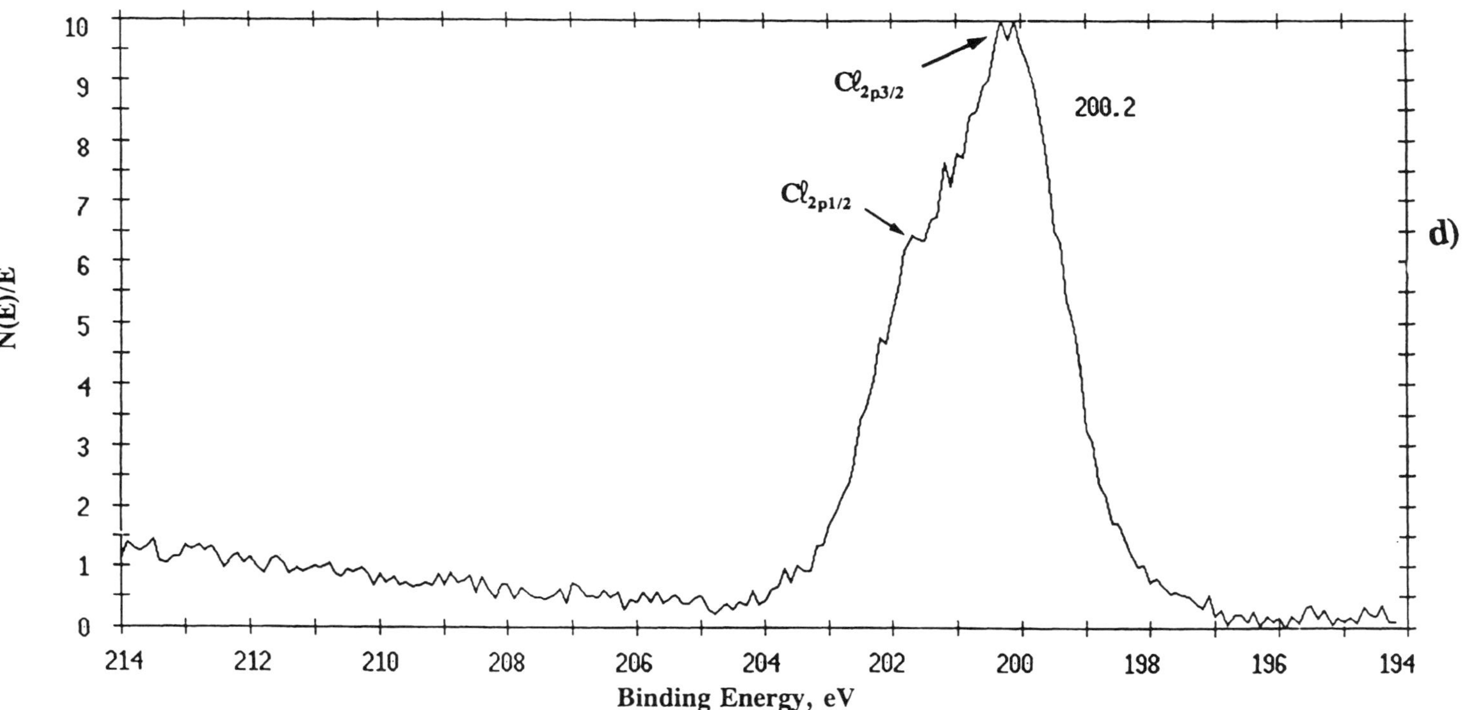

Figure 6d. XPS spectrum of chlorinated SBR (7 wt% TCI). Chlorine spectrum.

improved bond strengths resulting from chlorination of natural or styrene–butadiene rubber with TCI were ascribed to the enhanced wettability of the treated surface by the adhesive. This conclusion was based on the large reduction in contact angle of rubber surfaces after chlorination and on the fact that the surface layer of the chlorinated rubber contained less than one chlorine atom for each available C=C bond. In the styrene–butadiene rubber used in this study, we also found an improved wettability of rubber after chlorination but the chemical interaction due to the presence of carboxyl groups seems to be quantitatively more relevant to the adhesion properties.

In summary, the improved adhesion of SBR obtained by chlorination with TCI can be explained by the enhanced contribution of mechanical, thermodynamical and chemical adhesion components. Nevertheless, the nature of the chlorination process of SBR needs to be clarified.

The increase of the time of chlorination of rubber with TCI (10 min–6 h) does not significantly modify the T-peel strength of SBR/polyurethane adhesive joints (for 2 wt% TCI, T-peel strength between 7.3 and 8.4 kN m^{-1} is obtained) [9, 31]. SEM micrographs in Fig. 7, however, show a larger degree of surface heterogeneity and more pronounced microcracks in the rubber surface as the time of chlorination increases. Thus, the mechanical interlocking with the polyurethane adhesive should be improved and higher adhesion will be obtained. On the other hand, IR spectra (Fig. 8) show a more noticeable removal of zinc stearate and a somewhat less elimination of paraffin wax when the time of chlorination increases, but the relative intensity of bands due to carboxyl and chlorinated groups remains unchanged. Furthermore, the advancing contact angles (ethane diol, 25°C) do not vary greatly with the time of chlorination (40–43°). These experimental results indicate that the chlorination reaction of TCI with SBR is fast (it takes less than one hour) and the slightly improved adhesion obtained on increasing the time of halogenation should be ascribed mainly to the removal of abhesive substances from the rubber surface and to the formation of microcracks. No noticeable changes in surface polarity and surface energy are produced by increasing the time of chlorination.

Finally, the increase of TCI concentration does not produce a pronounced enhancement of T-peel strength (1 wt% TCI: 7.2 kN m^{-1}; 2 wt% TCI: 7.7 kN m^{-1}; 5 wt% TCI: 6.8 kN m^{-1}), and only a small concentration of TCI seems to be sufficient to reach the maximum adhesion of rubber. IR spectra (Fig. 5) and XPS measurements (Table 2), however, show a greater degree of reaction of the chlorination agent with the SBR surface when the TCI concentration is increased, and, at the same time, the surface roughness is more pronounced. Consequently, an increase of T-peel strength for a higher TCI concentration should be expected. However, there is no significant variation in surface energy of the rubber when the TCI percentage is increased. For instance, the advancing contact angle of ethane diol at 25°C varies between 47 and 39° for amounts of 1 wt% and 7 wt% TCI treated rubber samples, respectively. These results demonstrate that once the reaction between TCI and the SBR surface is completed (at low percentages of TCI), the increase in the concentration of TCI causes an excess of chlorination agent on the surface which should facilitate the reaction of TCI with the SBR bulk, giving a thicker layer of chlorinated SBR. This accounts for the invariance of T-peel strength and advancing contact angle when the percentage of TCI

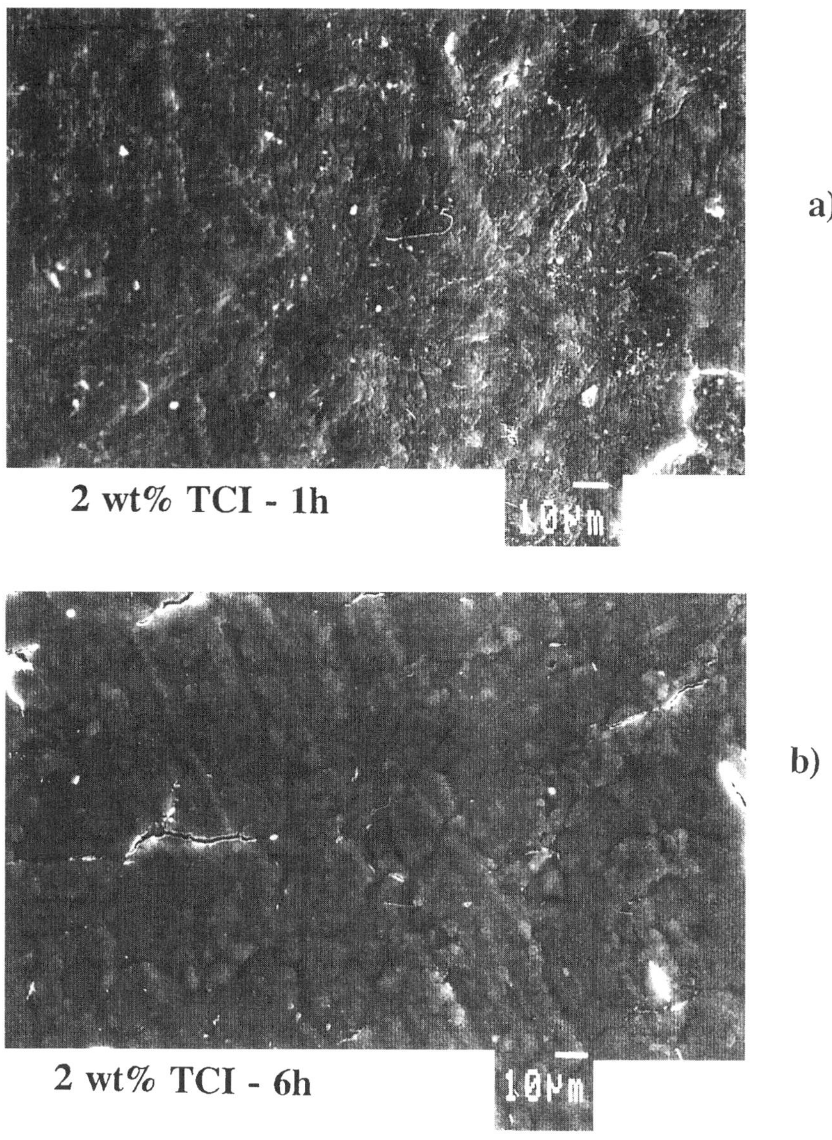

Figure 7. SEM micrographs of 2 wt% TCI treated SBR. Influence of the time of chlorination: a) 1 h; b) 6 h.

applied to the rubber is increased and, at the same time, gives an explanation for the IR spectra of Fig. 5 (this technique analyzes about 5 μm depth). A further confirmation of this hypothesis is obtained from DSC experiments: for untreated rubber the glass transition temperature (T_g) is $-40\,^{\circ}$C, whereas for the chlorinated rubber surface with 7 wt% TCI, the T_g is $-52\,^{\circ}$C. This diminution of T_g in the chlorinated SBR when noticeable amounts of TCI are used can be ascribed to the modification of the

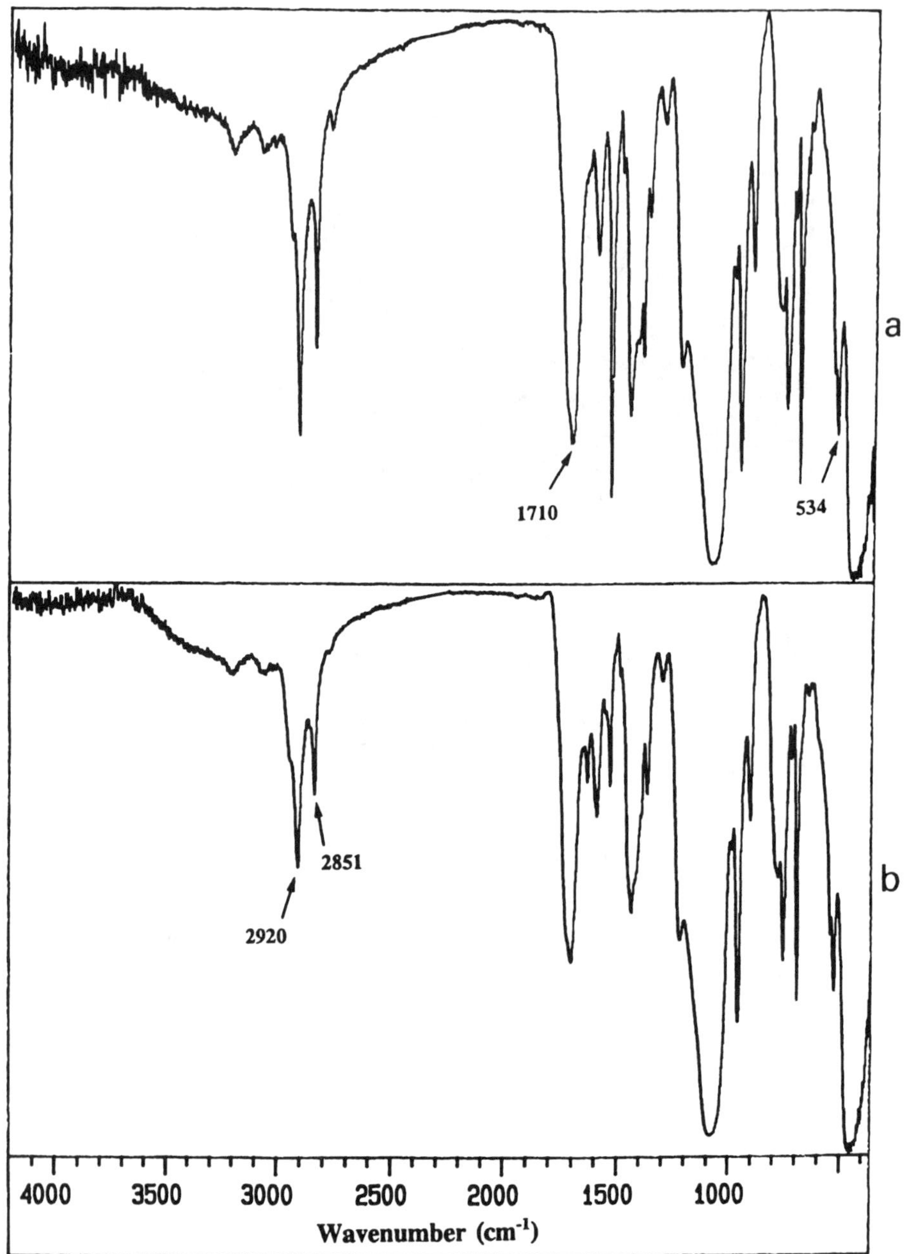

Figure 8. IR spectra of 2 wt% TCI treated SBR. Influence of the time of chlorination: a) 1 h; b) 6 h. Arrows show how the increase of time reduces the relative intensity of paraffin wax but does not affect the intensity of C—Cl and C=O groups.

properties of SBR by the surface treatment, in such a way that the crosslink density of the surface of the rubber increases and, consequently, a hardening of the material and increase in its durometer value is produced.

4. CONCLUSIONS

The following conclusions are drawn from this study:

1. Physical treatments of SBR (ultrasonic cleaning, solvent wiping, abrasion) do not produce a noticeable increase of adhesion to polyester–urethane adhesives because, although a certain surface roughness is created, the abhesive substances (zinc stearate, paraffin wax) are not sufficiently removed from the rubber surface. Between the different treatments studied, abrasion provides the best results.
2. Chlorination of SBR with trichloroisocyanuric acid produces a noticeable improvement in T-peel strength, due to the enhanced contribution of mechanical (surface roughness, microcracks), thermodynamical (increase of polar contribution to the surface energy) and chemical (removal of abhesive substances, creation of polar groups) surface modifications.
3. Oxidized species of >C=O, −C−OH and −COR type are responsible for the strong adhesion between the chlorinated SBR surface and the polyurethane adhesive. The chlorination reaction is fast and requires only a small concentration of chlorination agent. When the amount of TCI is increased, the chlorination reaction progresses from the exterior to the internal rubber bulk creating a thicker layer of chlorinated rubber but this does not contribute to enhanced adhesion.

Acknowledgements

The authors wish to thank Prof. John Dillard (Department of Chemistry, Virginia Polytechnic Institute, Blacksburg, VA, USA) for carrying out the XPS measurements. Financial support from CICYT (Project no. MAT 92-0522) is gratefully acknowledged.

REFERENCES

1. M. Morton, *Rubber Technology*, pp. 20–57. Van Nostrand Reinhold, New York (1987).
2. A. J. Kinloch, *Adhesion and Adhesives. Science and Technology*, pp. 127–128. Chapman and Hall, London (1987).
3. A. H. Landrock, *Adhesives Technology Handbook*. Noyes, New Jersey (1985).
4. B. P. Spearman and J. D. Hutchinson, *Adhesives Age* **17** (4), 30 (1974).
5. D. Pettit and A. R. Carter, *J. Adhesion* **5**, 333 (1973).
6. J. M. Martín-Martínez, J. C. Fernández-García, F. Huerta and A. C. Orgilés-Barceló, *Rubber Chem. Technol.* **64**, 510 (1991).
7. J. S. A. Langerwerf, *Schuh-Technik* **6**, 731 (1973).
8. D. Oldfield and T. E. F. Symes, *J. Adhesion* **16**, 77 (1983).
9. J. C. Fernández-García, A. C. Orgilés-Barceló and J. M. Martín-Martínez, *J. Adhesion Sci. Technol.* **5**, 1065 (1991).
10. A. R. Carter and D. Pettit, in: *Adhesives in Footwear*, p. 43. SATRA Intern. Conf. Boston, Massachusetts (1976).
11. J. S. A. Langerwerf, *Technicuir* **3**, 79 (1973).
12. R. F. Wegman (Ed.), *Surface Preparation Techniques for Adhesive Bonding*. Noyes, New Jersey (1989).
13. J. D. Minford (Ed.), *Treatise on Adhesion and Adhesives*, Vol. 7, Ch. II, pp. 231–232. Marcel Dekker, New York (1991).
14. R. S. Whitehouse, *Synthetic Adhesives and Sealants*. John Wiley, New York (1987).

15. T. P. Ferrandiz-Gómez and J. M. Martín-Martínez, unpublished results (1992).
16. W. J. Feast and H. S. Munro (Eds), *Polymer Surfaces and Interfaces*. John Wiley, New York (1989).
17. D. Pettit and A. R. Carter, Patent 1 278 258. United Kingdom (1972).
18. J. S. A. Langerwerf, *Centrum voor Schoentechniek van het Instituut voor Leder en Schoenen TNO*, Report 137-71S (1972).
19. C. W. Extrand and A. N. Gent, *Rubber Chem. Technol.* **61**, 691 (1988).
20. A. R. Carter, D. Pettit and J. S. A. Langerwerf, Patent 4 110 495. USA (1978).
21. A. I. V. Pyatravichyus, V. L. Yankauskaite, V. L. Rayatskas, M. M. Kakaev and E. Samarskis, *Izvestiya VUZ, Teknologiya Legkoi Promyshlennosti* **5**, 79 (1988).
22. A. I. V. Pyatravichyus, Patent SU 883133. URSS (1981).
23. D. Hace, V. Kovacevic, D. Manojlovic and I. Smit, *Angew. Makromol. Chemie* **176**, 161 (1990).
24. E. Schoenberg, H. A. Marsh, S. J. Walters and W. M. Satman, *Rubber Chem. Technol.* **52**, 526 (1979).
25. R. Vukov, *Rubber Chem. Technol.* **57**, 275 (1984).
26. M. L. Poustma, *J. Am. Chem. Soc.* **87**, 2172 (1965).
27. E. C. Juenge, D. A. Beal and W. P. Duncan, *J. Org. Chem.* **35**, 719 (1970).
28. E. C. Juenge and D. A. Beal, *Tetrahedron Lett.* **55**, 5819 (1968).
29. F. M. Fowkes, *J. Phys. Chem.* **67**, 2538 (1963).
30. N. Pastor-Sempere, Master Thesis. University of Alicante, Spain (1993).
31. J. M. Martín-Martínez, J. C. Fernández-García and A. C. Orgilés-Barceló, *J. Adhesion Sci. Technol.* **6**, 1091 (1992).
32. I. Skeist (Ed.), *Handbook of Adhesives*, pp. 67–69, 3rd edn. Van Nostrand, New York (1990).

Polymer Surface Modification: Relevance to Adhesion, pp. 401–416
K. L. Mittal (Ed.)

Surface modification of drawn gel-cast ultra-high molecular weight polyethylene films

M. S. SILVERSTEIN* and J. SADOVSKY

Department of Materials Engineering, Technion — Israel Institute of Technology, Haifa 32000, Israel

Revised version received 9 February 1995

Abstract—Gel-spun ultra-high molecular weight polyethylene (UHMWPE) fibers have superior properties but their use in composite material applications is limited by their poor adhesion to polymer matrices. Previous studies have shown that etching improves the adhesion of epoxy to the fibers, but leads to a reduction in mechanical properties. The purpose of this research was to use uniaxially drawn gel-cast UHMWPE films as a model system since both films and fibers have a highly oriented fibrillar structural hierarchy. Etching has detrimental effects on the mechanical properties and crystallinity of these very thin films. The small amount of carbonyl and carboxyl groups added to the surface through etching raises the film's surface tension and enhances wetting by epoxy. Even though the unmodified film cannot be bonded with epoxy, the interlaminar shear strength between epoxy and the etched films approaches the cohesive strength of the epoxy. A combination of interfacial and UHMWPE cohesive failures is observed. The increase in adhesion is attributed to the slight increase in surface oxygen.

Keywords: UHMWPE; etching; chromic acid; plasma; gel; film; surface tension; epoxy.

1. INTRODUCTION

Gel-spun ultra-high molecular weight polyethylene (UHMWPE) fibers are highly oriented and highly crystalline and thus possess many outstanding properties that are desirable in composite materials [1]. These desirable properties include superior toughness, chemical resistance, and biocompatibility. Unfortunately, the inclusion of UHMWPE fibers in composite materials has been limited by their poor adhesion to polymer matrices [2]. Both chemical and plasma etching can roughen the surface and add polar groups to the surface of non-polar PE [3, 4]. Roughening can enhance adhesion through the increase in contact area and the formation of mechanical interlocks with the polymer matrix. The creation of polar groups can enhance adhesion through the increase in surface tension and, in addition, through the formation of sites that can chemically bond to the polymer matrix. The proposed existence of a weak boundary layer (WBL) and its removal through etching have also been mentioned as

*To whom correspondence should be addressed.

contributing to improved adhesion. The removal of a boundary layer is important for processing techniques, such as gel spinning, that can leave a residue of solvents, processing aids, and low molecular weight polymer ejected from the bulk during crystallization.

1.1. Surface modification of PE

Chemical etching of PE has been used commercially and studied intensely for both low-density PE (LDPE) and high-density PE (HDPE). Although the results of etching HDPE and LDPE vary, some general conclusions can be drawn. Chromic acid, the commercially important etchant for PE, attacks the PE molecule through hydrogen abstraction, preferentially in the amorphous region [3, 4]. The etched surface is significantly roughened and has a considerable amount of oxygen in the form of hydroxyl, ether, carbonyl, and carboxyl groups. These changes to the surface have yielded increased wettability and a ten-fold or more increase in the shear strength of bonds with epoxy [3, 5, 6]. There has been considerable discussion as to the origin of this improvement in shear strength, whether it can be associated with the roughened topography, the presence of oxygen, or the removal of a WBL [3, 4]. The presence of a WBL has been discounted in many cases and has been demonstrated to exist in other specific cases [4]. The oxidation of PE with chromic acid has also been used as an initial step for more complex surface modifications of PE involving grafting [7].

Plasma treatments are also commercially important for PE surface modification. Plasma exposure can produce many simultaneous reactions including polymer degradation, polymer crosslinking, surface oxidation, and the removal of a WBL. Both crosslinking and surface oxidation have been demonstrated to be significant in plasma-etched PE, which exhibits improved wettability and bondability [3, 4, 8]. In some instances, the improvement in wettability was greater with oxygen and nitrogen plasmas than with helium plasmas, although bondability was improved with all gases [3, 9–11]. The presence of hydroxyl, ether, carbonyl, and carboxyl groups on the surface of plasma-modified PE has been demonstrated and the effects of these groups on adhesion have been investigated [12–14].

1.2. Surface modification of UHMWPE

Previous work has investigated the effects of various chemical etchants on gel-spun UHMWPE fibers and has shown that the adhesion of epoxy to the fibers could be significantly improved through chromic acid etching [15]. Chromic acid etching removed the smooth WBL observed on the fiber surface, exposed a highly oriented fibrillar structural hierarchy, and increased the amount of oxygen bonded to the UHMWPE surface [16]. The improvement in fiber–epoxy adhesion, however, was attained at the expense of a significant reduction in fiber mechanical properties [17]. After the removal of the WBL, the topography and fiber diameter remained unchanged with the etching time, in spite of the continued decrease in fiber mechanical properties [17]. The decrease in mechanical properties was attributed to the embrittlement of the fibers through polymer degradation and the production of failure-inducing flaws. It was difficult to characterize the surfaces of these small (20–30 μm) diameter, irregularly shaped fibers in a straightforward manner [16, 18].

A great deal of work has focused on the surface modification of gel-spun UHMWPE fibers through plasma etching. The interlaminar shear strength (ILSS) of an UHMWPE composite was shown to increase with the plasma etching time, while the tensile strength of the fibers decreased [19]. Surface oxidation was found to provide the major contribution to the ILSS at short treatment times, although surface roughening and crosslinking also played a significant role. The improvement in ILSS has also been related to both the removal of a WBL and the incorporation of oxygen onto the surface [2]. Plasma etching in mild conditions produced no change in the topography while yielding a significant amount of surface oxygen that could then be directly connected to the improvement in wetting and interfacial shear strength [20–22]. Ether, hydroxyl, carbonyl, and carboxyl functionalities have been observed [23, 24]. In all of these studies it was difficult to directly characterize the effects of plasma treatment on fiber wettability.

The wettability of planar surfaces is easier to characterize using techniques such as contact angle measurements. A uniaxially oriented gel-cast UHMWPE film can be used as a planar model for a gel-spun UHMWPE fiber for the initial surface modification studies. Casting a very dilute UHMWPE gel with minimal entanglements can provide a film that can be highly oriented by drawing to very high draw ratios at temperatures somewhat below the melting point [25–30]. The purpose of this research was to study the effects of both chemical and plasma etching on drawn gel-cast UHMWPE films. This research describes the effects of surface modification on the surface chemistry, surface topography, wettability, bonds with epoxy, and bulk properties.

2. EXPERIMENTAL

2.1. Film processing and modification

The films were made by gel-casting a 3×10^6 g/mol molecular weight UHMWPE powder (Hostalen GUR 402, Hoechst) from decalin (decahydronaphthalene). The concentration of UHMWPE in decalin was 0.5 wt%. The concentration of antioxidant (Plastanox 2246, Cyanamide) was 0.5 wt% based on the weight of UHMWPE. The mixture was heated at 160°C with stirring and yielded a transparent gel after approximately 45 min. The gel was cast into a glass dish and left in an oven at 65°C for 24 h. The gel was immersed in ethanol for 2 h to extract decalin, rinsed with distilled water, and then dried in a vacuum oven at 25°C for 24 h. The resulting films were uniform and 60–120 μm thick and will be referred to here as the 'as-cast' films.

The as-cast films were uniaxially oriented to a draw ratio of 20 at a temperature somewhat below the melting point. Strips 1.7 cm wide were cut from the film and hung vertically from their top edge in an oven at 120°C. A 200 g weight was attached to the bottom edge of the strip and the strip was allowed to draw until it had reached a draw ratio of 20. The resulting films were approximately 1.1 cm wide and 10–20 μm thick and will be referred to as the 'unmodified' films.

The surface was modified by either chemical or plasma etching. Chemical etching was carried out in chromic acid (CA) at 25°C for 4 h. The etched films were then washed with distilled water, as described previously [17]. Oxygen plasma etching was carried out at 50 W and 67 Pa for 10 min using a commercial plasma etcher (Jupiter III, March Instruments). The plasma reactor system consisted of a parallel-plate electrode radio frequency (13.56 MHz) plasma unit with a 3 cm gap between the electrodes. The reactor was evacuated with a vacuum pump (Alcatel AC-2012); the temperature of the anodized aluminum parallel-plate electrodes was maintained at 20°C with a circulating liquid cooler (Neslab RTE-100); and the gas flow was measured by a calibrated rotameter. The unmodified film was placed on the bottom electrode and the reactor was evacuated to 5 Pa, purged with argon, and then evacuated to 2.5 Pa. The film was then exposed for 10 min to an oxygen plasma at 50 W. The reactor was then evacuated to 2.5 Pa, purged with argon, and then opened to the atmosphere.

The epoxy (DER324, Dow Chemical) was used without a curing agent for contact angle measurements and with a curing agent for the ILSS measurements. The curing agent (ethylene diamine) was added to the epoxy in a mass ratio of 1 to 10.

2.2. Film characterization

The topography of the films was examined in a scanning electron microscope (SEM) (JEOL JSM-840). The films were coated with a 0.02 μm layer of evaporated gold before introducing them into the SEM. The melting point and degree of crystallinity were characterized through differential scanning calorimetry (DSC) (Mettler TC 10A). The film was heated from room temperature to 160°C at 10°C/min and the crystallinity was calculated assuming a heat of fusion of 280 J/g [31]. The crystalline orientation in the films was characterized through wide-angle X-ray scattering (WAXS) (PW 1840, Philips) using a Co K_α source (0.179 nm wavelength) without a monochromator.

The modulus and strength were determined in uniaxial tension using a tensile tester (JJ T 5002, Lloyd Instruments). Strips of film 1.5 cm in length were strained at 5 cm/min. The chromic acid-etched films were too brittle to yield 1.5 cm strips and so 1 cm strips were tested (at the same crosshead speed and thus a higher strain rate). The tensile modulus was taken from the steepest slope in the initial part of the stress–strain curve and the tensile strength from the maximum force attained. At least four samples were tested for each measurement.

Electron spectroscopy for chemical analysis (ESCA) was used to characterize the surface chemistry, using an Al K_α source and a 40° angle of incidence (Physical Electronics 555 ESCA/Auger, Perkin Elmer). The overall spectrum was taken at low resolution and the spectra of elements of interest were taken at high resolution.

The spectra were deconvoluted in order to describe more accurately the contributions of different bonds to the overall peak. The total spectrum is described by the sum of mixed Gaussian/Lorentzian functions, each representing a particular spectrum peak. An individual spectrum is described by equation (1) [32]. The peak heights (p) of each spectrum were readily solved by a curve-fit program, given the known binding energy at the peak (x_0) and assuming a consistent shape for all the spectra. This shape is set by keeping two parameters constant: β, which is approximately one half the full width at half-maximum (FWHM), and M, the tendency for the spectrum to be Lorentzian (where $M = 0$ describes a Gaussian spectrum and $M = 1$ describes a Lorentzian spectrum). From curve-fitting spectra with minimal oxygen, the C 1*s* spectrum peak shape was described by setting $M = 0.6 \pm 0.1$ and $\beta = 0.9 \pm 0.1$. The carbon–carbon peak was taken to be 285 eV with a 1.5 eV shift for every carbon–oxygen bond [16]. A shift of −0.8 was used to describe a carbon–silicon bond [33].

$$f(x) = \frac{p}{\left\{1 + M\left[(x - x_0)/\beta\right]^2\right\} \exp\left\{[1 - M][\ln(2)]\left[(x - x_0)/\beta\right]^2\right\}}. \tag{1}$$

Transmission Fourier transform infrared (FTIR) spectroscopy (IFS48, Bruker) was used to provide more information regarding the chemical make-up of the films. The ratios of peaks representing different functional groups were compared with the C—C peak at 1469 cm^{-1}. Peaks representing ether, carbonyl, and carboxyl were taken at 1110, 1740, and 2640 cm^{-1}, respectively [16].

The surface tension (γ_s) of the UHMWPE films was determined through an advancing droplet contact angle technique [4] with an accuracy of 1° using a Kernco goniometer. Assuming that the interactions with the epoxy (without curing agent) are primarily dispersive [4], the γ_s value of the UHMWPE film can be determined by the following expression [4]:

$$\gamma_s = 0.25\gamma_1 (1 + \cos\theta)^2, \tag{2}$$

where γ_1, the surface tension of the epoxy, is 46.2 mN/m [4] and θ is the contact angle between the epoxy droplet and the film. The droplets were ellipsoidal with the major axis in the draw direction and the contact angle was measured in that direction. Typical droplets were photographed using reflection optical microscopy (ROM) (Zeiss Axiophot).

The adhesion of epoxy to UHMWPE was characterized using the Al/epoxy/UHMWPE/epoxy/Al (five-layer) ILSS specimen illustrated schematically in Fig. 1. A 2 cm × 0.2 cm aluminum strip was bonded to each side of a 1 cm × 0.2 cm UHMWPE film with epoxy. The epoxy (with curing agent) was applied to the aluminum strips and the UHMWPE film placed between the strips. The excess epoxy

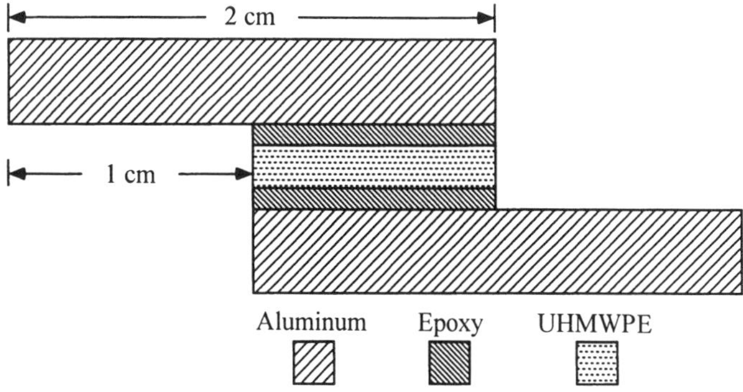

Figure 1. Schematic diagram of the five-layer ILSS sandwich specimen.

was allowed to flow out under the application of a 200 g weight. The epoxy was cured in an oven at 80°C for 5 h under a 200 g weight. The specimen was removed from the oven and the ILSS tested at room temperature. As illustrated in Fig. 1, the upper aluminum strip protrudes 1 cm to the left of the five layers and the bottom aluminum strip protrudes 1 cm to the right. One protruding aluminum strip was clamped in the upper jaw of a tensile tester (JJ T 5002, Lloyd Instruments), the other protruding aluminum strip was clamped in the lower jaw, and the strips were pulled apart at 5 cm/min. The failure surfaces were coated with gold as described previously and examined in the SEM.

3. RESULTS AND DISCUSSION

The thickness of the unmodified UHMWPE films (10–20 μm) is of the same order of magnitude as the UHMWPE fiber diameter. Previous research has indicated that this thickness is not sufficient to isolate the bulk from the effects of surface modification [15]. Unlike the fiber, the unmodified film in the SEM micrograph of Fig. 2a is not covered by a smooth WBL. The fibrils 2–3 μm in diameter on the film surface are similar to those seen in the fiber after its WBL was removed by chromic acid etching. The film has the typical corduroy appearance of a highly oriented fibrillar structural hierarchy [16]. The fibrillar topography is relatively unaffected by chromic acid etching (Fig. 2b). The surface seems less three-dimensional and more planar and the fibrils seem more closely bonded together after plasma etching (Fig. 2c). This change in topography may reflect not only reactions with the oxygen in the plasma which may preferentially remove fibrils protruding from the plane, but also heat generated at the film surface from the impact of high-energy species and crosslinking from the plasma's ultraviolet radiation. In addition, the surface of the oxygen plasma-etched film is damaged by gouges which seem to run perpendicular to the fiber axis across individual fibrils (Fig. 2c).

3.1. Bulk properties

The high narrow peaks in the wide-angle X-ray scattering pattern (Fig. 3) represent the highly oriented crystals produced by drawing the film. The as-cast film, in contrast, exhibited small, wide WAXS peaks representing a random arrangement of crystals (not shown) [34]. The 2θ peak at 25.2° yields a spacing of 0.41 nm which is assigned to the (110) reflection; the peak at 28.0° yields a spacing of 0.37 nm which is assigned to the (200) reflection. The a and b axes calculated from these WAXS peaks, 0.74 nm and 0.49 nm, are similar to the literature values [26, 31]. The c axis, the direction of the polymer chains, lies in the plane of the film and does not produce a scattering peak. Chromic acid etching does have some effect on crystallinity, as seen in Fig. 3. While the 2θ peaks are not shifted, their relative heights are changed. As the topography is relatively unaltered on chromic acid etching (Fig. 2), this change in scattering seems to reflect some change in crystallinity, although the exact implications of the change are not clear.

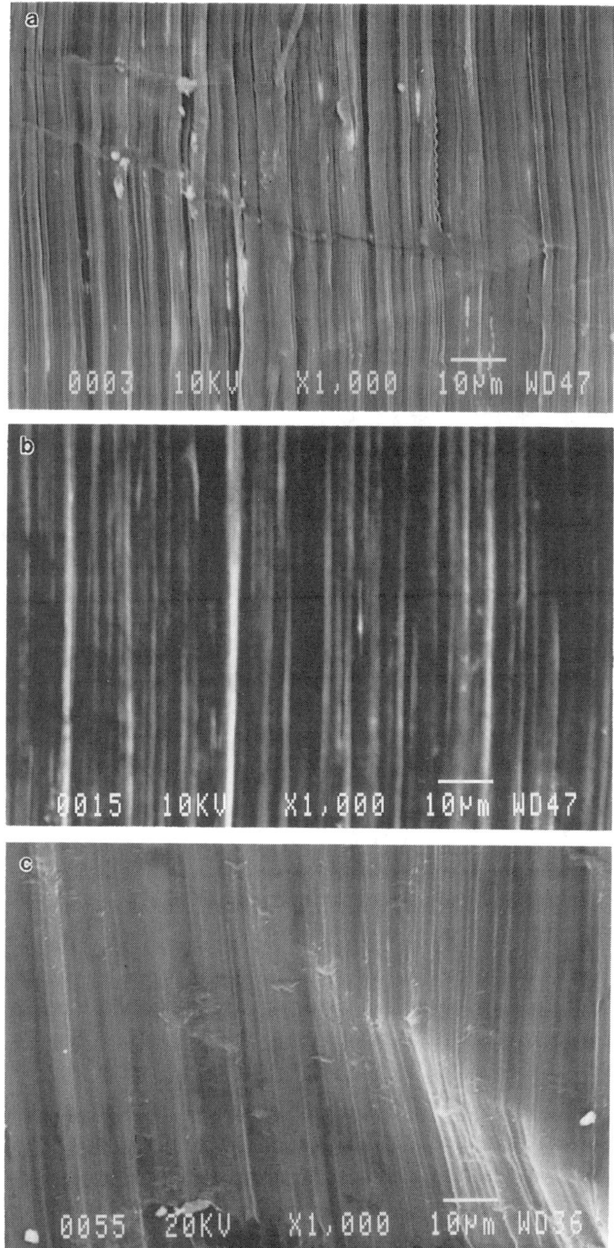

Figure 2. SEM micrographs of UHMWPE films. (a) Unmodified; (b) chromic acid etched; (c) oxygen plasma etched.

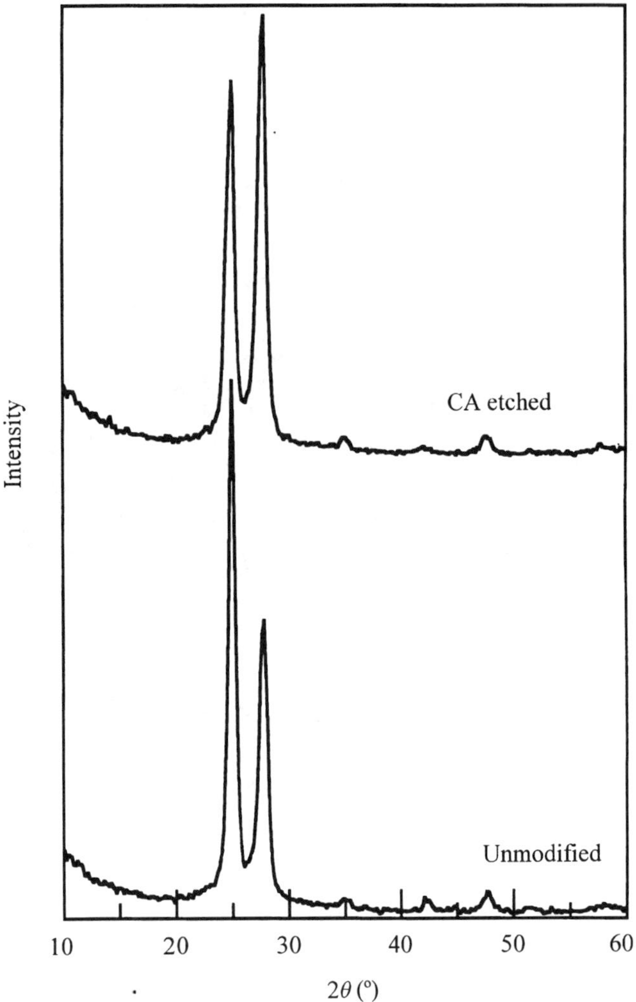

Figure 3. WAXS spectrum from unmodified and chromic acid-etched UHMWPE films.

More quantitative information regarding the effect of etching on crystallinity is provided by differential scanning calorimetry (DSC). The melting point of the as-cast UHMWPE is 136°C and increases with the draw ratio [34]. The unmodified film with a draw ratio of 20 exhibits a DSC thermogram with two peaks: the smaller at 136°C and the larger at 142°C. This double peak indicates the presence of both crystals that have been oriented by drawing and crystals that have not. The crystallinity of 76% is determined by taking the area under both peaks. The DSC thermograms of the etched films show a single peak. The crystallinity decreases by a least 10% on etching, with a corresponding decrease in the melting point (Table 1).

Etching has an even greater influence on the film's mechanical properties. The drawn UHMWPE has a tensile modulus of 2.1 GPa and a tensile strength of 270 MPa. These mechanical properties are more than an order of magnitude higher than those

Table 1.
Bulk properties of UHMWPE films

	Unmodified	Etched	
		CA	O_2 plasma
Melting point (°C)	142.3[a] ± 0.6	141 ± 2	138
Crystallinity (%)	76 ± 5	54	68
Tensile modulus (GPa)	2.1 ± 0.5	1.4 ± 0.4	—
Tensile strength (MPa)	270 ± 30	200 ± 80	—

[a] With a smaller peak at 136°C.

of the as-cast film [34] but are an order of magnitude lower than what can be achieved with optimized casting and drawing [26–28]. The tensile modulus and tensile strength decrease by one-third and one-quarter, respectively, on chromic acid etching (Table 1). The decrease in modulus on etching occurred in spite of the fact that the etched samples were tested at a higher strain rate which is expected to yield an increase in modulus [35]. These deleterious effects of etching on the film's mechanical properties are quite similar to those observed for UHMWPE fibers [17]. The oxygen plasma-etched films were so brittle that their mechanical properties could not be evaluated.

The unmodified films failed in tension through fibril splitting and delamination with tears running in the draw direction. This type of failure reveals a highly oriented fibrillar structural hierarchy throughout the bulk, similar to that seen in the UHMWPE fibers [15, 16]. The toughness of fibers with oriented fibrillar structural hierarchies has been attributed to this energy-absorbing fibril splitting and delamination failure mechanism [15, 36]. The etched films fail in a more brittle manner and do not undergo the same splitting and delamination. The decreases in tensile modulus and strength are caused by polymer degradation, the production of failure-inducing flaws, and the decrease in crystallinity. Both the crystallinity and the mechanical properties of these films are clearly affected by a surface modification which was not supposed to affect the bulk. The significant effect of surface modification on the bulk originates in their meager 20–30 μm thickness. Any improvement in adhesion will be offset by the loss in bulk properties.

3.2. Surface properties

The unmodified film shows a minimal surface oxygen content, 3.1 at.% as determined by ESCA (Table 2) and reflected in the C 1*s* ESCA spectrum in Fig. 4. Deconvolution of the C 1*s* spectrum yields hydroxyl/ether, carbonyl, and carboxyl peaks (not shown) whose areas ($A_{286.5}$, A_{288}, and $A_{289.5}$, respectively) each contribute from 1.2 to 1.7% to the total area of the C 1*s* peak (A_T), as seen in Table 2. The total area of the oxygen-related peaks is thus consistent with a 3.1 at.% oxygen content. The low-resolution ESCA scan of the chromic acid-etched film indicates that there is 20.7 at.% oxygen on the surface. This figure, however, is misleading, owing to the contamination of the surface with 8.5 at.% silicon (Table 2). According to the Si 2*p* ESCA spectrum (not shown), the silicon was also bound to oxygen. It is reasonable to assume that

Table 2.
Surface properties of UHMWPE films

	Unmodified	Etched	
		CA	O_2 plasma
O (at.%)	3.1	20.7	—
Si (at.%)	—	8.5	—
$A_{286.5}/A_T$	1.7	0.0	—
A_{288}/A_T	1.3	3.9	—
$A_{289.5}/A_T$	1.2	2.4	—
H_{2640}/H_{1469}	0.05	0.02	—
θ (degrees)	27 ± 2	16 ± 2	15 ± 2
γ_s (mN/m)	41 ± 1	44 ± 1	45 ± 1

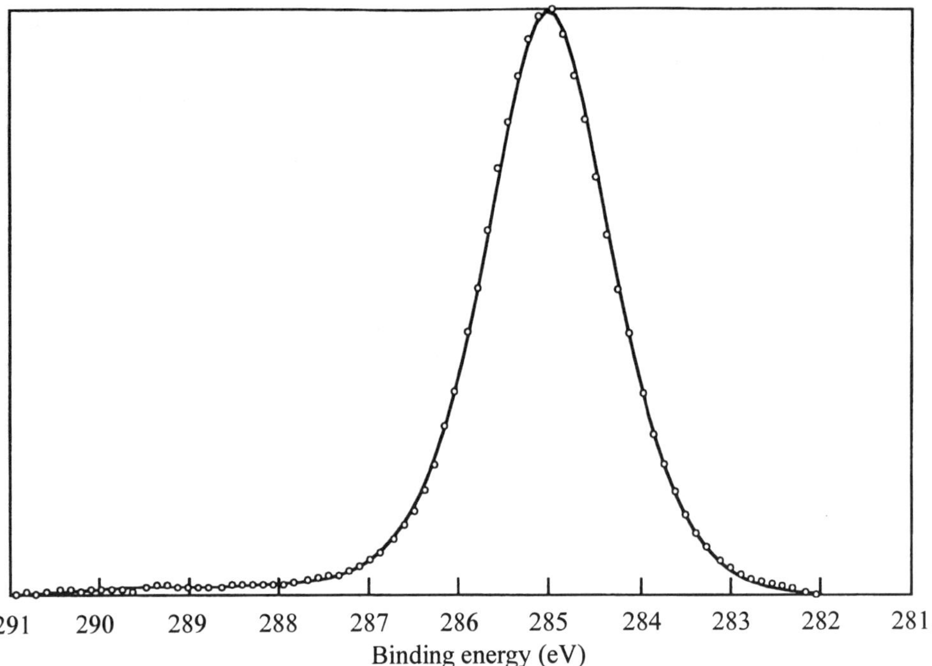

Figure 4. C 1*s* ESCA spectrum of an unmodified UHMWPE film.

a great deal of the oxygen in the spectrum results from Si contamination and is not bound to the carbon in UHMWPE. Etching the as-cast film has shown only a small increase in oxygen (there was no silicon contamination) [34] and so a similarly small increase in oxygen bound to UHMWPE is expected here for the drawn films.

More quantitative information can be generated from the deconvoluted C 1*s* ESCA spectrum of a chromic acid-etched film in Fig. 5. The area of the C—Si peak at 284.2 eV contributes 14.5% to the total area. The area of the hydroxyl/ether peak is approximately zero. The carbonyl peak contributes 3.9% to the total area and the

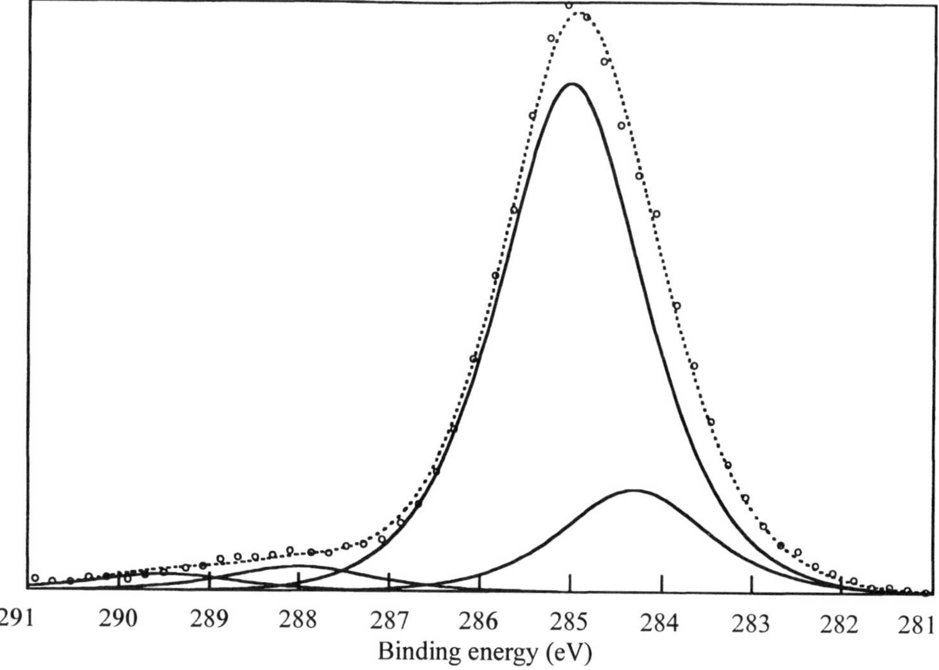

Figure 5. Deconvoluted C 1*s* ESCA spectrum of chromic acid-etched UHMWPE film. (—) Individual peaks; (- - -) total spectrum.

carboxyl peak 2.4%, while the hydroxyl/ether peak is not significant. The deconvolution indicates, as expected, that there is only a small increase in the oxygen bound to UHMWPE. These results are quite different from those for chromic acid-etched UHMWPE fibers, which exhibited a significant oxygen content [16].

The increase in oxygen content is confirmed by the FTIR spectra of the unmodified and chromic acid-etched films in Fig. 6. The FTIR spectrum of the unmodified film is typical of UHMWPE and does not exhibit significant peaks representing oxygen-containing groups at 1110 and 1740 cm^{-1}. There is a small peak representing carboxyl groups at 2640 cm^{-1} with a 2640 cm^{-1} to 1469 cm^{-1} peak height ratio (H_{2640}/H_{1469}) of 0.02 indicating the lack of oxygen bound to the UHMWPE. The peaks at 1110 and 1740 cm^{-1} are still barely discernible in the spectrum of the chromic acid-etched film (Fig. 6). The peak at 2640 cm^{-1}, however, is more pronounced than previously and H_{2640}/H_{1469} more than doubles (Table 2). This confirms the conclusions drawn from ESCA. The amount of oxygen bound to the UHMWPE film increases with chromic acid etching but remains small compared with that observed in the chromic acid-etched fibers [16].

A typical epoxy droplet used for contact angle measurements is seen in Fig. 7. The droplet has a sectioned ellipsoidal shape with the major axis in the draw direction and the curved edges sectioned by lines in the draw direction. The oriented fibrillar topography limits the spread of the droplet perpendicular to the draw direction and enhances the spread of the droplet in the draw direction. Owing to the distorted shape of the droplet, the contact angle measurement and calculation of surface tension can

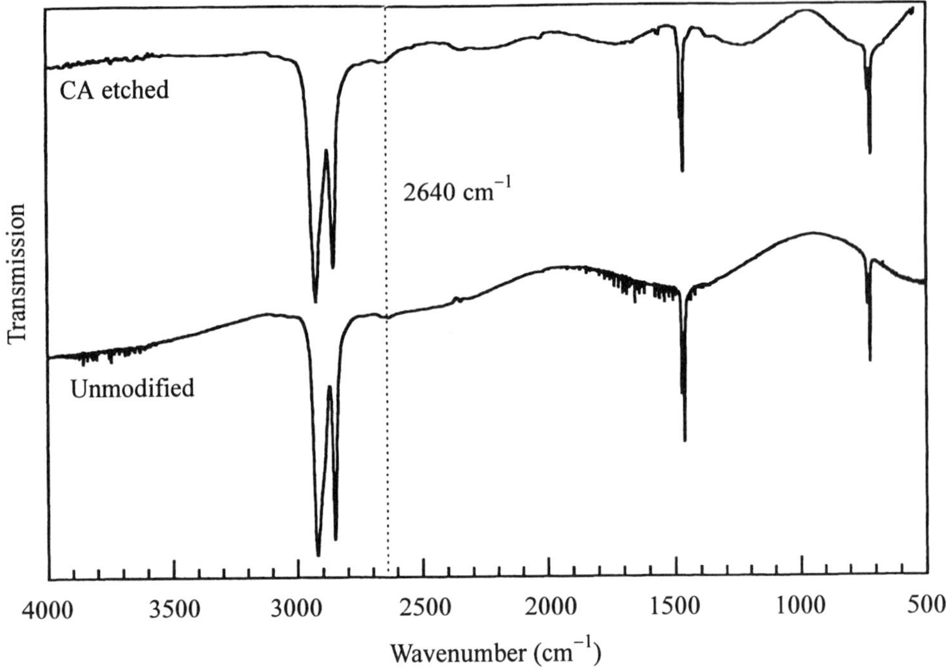

Figure 6. FTIR spectra of unmodified and chromic acid-etched UHMWPE films.

Figure 7. ROM micrograph of an epoxy droplet on an unmodified UHMWPE film.

only be used as a relative measure for the comparison of similar topographies and not for absolute values. The contact angle was measured in the draw direction and yielded reproducible results.

A decrease in contact angle on chromic acid and oxygen plasma etching is observed, as seen in Table 2. This decrease in contact angle can reflect both the introduction of oxygen-containing groups onto the surface and the roughening of the surface [4]. The unmodified and chromic acid-etched film topographies are quite similar (Figs 2a and 2b) and so are not expected to contribute to variations in contact angle. The plasma-etched film is smoother than the unmodified film (Fig. 2c) and so an increase in contact angle would be expected from topographical considerations [4]. The decreases in contact angle on etching may thus be largely attributed to changes in the chemical groups on the surface. The surface tension in Table 2 is calculated from the contact angle data using equation (2). The increase in surface tension on chromic acid etching from 41 to 44 mN/m (Table 2) reflects the slight increase in oxygen bonded to UHMWPE on the film surface as indicated by ESCA and FTIR.

3.3. Adhesion to epoxy

The unmodified film did not adhere to epoxy and an ILSS test specimen could not be produced. The improved wetting by epoxy and the presence of groups on the surface which may react with the epoxy during curing are expected to enhance epoxy–UHMWPE adhesion. The etched films did adhere to epoxy and the interlaminar shear strengths in Table 3 are close to the measured epoxy cohesive strength of 6 MPa. The ILSS failure seen in Fig. 8 for an oxygen-etched UHMWPE film is a combination of epoxy–UHMWPE interfacial failure and UHMWPE cohesive failure. Both interfacial and UHMWPE cohesive failure can be seen on each of the fracture surfaces ($a1$ and $a2$) in Fig. 8a. The UHMWPE cohesive failure is seen at a higher magnification in Figs 8b and 8c for $a1$ and $a2$, respectively. The UHMWPE strips pulled out by the shear failure are typical of the underlying structural hierarchy that still exists in the surface-modified films.

Etching has a larger effect on adhesion than the characterization of the changes in surface chemistry and topography would portend. The changes to the UHMWPE film surface and their ramifications on adhesion should be investigated further.

Table 3.
Adhesion of epoxy to UHMWPE films

	Unmodified	Etched	
		CA	O_2 plasma
Failure strength (MPa)	—	6 ± 2	5.4 ± 0.7
Type of failure	No adhesion	Combination[a]	Combination[a]

[a] Combination of epoxy–UHMWPE interfacial and UHMWPE cohesive.

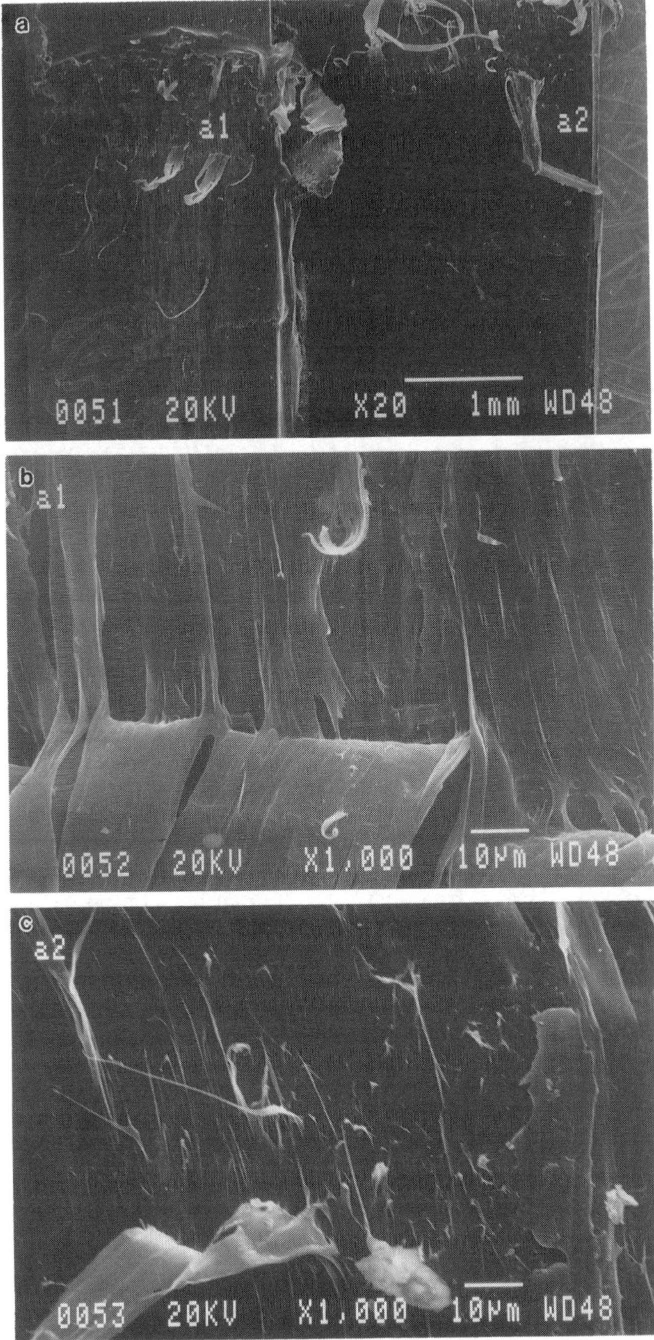

Figure 8. SEM micrographs of the fracture surfaces ($a1$ and $a2$) from an epoxy–UHMWPE ILSS specimen. The UHMWPE film had been etched in an oxygen plasma. (a) Low magnification of both $a1$ and $a2$; (b, c) high magnification of $a1$ and $a2$, respectively.

4. CONCLUSIONS

The drawn gel-cast UHMWPE films are suitable models for the gel-spun UHMWPE fibers and exhibit similar behavior when etched.

- The film has a highly oriented fibrillar structural hierarchy with fibrils 2–3 μm in diameter.
- The degree of crystallinity and melting point both decrease on etching, as do the tensile modulus and tensile strength. The significant decrease in mechanical properties is related to the polymer degradation, the production of failure-inducing flaws, and the decrease in crystallinity. The influence of surface modification on the bulk properties is attributed to the meager film thickness (10–20 μm).
- Etching slightly increases the oxygen content on the film surface through the formation of carbonyl and carboxyl groups. This increase in surface oxygen increases the film's surface tension and enhances its wetting with epoxy.
- Although the unmodified film cannot be bonded with epoxy, the interlaminar shear strength between epoxy and the etched films approaches the cohesive strength of the epoxy. The failure is a combination of interfacial and UHMWPE cohesive failure. This increase in adhesion seems to result from the slight increase in surface oxygen.

Acknowledgement

We gratefully acknowledge the support of the Technion Fund for the Promotion of Research.

REFERENCES

1. P. J. Barham and A. Keller, *J. Mater. Sci.* **20**, 2281 (1985).
2. B. Tissington, G. Pollard and I. M. Ward, *J. Mater. Sci.* **26**, 82 (1991).
3. S. Wu, *Polymer Interface and Adhesion*. Marcel Dekker, New York (1982).
4. A. J. Kinloch, *Adhesion and Adhesives*. Chapman & Hall, New York (1987).
5. D. Briggs, D. M. Brewis and M. B. Konieczo, *J. Mater. Sci.* **11**, 1270 (1976).
6. D. Briggs, in: *Surface Analysis and Pretreatment of Plastics and Metals*, D. M. Brewis (Ed.), p. 199. Macmillan, New York (1982).
7. D. E. Bergbreiter and J. Zhou, *J. Polym. Sci., Part A: Polym. Chem.* **30**, 2049 (1992).
8. Y. Yao, X. Liu and Y. Zhu, *J. Appl. Polym. Sci.* **48**, 57 (1993).
9. J. Behnisch, A. Holländer and H. Zimmermann, *J. Appl. Polym. Sci.* **49**, 117 (1993).
10. L. Dorn, J. Gärtner and M. Rasche, *Kunststoffe* **76**, 249 (1986).
11. W. L. Wade, Jr, R. J. Mammone and M. Binder, *J. Appl. Polym. Sci.* **43**, 1589 (1991).
12. R. Foerch, G. Kill and M. J. Walzak, *J. Adhesion Sci. Technol.* **7**, 1077 (1993).
13. R. Foerch, J. Izawa and G. Spears, *J. Adhesion Sci. Technol.* **5**, 549 (1991).
14. S. Sapieha, J. Cerny, J. E. Klemberg-Sapieha and L. Martinu, *J. Adhesion* **42**, 91 (1993).
15. M. S. Silverstein and O. Breuer, *J. Mater. Sci.* **28**, 4718 (1993).
16. M. S. Silverstein, O. Breuer and H. Dodiuk, *J. Appl. Polym. Sci.* **52**, 1785 (1994).
17. M. S. Silverstein and O. Breuer, *J. Mater. Sci.* **28**, 4153 (1993).
18. M. S. Silverstein and O. Breuer, *Polymer* **34**, 3421 (1993).
19. D. W. Woods and I. M. Ward, *Surface Interface Anal.* **20**, 385 (1993).
20. D. N. Hild and P. Schwartz, *J. Adhesion Sci. Technol.* **6**, 879 (1992).
21. D. N. Hild and P. Schwartz, *J. Adhesion Sci. Technol.* **6**, 897 (1992).

22. F. P. M. Mercx, *Polymer* **35**, 2098 (1994).
23. N. Inagaki, S. Tasaka, H. Kawai and Y. Kimura, *J. Adhesion Sci. Technol.* **4**, 99 (1990).
24. D. A. Biro, G. Pleizier and Y. Deslandes, *J. Appl. Polym. Sci.* **47**, 883 (1993).
25. P. Smith, P. J. Lemstra and H. C. Booij, *J. Polym. Sci., Polym. Phys.* **19**, 877 (1981).
26. P. Smith, P. J. Lemstra, J. P. L. Pijpers and A. M. Kiel, *Colloid Polym. Sci.* **259**, 1070 (1981).
27. P. J. Lemstra and P. Smith, *Br. Polym. J.* **12**, 212 (1980).
28. P. Smith and P. J. Lemstra, *Colloid Polym. Sci.* **258**, 891 (1980).
29. Y. Sakai, K. Umetsu and K. Miyasaka, *Polymer* **34**, 318 (1993).
30. B. E. Krisyuk, V. A. Marikhin, L. P. Myasnikova and N. L. Zaalishvili, *Int. J. Polym. Mater.* **22**, 161 (1993).
31. R. P. Quirk and M. A. A. Alsamarraie, in: *Polymer Handbook*, 3rd edn, J. Brandrup and E. H. Immergut (Eds), p. V/15. John Wiley, New York (1989).
32. P. M. Sherwood, in: *Practical Surface Analysis by Auger and X-ray Photoelectron Spectroscopy*, D. Briggs and M. P. Seah (Eds), p. 445. John Wiley, New York (1983).
33. G. Beamson and D. Briggs, *High Resolution XPS of Organic Polymers*. John Wiley, New York (1992).
34. M. S. Silverstein and J. Sadovsky, in preparation.
35. P. Schwartz, A. N. Netravali and S. Sembach, *Textile Res. J.* **56**, 502 (1986).
36. J. X. Li, M. S. Silverstein, A. Hiltner and E. Baer, *J. Appl. Polym. Sci.* **44**, 1531 (1992).

Polymer Surface Modification: Relevance to Adhesion, pp. 417–427
K. L. Mittal (Ed.)

Modification of polyamide fiber surfaces by micro-organisms

ELENA V. PISANOVA* and SERGE F. ZHANDAROV

Metal–Polymer Research Institute, Belarus Academy of Sciences, 32a, Kirov Street, Gomel 246652, Republic of Belarus

Revised version received 7 September 1994

Abstract—The properties of some polyamide fibers — polycaproamide (PCA), poly-*p*-amidobenzimidazole (PABI), poly-*p*-phenyleneterephthalamide (PPTA), poly-*m*-phenyleneisophthalamide (PPIA), and polyimide (PI) — treated by different micro-organisms that are abundant in nature (the bacteria *Bacillus* and *Pseudomonas*, and the fungi *Aspergillus*) were studied. The interfacial bond strength between these fibers (untreated and exposed to micro-organisms) and thermoplastic polymers such as polyethylene (PE), polycarbonate (PC), and polysulfone (PSF) was investigated by the single-filament composite and pull-out techniques. If the treatment time is not very long (1–3 weeks), the tensile strength of the fibers does not deteriorate. However, the topography of the surface undergoes considerable changes in some cases. The result of the treatment depends substantially on the chemical nature of the nutrient as well as on the micro-organism species. Two main avenues for fiber modification are possible: (1) decomposition of macromolecules in the surface layer, resulting in surface growth; and (2) adsorption of the metabolic products, which heal surface flaws and alter the chemical nature of the surface. The increase in the interfacial shear strength (IFSS) for the modified fiber–thermoplastic matrix systems was from 15% for the PCA/PE composite to 80% for the PABI/PSF and PABI/PC pairs. This increase was especially marked when a sublayer of a substance specially added to the nutrient, having better compatibility with the matrix, was formed by micro-organisms on the fiber surface.

Keywords: Polyamide fibers; biochemical treatment; adhesion; thermoplastic matrices; surface modification; surface topography.

1. INTRODUCTION

Many micro-organisms can affect polymer materials, causing their degradation [1]. Because a characteristic property of polymer fibers is the dense packing of oriented macromolecules which prevents the enzymes from penetrating inside the fiber, active micro-organisms cause changes only in the surface layer. This fact can be used for target-oriented modification of the surface of organic fibers employed as a reinforcement in composite materials production.

*To whom correspondence should be addressed.

Our previous studies revealed that the biochemical treatment of polyamide reinforcing fibers enhanced the bond strength at the fiber–matrix interface [2] and, in this way, improved the strength properties of the plastics obtained [3]. This obviously results from alterations in the structure of the fiber surface layers. The aim of the present work was to investigate in detail the chemical structure and topography of the fiber surface treated by micro-organisms.

2. EXPERIMENTAL

The following organic fibers which are recognized as advanced reinforcements for polymer composite materials — poly-*p*-amidobenzimidazole (PABI), poly-*p*-phenyleneterephthalamide (PPTA), poly-*m*-phenyleneisophthalamide (PPIA), polycaproamide (PCA), and polyimide (PI) — were chosen for the tests. The fibers were exposed to micro-organisms abundant in nature: the bacteria *Bacillus* and *Pseudomonas*, and fungi *Aspergillus*. Adaptive strains capable of using polyamides as the only source of carbon and nitrogen were applied. The treatment consisted in keeping the fibers in aqueous medium with micro-organisms for 7–21 days in normal conditions and under moderate aeration. Trace amounts of some ions vital for the micro-organisms (Fe^{3+}, Ca^{2+}, Al^{3+}, Na^{+}, Mg^{2+}) were added to the medium. In some cases, organic compounds which do not contain nitrogen (e.g. polyvinyl alcohol) were chosen as nutrients. In this way, the micro-organisms were forced to consume nitrogen from the fibers. After the treatment the fibers were sterilized, rinsed in hot distilled water, and then dried. Control fibers underwent identical treatment in the nutrient medium but in the absence of micro-organisms.

The fiber surface was investigated with a scanning electron microscope (SEM; JSM-50A, JEOL, Japan). The topography of the surface was studied with the secondary electron emission technique using the SEM with a personal computer attached [4]. The investigation of the fibers was also carried out using infrared spectroscopy (UR-20 spectrophotometer, Germany) and optical spectroscopy (LMA-10 laser microstructural analyzer, Germany). The degree of the fibers swelling in dimethylformamide (DMFA) was determined as the ratio of the weight of the swelled sample to its initial weight. The fiber tensile strength was measured using an F0-1C tensile machine (VEB Werkstoffprüfmaschinen, Germany) at a gauge length of 20 mm and a loading rate of 0.8 g/s.

The adhesional bond strength in the single fiber/thermoplastic matrix systems was determined by two techniques: (1) the three-fiber method (a version of the well-known pull-out technique) which we had modified for thermoplastic polymers [2, 5]. This technique is based on the measurement of the force required for pulling the fiber out of a polymer droplet of known size; and (2) the single-filament-composite (SFC) test, which involves stretching a matrix bar along the axis of the single fiber embedded in it and after the fiber fragmentation is completed, analyzing the final fragment distribution. We employed the testing procedure specially developed by us for investigation of thermoplastic polymers [6] consisting in the specimen stretching until a neck having the maximum possible length is formed.

3. RESULTS AND DISCUSSION

It is well known [7] that the process of fiber interaction with micro-organisms follows a multistage course. At the initial stage, the adsorption of micro-organisms by the fiber surface and their fixation on it are observed. Within 3–4 weeks, as a rule, there are no distinct changes in the fiber surface. However, the application of adaptive strains can considerably accelerate this process. The fiber strength does not normally deteriorate during this period of biochemical treatment. Rather, our investigations of the strength of PCA single fibers treated by various micro-organisms demonstrated that, for short exposure times (1–2 weeks), the fiber tensile strength increased by 6–23% (Table 1). A similar strength improvement was also observed for fibers of other chemical natures, e.g. PABI. We studied PABI fibers especially thoroughly to clarify the mechanism of fiber strengthening.

As one can see in Table 2, the biochemical treatment of PABI fibers results in an increase in the mean strength of the fibers; it also reduces the scatter in the strength values. A comparison of experimentally obtained strength distributions for untreated (Fig. 1a) and treated (Fig. 1b) fibers demonstrates a narrowing of the distribution and a shift of its maximum to the right, i.e. the observed strengthening is a result of the 'disappearance' of fibers having the least strength. In this case, the strength

Table 1.
Characteristics of untreated and modified PCA fibers

Micro-organism	Tensile strength (GPa)	Elongation to break (%)	Bond strength in the PCA/HDPE system (MPa)[a]
—	1.52	16.0	13.1
B. subtilis	1.61	18.5	15.8
P. putida	1.67	17.9	15.0
B. brevis	1.71	18.9	12.1
B. vulgaris	1.74	17.8	21.0
B. megaterium	1.74	18.2	17.6
B. mesentericus	1.79	18.3	15.9
B. cereus	1.79	20.2	20.0
B. cereus[b]	1.23	15.5	19.9

B. = *Bacillus*; *P.* = *Pseudomonas*. Treated for 2 weeks.
[a]Thermal treatment conditions: 180°C for 15 min.
[b]The treatment time was 8 weeks.

Table 2.
Strength distributions of untreated and modified PABI fibers

PABI fiber	No. of tests (MPa)	Mean strength (MPa)	SD (MPa)	Weibull parameters			
				ρ_1	σ_1 (MPa)	ρ_2	σ_2 (MPa)
Untreated	120	3760	451	11.6	3622	24.7	4364
Modified	120	4060	239	37.0	3911	24.0	4431

SD = Standard deviation.

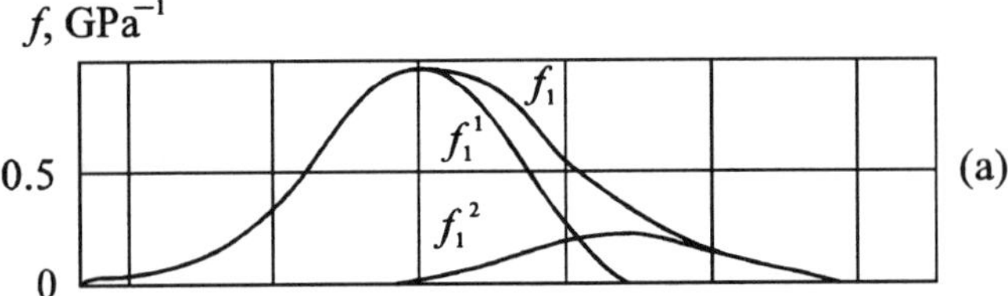

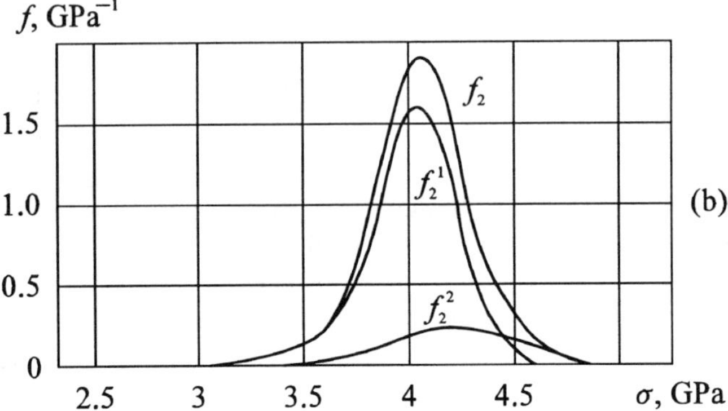

Figure 1. Experimental strength distribution of untreated (a) and modified (b) PABI fibers presented in the form of the sum of two Weibull distributions: $f_1 = 0.84 f_1^1 + 0.16 f_1^2$; $f_2 = 0.84 f_2^1 + 0.16 f_2^2$. f = Probability density (GPa^{-1}); σ = fiber strength (GPa).

improvement can be accounted for in terms of fiber defects healing as a result of the adsorption of products of micro-organism metabolism on the fiber surface [8]. Both experimental probability density curves (f_1 for untreated fibers and f_2 for treated ones) can be described with good accuracy as linear combinations of two Weibull distributions (Fig. 1 and Table 2), both of which appear in linear combinations with equal weight coefficients:

$$f_1 = 0.84 f_1^1 + 0.16 f_1^2;$$
$$f_2 = 0.84 f_2^1 + 0.16 f_2^2.$$

Table 2 demonstrates that PABI single fibers can be divided into two types distinguished by their different behaviors under biochemical treatment. The smaller portion — the strongest fibers ($\simeq$16% of the total number of fibers) — does not show a change in fiber strength in this process; the Weibull parameters of the f_1^2 and f_2^2 distributions practically coincide, so f_1^2 and f_2^2 can be considered as identical distributions ($f_1^2 \approx f_2^2$). On the contrary, the Weibull parameters of the f_1^1 and f_2^1 distributions differ substantially, demonstrating a marked increase in the strength of the corresponding fibers ($\simeq$84% of the total number of fibers) resulting from the treatment. Proceeding from this fact, we suggest the existence of two modes of PABI fiber failure, one controlled by surface flaws and the other by the internal fiber structure.

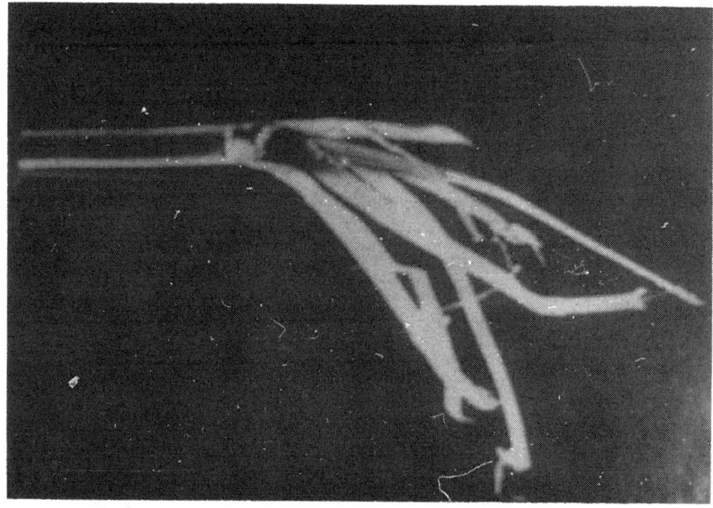

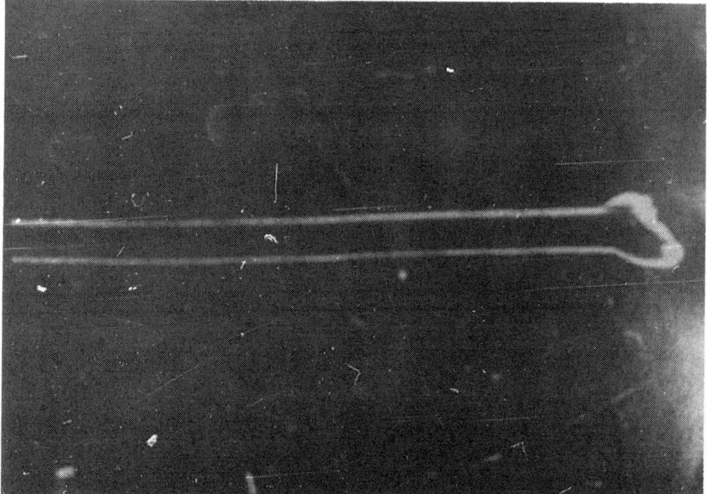

Figure 2. The failure of untreated (a) and modified (b) PABI fibers.

As is shown elsewhere [9], the characteristics of the supermolecular structure of PABI fibers enables the enzymes to penetrate the space between the fibrils. The partial hydrolysis of internal macromolecules results in interfibrillar bonding. Additional evidence for this is the difference in the failure modes between untreated and treated fibers (Fig. 2). The untreated fibers undergo fibrillation (Fig. 2a), which is a characteristic of highly anisotropic fibers; the treated fibers fail via transversal crack growing (Fig. 2b) showing a high degree of interfibrillar interaction [10].

There is no surface damage visible on scanning electron microphotographs of the fibers treated for a moderate period. At the same time, outgrowths of various shapes

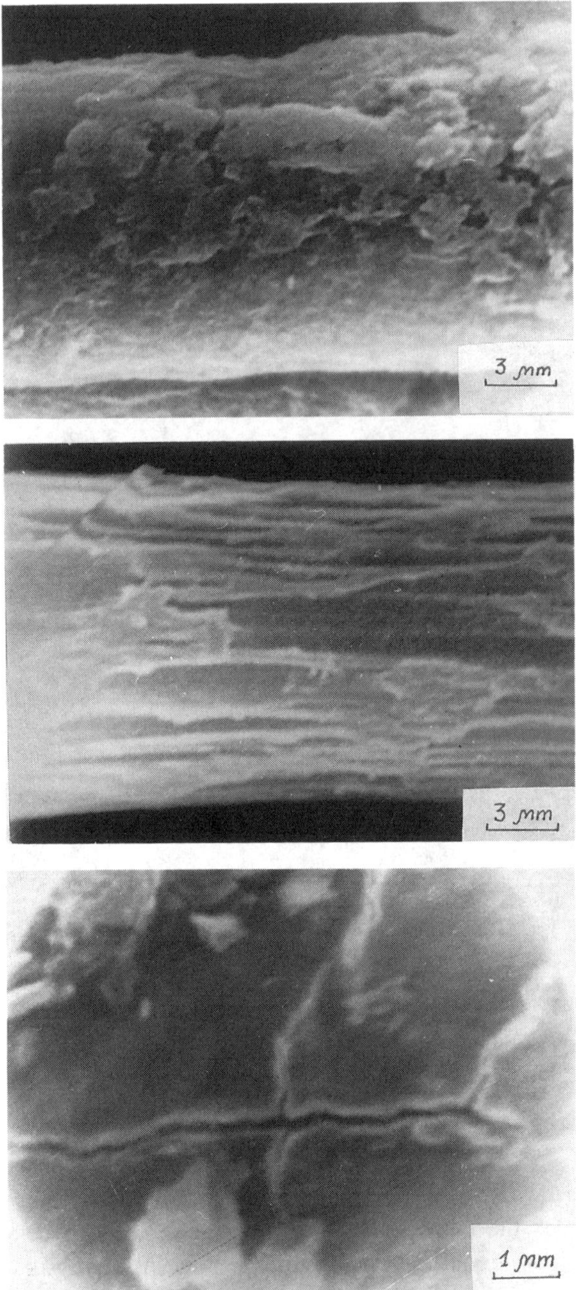

Figure 3. PI fibers modified by *Bacillus cereus* (a) and *Pseudomonas putida* (b, c) bacteria. Exposure time: 2 weeks (a, b); 8 weeks (c).

and sizes are observed (Fig. 3). The reason for their formation can be the adsorption of the products of micro-organism metabolism on the fiber [9, 11, 12] as well as biochemical destruction of the macromolecules in the surface layer which alters its supermolecular structure.

A more detailed study of the surface of the fibers was carried out using the secondary electron emission technique [4]. Such statistical parameters as the relative surface area (RSA) (defined as the ratio of the real area to the nominal one), the number of extrema (peaks and pits) per unit area, and the mean degree of inclination of these irregularities with respect to the fiber axis were determined. The PABI fibers were treated by the bacteria *Pseudomonas putida*, and the PI fibers by *Bacillus cereus* for 14 days. In both cases, the treatment causes a decrease in the RSA (by 2% for PABI and by $\simeq$30% for PI fibers). At the same time, the number of extrema for treated PABI fibers is six times greater than that for untreated samples, and three times greater for PI fibers (Table 3); the asperities themselves become smoother.

Thus, all the results that we have obtained are in good mutual agreement and show that at the first stage of the microbiological treatment, the cavities, pores, and micro-cracks on the fiber surface are filled with the products of micro-organism metabolism (Figs 3a and 3b), resulting in healing of the defects and consequently in an increase of the fiber strength.

Longer treatment causes a natural decrease of the strength of the fibers (Table 1). Small cracks appear on the fiber surface (Fig. 3c) similar to those observed by Ermilova [7] and Watanabe [13]. Hence, it is very important to determine the optimum time of exposure so as not to deteriorate the strength of the reinforcing fibers and to alter substantially the nature of the surface thus improving the interaction between the matrix and the reinforcement at the interface.

The mechanism of biochemical degradation of the polyamides is similar to the enzymatic splitting of proteins [9, 13]:

$$-C_6H_4-NH-\underset{\underset{O}{\|}}{C}-C_6H_4- \xrightarrow{H_2O} -C_6H_4-NH_2 + HO-\underset{\underset{O}{\|}}{C}-C_6H_4-$$

The amino groups produced are consumed by the micro-organisms; as for carboxyl groups, they remain on the fiber surface [9]. Depending on the composition of

Table 3.
Characteristics of the fiber surface topography

		Topography characteristics		
Fiber	Micro-organism	Relative surface area	Number of extrema per unit area (μm^{-2})	Mean inclinanation of irregularities (degrees)
PABI	—	1.052	1.5	17.6
PABI	*Pseudomonas putida*	1.032	8.9	14.1
PI	—	1.412	0.5	42.6
PI	*Bacillus cereus*	1.066	1.4	19.5

the medium, they may react with different substances present in the solution. For instance, if ions of metals (Ca, Na, Fe, Mg, Al) were added to the nutrient liquid, they would be deposited on the PABI surface. Metal ions are easily detectable with a laser microanalyzer. In the spectra of untreated and control fibers the lines of metals are absent, but biochemically treated fibers give intense signals corresponding to the spectrum of the added metal. If an alcohol is present in the nutrient medium, the etherification reaction proceeds. The fragments of polyvinyl alcohol molecules, interacting with carboxyl groups on the fiber surface, produce esters according to the following scheme:

$$-C_6H_4-\underset{\underset{\displaystyle O}{\|}}{C}-OH \;+\; HO-\overset{|}{\underset{\underset{\displaystyle |}{CH_2}}{\underset{|}{CH}}} \xrightarrow{-H_2O} -C_6H_4-\underset{\underset{\displaystyle O}{\|}}{C}-O-\overset{|}{\underset{\underset{\displaystyle |}{CH_2}}{\underset{|}{CH}}}$$

This mechanism is verified by the appearance of absorption bands at 1028, 1244, and 1780 cm^{-1} (assigned to ester groups) in the IR spectrum of PABI fibers treated in the presence of polyvinyl alcohol. The presence of chemically bound polyvinyl alcohol molecules on the fiber surface is shown by the peaks at 1090 cm^{-1} ($\nu_{C=O}$) and 1420 cm^{-1} (δ_{C-H}) [14] characteristic of PVA. At the same time, the intensity of amide I and amide II bands (at 1665 and 1535 cm^{-1}) decreases (Fig. 4).

The result of the biochemical treatment of the fiber depends not only on the species of the micro-organisms used and the nutrient medium composition, but also, to a great extent, on the chemical nature of the fiber itself. Fibers of similar chemical

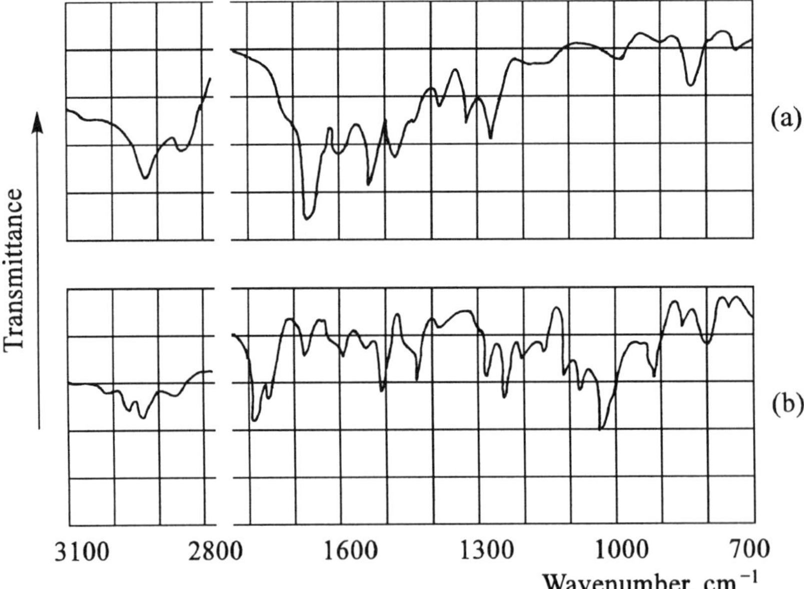

Figure 4. Infrared spectra of untreated PABI fibers (a) and PABI fibers modified in the presence of polyvinyl alcohol (b).

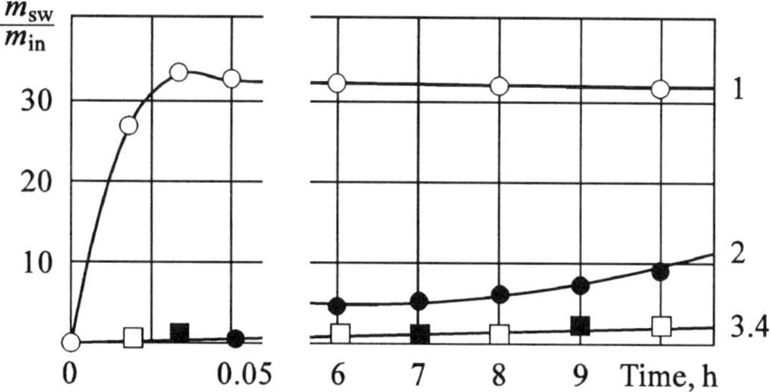

Figure 5. The kinetics of swelling of PABI (1, 2) and PPTA (3, 4) fibers treated by micro-organisms (1, 3) and control fibers (2, 4). m_{sw} = mass of the swelled sample; m_{in} = initial mass of the sample.

nature under the effect of the same strain of micro-organism often revealed quite different behavior. So, PABI fibers treated with *Bacillus* bacteria were seen to swell fast in DMFA (Fig. 5), but PPTA and PPIA practically did not swell. The likely reason for this is the particular supermolecular structure of PABI without regions with three-dimensional order [10]; this fact promotes the diffusion of enzymes into the interfibrillar space [9]. In the case of the more ordered PPTA and PPIA structures, the biochemical modification is restricted to the surface layer only.

The alteration of the chemical nature of the fibers due to the microbiological treatment, as a rule, results in changes in the adhesion between these fibers and various matrices (Tables 1 and 4). Drzal [15], while studying the interaction of epoxy resins with fibers, demonstrated that the SFC test was not really applicable to Kevlar fibers. When these are being stretched, fibrillation occurs rather than brittle failure, with subsequent mutual sliding of the fibrils, and the fragment lengths cannot be measured. We observed the same effect when we tried to determine by the SFC technique the interfacial shear strength in PSF- and PC-based composites with PABI, PI, PPTA, and PPIA reinforcements. One may conclude that in this case the adhesional bond strength at the interface exceeds the cohesional fiber strength; so, it is impossible to measure it using the SFC test. However, for the case of low adhesion, e.g. for the PCA/polyethylene pair, the SFC test is quite valid. The data presented in Table 1

Table 4.
Properties of the untreated and modified PI and PABI fibers

Fiber	Micro-organism	Diameter (μm)	Mean strength (GPa)	Bond strength in the fiber/PC system (MPa)[a]
PI	—	15.7	0.78	39.3
PI	*Caulobacter bacteroides*	15.7	0.72	51.4
PABI	—	13.5	3.76	45.7
PABI	*Pseudomonas putida*	13.5	3.67	57.0
PABI	*Bacillus cereus*[b]	13.5	4.06	86.1

[a]Thermal treatment conditions: 275°C for 15 min.
[b]In a nutrient medium containing polyvinyl alcohol.

were obtained by the SFC technique and those in Table 4 by the three-fiber pull-out method.

As can be seen in Table 4, the adhesional bond strength increases with the microbiological treatment of the fibers. This effect is especially pronounced when a grafted polymer layer is formed on the fiber surface. This layer containing ester groups likely enhances affinity to the polycarbonate and plays the role of an interphase.

4. CONCLUSION

The biochemical (microbiological) treatment is a simple and efficient way of altering the surface properties of the reinforcing fibers in composite materials. Because this process is very sensitive to the micro-organism strain, the chemical nature of the fiber, and the composition of the nutrient medium, it enables a wide-ranging, target-oriented variation of the fiber surface properties, without degradation of the strength of the fibers, to be obtained. In many cases, the treatment with micro-organisms results in an improvement of the adhesional bonding between fibers and thermoplastic matrices, thus providing a way to obtain reinforced plastics with improved characteristics [3, 11].

Acknowledgement

We would like to acknowledge Dr M. B. Vainshtein (Institute of Biochemistry and Physiology of Micro-organisms, Russian Academy of Sciences) for carrying out the microbiological treatment of the fibers.

REFERENCES

1. M. N. Rotmistrov, P. I. Gvozdyak and S. S. Stavskaya, *Microbiological Destruction of Synthetic Organic Substances*. Naukova Dumka, Kiev (1975).
2. A. I. Sviridenok, T. K. Sirotina and E. V. Pisanova, *J. Adhesion Sci. Technol.* **5**, 229–237 (1991).
3. A. I. Sviridenok, V. V. Meshkov, E. V. Pisanova and T. K. Sirotina, in: *ACHEMA '91: Proceedings of the International Meeting on Chemical Engineering and Technology*, Vol. 1, p. 284. DECHEMA, Frankfurt am Main, Germany (1991).
4. A. Ya. Grigoryev, N. K. Myshkin, N. F. Semenyuk and O. V. Kholodilov, *Friction and Wear* (*USSR*) No. 5, 793–798 (1988).
5. V. A. Dovgyalo, S. F. Zhandarov and E. V. Pisanova, *Mech. Composites* (*USSR*) No. 1, 9–12 (1990).
6. E. V. Pisanova, S. F. Zhandarov and V. A. Dovgyalo, *Polym. Composites* **15**, 147–155 (1994).
7. I. A. Ermilova, *Theoretical and Practical Basis of the Microbiological Destruction of Chemical Fibers*. Nauka, Moscow (1991).
8. O. A. Novikova and V. P. Sergeev, *Modification of Reinforcing Fibers in Composite Materials*. Naukova Dumka, Kiev (1989).
9. A. I. Sviridenok, T. K. Sirotina and E. V. Pisanova, *High Molecular Weight Compounds* (*USSR*), *Ser. B* **31**, 571–576 (1989).
10. K. E. Perepelkin, *Structure and Properties of Fibers*. Nauka, Moscow (1985).
11. A. I. Sviridenok, T. K. Sirotina and V. V. Meshkov, *USSR Acad. Rep.* **298**, 666 (1988).
12. A. Yu. Lugauskas, L. I. Levinskajte, D. K. Lukshajte and D. N. Pechulite, *Plast. Mater.* (*USSR*) No. 2, 24–28 (1991).
13. T. Watanabe, *J. Jpn. Soc. Fiber Sci. Technol.* **43**, 192–197 (1987).

14. I. Dechant, R. Danz, W. Kimmer and R. Schmolke, *Ultrarotspectroskopische Untersuchungen an Polymeren*. Akademie-Verlag, Berlin (1972).
15. L. T. Drzal, *15th Natl. SAMPE Techn. Conf., Azusa, Calif.* **15**, 190–201 (1983).

Part 4

General Papers

Polymer Surface Modification: Relevance to Adhesion, pp. 431–454
K. L. Mittal (Ed.)

Physico-chemical properties of surface-modified polymers

K. GRUNDKE,* H.-J. JACOBASCH, F. SIMON and ST. SCHNEIDER
Institute of Polymer Research Dresden, 01005 Dresden, P.O.B. 120411, Germany

Revised version received 1 July 1994

Abstract—Contact angle, electrokinetic, and X-ray photoelectron spectroscopic (XPS) measurements have been used to study the surface properties of flame- and oxygen plasma-pretreated polypropylene/ethylene-propylene-diene monomer rubber (PP–EPDM) blends and of ethylene vinyl acetate (EVA) copolymers grafted with carboxyl group-containing monomers. The contact angles of pure test liquids (water, methylene iodide, and ethylene glycol) were used to calculate the dispersive and polar components of the surface free energy according to Owens and Wendt, and the acid–base parameters according to Van Oss and co-workers. In addition, the acid–base properties of the differently pretreated polymers could be evaluated quantitatively by measuring the zeta potential vs. the pH in a 10^{-3} mol/l KCl solution. The zeta potential measurements show that oxygen plasma-treated PP–EPDM and grafted EVA indicate an acidic surface character, whereas the flame-treated PP–EPDM blends possess both acidic and basic surface groups. The basic surface character of flame-treated PP–EPDM injection-moulded sheets could be enhanced by the presence of sterically hindered amine light stabilizers in the blend. This increase in the basic surface character was not only proved by zeta potential measurements, but also by the contact angle method according to Van Oss and co-workers. These results correlate with an increase of the oxygen content in the surface region and the occurrence of nitrogen-containing functional groups detected by XPS. The plasma-treated surface region of PP–EPDM blends contained an increased amount of carboxyl group-containing species (O=C–O). Flame-treated surfaces with additional light stabilizers in the blend indicated an increased concentration C–OH groups together with protonated nitrogen in the surface region. It was found that the adhesion strength of water-based primers was higher at these surfaces. A general interrelation between the acidic and basic parameters determined by zeta potential measurements, on the one hand, and the acidic and basic parameters determined by contact angle measurements, on the other hand, could not be found. A direct correlation was found between the increasing acidic character of EVA grafted with different amounts of carboxyl group-containing monomers and the decrease in the receding contact angle.

Keywords: Polypropylene/ethylene-propylene-diene monomer rubber blends; grafted ethylene vinyl acetate copolymers; plasma treatment; flame treatment; contact angle; zeta potential; XPS; acid–base interactions; painting; adhesion.

1. INTRODUCTION

Several techniques have been developed for modifying polymer surfaces to improve their wettability and adhesion properties without altering their bulk properties. These

---

*To whom correspondence should be addressed.

include chemical modifications (grafting) [1, 2] and exposure to flames, and low-pressure non-equilibrium plasma discharge [3–6], which offer an attractive way to increase the polarity of the polymer surface by creating reactive surface groups.

The objective of the present paper is to present recent results obtained on the characterization of these modified polymer surfaces by means of physico-chemical methods. Table 1 summarizes some of these methods and the parameters which can be measured.

By applying semi-empirical approaches, conclusions can be drawn concerning the intermolecular forces resulting from polymer surfaces. Table 2 summarizes inter-atomic and intermolecular forces at polymer surfaces, their range, the interaction energy to be expected, and the physico-chemical parameters characterizing these forces.

The highest interaction energies can be achieved by chemical reactions at the interface leading to covalent chemical bonds and Coulomb forces. These interactions require highly reactive functional groups at the surface. The efficiency of such reactive sites at polymer surfaces for improving adhesion properties is probably limited by

Table 1.
Physico-chemical methods for characterizing polymer surfaces [15]

Method	Parameters		Information obtained concerning
	Measured	Calculated	
Inverse gas chromatography	Retention volume of gases	Adsorption free energy	Dispersion forces, acid–base interactions
Wetting measurements	Contact angles of test liquids	Surface free energy	Dispersion forces, acid–base interactions
Electrokinetic measurements	Zeta potential of solids in electrolyte solutions	Adsorption free energy, pK_A, pK_B	Dispersion forces, acid–base interactions, Coulomb forces, displacement forces
Quantitative adsorption measurements	Adsorbed amount of gases or solutes	Adsorption free energy	Displacement forces

Table 2.
Interatomic and intermolecular forces at polymer surfaces [15, 16]

Type	Range	Interaction energy (kJ/mol)	Parameter determined
Covalent bond	Short	63–920	Reactivity of functional groups
Coulomb interaction	Short	335–1050	Charge of functional groups
Dispersion (London) forces	Long	$\leqslant 40$	Hamaker constant
Acid–base interactions	Short	$\leqslant 50$	
Brönsted theory			pK values
Lewis theory			Acceptor and donor numbers according to Gutmann, Drago constant

their tendency to be quickly neutralized by atmospheric contaminants [7]. Improved adhesion can also be achieved by acid–base interactions. These forces are due to specific interactions between suitable, less reactive, functional groups of each partner at the interface. Today, it is assumed that the majority of the specific interactions across polymer interfaces are those of the acid–base type in the most general Lewis sense [8]. For that reason, it is preferable to determine the 'reactivity' of the polymer surface by detecting the presence of acidic or basic sites on a surface.

Inverse gas chromatography and quantitative adsorption measurements (see Table 1) are restricted to high surface area solids and have been applied mainly to powders or fibres. These methods are not applicable to macroscopic specimens.

Despite their incomplete theoretical background and many experimental difficulties, contact angle measurements have been used extensively to elucidate the specific surface-chemical changes imparted by surface treatments of macroscopic specimens. The contact angles of non-polar and polar test liquids are used to determine the dispersive and polar components of the surface free energy of the polymer surface by applying various empirical relations (Table 3). The widely used method of Owens and Wendt [9] is based on the controversial assumption that the polar interactions can be calculated by a geometric mean mixing rule [22]. A more recently developed approach by Van Oss *et al.* [10] postulates that the surface free energy can be divided into a Lifshitz–van der Waals component and an electron-acceptor (Lewis acid) and electron-donor (Lewis base) parameter. It is not yet clear if this empirical approach (see Table 3) can be confidently used to characterize the acid–base properties of polymers.

In addition to contact angle measurements, this paper also presents electrokinetic measurements to study the acid–base properties of surface-modified polymers. Zeta

Table 3.
Empirical relations between the contact angle and the surface free energy of the liquid and the solid

Approach	Equation
Neumann [17]	$\cos\Theta = \dfrac{(0.015\gamma_s - 2)(\gamma_s\gamma_l)^{1/2} + \gamma_l}{\gamma_l[0.015(\gamma_s\gamma_l)^{1/2} - 1]}$
Owens–Wendt [9] Kaelble [18]	$(1+\cos\Theta)\gamma_l = 2\left[(\gamma_s^d\gamma_l^d)^{1/2} + (\gamma_s^p\gamma_l^p)^{1/2}\right]$
Wu [19]	$(1+\cos\Theta)\gamma_l = 4\dfrac{\gamma_l^d\gamma_s^d}{\gamma_l^d+\gamma_s^d} + 4\dfrac{\gamma_l^p\gamma_s^p}{\gamma_l^p+\gamma_s^p}$
Fowkes [20]	$(1+\cos\Theta)\gamma_l = 2(\gamma_s^d\gamma_l^d)^{1/2}$
Van Oss, Good, Chaudhury [10]	$(1+\cos\Theta)\gamma_l = 2\left[(\gamma_s^{LW}\gamma_l^{LW})^{1/2} + (\gamma_s^+\gamma_l^-)^{1/2} + (\gamma_s^-\gamma_l^+)^{1/2}\right]$

γ_l, γ_s: Surface free energies of the liquid and solid; γ_l^d, γ_s^d: dispersion terms of the liquid and solid; γ_l^p, γ_s^p: polar terms of the liquid and solid; γ_l^{LW}, γ_s^{LW}: Lifshitz–van der Waals terms of the liquid and solid; γ_l^+, γ_s^+: electron-acceptor terms of the liquid and solid; γ_l^-, γ_s^-: electron-donor terms of the liquid and solid.

potential measurements have been used for a long period to investigate the electrical properties of colloidal particles [11]. By applying the streaming potential method, it is possible to characterize the electrical properties of macroscopic specimens, such as plates or foils. Using an automatic streaming potential measuring device, polymer surfaces can be characterized with respect to acidic or basic groups [12–14].

The intermolecular forces, determined by contact angle and electrokinetic measurements, are related to the chemical structure of the surface region investigated by X-ray photoelectron spectroscopy (XPS).

2. EXPERIMENTAL

2.1. Methods

2.1.1. Contact angle measurements. The contact angles were determined by the sessile drop method using the G 40 contact angle measuring system of Krüss GmbH. Two techniques were used to adjust the liquid droplets at the macroscopic polymer sheets. In the first case, the liquid droplet was adjusted with a microsyringe at the polymer sheet without holding the syringe in the droplet. The contact angle Θ was measured when the liquid droplet rested on the surface as observable with the eyes. The angles observed in this way are of the advancing type. Maximum advancing angles Θ_a were determined while holding the micrometer syringe in the liquid droplet and adding more liquid to the droplet until a steady value of Θ_a was obtained. To measure the receding angles Θ_r, liquid was withdrawn from the droplet by means of the micrometer syringe. These two kinds of angles are termed 'recently advanced' and 'recently retreated' [21]. At least five liquid droplets were placed on each polymer surface. The contact angles reported in this paper are mean values of at least five measurements per polymer sheet for each solid/liquid combination. The relevant properties of the liquids used are given in Table 4.

The contact angles of pure test liquids (water, methylene iodide, and ethylene glycol) were used to calculate the dispersive and polar components of the surface free energy according to Owens and Wendt [9] and the acid–base parameters according to Van Oss *et al.* [10] (see equations in Table 3).

Table 4.
Surface free energy parameters (in mJ/m^2) of the test liquids used for contact angle measurements

Liquid	γ_l	γ_l^{d} [a]	γ_l^{LW} [b]	γ_l^{+} [b]	γ_l^{-} [b]
Water	72.8	21.8	21.8	25.5	25.5
Ethylene glycol	48.0	29	29	1.92	47.0
Methylene iodine	50.8	48.5	50.8		

[a] [21].
[b] [20].

2.1.2. Zeta potential measurements. The electrical properties of polymer surfaces can be investigated by electrokinetic methods. They provide information on the zeta potential, ζ, which is the electrical potential at the shear plane between a charged surface and a liquid, moving with respect to each other. The origin of charges at polymer surfaces in contact with aqueous electrolyte solutions may be attributed to the dissociation of acidic or basic surface groups and the preferential adsorption of cations or anions in competition with the adsorption of water (Fig. 1). Because of the stronger adsorption of anions in neutral 1 : 1 electrolytes, the zeta potential is negative for most polymer surfaces.

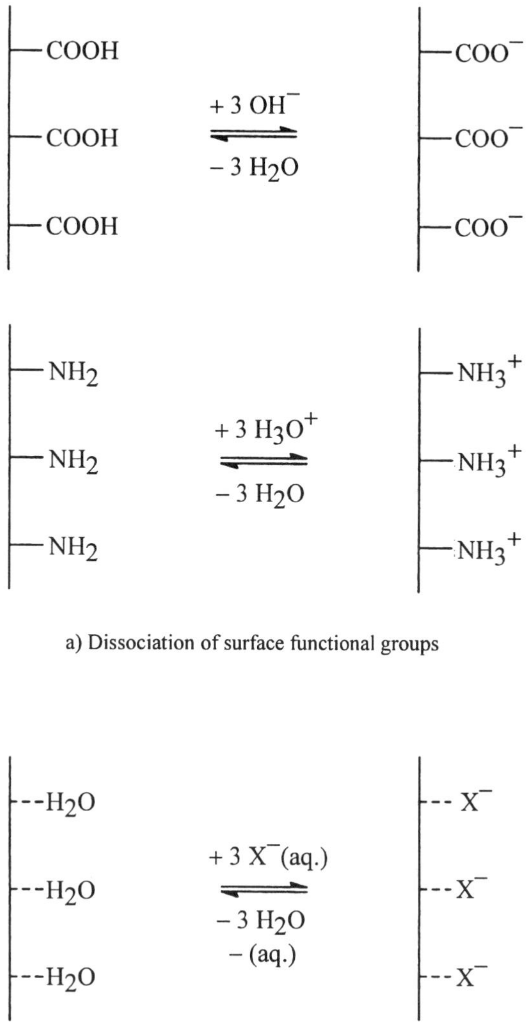

$$\begin{matrix} -COOH \\ -COOH \\ -COOH \end{matrix} \quad \underset{-3\,H_2O}{\overset{+3\,OH^-}{\rightleftharpoons}} \quad \begin{matrix} -COO^- \\ -COO^- \\ -COO^- \end{matrix}$$

$$\begin{matrix} -NH_2 \\ -NH_2 \\ -NH_2 \end{matrix} \quad \underset{-3\,H_2O}{\overset{+3\,H_3O^+}{\rightleftharpoons}} \quad \begin{matrix} -NH_3^+ \\ -NH_3^+ \\ -NH_3^+ \end{matrix}$$

a) Dissociation of surface functional groups

$$\begin{matrix} \text{--}H_2O \\ \text{--}H_2O \\ \text{--}H_2O \end{matrix} \quad \underset{\substack{-3\,H_2O \\ -\,(aq.)}}{\overset{+3\,X^-(aq.)}{\rightleftharpoons}} \quad \begin{matrix} \text{--}X^- \\ \text{--}X^- \\ \text{--}X^- \end{matrix}$$

b) Preferential adsorption of dehydrated anions

Figure 1. Mechanism of electrochemical double-layer formation at the polymer/electrolyte interface.

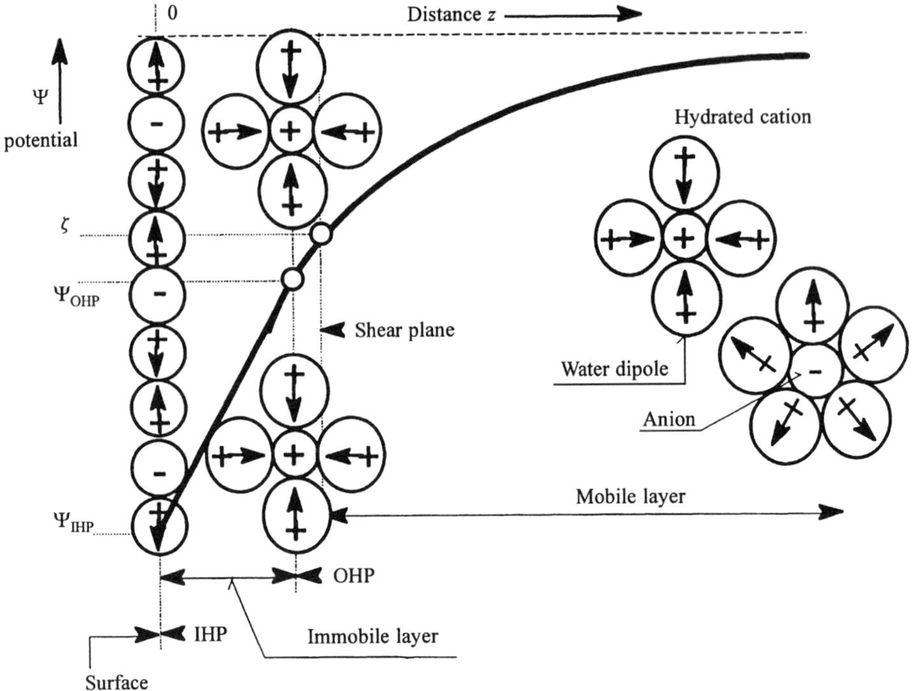

Figure 2. Build-up of the electrochemical double layer according to the Gouy–Chapman–Stern–Grahame model.

According to Stern [23], the charges at a solid surface are compensated by charges attributed partly to ions strongly adsorbed at the interface and partly to ions situated at a greater distance from the interface due to their thermal movement (Fig. 2).

This electrical double layer is divided into an immobile or Stern layer and a diffuse layer. Externally applied electrical or mechanical forces cause relative movement between the fixed part and the diffuse part of the double layer in electrokinetic experiments. By measuring the zeta potential as a function of the known properties of the bulk liquid phase (pH value, ionic strength) and applying the model of the electrical double layer by Börner and Jacobasch [13], any solid substance can be characterized with respect to acidic or basic groups at the surface and the occurrence of dispersion or Coulomb forces [12–14].

While the electrokinetic properties of colloidal particles have been widely investigated by electrophoretic mobility measurements, the characterization of the surface of macroscopic specimens is still rather uncommon. An automatic device, commercially available as Electrokinetic Analyser EKA of Anton Paar KG, Graz, Austria, allows a rapid and reproducible measurement of the zeta potential by streaming potential measurements. This electrokinetic method is characterized by the fact that an electrical potential (streaming potential) is measured as a function of the pressure decay in a single capillary or capillary bundle when an electrolyte solution is pumped through it. By using various measuring cells, the zeta potential of coarse polymer particles and

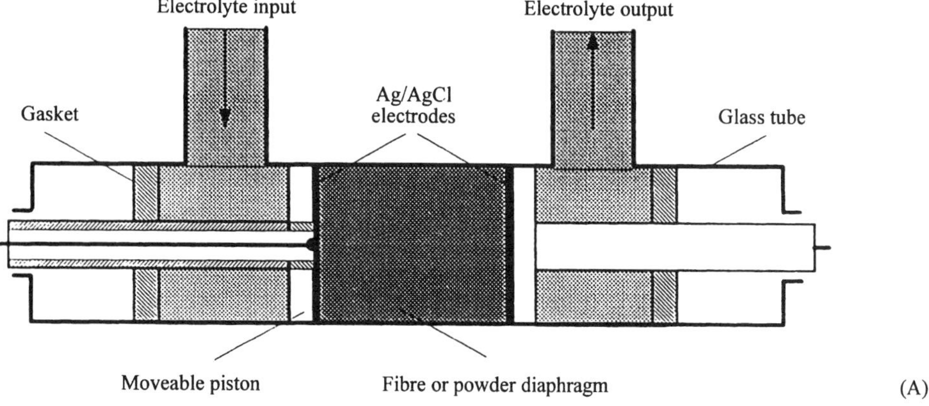

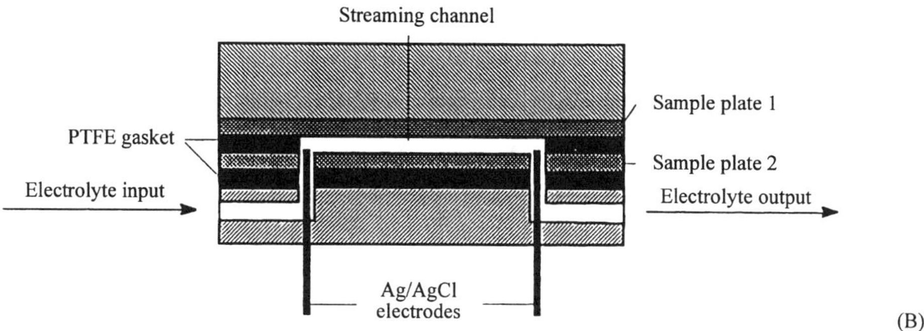

Figure 3. Measuring cells of the electrokinetic analyser for determining the zeta potential of coarse polymer particles (granulated material) (A) and polymer sheets (B).

of injection-moulded polymer plates and sheets can be determined [24] (see Fig. 3). Details of this experimental technique have been described elsewhere [12–15, 25].

The zeta potential values of the polymer surfaces investigated in the present study were determined in 10^{-3} mol/l KCl solutions of varying pH values. Generally, the presence of acidic or basic functional groups corresponds to functions $\zeta = \zeta(\text{pH})$, as can be seen schematically in Fig. 4A. The increase in negative zeta potential with increasing pH is due to the increased dissociation of acidic surface groups. In the case of basic groups, the number of positively charged groups increases with decreasing pH. Complete dissociation of acidic or basic functional groups is related to the plateau in the ζ–pH curve. Non-polar polymers yield ζ–pH plots without plateaus due to the lack of dissociable groups. The increase in negative zeta potential with pH is caused by the increasing adsorption of hydroxyl (OH^-) ions (see Fig. 4B.)

According to Hunter [11], the pK values of dissociating functional groups of solids can be calculated from ζ–pH plots. Mathematical modelling of the electrokinetic

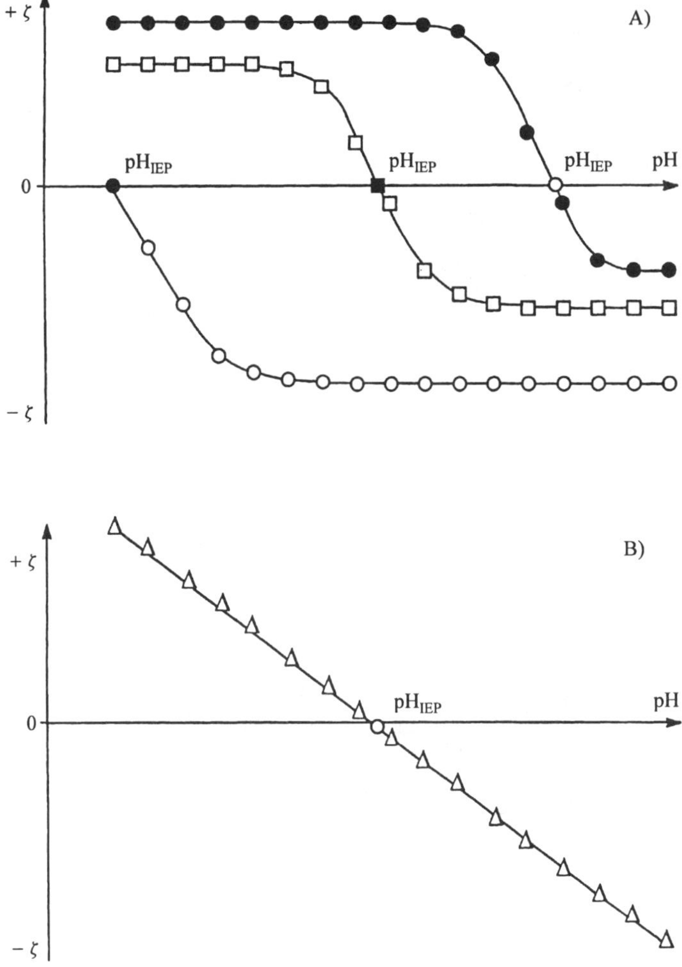

Figure 4. Schematic representation of zeta potential vs. pH plots of solids. (A) ●, Dissociable basic functional groups; □, dissociable acidic and basic functional groups (amphoteric behaviour); ○, dissociable acidic functional groups. (B) △, Non-polar surface without dissociable functional groups.

double layer [13], by means of the three parameters zeta potential ζ, bulk concentration of electrolyte ions c_i^{bulk}, and pH as an expression of the bulk concentration of H^+ and OH^- ions, respectively, permits calculation of the adsorption potentials for all ionic species (Φ_i) of electrolyte solutions (e.g. K^+, Cl^-, H^+, OH^- ions in the case of potassium chloride solutions), the charge densities (σ^{k}) in the two Helmholtz planes and the diffuse layer, and the integral capacity of the Stern layer (C^{SP}). Figure 5 shows that the adsorption potentials of H^+ and OH^- ions correspond to the dissociation constants of basic and acidic surface groups (pK_{B} and pK_{A} values, respectively).

If acidic or basic groups are present at the polymer surface and the build up of the electrical double layer is governed by dissociation processes, then the adsorption potential of H^+ and the corresponding pK_{B} value are parameters describing the basic

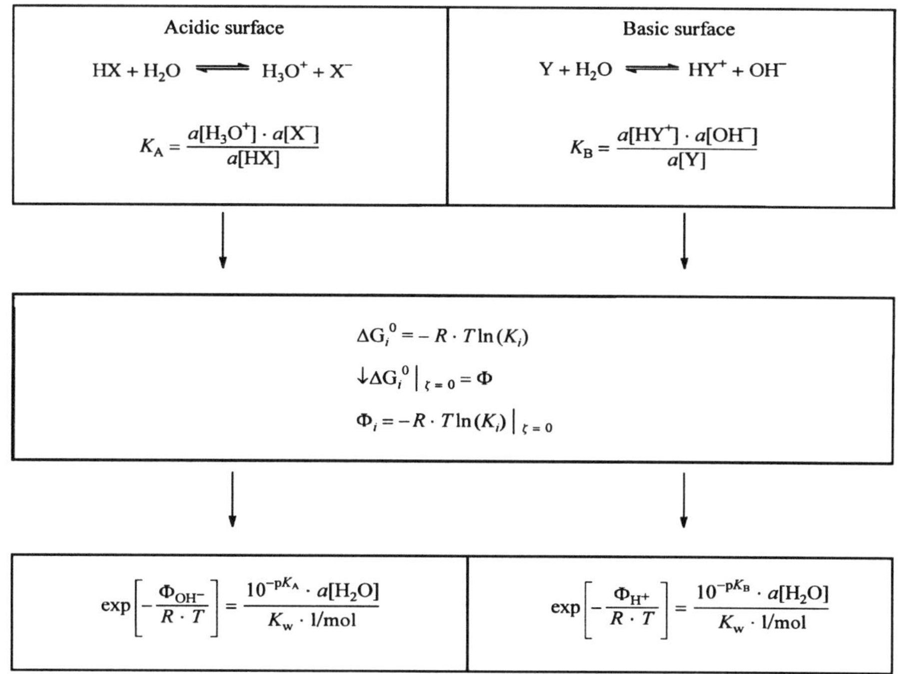

Figure 5. Relations between the adsorption potentials Φ_i of H^+ and OH^- ions and the dissociation constants of basic and acidic surface groups (pK_B and pK_A values).

character of the polymer surface. The adsorption potential of OH^- and the corresponding pK_A value are parameters describing the acidic character of the surface. Table 5 gives pK values of strong to weak acids or bases and the corresponding adsorption potentials. Thus, it is possible to distinguish between acidic and basic functional groups with varying acid or base strength [13, 26].

2.1.3. X-ray photoelectron spectroscopy (XPS). The surface-chemical structure was determined by XPS using a VG Escalab 220i surface analytical instrument with non-monochromatized Mg-$K_{\alpha 1,2}$ radiation (300 W at 15 kV). The base pressure in the analytical system was below 10^{-8} mbar. All spectra were taken at an electron take-off angle of 90° unless otherwise specified in the text. The measured XPS spectra were not smoothed prior to deconvolution. No charge compensation was applied. All spectra were referenced to the C 1*s* peak for C_xH_y, which was assigned a value of 285.00 eV. The curve-fitting ECLIPSE routine was used, which allows a variation of parameters such as the Gaussian/Lorentzian ratio, the full width at half-maximum, the peak positions, and the height of the peak. The curve-fitting quality was evaluated by the χ^2 convergence. The ratios O/C and N/C are the ratios of their peak areas in the XPS spectrum.

2.1.4. Flame and low pressure plasma treatments. Injection-moulded sheets of PP–EPDM were flame-treated in an automatic device developed by AKZO. The

Table 5.
pK values and corresponding adsorption potentials Φ_i [12]

	pK_A	Φ_{OH^-} (kJ/mol)	Φ_{H^+} (kJ/mol)	pK_B	
Very strong acid	−3	−106.0	+7.10	−17	Very weak base
	−2	−100.3	+1.45	+16	
	−1.74	−98.84	−0.02	+15.74	
Strong acid	−1	−94.65	−4.21	+15	Weak base
	0	−89.00	−9.86	+14	
	+1	−83.35	−15.51	+13	
	+2	−77.69	−21.16	+12	
	+3	−72.04	−26.82	+11	
	+4	−66.39	−32.47	+10	
	+4.5	−63.56	−35.30	+9.5	
Moderately strong acid	+5	−60.74	−38.12	+9	Moderately strong base
	.	.	.	.	
	.	.	.	.	
	+8	−43.78	−55.08	+6	
	+9	−38	−60.74	+5	
Weak acid	+10	−32.47	−66.39	+4	Strong base
	.	.	.	.	
	.	.	.	.	
	+15	−4.21	−94.65	−1	
	+15.74	−0.02	−98.84	−1.74	
Very weak acid	+16	+1.45	−100.30	−2	Very strong base
	+17	+7.10	−106.00	−3	

flame was obtained using a stoichiometric oxygen (air)/butane mixture (oxygen content 22%). The polymer sheets passed the flame with a velocity of 400 mm/s. The distance from the polymer surface to the upper planar flame front was maintained at 3 cm.

Low pressure plasma treatments were carried out using commercial apparatus of Technics Plasma GmbH and Plasma Electronic GmbH. The plasma parameters were as follows: excitation frequency 13.56 MHz or 2.45 GHz; pressure 0.3 mbar. The treatment time ranged from 10 to 60 s and the power was 200 W.

The effect of pretreating injection-moulded PP–EPDM sheets on their paintability was investigated by a twist-o-meter test [27, 28] after coating the PP–EPDM with water-based primers.

2.1.5. Chemical modification. Finely granulated ethylene vinyl acetate (EVA) copolymer was chemically modified with carboxyl group-containing monomers by solid-state grafting. This technique was developed by Buna AG [29]. The reactive monomers used were acrylic acid, maleic anhydride, and mixtures of these monomers. The commercially available products were applied as compatibilizers in polymer blends.

Table 6.
The polymer materials investigated

Trade name of material	Manufacturer	Composition
Keltan TP 2600	DSM	PP–EPDM
Keltan TP 0550		PP–EPDM
Hostalen PPR 8018 A	Hoechst	PP–EPDM
Hostalen PPR 8018 A, HL 25		PP–EPDM, HALS additive[a]
Hostalen PPN 8018 B		PP–EPDM
Hostalen PPN 8018 B, HL 25		PP–EPDM, HALS additive[a]
Chimasorb 944 FL	Ciba Geigy	Oligomeric HALS additive
Tinuvin 770 DF	Ciba Geigy	Monomeric HALS additive
Scona® TPEV	Buna	Grafted EVA (14 wt% vinyl acetate)[b]

[a] HALS = Hindered amine light stabilizer (standard additive).
[b] Solid-state grafting with carboxyl group-containing monomers (acrylic acid, maleic anhydride, and their mixtures).

2.2. *Materials*

The polymeric materials studied in this work were commercially available polypropylene/ethylene-propylene-diene monomer rubber (PP–EPDM) blends (Keltan of DSM and Hostalen of Hoechst) and chemically modified EVA copolymers. EPDM is an elastomeric terpolymer of ethylene, propylene, and a non-conjugated diene. Injection-moulded sheets (120×50×2 mm) of PP–EPDM were pretreated by flame and oxygen plasma. Hindered amine light stabilizers (HALS) were added as standard additives to Hostalen.

1 K Hydroprimer H0014 and 2 K Hydroprimer 65258 were applied to coat the PP–EPDM. The primers are synthetic resin dispersions of Wörwag GmbH. EVA was chemically modified as finely granulated material. In some cases, this modified granular material was melted and pressed against PTFE. Further information about the materials is given in Table 6.

3. RESULTS AND DISCUSSION

3.1. *Surface characterization of flame- and plasma-treated PP–EPDM blends*

Because of the low cost and the possibility of recycling, polyolefins are increasingly used in blends with other polymers. The low surface free energy and the absence of reactive functional groups make PP–EPDM blends, used as automotive bumpers, very difficult to paint. The plastics processing industry uses pretreatment methods to create functional groups at the surface. In addition to flame treatment, low pressure plasma treatment has been increasingly utilized.

Contact angle measurements with water and methylene iodide were used to calculate the dispersive and polar components of the surface free energy of untreated and differently treated PP–EPDM samples. The results are summarized in Table 7, together

Table 7.
Dispersive (γ_s^d) and polar (γ_s^p) components of the surface free energy (γ_s) of differently pretreated PP–EPDM injection-moulded sheets and of cured primer coatings (determined according to Owens and Wendt [9])

Substrates	γ_s^d (mJ/m^2)	γ_s^p (mJ/m^2)	γ_s (mJ/m^2)	γ_s^p/γ_s (mJ/m^2)
Keltan TP 2600 (untreated)	25.2	0.5	25.7	0.02
Flame-treated	25.9	9.0	34.9	0.26
Plasma-treated[b]	23.3	5.4	28.7	0.19
Keltan TP 0550 (untreated)	26.0	1.1	27.1	0.04
Flame-treated	32.4	7.3	39.7	0.18
Hostalen 8018 A (untreated)	25.0	0.5	25.5	0.02
Flame-treated	30.3	2.5	32.8	0.10
Flame-treated (HALS)[a]	31.4	6.7	38.1	0.17
Plasma-treated[b]	24.1	7.2	31.3	0.23
Plasma-treated[b] (HALS)[a]	27.2	7.0	34.2	0.20
Hostalen 8018 B (untreated)	26.7	0.6	27.3	0.02
Flame-treated	31.8	3.8	35.6	0.10
Flame-treated (HALS)[a]	27.2	17.4	44.6	0.39
1 K Hydroprimer	25.5	4.8	30.3	0.16
2 K Hydroprimer	27.0	5.0	32.0	0.16

[a] HALS additives were additionally added to the PP–EPDM blend.
[b] 2.45 GHz, 0.3 mbar, 200 W, 30 s.

with the surface free energy components, calculated from contact angle measurements of cured primers which were used to paint the PP–EPDM.

These results confirm that the polar component of the surface free energy is increased by flame or plasma treatment of non-polar PP–EPDM surfaces. When hindered amine light stabilizers (HALS) were added to the blend, the effect of the flame treatment could be enhanced. This phenomenon was not observed after plasma treatments.

The geometric mean mixing rule was applied to the system PP–EPDM/water-based primer and interfacial energies were calculated from the dispersive and polar components of the surface free energies of the adhesion partners (Table 8). The results of adhesion strength measurements were compared with these interfacial energies. The Epprecht twist-o-meter test [31] was adopted to measure the strength of adhesion as the maximum torsional stress exerted to separate a cylindrical body which was glued on the primer-coated PP–EPDM samples. A two-component adhesive was used (resin: Araldite AW 106; hardener: HV 953 U, 10:4) [27].

High interfacial energies should give a low adhesion strength, but the opposite was observed for flame-treated PP–EPDM with different polarities (see Table 8). In the following, it is shown that the increase in adhesion strength can be due to acid–base interactions determined quantitatively by zeta potential measurements.

Figure 6 presents zeta potential (ζ) vs. pH plots for untreated PP–EPDM injection-moulded sheets and for flame- and plasma-treated PP–EPDM. Obviously, the ζ–pH

Table 8.
Comparison between the interfacial energies (γ_{12}) calculated from the surface free energies of the adhesion partners (differently treated PP–EPDM and cured primer coatings) and the torsional stress values determined by a twist-o-meter test [27]

Adhesion system	Interfacial energies γ_{12} (mJ/m^2)	Adhesion strength (torsional stress)[a] (MPa)
Keltan TP 2600/ 1 K Hydroprimer		
Untreated PP–EPDM	2.0	0 (no adhesion)
Flame-treated	0.7	8.0 ± 0.9
Plasma-treated	1.0	8.0 ± 1.0
Keltan TP 2600/ 2 K Hydroprimer		
Untreated PP–EPDM	2.0	0 (no adhesion)
Flame-treated	0.7	10.0 ± 0.8
Plasma-treated	0.8	9.4 ± 1.2
Hostalen 8018 B/ 1 K Hydroprimer		
Flame-treated	0.4	9.0 ± 1.4
Flame-treated (HALS)	4.0	16.8 ± 2.1
Hostalen 8018 B/ 2 K Hydroprimer		
Flame-treated	0.3	11.1 ± 2.8
Flame-treated (HALS)	3.8	18.7 ± 1.6

[a] [27] and [28].

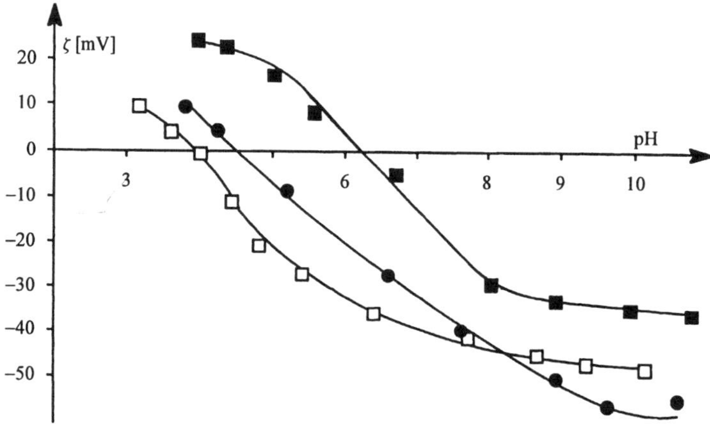

Figure 6. Plot of the zeta potential vs. pH: PP–EPDM (Keltan TP 2600) injection-moulded sheets investigated by streaming potential measurements in 10^{-3} mol/l KCl solution. (●) Untreated; (■) flame-treated; (□) oxygen plasma-treated.

plot of untreated PP–EPDM yields a curve which is typical of a non-polar polymer surface (see Fig. 4). The flame-treated sheet seems to have an amphoteric surface because of the S-shaped curve (cf. Fig. 6 with Fig. 4) and a shift of the isoelectric point (IEP) to higher pH values. The shift of the IEP to lower pH in the case of the plasma-treated sheets is evidence for acidic surface groups. These changes in the ζ–pH plots were observed for all PP–EPDM samples investigated after flame and plasma treatments.

Figure 7 summarizes the ζ–pH plots for several PP–EPDM samples after the flame treatment, indicating an amphoteric surface character because of the S-shaped curves and a shift of the IEP to higher pH values.

Figure 8 shows plots for oxygen plasma-treated PP–EPDM (Hostalen 8018 A) typical of an acidic surface character because of the shift of the IEP to lower pH in

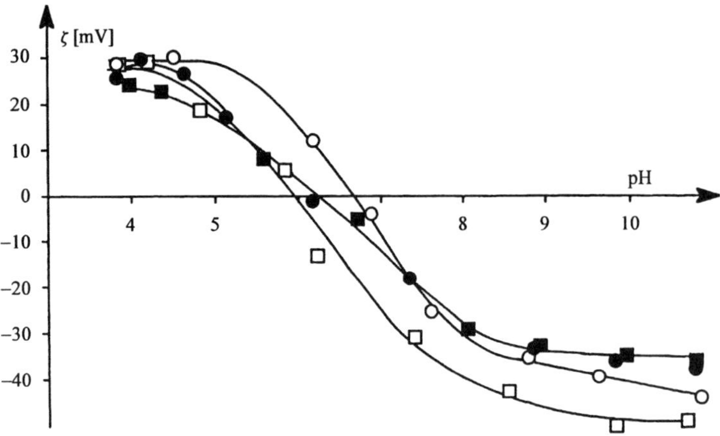

Figure 7. Plot of the zeta potential vs. pH: flame-treated PP–EPDM materials investigated by streaming potential measurements in 10^{-3} mol/l KCl solution. (○) Hostalen PPN 8018 B (HALS); (●) Hostalen PPN 8018 B; (■) Keltan TP 2600; (□) Keltan TP 0550.

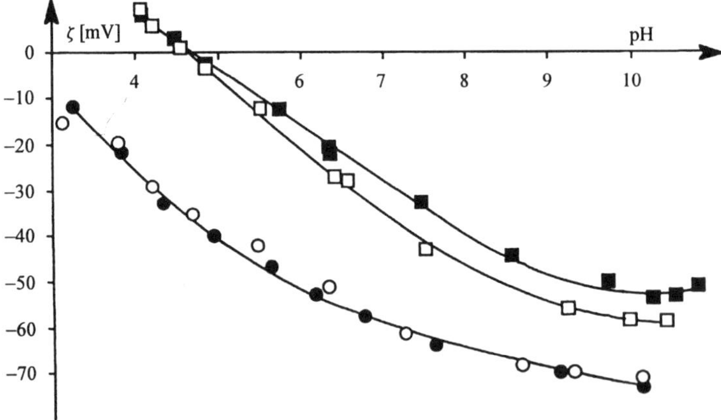

Figure 8. Plot of the zeta potential vs. pH: untreated and oxygen plasma-treated PP–EPDM materials investigated by streaming potential measurements in 10^{-3} mol/l KCl solution. (□) Untreated Hostalen PPR 8018 A; (■) untreated Hostalen PPR 8018 A (HALS); (○) oxygen plasma-treated Hostalen PPR 8018 A; (●) oxygen plasma-treated Hostalen PPR 8018 A (HALS).

Table 9a.
Surface parameters of differently treated PP–EPDM injection-moulded sheets

Sheet material	Adsorption potentials calculated from ζ–pH plots (kJ/mol)		Surface free energy components determined by contact angle measurements (mJ/m^2)				Elemental surface composition determined by XPS	
	Φ_{H^+}	Φ_{OH^-}	γ_s^d	γ_s^p [a]	γ_s^- [b]	γ_s^+ [b]	O/C	N/C
Keltan TP 2600								
Untreated	−28.9	−56.9	25.3	0.4	0.5	0.09	0.034	0
Plasma-treated	−26.3	−62.8	23.3	5.4	11.6	0.1	0.149	0
Flame-treated	−49.9	−60.3	25.9	9.0	30.3	2.6	0.154	0.013
Hostalen PPN 8018 B								
Untreated (HALS)	−34.7	−47.7	30.5	1.3				
Flame-treated (HALS)	−48.1	−52.3	27.2	17.4	31.1	0.04	0.211	0.026
Flame-treated (without HALS)	−40.3	−60.0	31.8	3.8	0.5	3.5	0.142	0.013
Hostalen PPR 8018 A								
Untreated (HALS)	−35.5	−51.3	22.9	1.2			0.022	0
Plasma-treated (HALS)	−29.3	−60.2	27.2	7.0			0.273	0.007
Plasma-treated (without HALS)	−28.2	−60.6	24.1	7.2			0.323	0.003
Flame-treated (HALS)	−47.9	−55.8	31.4	6.7	10.1	0.04	0.125	0.014
Flame-treated (without HALS)	−47.2	−59.1	30.3	2.5	1.4	0.8	0.109	0.012

[a] Determined according to Owens and Wendt [9].
[b] Determined according to Van Oss *et al.* [10].

comparison with the untreated surface. Similar ζ–pH plots have been reported in the literature [7] for polystyrene plates which were treated by oxygen or NH_3 RF plasma discharge and characterized by streaming potential measurements.

The results of quantitative evaluation of the thermodynamic parameters from the ζ–pH plots are summarized in Table 9a, together with the results of contact angle and X-ray photoelectron spectroscopic measurements.

Φ_{H^+} and Φ_{OH^-} calculated for the untreated PP–EPDM can be due to the adsorption of ions at the non-polar PP–EPDM surface which is in contact with an aqueous electrolyte solution. It can be seen that the plasma treatment of PP–EPDM results in an increase of the parameter Φ_{OH^-} (that means Φ_{OH^-} becomes more negative). This can be due to the dissociation of acidic surface groups in addition to the adsorption of ions at non-polar surface groups. Φ_{H^+} is not changed. Flame treatments change both Φ_{OH^-} and Φ_{H^+}, which means that both acidic and basic surface groups are formed. The changes in the acid–base character of pretreated surfaces correlate with an increase of the oxygen and nitrogen concentrations in the surface region, determined by XPS, and an increase in the polar component of the surface free energy calculated according to the method of Owens and Wendt from contact angle measurements using water and methylene iodide (see Table 9b). A direct correlation between the acidic and basic parameters determined by zeta potential measurements, on the one hand, and the acidic and basic parameters calculated from contact angle measurements, on the other hand, could not be found (Tables 9a and 9b). It should be noted that the acid–base properties determined by zeta potential measurements can be due to the dissociation of functional surface groups according to the Brönsted theory, whereas the Van Oss, Good, and Chaudhury approach determines the electron donor–electron acceptor character of a given surface according to the Lewis concept of acids and bases.

There are indications from XPS measurements that the acidic surface character of plasma-treated PP–EPDM determined by zeta potential measurements is connected

Table 9b.
Contact angles of the test liquids used to calculate the surface free energy parameters according to Owens and Wendt [9] and Van Oss *et al.* [10]

PP–EPDM	Contact angle (degrees)		
	Water	Methylene iodide	Ethylene glycol
Keltan TP 2600 (untreated)	103 ± 1	65 ± 1	76 ± 2
Flame-treated	76 ± 4	57 ± 5	75 ± 4
Plasma-treated	86 ± 6	64 ± 4	72 ± 4
Hostalen 8018 A			
Flame-treated	89 ± 2	53 ± 1	56 ± 2
Flame-treated (HALS)	77 ± 1	48 ± 3	53 ± 2
Hostalen 8018 B			
Flame-treated	84 ± 3	50 ± 5	38 ± 4
Flame-treated (HALS)	61 ± 5	51 ± 3	49 ± 3

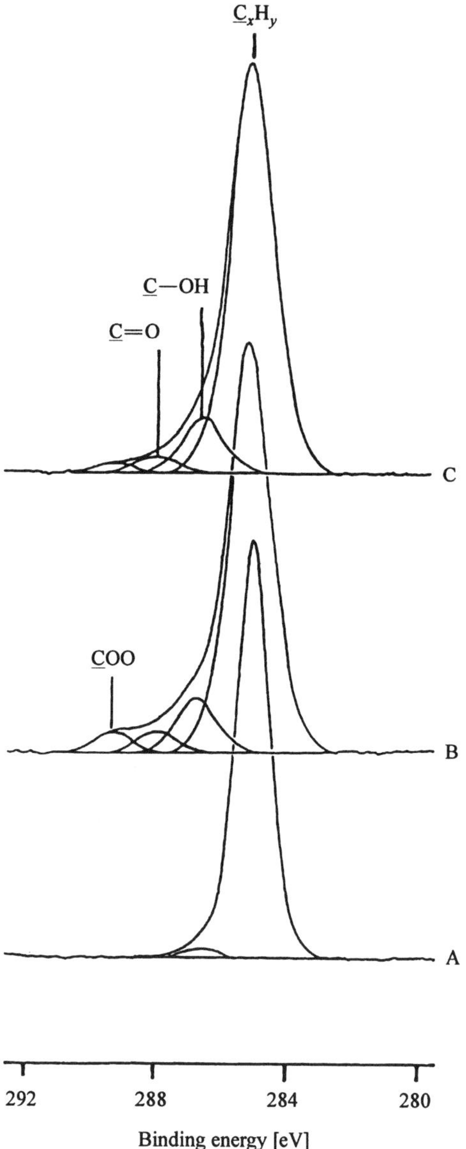

Figure 9. Curve-fitted C 1*s* XPS spectra: (A) untreated PP–EPDM surface (Hostalen PPR 8018 A HALS); (B) after oxygen plasma treatment; (C) after flame treatment.

with an increased concentration of carboxyl groups in the surface region in contrast to flame-treated surfaces (Fig. 9).

The basic character of the flame-treated surfaces could be strengthened if light stabilizers (HALS) were added to the PP–EPDM blend. The increase in the Φ_{H^+} value and especially the increase in the γ_s^- parameter (Table 9a) are indications of this behaviour. XPS measurements show that these changes are connected with the occurrence of protonated nitrogen in the surface region (Fig. 10).

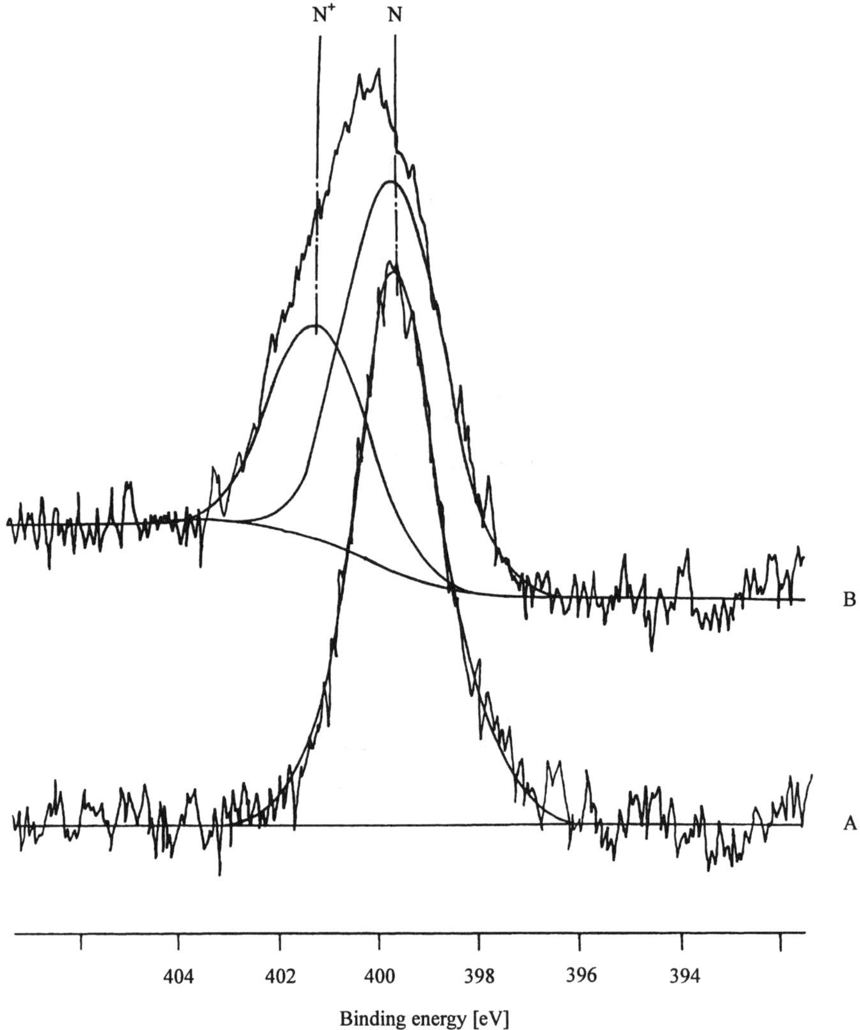

Figure 10. Curve-fitted N 1*s* XPS spectra of the PP–EPDM surface (Hostalen PPR 8018 A HALS): (A) after plasma treatment; (B) after flame treatment.

Figure 11 shows the ζ–pH plots for cured water-based primers which are industrially applied in the painting process of automotive bumpers pretreated by flames or low pressure plasma discharge.

These curves are typical for acidic surfaces. The calculated Φ_{OH^-} values are −67.8 kJ/mol (1 K Hydroprimer) and −71.2 kJ/mol (2 K Hydroprimer). According to Table 5, these adsorption potentials correspond to strong acidic groups at the surface.

Thus, the increase in the basic character of flame-treated PP–EPDM surfaces due to the addition of HALS should result in stronger acid–base interactions with the acidic water-based primers. Whereas surface energetic considerations based on contact angle measurements were not able to predict the strong interaction between flame-treated

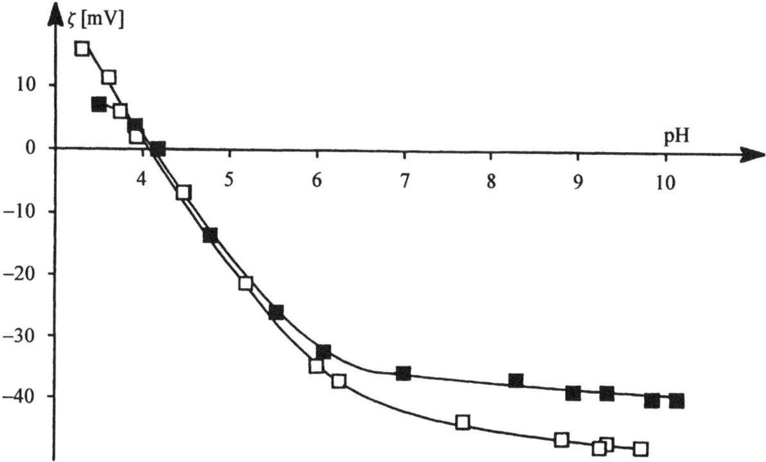

Figure 11. Plot of the zeta potential vs. pH: cured water-based primers investigated by streaming potential measurements in 10^{-3} mol/l KCl solution. (○) 1 K Hydroprimer; (□) 2 K Hydroprimer.

Hostalen 8018 B (HALS) and the primers (see Table 8), acid–base interactions determined by zeta potential measurements explain the good adhesion of these systems.

3.2. Surface characterization of EVA copolymers grafted with carboxyl group-containing monomers

Non-polar EVA copolymers were chemically modified by solid-state grafting with carboxyl group-containing monomers [29]. It is important to know the efficiency of this chemical modification technique because these materials are applied as compatibilizers in polymer blends. The unmodified EVA copolymer was a granulated material whose morphology was not changed by the grafting process. Thus, granular EVA materials grafted with different reactive monomers had to be characterized by physico-chemical methods. At first, the wetting behaviour of the coarse powders was studied by means of a capillary penetration technique [30]. It was found that the majority of the grafted EVA powders were completely wetted by selected polar (benzyl alcohol) and non-polar (α-bromonaphthalene) test liquids, in contrast to the untreated EVA. From these measurements it could be concluded that the grafted powders had higher surface free energies due to the presence of polar surface groups. However, it was not possible to differentiate between the grafted EVA copolymers. Figure 12 shows ζ–pH plots of EVA powders grafted with increasing concentration of maleic anhydride (MA).

The measuring cell was filled with coarse polymer particles, as shown in Fig. 3; the electrolyte solution was pumped through this capillary system and the streaming potential was measured. As can be seen from Fig. 12, the curve of the untreated EVA, which shows a ζ–pH plot typical of a non-polar polymer surface, is clearly changed by the modification with MA. The isoelectric points were shifted to lower pH values and plateaus could be detected. These curves are the typical ζ–pH plots of surfaces with acidic functional groups (cf. Fig. 12 with Fig. 4). Figure 13 shows ζ–pH plots

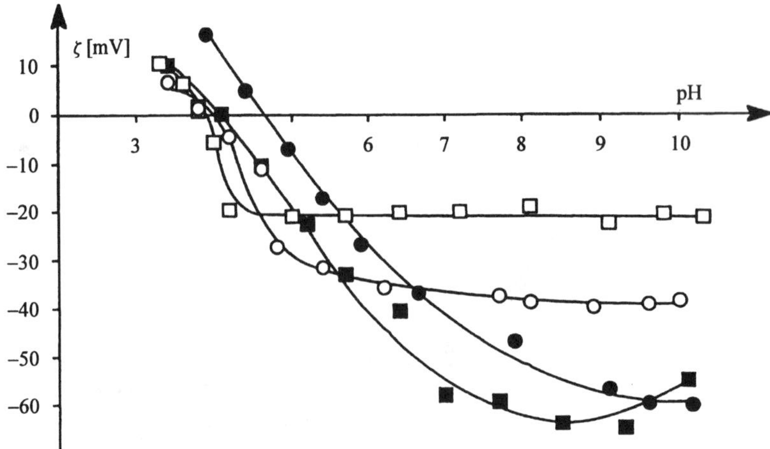

Figure 12. Zeta potential vs. pH of non-grafted and MA-grafted EVA powders (TPEV-Buna) in 10^{-3} mol/l KCl solution. EVA grafted with increasing amount of MA: (●) non-grafted EVA; (■) 0.1 MA; (□) 0.5 MA; (○) 1.0 MA (MA = maleic anhydride).

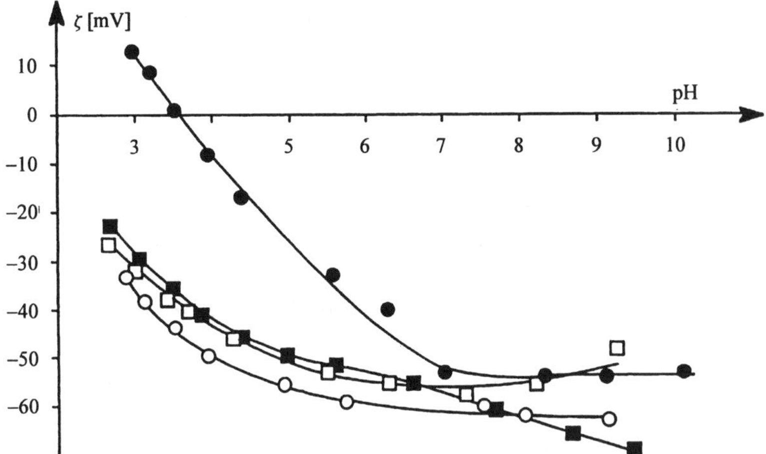

Figure 13. Plot of the zeta potential vs. pH: EVA copolymers differently modified by solid-state grafting with carboxyl group-containing monomers investigated by streaming potential measurements in 10^{-3} mol/l KCl solution. EVA grafted with: (●) MA; (○) MA : AA = 2 : 1; (■) MA : AA = 1 : 1; (□) MA : AA = 1 : 2 (MA = maleic anhydride, AA = acrylic acid).

of EVA powders grafted with monomer mixtures of MA and acrylic acid. It can be seen that the acidic surface character could be enhanced additionally by grafting with these monomer mixtures.

Thus, differences in the acidic character of the grafted EVA can be detected by means of zeta potential measurements. Table 10 summarizes the adsorption potentials of H^+ and OH^- ions calculated from ζ–pH plots of EVA powders grafted with different amounts of reactive monomers, together with the advancing and receding water contact angles measured on sheets which were made from the powders by moulding against PTFE. The samples are arranged according to increasing Φ_{OH^-}, and this corresponds to decreasing pK_A values.

Table 10 shows a correlation between increasing acidic character of the powders and a decrease in the receding contact angles of water (see Fig. 14).

Table 10.
Acid–base parameters of grafted EVA copolymer powders together with advancing and receding contact angles of water measured on sheets which were made from these powders by moulding against PTFE

Grafted EVA sample	Acid–base parameters of grafted EVA powders determined by ζ–pH plots				Advancing Θ_a and receding Θ_r contact angles of water		
	Φ_{H^+} (kJ/mol)	pK_B	Φ_{OH^-} (kJ/mol)	pK_A	Θ_r (degrees)	Θ_a (degrees)	$\Delta\Theta$
EVA grafted with MA[a]	−24.5	11.4	−62.2	4.7	83 ± 8	102 ± 6	19
EVA grafted with MA/styrene, ratio 2 : 1	−24.5	11.5	−65.1	4.2	73 ± 11	90 ± 5	17
EVA grafted with MA/styrene, ratio 1 : 1	−20.4	12.1	−66.2	4.0	72 ± 4	102 ± 3	30
EVA grafted with MA[b]	−19.8	12.2	−70.1	3.3	66 ± 9	93 ± 5	27
EVA grafted with MA and AA, ratio 2 : 1	−19.5	12.3	−72.0	3.0	58 ± 6	80 ± 8	22
EVA grafted with MA and AA, ratio 1 : 2	−20.1	12.2	−73.4	2.8	61 ± 6	92 ± 4	31

[a] EVA grafted with half the amount of MA as compared with sample[b].

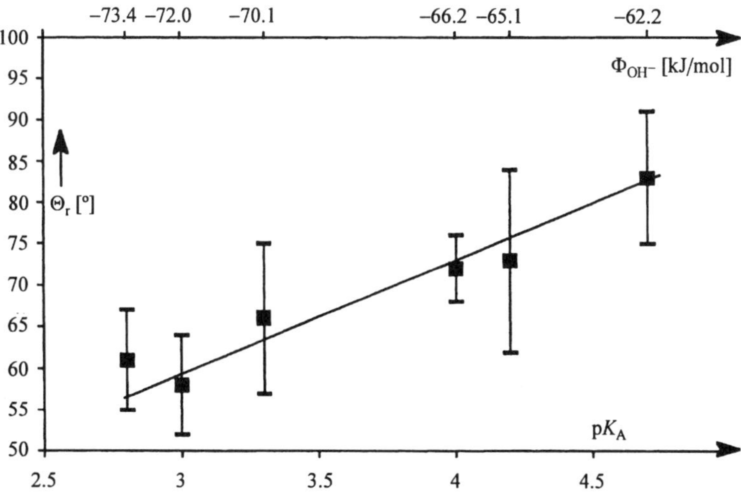

Figure 14. Receding water contact angles (Θ_r) measured on sheets which were made from grafted EVA powders by moulding against PTFE vs. the adsorption potential Φ_{OH^-} and the pK_A value of the grafted powder surfaces determined by streaming potential measurements.

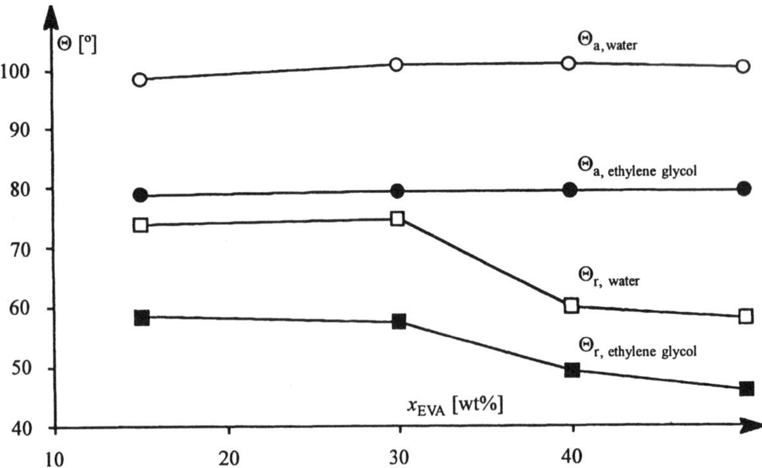

Figure 15. Advancing (Θ_a) and receding (Θ_r) contact angles as a function of the composition of PP–EVA blends.

Thus, it can be concluded that the receding contact angle of water seems to be sensitive to the presence of acidic functional groups in the surface region. As can be seen from Table 10, the standard deviations of the mean contact angles were high, indicating heterogeneous surfaces as a result of the compression moulding of the powders. However, further contact angle measurements on injection-moulded sheets of polypropylene–EVA blends with varying amounts of grafted EVA (Fig. 15) seem to confirm the observation that the receding contact angle of polar test liquids is sensitive to changes in the concentration of carboxyl group-containing species in the blend. Figure 15 also shows that the advancing contact angle was not affected. A possible explanation for this behaviour is the rearrangement of the polar groups in contact with polar liquids which can be detected by a decrease in the receding contact angle [31]. The maximum standard deviation was $\pm 4°$ for the mean advancing contact angles and $\pm 8°$ for the mean receding contact angles.

4. CONCLUSIONS

Electrokinetic measurements are, in addition to contact angle and spectroscopic investigations, suitable techniques for studying the surface characteristics of polymers treated with flame and low pressure plasma discharge or grafted with reactive monomers. The streaming potential method, which was used to measure the zeta potential of polymer sheets and coarse polymer particles (granular material), provides a tool for studying the acid–base properties of polymer surfaces. By measuring the zeta potential vs. the pH in a 10^{-3} mol/l KCl solution, the dissociation constants of acidic or basic functional groups can be determined. Thus, the acid–base properties of surface-modified polymers can be quantitatively evaluated by zeta potential measurements.

The changes in the acid–base character detected by electrokinetic measurements correlated with the elemental composition of the surface region investigated by XPS. The acidic and basic parameters of the surface free energy calculated by the approach of Van Oss and co-workers did not generally correlate with the acidic and basic parameters calculated by zeta potential measurements.

A correlation was found between increasing acidity of grafted EVA copolymer powders determined by zeta potential measurements and the decrease in the receding water contact angle on sheets formed by compression moulding of these powders.

Acknowledgements

We would like to thank the AIF and BMFT for their financial support for this work and the various companies, especially Hoechst, DSM, Mercedes Benz, Pebra, Wörwag, Herberts, Ciba Geigy, and Buna, for making available the materials investigated.

REFERENCES

1. D. Briggs, in: *Surface Analysis and Pretreatment of Plastics and Metals*, D. M. Brewis (Ed.), pp. 199–226. Applied Science Publishers, London (1982).
2. J. Konar, A. K. Sen and A. K. Bhowmick, *J. Appl. Polym. Sci.* **48**, 1579–1585 (1993).
3. F. Garbassi, M. Morra, E. Occhiello, L. Barino and R. Scordamaglia, *Surface Interface Anal.* **14**, 585–589 (1989).
4. E. Occhiello, M. Morra, G. Morini, F. Garbassi and P. Humphrey, *J. Appl. Polym. Sci.* **42**, 551–559 (1991).
5. E. Papirer, D. Y. Wu and J. Schultz, *J. Adhesion Sci. Technol.* **7**, 343–362 (1993).
6. E. M. Liston, L. Martinu and M. R. Wertheimer, *J. Adhesion Sci. Technol.* **7**, 1091–1127 (1993).
7. P. G. Rouxhet, A. Doren, J. L. Dewez and O. Heuschling, *Prog. Org. Coat.* **22**, 327–344 (1993).
8. K. L. Mittal and H. R. Anderson, Jr (Eds), *Acid–Base Interactions: Relevance to Adhesion Science and Technology*. VSP, Zeist, The Netherlands (1991).
9. D. K. Owens and R. C. Wendt, *J. Appl. Polym. Sci.* **13**, 1741 (1969).
10. C. J. van Oss, M. K. Chaudhury and R. J. Good, *Chem. Rev.* **88**, 927–941 (1988).
11. R. J. Hunter, *Zeta-Potential in Colloid Science — Principles and Applications*. Academic Press, London (1981).
12. H.-J. Jacobasch and J. Schurz, *Österr. Chem. Zeitschr.* **7–8**, 164 (1987).
13. M. Börner and H.-J. Jacobasch, *Proc. Symp. Electrokinetic Phenomena*, pp. 231–264. Institute of Polymer Research, Dresden (1987).
14. H.-J. Jacobasch, *Angew. Makromol. Chem.* **128**, 47 (1984).
15. H.-J. Jacobasch, *Makromol. Chem., Macromol. Symp.* **75**, 99–113 (1993).
16. W. Gutowski, in: *Fundamentals of Adhesion*, L.-H. Lee (Ed.), pp. 87–133. Plenum Press, New York (1991).
17. A. W. Neumann, *Adv. Colloid Interface Sci.* **4**, 105 (1974).
18. D. H. Kaelble, *Physical Chemistry of Adhesion*, pp. 153–170. Wiley–Interscience, New York (1971).
19. S. Wu, *Polymer Interface and Adhesion*. Marcel Dekker, New York (1982).
20. F. M. Fowkes, *J. Adhesion Sci. Technol.* **1**, 7 (1987).
21. R. J. Good, *J. Adhesion Sci. Technol.* **6**, 1269–1302 (1992).
22. J. K. Spelt and A. W. Neumann, *Langmuir* **3**, 588 (1987).
23. O. Stern, *Z. Elektrochem.* **30**, 508 (1924).
24. J. Schurz, Ch. Jorde, V. Ribitsch, H.-J. Jacobasch, H. Körber and R. Hanke, *GIT Z. Laboratoriums-technik* **30**, 98 (1986).
25. H.-J. Jacobasch, *Oberflächenchemie Faserbildender Polymerer*. Akademie-Verlag, Berlin (1984).

26. H.-J. Jacobasch, K. Grundke, St. Schneider and F. Simon, *J. Adhesion* **40** (1994) (in press).
27. Unpublished results of U. Schatton, Mercedes-Benz (1993).
28. W. Boller and A. G. Epprecht, *Farbe Lack* **80**, 936–941 (1974).
29. J. Gerecke, D. Wulff and K. Thiele, Pat. DD 300977 A7 (1989).
30. K. Grundke, M. Börner and H.-J. Jacobasch, *Colloids Surfaces* **58**, 47–59 (1991).
31. F. J. Holly and M. F. Refojo, *J. Biomed. Mater. Res.* **9**, 315–326 (1975).

Polymer Surface Modification: Relevance to Adhesion, pp. 455–478
K. L. Mittal (Ed.)

Surface characterizations of modified polyethylene pulp and wood pulps fibers using XPS and inverse gas chromatography

HALIM CHTOUROU,[1] BERNARD RIEDL[1,*] and BOHUSLAV VACLAV KOKTA[2]

[1]*Centre de Recherche en Sciences et Ingénierie des Macromolécules, Département des Sciences du Bois et de la Forêt, Faculté de Foresterie et de Géomatique, Université Laval, Ste-Foy, Québec, G1K 7P4 Canada*

[2]*Centre de Recherche en Pâtes et Papiers, Université du Québec à Trois-Rivières C.P.500, Trois-Rivières, Québec, G9A 5H7 Canada*

Revised version received 1 August 1994

Abstract—The fiber/fiber interface influences strongly the mechanical properties of a composite paper material. Polyethylene (PE) pulp fiber was treated, for surface modification to improve adhesion with conventional paper fiber, using ozone in aqueous medium, and fluorinated gases. X-ray photoelectron spectroscopy (XPS) was used to analyze the surface of conventional paper fiber as well as the variation of the chemical composition and the O/C atomic ratio, within the surface level of the PE pulp fiber after the treatments. The thermodynamic surface characteristics of the fiber were then determined using the inverse gas chromatography (IGC) technique at infinite dilution. Following Fowkes' approach, acid/base surface characteristics were obtained. The variation of surface thermodynamic properties due to treatments were compared to the variation of the chemical compositions as obtained by XPS.

Keywords: PE pulp fiber; ozonation; fluorination; lignocellulosic pulps; inverse gas chromatography; X-ray photoelectron spectroscopy; surface characteristics.

1. INTRODUCTION

Synthetic polymers are playing an increasingly important role in the manufacture of materials originally based on natural polymers. Polyethylene (PE) and polypropylene (PP) pulps, a new application for synthetic polyolefins, are designed to be blended in all proportions with conventional wood pulps and made into papers using conventional paper-making equipment. The resulting papers should differ significantly from pure cellulosic or synthetic papers.

*To whom correspondence should be addressed.

In this application, and in any other multicomponent material, adhesion is the main factor controlling the performance of the resulting product. The nature of adhesion is strongly dependent on the surface properties of the components.

The chemical composition of a surface can be characterized as to its atomic composition by using XPS [1, 3]. In adhesion science, surfaces are characterized by a thermodynamic parameter, called surface tension or surface energy (γ_S). For years it was admitted that the surface energy is the sum of two components: a dispersive component (γ_S^D), due to London dispersion intermolecular interactions; and a non-dispersive, (γ_S^P), polar component due to all others intermolecular interactions [4].

Through wetting/adsorption theory, and contact angle techniques, which were widely used to study adhesion phenomena, these two components of surface energy were very much used to predict the work of adhesion between two phases (solid/solid or solid/liquid). However this assumption, ($\gamma_S = \gamma_S^D + \gamma_S^P$), was not sufficient to predict the overall adhesion and the interphase interactions. In addition, techniques used such as contact angles are mostly limited to smooth and hydrophobic surfaces.

In the last few years, Inverse Gas Chromatography (IGC) has been widely used as new a technique to characterize solid surface energy, not only for regular surfaces but also for rough hydrophilic surfaces [5, 6]. Acidity and basicity of surfaces could be well estimated through this (IGC) technique, and it appears that the specific interactions, responsible for γ_S^P term, could be better described using the concept of acid/base or acceptor/donor interactions [7–9].

As the PE pulp fiber used in this work is designed to be blended with lignocellulosic fiber for composite paper making, we characterize the thermodynamic properties of the lignocellulosic pulps fibers, and the PE pulp fiber treated with ozone [1, 2], and fluorine gases [3] for surface modification, in order to predict interfiber adhesion. We also present XPS results on the PE pulp fibers [1, 3] and on the lignocellulosic pulps fibers, to look for any correlation between XPS and IGC results.

2. MATERIALS AND METHODS

2.1. Materials

In this study, two different kinds of pulps, *synthetic* polyethylene and *natural* lignocellulosic pulps, were used for surface characterizations.

2.1.1. PE pulp fiber. This was received from DuPont company as folded, thick and low density sheets, and was modified by ozone in aqueous solution, and by fluorine gas treatment. This experimental procedure used for these surface treatments of the PE pulp fiber has been extensively described in previous papers [1–3].

2.1.2. Lignocellulosic pulps fibers. Explosion and Kraft pulps obtained from hardwood aspen and birch, respectively, were the two kinds of wood pulps fibers used in this work. In this work we are not concerned with the cooking conditions of these

Table 1.
Characteristics of injected probes[‡] [17]

Probes	a (Å^2)	γ_L^D (mJ/m^2)	$DN^{\ddagger}$ (kcal/mol)	$AN^{\ddagger}$ (arbitrary unit)	Specific characteristics
C_6H_{14}	51.5	18.4	—	—	Neutral
C_7H_{16}	57.0	20.3	—	—	
C_8H_{18}	62.8	21.3	—	—	
C_9H_{20}	68.9	22.7	—	—	
C_6H_6	46.0	26.7	0.1	8.2	Acidic
CH_3NO_2	37.8	26.2	2.7	20.5	
CH_2Cl_2	31.5	27.6	—	20.4	
$CHCl_3$	44.0	25.9	—	23.1	
T.H.F.	45.0	22.5	20.0	8.0	Basic
D.E.E.	47.0	15.0	19.2	3.9	
Acetone	42.5	16.5	17.0	12.5	Amphoteric
Ethyl acetate	48.0	19.6	17.1	9.3	

T.H.F. and D.E.E. are tetrahydrofuran and diethylether, respectively. The values of a, the surface area of the probes (Å^2), are reported in different references (see [8]).

pulps. As these pulps are designed to be blended with the PE pulp to make composite papers, only the surface characteristics of these pulps, influencing the inter-fiber adhesion, were studied.

2.1.3. Probes. The probes used in this study (Table 1) were obtained as chromatographically pure materials from Aldrich and were used without any further purification.

2.2. Methods

2.2.1. Inverse gas chromatography. Chromatographic measurements at infinite dilution were carried out with a Hewlett Packard 5700A apparatus equipped with dual hydrogen flame ionization detectors maintained at 300°C. To ensure flash vaporisation, the temperature at the injection ports of the chromatograph was around 300°C, which is 50°C above the highest boiling point of the alkane probes. The probes used to evaluate the dispersive component of the surface energy of the stationary phases were n-alkanes ranging from n-hexane to n-nonane.

The stationary phases were packed into copper tubing of 1.2 m length and 4.0 mm internal diameter. To maintain the temperature of the column constant, which was controlled within ±0.5°C and monitored with an Omega digital thermometer, a circulating water bath from Julabo (Model UC-5b) was used. Nitrogen was used as the carrier gas, and methane was used to determine the dead volume. The amount of probes injected in the column through the injection port, in the form of vapour, was small enough and at very low concentration in the carrier gas. A 1 μl syringe was used. A small amount, 0.1 μl, from the liquid probes was taken up in the syringe needle, and then pushed out completely from the syringe (horizontally). The end of the needle was carefully observed, and when the probe was nearly completely evaporated, as evidenced by the presence of a very small drop, the piston was pulled a little bit and injected into the injection port. Thus, the very small amount injected allowed us to assume that only probe-adsorbent interactions were studied, the adsorbed molecules being sufficiently far apart to neglect their mutual interactions.

The IGC methodology is hardly new by now. The details has been reported in the literature [6–16]. However, the following should be useful for some readers.

By assuming that the retention mechanism is due only to surface adsorption, the net retention volume (V_N), the fundamental parameter in IGC measurements, which is the volume of the carrier gas required to elute the probes from the column, is given by the following expressions:

$$V_N = K_S A = Q(t_r - t_m), \tag{1}$$

where K_S is the surface partition coefficient of the given probe between the stationary and the mobile phases, A is the surface area (m^2) of the stationary phase, equal to the specific surfaces area multiplied by the sample weight, t_r is the retention time of the injected probe through the column, which is the main parameter, obtained by the IGC, t_m is the retention time of methane, measured as t_r from the peaks maxima. Q is the corrected flow rate of the carrier gas at column temperature and at 760 mm/Hg obtained as follows [10]:

$$Q = Q_0 J \frac{T_c}{T_a}\left(1 - \frac{P_w}{P_a}\right), \tag{2}$$

where Q_0 is the measured flow rate (ml/min, Table 2), T_c and T_a are the experimental and the ambient temperatures (K), respectively, P_a and P_w are the atmospheric and the saturated vapour pressures of water at ambient temperature, and J is the James–Martin compression correction term determined as follows, where P_1 is equal to P_a plus the pressure drop in the column:

$$J = \frac{3}{2}\,\frac{1 - \left(\frac{P_1}{P_a}\right)^2}{1 - \left(\frac{P_1}{P_a}\right)^3}. \tag{3}$$

The specific net retention volume (V_g^0), at 0°C and per gram of adsorbent is given by the following equation [11], where W is the weight of the stationary phase:

$$V_g^0 = \frac{273.15}{T_c}\,\frac{V_N}{W}. \tag{4}$$

The standard free energy, the standard enthalpy, and the standard entropy of adsorption, ΔG_A^0, ΔH_A^0 and ΔS_A^0, respectively, are given by the following expressions [10]:

$$-\Delta G_A^0 = RT \ln V_g^0 + C = RT \ln\left(K_s \frac{P_{s,g}}{\Pi_s}\right), \tag{5}$$

Table 2.
Column description, $-\Delta G_A^0(-CH_2-)$ and γ_S^D obtained from IGC measurements

Samples	W (g)	A (m^2/g)	Q_0, (60°C) (ml/min)	J	Q, (60°C) (ml/min)	$-\Delta G_A^0(-CH_2-)$ (kJ/mol) 50/60/70°C	γ_S^D (mJ/m^2) 50/60/70°C
Untreated PE	1.48	2.68	27.66	0.95	28.57	2.4/2.2/2.1	32.7/28.6/26.0
Ozonated PE (2 h)	2.37	2.20	21.86	0.94	22.17	2.6/2.4/2.2	37.1/33.1/27.6
Ozonated PE (3 h)	1.48	1.83	33.04	0.89	31.94	2.3/2.2/2.0	28.9/27.7/23.6
Fluorinated PE (#1)	2.00	2.49	18.18	0.93	18.21	2.4/2.3/2.1	32.8/29.7/26.6
Fluorinated PE (#2)	1.48	2.16	23.70	0.93	23.90	2.3/2.2/2.1	31.3/27.6/24.7
Explosion pulp	2.62	1.30	22.56	0.99	24.11	1.3/1.1/0.9	9.8/6.9/4.8
Kraft pulp	1.92	1.60	19.52	0.98	20.72	1.1/1.0/1.0	6.8/5.7/5.4

W is the sample weight, and A is its specific surface area.

$$-\Delta H_A^0 = R\,\frac{d\left(\ln V_g^0\right)}{d\left(\frac{1}{T}\right)}, \tag{6}$$

$$-\Delta S_A^0 = \frac{\Delta G_A^0 - \Delta H_A^0}{T}, \tag{7}$$

where $P_{s,g}$ is the adsorbate vapour pressure in the gaseous standard state, equal to 101 kN/m^2, and Π_s is the spreading pressure of the adsorbed film to a reference gas phase state defined by the pressure $P_{s,g}$ of the solute, equal to 0.338 mN/m (reference state of De Boer [12].

A linear variation of $-\Delta G_A^0$ (or $RT \ln V_N$) as a function of the number of carbon atoms in the non-polar probes (n-alkanes) is a common observation. The free energy of adsorption corresponding to one methylene group, $-\Delta G_A^0(-CH_2-)$, was obtained through the injection of a homologous series of n-alkane probes:

$$-\Delta G_A^0(-CH_2-) = RT \ln \frac{V_N(C_{n+1}H_{2n+4})}{V_N(C_nH_{n+2})}. \tag{8}$$

The London dispersive component (γ_S^D) of the surface energy of the stationary phase is calculated by the following equation [10]:

$$\gamma_S^D = \frac{1}{4}\,\frac{\Delta G_A^0(-CH_2-)^2}{\gamma(CH_2)N^2a^2}, \tag{9}$$

where N is Avogadro's number, a is the area of an adsorbed methylene group, and $\gamma(CH_2)$ is the surface energy of pure methylene group surface [10], $\gamma(CH_2) = 35.6 + 0.058\,(293 - T)$, in mJ/m^2.

St-Flour and Papirer [13] found that linearity is usually obtained when plotting $RT \ln(V_N)$ as a function of $\ln(P_0)$, in the case of n-alkane probes. In addition Schultz *et al.* [7] also found a linear relation by plotting $RT \ln(V_N)$ as a function of $a(\gamma_L^D)^{1/2}$, where a is the surface area of the probe and γ_L^D is its dispersive component of surface energy in the liquid state. These correlations are observed since n-alkanes at infinite dilution behave nearly as ideal gases.

In order to determine the contribution of the surface specific interactions to the total surface energy, it is necessary to inject polar probes in the column, in addition to n-alkanes. By the assumption that alkanes exchange only dispersive interactions and also assuming that dispersive and polar components of surface energy are additive, the alkane line may be taken as a reference for the determination of the dispersive component for polar probes. The difference of ordinates between the alkane straight line and the polar probe gives ΔG_A^{SP}, corresponding to specific interactions.

As noted by Fowkes *et al.* [14–16], the specific interactions are all acid–base types. Schultz *et al.* [7] and Martin [8] characterized solid surfaces by an acidic constant K_A and a basic constant K_B, using Gutmann's [17] acid–base concepts, and proposed the following relation:

$$-\Delta H_A^{SP} = K_A DN + K_B AN, \tag{10}$$

where ΔH_A^{SP} is the enthalpy of adsorption corresponding to the specific interactions, K_A and K_B are, respectively, equivalent to AN (acceptor number) and DN (donor number) of molecules. An acid is able to attract electrons while a base releases electrons. The DN, expressed in kilocalories per mole, is the molar enthalpy of interaction between a base and a reference acceptor $SbCl_5$, in a dilute solution of 1,2-dichloroethane. Unlike the DN, the AN is an arbitrary unit set at zero for the shift induced by hexane and at 100 for $SbCl_3$ in a dilute solution of 1,2-dichloroethane [17]. As DN and AN are expressed in different units, recently Riddle and Fowkes [18] redefined a new AN^* on the same scale and with the same units as DN, taking into consideration the amphoteric character of molecules. In this work we have considered DN and AN even if they do not have the same units. However, subsequent analysis and discussion will be done and published taking into consideration AN^* rather than AN.

2.2.2. X-ray photoelectron spectroscopy. The XPS apparatus used for the pulp surface characterization was an ESCALAB Mk II spectrometer, fitted on a Microlab system from Vacuum Generators and equipped with a non-monochromatized dual Mg–Al anode X-ray source. Kinetic energies were measured by a hemispherical electrostatic analyzer ($r = 150$ mm) in the constant pass energy (20 eV) mode with a resolution of 1.1 eV. The base pressure was 10^{-8} to 10^{-6} Torr. The PE fibers, contained in a stainless steel cup, were introduced into the vacuum system. Before analysis, samples were cooled for 20 min at a temperature of about −80°C. This yields minimal sample degradation. All analysis were done with MgK_α source at 300 W.

3. RESULTS AND DISCUSSION

In this study, results obtained on the untreated, the ozone-treated (2 and 3 hours) [1, 2], the fluorinated (levels 1 and 2, the two degrees of treatments done graciously by Air Products Inc. [3]) PE pulp fibers, and the lignocellulosic pulps fibers are discussed. The specific surface area of these pulps, as obtained through nitrogen adsorption isotherms and the BET equations [19], are presented in Table 2. A clear decrease in the specific surface area of the PE fiber is noted after the ozone treatment and the fluorination. The specific surface area decreased from 2.7 m^2/g to 1.8 and 2.2 m^2/g after ozonation (3 h) and fluorination (#2), respectively. The treatment should increase the polar component of the surface energy of the PE pulp fiber. More specific interactions (acid/base type) should be generated at the surface level of each fiber and also between fibers close to each other. These interactions should decrease the dimensions of the pores and the voids, and consequently decrease the specific surface are of the PE pulp fiber.

It should be noted that the specific surface area of the PE pulp fiber measured here was smaller than that reported in the literature [20], 5–20 m^2/g. Whereas the specific surface areas of the Explosion (1.3 m^2/g) and the Kraft pulps fibers (1.6 m^2/g), were relatively comparable with what is reported in the literature [21–22] for wood pulps fibers using krypton. Using krypton, instead of nitrogen, Kamdem and Riedl [21] found a specific surface area of 1.6 m^2/g for a chemico-thermo-mechanical pulp (CTMP) fiber. In addition Gurnagul and Gray [22] measured a specific surface area of 1.5 m^2/g for a thermo-mechanical pulp (TMP) fiber, using krypton.

3.1. IGC results

3.1.1. Dispersive component of the surface energy, γ_S^D. Table 2 shows $-\Delta G_A^0$ ($-CH_2-$), the free energy of adsorption of one methylene group, and the London dispersive components (γ_S^D) of the surface energies of the fibers, calculated through equations (8) and (9), respectively. The free energy of adsorption of one methylene group does not change significantly with the surface treatment of the PE fiber. It was about 2.4, 2.2 and 2.1 kJ/mol, at 50, 60 and 70°C, respectively, for all PE pulps fibers. For cellulosic pulps fibers these values were almost half, 1.2, 1.1 and 1.0 kJ/mol, at the same temperatures, respectively.

Kamdem and Riedl [21] reported a value of 2.6 kJ/mol for the $-\Delta G_A^0(-CH_2-)$, at 25°C, on CTMP pulp fiber. Also, at this temperature, the following values have been reported in the literature, 2.9 kJ/mol for the Kraft papers [23], 2.7 kJ/mol for the TMP fiber [24], 2.99 kJ/mol for cotton cellulose [10, 24], and 2.95 kJ/mol for cellulose paper [25]. A higher interaction between the vapour probe and the sample surface should occur as the sample temperature decreases. The values of $-\Delta G_A^0(-CH_2-)$ found in this work for the Explosion and the Kraft pulps fibers were relatively lower than those mentioned above and this is due to the fact that we worked at 50–70°C and not at 25°C.

γ_S^D also shows a decrease as the temperature of the column (the stationary phases) increased. The lignocellulosic pulps fibers showed dispersive components of surface energies much lower than those for the PE pulps fibers. The surface treatments of the PE pulp fiber did not result in large changes in γ_S^D. However, this is not enough to characterize the effect of these treatments on the complete surface energy of the PE fiber, since polar interactions must also be characterized as follows.

3.1.2. Specific interactions using Schultz's method. Figures 1 to 4 presenting $RT \ln V_N$ versus $a(\gamma_L^D)^{1/2}$, at 60°C as an example, show linear relations for n-alkane probes, as shown by Schultz *et al.* [7]. All other polar probes are on or over these n-alkane reference lines. The vertical distances between polar probe ordinates ($RT \ln V_N$) and the disperse interaction reference lines give ΔG_A^{SP} for each probe and are related to the polar nature of the surface of the stationary phases. As an example, ΔG_A^{SP} values were higher with the fluorinated fiber (#2) than with the ozone-treated fiber (3 h) and the untreated fiber.

It can be concluded from these figures, on a qualitative basis, that: 1) the DuPont PE pulp fiber has relatively no acidic (acceptor) and no basic (donor) characteristics; 2) the fluorinated pulp fiber has stronger basic characteristics and also stronger acidic characteristics — it could be called amphoteric; and 3) the ozone-treated pulp fiber has lower basic and acidic characteristics than fluorinated fiber.

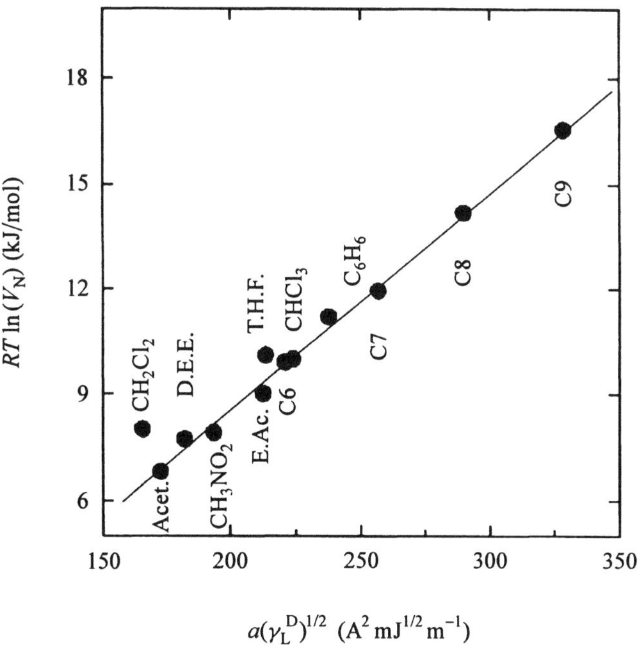

Figure 1. Plot of $RT \ln(V_N)$ versus $a(\gamma_L^D)^{1/2}$ for the *n*-alkanes and the polar probes at 60°C, on the untreated PE pulp fiber.

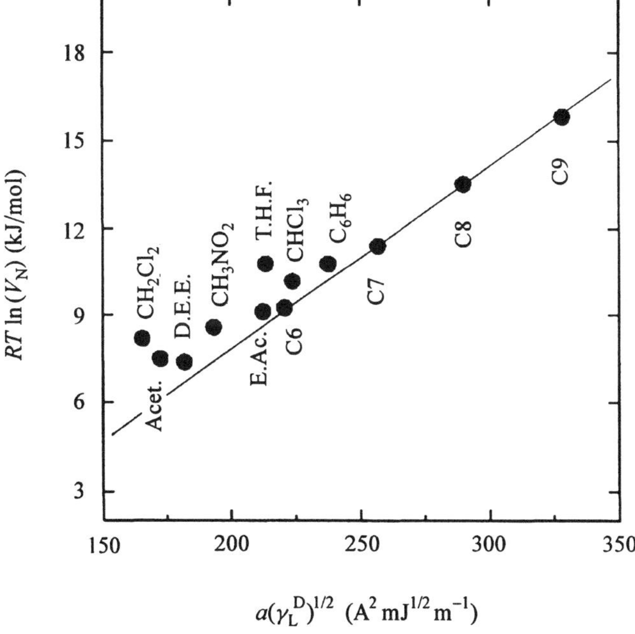

Figure 2. Plot of $RT \ln(V_N)$ versus $a(\gamma_L^D)^{1/2}$ for the *n*-alkanes and the polar probes at 60°C, on the ozonated PE pulp fiber (3 h).

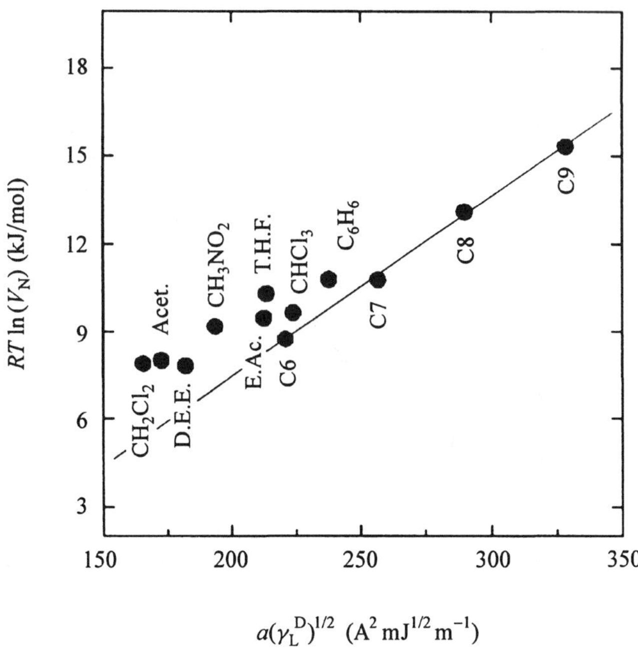

Figure 3. Plot of $RT \ln(V_N)$ versus $a(\gamma_L^D)^{1/2}$ for the *n*-alkanes and the polar probes at 60°C, on the fluorinated PE pulp fiber (#2).

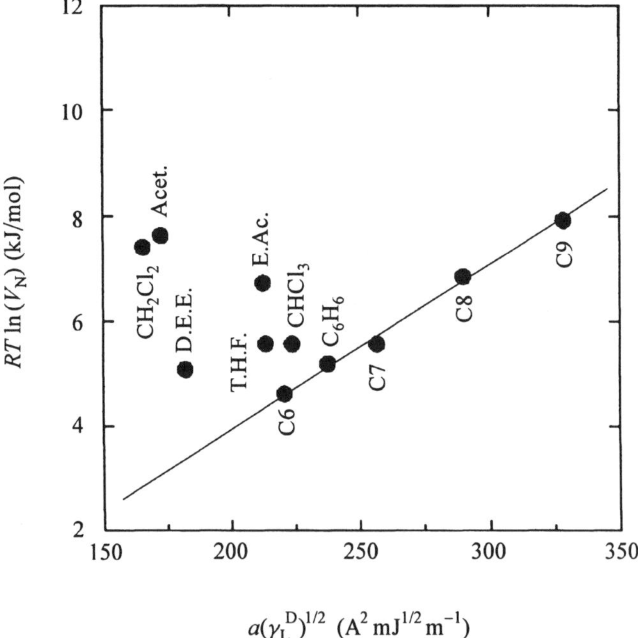

Figure 4. Plot of $RT \ln(V_N)$ versus $a(\gamma_L^D)^{1/2}$ for the *n*-alkanes and the polar probes at 60°C, on the Explosion pulp fiber.

3.1.3. Surface thermodynamic functions of adsorption. Within the range of temperatures used in this study (50–70°C), the plots of $\ln(V_g^0)$ versus $1/T$ for all probes and stationary phases are linear, confirming the validity of equation (6). As an example, Fig. 5 shows this result on the untreated PE pulp fiber.

Linearity also is observed for $\ln(V_g^0)$ versus the number of carbon atoms (n-alkanes). This is illustrated in Fig. 6, as an example.

Tables 3 shows $-\Delta H_A^0$, (equation (6)), for the different probes on the untreated, the ozonated, the fluorinated PE pulps fibers, and the lignocellulosic pulps fibers, obtained from their respective $\ln V_g^0$ versus $1/T$ slopes. The incremental $-\Delta H_A^0$ per $-CH_2-$ (the slope of the curve $-\Delta H_A^0$ versus the number of carbon atoms) is equal to 6.8, 9.7, 7.2, 9.9, 7.6, 9.2 and 3.7 kJ/mol for the untreated PE, the ozonated PE (2 and 3 hours), the fluorinated PE (levels 1 and 2), the Explosion and the Kraft pulps fibers, respectively. The $-\Delta H_A^0$ are plotted in Figs 7 to 11, as a function of $a(\gamma_L^D)^{1/2}$. A linearity is always observed between the n-alkanes, for the untreated, the ozonated, the fluorinated PE pulps fibers, and the lignocellulosic pulps fibers.

It should be noted that when the CH_3NO_2, an acidic probe, was injected into the column containing the Explosion pulp fiber, no response was recorded by the IGC detector. This probe probably adsorbed and desorbed with slow kinetics, such that no

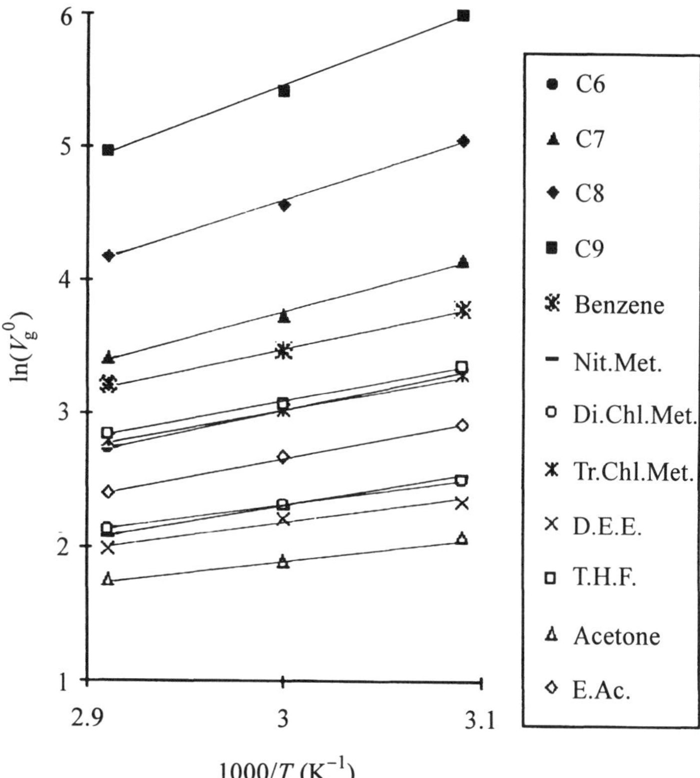

Figure 5. Plot of $\ln(V_g^0)$ versus $1/T$ for the n-alkanes and the polar probes on the untreated PE pulp fiber.

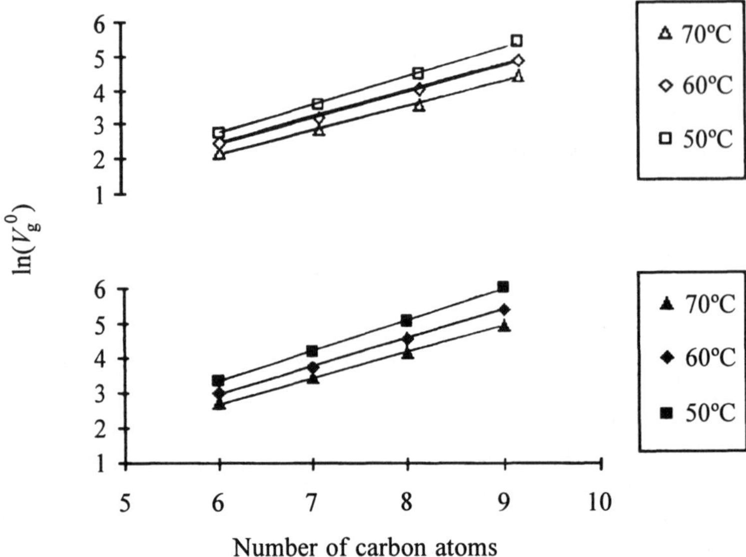

Figure 6. Plot of ln (V_g^0) versus number of carbon atoms of the n-alkanes, as a function of the column temperature, on the untreated (full symbols) and the fluorinated (empty symbols, level 1) PE pulp fiber.

Table 3.
Standard enthalpy for the zero-coverage adsorption of the probes on the stationary phases

	$-\Delta H_A^0$ (kJ/mol)						
Probes	Untreated PE	Ozonated PE (2 h)	Ozonated PE (3 h)	Fluorinated PE #1	Fluorinated PE #2	Explosion pulp	Kraft pulp
C_6	27.1	25.5	23.4	25.0	10.6	10.4	09.6
C_7	34.2	34.2	30.4	36.8	19.0	20.0	13.0
C_8	40.6	43.3	38.9	47.3	26.0	27.7	16.2
C_9	47.5	54.7	45.0	54.7	33.4	38.0	20.7
C_6H_6	29.4	29.7	26.4	32.7	22.8	27.3	12.3
CH_3NO_2	21.3	26.6	21.7	28.1	40.7	—	29.3
CH_2Cl_2	17.1	17.1	18.0	24.7	24.7	24.5	20.2
$CHCl_3$	27.4	26.0	30.4	29.7	28.5	25.8	15.5
D.E.E.	19.1	22.8	22.0	31.2	27.0	9.5	18.7
T.H.F.	25.4	29.7	27.8	31.6	32.3	24.4	20.2
Acetone	16.8	18.0	15.1	25.8	26.6	16.4	22.4
Ethyl acetate	23.6	26.6	28.9	30.4	37.3	24.2	20.7

peak was observed and this may be due to a highly basic character of the Explosion pulp fiber surface.

Tables 4 and 5 show $-\Delta G_A^0$ (equation (5)) and $-\Delta S_A^0$ (equation (7)), respectively. $-\Delta G_A^0$ is the energy necessary to change one mole of the vapor probe from its gaseous state at 1 atm. to a standard state of the adsorbent surface. A linearity

Table 4.
Standard free energy for the zero-coverage adsorption of the probes on the stationary phases, at 50/60/70 °C

	$-\Delta G_A^0$ (kJ/mol)						
Probes	Untreated PE	Ozonated PE (2 h)	Ozonated PE (3 h)	Fluorinated PE #1	Fluorinated PE #2	Explosion pulp	Kraft pulp
C_6	22.0/21.9/21.8	20.7/20.7/20.7	22.6/22.3/22.6	20.7/20.6/20.5	20.7/21.3/21.7	16.7/17.0/17.3	16.1/16.3/16.5
C_7	24.2/23.9/23.8	23.1/22.9/22.6	24.6/24.4/24.4	22.9/22.7/22.5	23.2/23.4/23.7	17.9/18.0/17.9	17.2/17.5/17.6
C_8	26.7/26.2/26.0	25.8/25.4/24.9	26.9/26.5/26.3	25.4/25.0/24.5	25.4/25.7/25.7	19.1/19.1/19.1	18.1/18.3/18.4
C_9	29.2/28.6/28.2	28.4/27.9/27.2	29.4/28.8/28.6	28.0/27.4/26.9	27.8/27.9/27.8	20.6/20.3/20.0	19.4/19.4/19.4
C_6H_6	23.3/23.2/23.2	22.1/22.0/22.0	23.9/23.8/24.0	22.2/22.1/22.0	23.0/23.4/23.4	17.8/17.6/17.4	16.5/16.6/16.9
CH_3NO_2	19.9/19.9/19.9	19.0/18.9/18.8	21.5/21.6/20.8	19.0/19.0/18.9	21.9/21.7/21.3	—/—/—	18.3/18.1/17.8
CH_2Cl_2	19.8/20.0/20.1	18.7/18.8/19.0	20.9/21.2/21.4	18.9/18.8/18.8	20.5/20.4/20.4	19.8/19.8/19.8	16.1/16.0/15.9
$CHCl_3$	21.9/22.0/22.0	20.9/20.9/20.9	23.1/23.2/22.8	20.7/20.4/20.3	22.2/22.2/22.1	18.1/18.0/17.8	16.3/16.5/16.5
D.E.E.	19.4/19.7/19.7	18.3/18.2/18.2	20.3/20.2/21.0	18.3/18.1/18.0	20.2/20.4/20.2	17.2/17.5/17.8	16.3/16.2/16.4
T.H.F.	22.1/22.1/22.1	21.3/21.3/21.2	23.9/23.8/23.9	21.2/21.1/21.0	22.8/22.9/22.7	18.0/18.0/17.8	17.1/17.1/17.0
Acetone	18.7/18.8/19.1	17.8/17.9/18.1	20.3/20.5/20.6	18.1/18.0/17.8	20.5/20.6/20.4	19.8/20.0/20.2	16.2/16.0/15.9
Ethyl acetate	20.9/21.0/20.9	19.9/19.8/19.8	22.2/22.2/22.2	20.1/20.0/19.8	22.2/22.0/21.7	19.2/19.1/19.0	16.5/16.5/16.4

Table 5.
Standard entropy for the zero-coverage adsorption of the probes on the stationary phases, at 50/60/70°C

Probes	$-\Delta S_A^0$ ($\times 10^{-2}$ kJ/mol K)						
	Untreated PE	Ozonated PE (2 h)	Ozonated PE (3 h)	Fluorinated PE #1	Fluorinated PE #2	Explosion pulp	Kraft pulp
C_6	15.2/14.7/14.3	14.3/13.9/13.5	14.2/13.7/13.4	14.2/13.7/13.2	09.7/09.6/09.4	08.4/08.2/08.1	07.9/07.8/07.6
C_7	18.0/17.4/16.9	17.7/17.1/16.6	17.0/16.4/16.0	17.9/17.3/16.8	13.0/12.7/12.4	11.7/11.4/11.1	09.3/09.1/08.9
C_8	20.8/20.0/19.4	21.4/20.6/19.9	20.4/19.6/19.0	21.8/21.0/20.3	15.9/15.5/15.1	14.5/14.1/13.6	10.6/10.4/10.1
C_9	23.7/22.8/22.0	25.7/24.8/23.9	23.0/22.2/21.4	25.6/24.6/23.8	18.9/18.4/17.8	18.1/17.5/16.9	12.4/12.0/11.7
C_6H_6	16.3/15.8/15.3	16.0/15.5/15.1	15.6/15.1/14.7	17.0/16.4/15.9	14.2/13.9/13.4	14.0/13.5/13.0	08.9/08.7/08.5
CH_3NO_2	12.7/12.4/12.0	14.1/13.7/13.2	13.3/13.0/12.4	14.6/14.1/13.7	19.4/18.7/18.0	—/—/—	14.7/14.2/13.7
CH_2Cl_2	11.4/11.1/10.8	11.1/10.8/10.5	12.0/11.8/11.5	13.5/13.1/12.9	14.0/13.6/13.1	13.7/13.3/12.9	11.2/10.9/10.5
$CHCl_3$	15.3/14.8/14.4	14.5/14.1/13.7	16.6/16.1/15.5	15.6/15.0/14.6	15.7/15.2/14.7	13.6/13.1/12.7	09.9/09.6/09.3
D.E.E.	11.3/11.6/11.3	12.7/12.3/12.0	13.1/12.7/12.5	15.3/14.8/14.3	14.6/14.2/13.7	08.2/08.1/08.0	10.8/10.5/10.2
T.H.F.	14.7/14.2/13.8	15.8/15.3/14.8	16.0/15.5/15.0	16.3/15.8/15.3	17.0/16.6/16.0	13.1/12.7/12.3	11.5/11.2/10.8
Acetone	11.0/10.7/10.4	11.1/10.8/10.5	11.0/10.7/10.4	13.6/13.1 /12.7	14.6/14.2/13.7	11.2/10.9/10.7	11.9/11.5/11.2
Ethyl acetate	13.8/13.4/13.0	14.4/13.9/13.5	15.8/15.3/14.9	15.6/15.1/14.6	18.4/17.8/17.2	13.4/13.0/12.6	11.5/11.2/10.8

should be obtained when plotting $-\Delta G_A^0$ with the number of carbon atoms for the n-alkane probes for a given stationary phase. The slope of $-\Delta G_A^0$ versus the number of carbon atoms is the incremental variation of the free energy of the methylene group, $-\Delta G_A^0(-CH_2-)$. This parameter, $-\Delta G_A^0$, reflects the interaction between the vapor probe and the surface molecules of the non-mobile phase as a function of the temperature. $-\Delta G_A^0$ did not change significantly with the temperature of the non-mobile phases (50–70°C). Values of $-\Delta G_A^0$ were generally lower in the case of the lignocellulosic pulps than in the case of the PE pulps fibers. The highest values of $-\Delta G_A^0$ were obtained with the C_8 and C_9 alkanes probes.

The entropy of adsorption, $-\Delta S_A^0$ (Table 5), for the n-alkanes on the various samples vary from about 75 to 235 J/mol K. $-\Delta S_A^0$ versus the number of carbon atoms of n-alkanes for the various samples should give linear relations. The slope of $-\Delta S_A^0$ versus the number of carbon atoms is the incremental variation of the entropy of the methylene group, $-\Delta S_A^0(-CH_2-)$. There is a clear decrease of about 4 J/mol K in the entropy of adsorption of a given probe, on the various samples, when the sample temperature increases from 50 to 60 and to 70°C. Especially for the n-alkane probes, the lowest values of entropy, $-\Delta S_A^0$, are obtained on the Kraft, the Explosion and the fluorinated (#2) PE pulps fibers, indicating the restricted degree of freedom due to the nature of these surfaces, which are probably rather polar. Similarly, there is a decrease in the entropy values of the n-alkanes as the degree of treatment (oxidation) of the PE fiber surface increases. This is clear especially for the fluorinated PE pulp (#2), compared with the untreated and the fluorinated PE (#1) pulps. De Boer [12] attributed the decrease of entropy values upon adsorption to the loss of one degree of translational freedom for the n-alkane probes.

On CTMP fiber, at 25°C, Kamdem and Riedl [21] obtained entropies of adsorption of 44.9 and 58.4 J/mol K for C_8 and C_9 alkanes. In our present study we found the values of 106 and 124 J/mol K, and 145 and 181 J/mol K, on the Kraft and the Explosion fibers, respectively, at 50°C.

3.1.4. Enthalpy of specific adsorptions and acidity/basicity of the surfaces. The enthalpies of adsorption corresponding to the specific interactions, ΔH_A^{SP}, are obtained from the difference of ordinates between the n-alkane straight lines, as in Figs 7 to 11, and the polar probes. These are summarized in Table 6 for all samples of fibers characterized in this study. Equation (10) can be written as:

$$\frac{-\Delta H_A^{SP}}{AN} = \frac{DN}{AN} K_A + K_B. \tag{11}$$

This equation, plotted in Figs 12 to 14, shows that $\Delta H_A^{SP}/AN$ versus DN/AN plots are almost linear for all samples. K_A and K_B can be determined from the slope and the intercept at the origin of the straight lines. Table 7 shows the resulting values, confirming the amphoteric characteristic of the fluorinated PE fiber, a more acid surface characteristic of the ozonated PE fiber, and the neutral surface characteristic (surface energy with only dispersive component, $\gamma_S^P \approx 0$) of the untreated PE pulp fiber. Fluorination (level 2) does give a very increase in acidic and basic characteristics

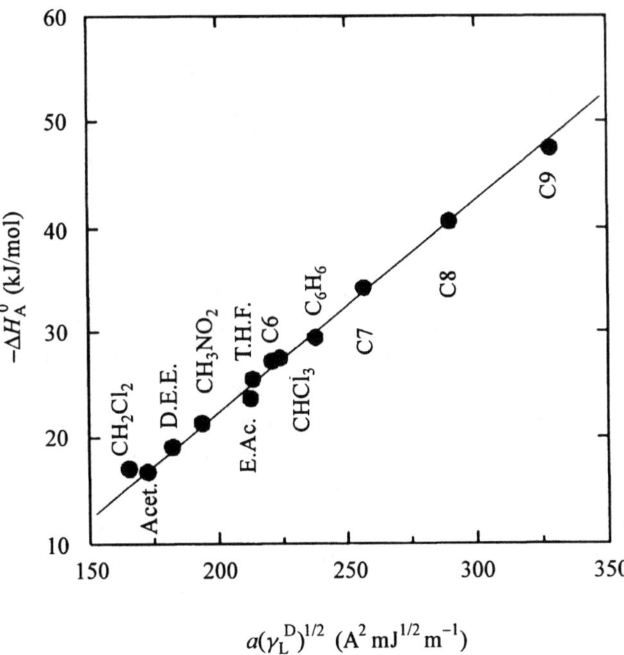

Figure 7. Plot of $-\Delta H_A^0$ versus $a(\gamma_L^D)^{1/2}$ for the *n*-alkanes and the polar probes, on the untreated PE pulp fiber.

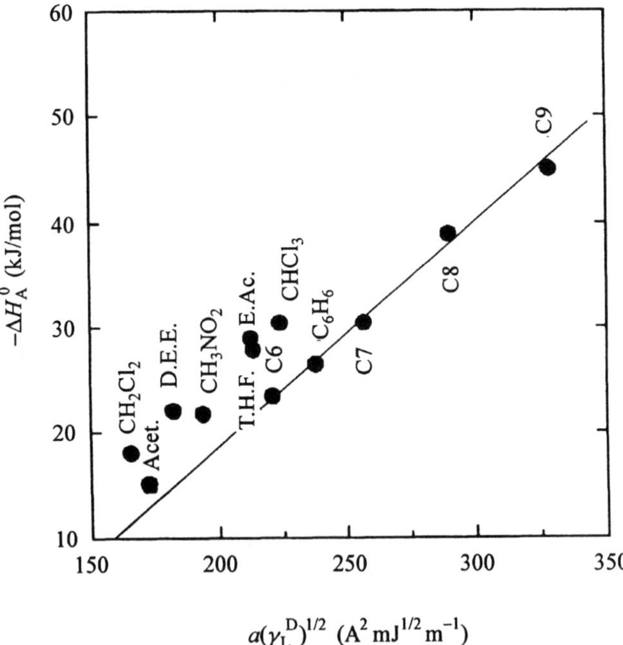

Figure 8. Plot of $-\Delta H_A^0$ versus $a(\gamma_L^D)^{1/2}$ for the *n*-alkanes and the polar probes, on the ozonated PE pulp fiber (3 h).

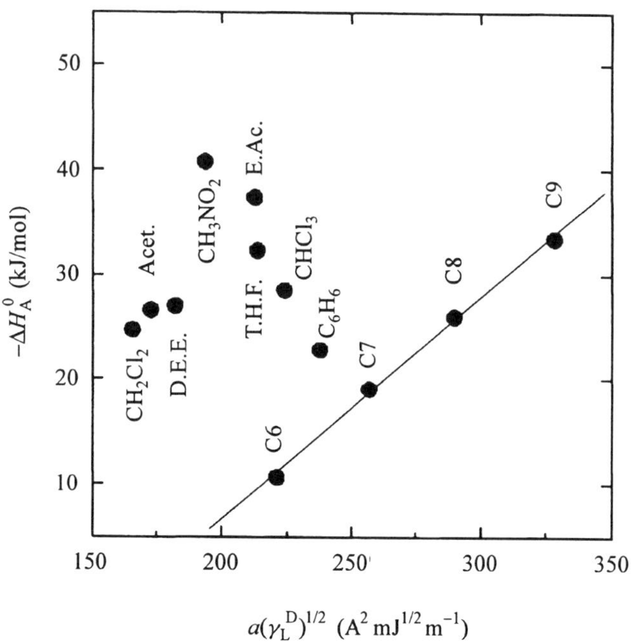

Figure 9. Plot of $-\Delta H_A^0$ versus $a(\gamma_L^D)^{1/2}$ for the n-alkanes and the polar probes, on the fluorinated PE pulp fiber (#2).

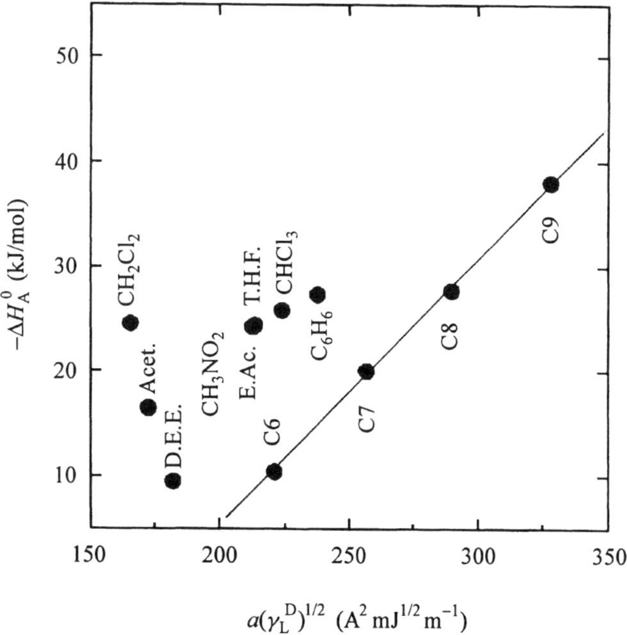

Figure 10. Plot of $-\Delta H_A^0$ versus $a(\gamma_L^D)^{1/2}$ for the n-alkanes and the polar probes, on the Explosion pulp fiber.

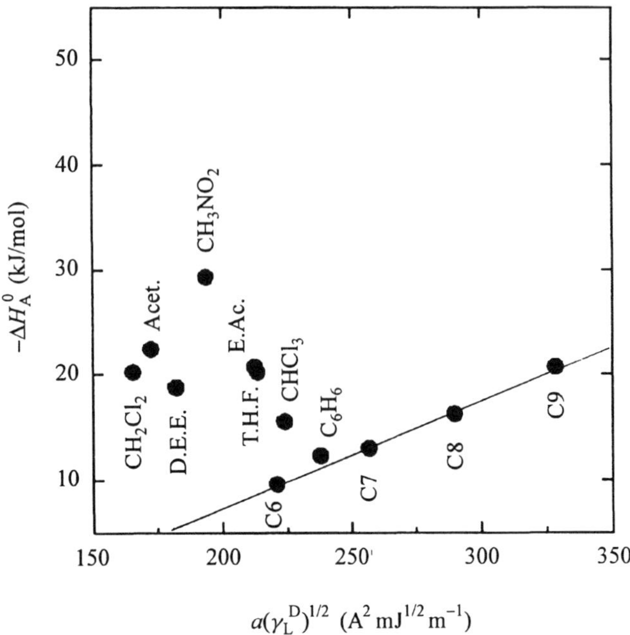

Figure 11. Plot of $-\Delta H_A^0$ versus $a(\gamma_L^D)^{1/2}$ for the *n*-alkanes and the polar probes, on the Kraft pulp fiber.

Table 6.
Specific enthalpies of adsorption of the polar probes on the surface of the stationary phases (kJ/mol)

	$-\Delta H_A^{SP}$						
Probes	Untreated PE	Ozonated PE (2 h)	Ozonated PE (3 h)	Fluorinated PE #1	Fluorinated PE #2	Explosion pulp	Kraft pulp
C_6H_6	0	0	0	3.1	8.3	12.5	1.3
CH_3NO_2	0	8.6	4.4	10.9	35.3	—	22.4
CH_2Cl_2	0	6.7	6.8	15.6	25.5	28.4	16.7
$CHCl_3$	0	0	6.5	4.2	16.6	14.6	5.9
D.E.E.	0	7.8	7.3	17.1	24.2	8.9	13.6
T.H.F.	0	6.5	6.2	12.0	22.8	15.9	11.5
Acetone	0	5.7	2.3	14.5	26.0	18.5	18.2
Ethyl acetate	0	3.6	7.3	11.1	28.0	15.9	12.1

of the untreated PE pulp fiber surface. Kraft pulp fiber has an amphoteric surface with moderate acid/base characteristics, whereas the Explosion pulp fiber has the strongest basic character and a relatively low acidic character.

3.2. XPS results

In [1] and [3], using XPS, we analyzed the surfaces of the untreated, the ozonated, and the fluorinated PE pulp fibers. In Table 8, we present a summary of this analysis

Table 7.
Acid/Base (Acceptor/Donor) characteristics (in arbitrary units)

Stationary phases	K_A	K_B
Untreated PE fiber	≈ 0	≈ 0
Ozonated PE fiber (2 h)	3.5	0.8
Ozonated PE fiber (3 h)	3.3	1.0
Fluorinated PE fiber (#1)	7.3	2.5
Fluorinated PE fiber (#2)	10.3	9.2
Explosion pulp fiber	2.4	12.1
Kraft pulp fiber	5.5	5.0

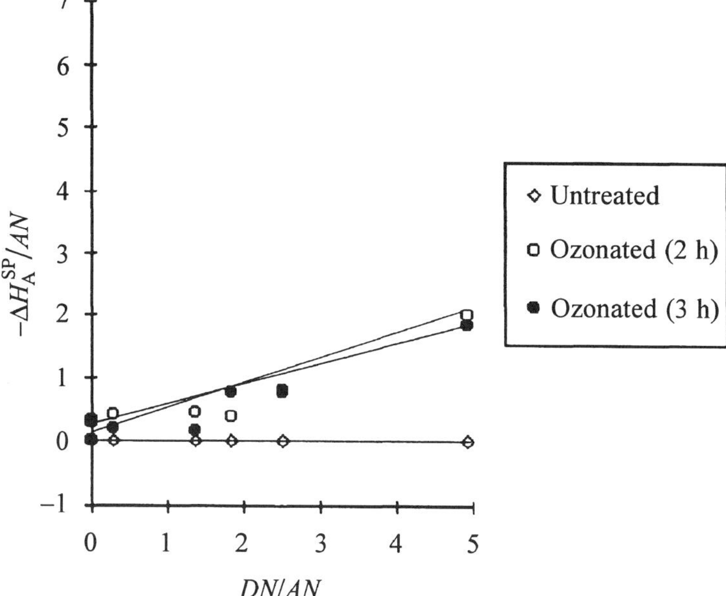

Figure 12. Plot of $-\Delta H_A^{SP}/AN$ versus DN/AN of the polar probes, for the untreated and the ozonated PE pulp fiber.

for the samples studied here, and we add new results obtained recently on Kraft and Explosion pulps.

As discussed in [1] and [3], the surface of the untreated PE fiber shows 7.5% of C2 component (related to −C−OH) and 0.9% of C3 component (C3[b] is a C3, related to O−C=O) [1, 3], arising from the poly(vinyl alcohol) and the poly(vinyl acetate) added during manufacturing as wetting agents. However IGC results still show neutral characteristics for the surface ($K_A \approx K_B \approx 0$). This may be due to the fact that these wetting agents do not induce much change in the surface energy of the fiber, probably because of their being immobilized in the crystal structure. The surface treatments are known to induce a loss of crystallinity enhancing interactions at the surface. IGC results related to specific interactions should be influenced by the loss/increase of availability and the distribution of the acceptor/donor groups at the surface.

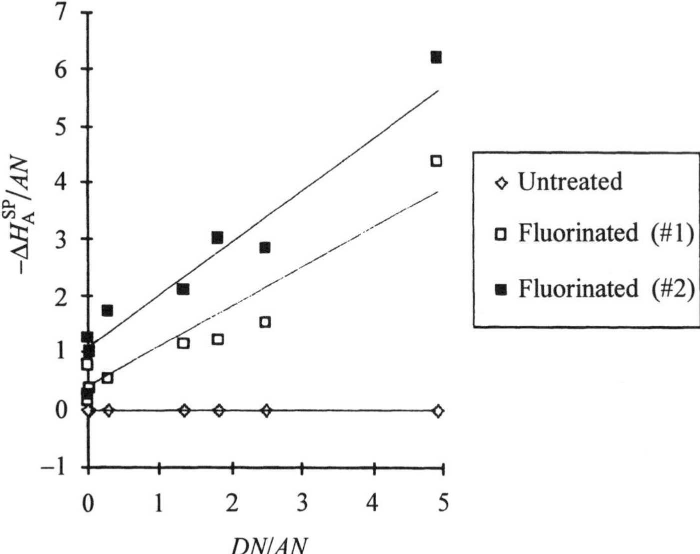

Figure 13. Plot of $-\Delta H_A^{SP}/AN$ versus DN/AN of the polar probes, for the untreated and the fluorinated PE pulp fiber.

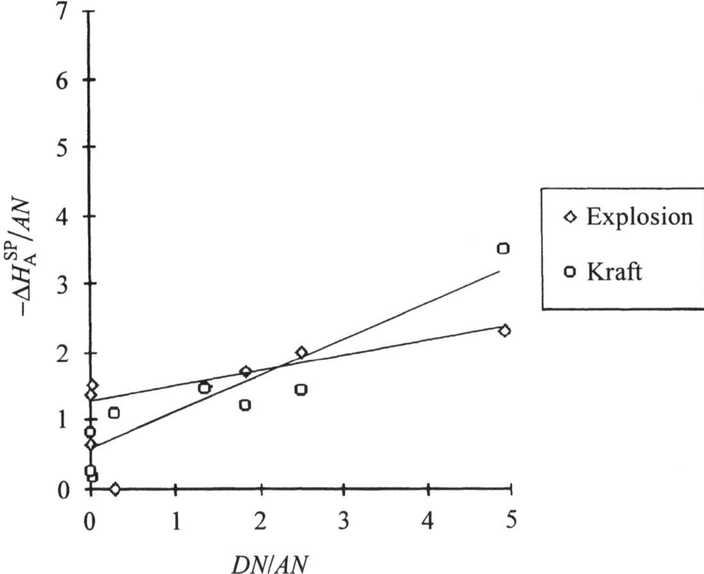

Figure 14. Plot of $-\Delta H_A^{SP}/AN$ versus DN/AN of the polar probes, for the Explosion and the Kraft pulp fibers.

The surface of the ozonated (2 h) PE pulp fiber shows less hydroxyl and more carboxyl components, C2 = 4.2% and $C3^b$ = 1.2%, than the untreated PE pulp fiber. The IGC results showed an increase in the acid/base characteristics of the surface. The acid/base constants were about 3.5/0.8, respectively, and this is in accordance

Table 8.
O/C and F/C atomic ratio, and fractional area for deconvoluted components from XPS C_{1s} and O_{1s} peaks

Samples	C_{1s} C1/C2/C3^a/C3^b(%)	O_{1s} O1/O2 (%)	O/C (%)	F/C (%)
PE untreated	91.6/7.5/–/0.9	44.0/56.0	9.9	—
Ozonated PE, 2 h	94.6/4.2/–/1.2	49.4/50.6	8.2	
Ozonated PE, 3 h	97.8/–/1.1/1.1	59.5/40.5	2.9	—
Fluorinated PE, #1	92.7/–/4.8/2.3	58.7/41.3	11.2	1.5
Fluorinated PE, #2	81.3/–/14.0/4.7	46.1/53.9	17.1	9.1
Explosion pulp	65.4/32.2/–/2.4	15.1/84.9	30.3	—
Kraft pulp	29.4/64.4/–/6.2	7.7/92.3	55.4	—

C3^a and C3^b are C3, related to −C=O and O−C=O, respectively. C3^a and C3^b have binding energies of 287.8 ± 0.2 eV and 289.0 ± 0.2 eV, respectively [1, 3].

with the hydroxyl over the carboxyl groups concentrations, 4.2/1.2, respectively. This again suggests etching of the crystalline areas of the PE which decreases the absolute surface concentration of oxygen moieties as shown by ESCA but increases their availability, as shown by IGC.

The surface of the ozonated (3 h) PE fiber shows 2.2% of C3 component (C3 is a sum of C3^a, related to −C=O, 1.1%; and C3^b, related to O−C=O, 1.1%) [1]. The loss of C2 component compared with the untreated and the two hours ozonated PE fiber is due to the loss of some oxygen-containing functionalities from the outermost layers, and may be due to the oxidation of −C−OH to −C=O or O=C−OH, as explained in [1]. However, the ozonated PE fiber surface showed heightened acidic characteristics ($K_A = 3.3$, $K_B = 1.0$). There is no significant changes in K_A and K_B between the two and the three hours ozonated PE pulp fibers, while there is some change in the chemical composition and a decrease in O/C atomic ratio.

On the fluorinated PE pulp fiber, there are no hydroxyl groups (C2 = 0%) at the surface, but the carbonyl and carboxyl groups increased, as the O/C and F/C atomic ratios increased. The surface of the first fluorinated PE fiber shows $\leqslant$ 7.1% of C3 component [3] (C3^a $\leqslant$ 4.8% and C3^b = 2.3%, related to −C=O and O−C=O, respectively). The surface of the second fluorinated PE fiber shows $\leqslant$ 18.7% of C3 component (C3^a $\leqslant$ 14.0% related to −C=O, and C3^b = 4.7% related to O−C=O) [3]. The first level of fluorination caused a large increase in K_A and a small increase in K_B, compared with the untreated PE fiber. The second fluorination level caused a large increase in K_B and a small increase in K_A, compared with the level one of fluorinated fiber.

It is really difficult to correlate the chemical composition of the surface of the PE pulps fibers and the acid/base constants obtained from the IGC analysis. The variation of the chemical compositions and the acid/base characteristics of the surface as a function of the treatment does not follows a simple regression function with two or three parameters, but is more complicated than that. As shown by Guillet [26], the IGC method is sensitive to the crystallinity of the surface. The diffusion level of the probes is higher in an amorphous phase than in a crystalline one. The crystallinity at the surface of the PE pulp fiber should be decreased especially with the ozone treatment, and this increases the sensitivity of the IGC to the presence of polar moieties at the surface. In addition, the distribution of the polar groups generated at the surface, the orientation of each group and each atom, the interactions between these groups, and also the interactions between these surface groups and the near-surface groups, should play an important role in the surface acid/base characteristics.

On a regular surface, such as that of an extruded polymeric fiber or film, ESCA and IGC results should correlate well. The PE pulp fibers used in this study have an irregular and a convented surface. They presented an apparent or *external* surface, the first 1–5 nm accessible to ESCA analysis, and a hidden or *internal* surface due to the presence of many crevices in the fiber structure. The surface probed by the solute or probe, in the mobile phase in the IGC analysis, should be both the *external* and the *internal* surfaces of the stationary phase.

For the lignocellulosic fiber, Kraft pulp has a higher O/C atomic ratio, and this is normal as cellulose and hemicellulose are more rich in oxygen, compared with lignin which is normally present at the surface of the Explosion pulp. Kraft pulp fiber surface has 64.4% of C2 and 6.2% of $C3^b$ components, and 7.7% of O1 and 92.3% of O2 components. So there are more hydroxyl groups at the surface and this suggests a more acidic character for this surface. The Explosion pulp fiber surface has 32.2% of C2 and 2.4% of $C3^b$ components, and 15.1% of O1 and 84.9% of O2 components (Table 8). In this case also there are more hydroxyl than carboxyl groups at the surface. But compared with the Kraft fiber, the O_{1s} peaks indicated more O1, related to $-C{=}O$ and $C{-}O{-}C$, on the Explosion pulp surface than the Kraft pulp surface, O1/O = 15.1 and 7.7%, respectively. IGC results showed more basic characteristics for the Explosion pulp fiber surface, $K_B = 12.1$, which is the highest value obtained in this study, and amphoteric characteristics for the Kraft pulp fiber surface. The Kraft pulp surface was almost two and a half times more acidic than the Explosion pulp surface which was almost two and a half time more basic than the Kraft pulp fiber surface.

Based on the XPS and IGC results, the quality of the adhesion between the lignocellulosic and the PE pulps fibers, before and after treatments, could be predicted. The strength properties of composite papers [27], which will be published soon, showed a very positive effect of the fluorination (#2), which increased significantly the inter-PE fiber adhesion and also the adhesion of the PE pulp fiber with the Kraft fiber, and especially with the Explosion pulp fiber.

4. CONCLUSIONS

IGC was used successfully to characterize the surface of a PE pulp fiber pretreated with wetting agents (PVA/PVAc) and modified with fluorine gas and ozone treatments. No acid/base characteristics were obtained on the surface of the PE pulp fiber pretreated with the wetting agents. Both treatments, the ozonation and the fluorination, created acid/base characteristics on the untreated PE pulp. The fluorinated pulp fiber (#2) showed the highest acidic and basic constants, which make it an amphoteric surface. Increasing the fluorination time, level 1 to level 2, was beneficial to the resulting surface which changed from acidic to amphoteric in nature. This was not the case with the ozonated surface.

IGC was also used successfully to characterize the surfaces of Explosion and Kraft pulps fibers. A large basic constant was obtained for the Explosion pulp surface, and a moderate acid/base constants were obtained for the Kraft pulp surface.

XPS results obtained at the surface of the PE pulp fiber showed a decrease in the O/C atomic ratio after the ozone treatment. These results are insufficient to predict the effect of a surface treatment on the surface energy and the interfiber adhesion. In addition to XPS, IGC is really recommended to characterize the surfaces and the effects of treatments, as well as to well predict the interfiber adhesion. While ESCA is sensitive to the chemical composition of the first 1–5 nm of the *external* surface of the PE pulp fiber, IGC seems to be sensitive to the chemical composition, the crystallinity and the availability of sites on the *external* as well as *hidden* surfaces.

REFERENCES

1. H. Chtourou, B. Riedl, B. V. Kokta, A. Adnot and S. C. Kaliaguine, *J. Appl. Polym. Sci.* **49**, 361 (1993).
2. H. Chtourou, B. Riedl and B. V. Kokta, *Polym. Degrad. Stab.* **43**, 149 (1994).
3. H. Chtourou, B. Riedl and B. V. Kokta, *J. Colloid Interface Sci.* **158**, 96 (1993).
4. F. M. Fowkes, *Ind. Eng. Chem.* **56** (12), 40 (1964).
5. H. P. Schreiber, M. R. Wertheimer and M. Lambla, *J. Appl. Polym. Sci.* **27**, 2269 (1982).
6. B. Riedl and P. D. Kamdem, *J. Adhesion Sci. Technol.* **6**, 1053 (1992).
7. J. Schultz, L. Lavielle and C. Martin, *J. Adhesion* **23**, 45 (1987).
8. C. Martin, PhD Thesis. Université de Haute Alsace, France (1988).
9. K. L. Mittal and H. R. Anderson, Jr. (Eds), *Acid–Base Interactions: Relevance to Adhesion Science and Technology*. VSP, Zeist, The Netherlands (1991).
10. G. M. Dorris and D. G. Gray, *J. Colloid Interface Sci.* **77**, 353 (1980).
11. D. G. Gray and J. E. Guillet, *Macromolecules* **5**, 316 (1972).
12. J. H. De Boer, *The Dynamic Character of Adsorption*, p. 113. Oxford Univ. Press (Clarendon), Oxford (1953).
13. C. St-Flour, PhD Thesis. Université de Haute Alsace, France (1981).
    C. St-Flour and E. Papirer, *Ind. Eng. Chem. Prod. Res. Dev.* **21** (2), 337 (1982).
14. F. M. Fowkes and M. A. Mostafa, *Ind. Eng. Chem. Prod. Res. Dev.* **17**, 3 (1978).
15. F. M. Fowkes, *Rubber Chem. Technol.* **57**, 328 (1983).
16. F. M. Fowkes, *J. Adhesion Sci. Technol.* **1**, 7 (1987).
17. V. Gutmann, *The Donor-Acceptor Approach to Molecular Interactions*. Plenum Press, New York (1978).
18. F. L. Riddle and F. M. Fowkes, *J. Am. Chem. Soc.* **112**, 3259 (1990).
19. S. G. Gregg and K. S. Sing, *Adsorption, Surface Area and Porosity*. Academic Press, New York (1982).
20. T. Rave, *Kirk-Othmer Encyclopedia of Chemical Technology*, M. Grayson (Ed.), Vol. 19, p. 425. Wiley, New York (1978).
21. D. P. Kamdem and B. Riedl, *J. Wood Chem. Technol.* **11**, 57 (1991).
22. N. Gurnagul and D. G. Gray, *J. Pulp Paper Sci.* **11**, J98 (1985).
23. N. Gurnagul and D. G. Gray, *Can. J. Chem.* **65**, 1935 (1987).
24. G. M. Dorris, PhD Thesis. McGill University, Canada (1979).
25. H. L. Lee and P. Luner, *Nordic Pulp and Paper Research J.* **2**, 164 (1989).

26. J. Guillet, in: *New Developments in Gas Chromatography*, J. H. Purnell (Ed.), p. 187. Wiley, New York (1973).
27. H. Chtourou, B. Riedl and B. V. Kokta (to be published).

Polymer Surface Modification: Relevance to Adhesion, pp. 479–489
K. L. Mittal (Ed.)

Enhancement of the monovalent cation perm-selectivity of Nafion by plasma-induced surface modification*

ZEN-ICHIRO TAKEHARA,** ZEMPACHI OGUMI, YOSHIHARU UCHIMOTO and KAZUAKI YASUDA

Division of Energy and Hydrocarbon Chemistry, Graduate School of Engineering, Kyoto University, Yoshida, Sakyo-ku, Kyoto 606-01, Japan

Revised version received 21 September 1994

Abstract—In this paper, we focus on improvement of the monovalent cation perm-selectivity of a perfluorinated cation-exchange membrane, Nafion 117, by depositing an anion-exchange layer using a plasma surface modification process. The anion-exchange layer was deposited from 4-vinylpyridine monomer vapor followed by quaternization with 1-bromopropane. The transference number of divalent cation (Fe^{2+}) through the membrane, t_{Fe}, decreased with increasing thickness of the plasma polymer layer at the expense of enhanced membrane resistance. A large interfacial resistance was observed between Nafion and the plasma polymer layer which was ascribed to the implantation of cationic species containing nitrogen. To avoid the formation of an interfacial layer, a novel method of plasma-induced surface modification was devised. After a Nafion 117 sheet was placed on an RF (radio-frequency) electrode and sputtered with an oxygen or argon plasma in order to produce active sites on the Nafion, 4-vinylpyridine or 3-(2-aminoethyl)aminopropyltrimethoxysilane vapor was introduced into the reactor to react with radical sites. t_{Fe} decreased with increasing RF power. t_{Fe} through Nafion modified with 3-(2-aminoethyl)aminopropyltrimethoxysilane was lower than that for Nafion modified with 4-vinylpyridine, probably due to its weak Si–C bond. Nafion treated by the plasma surface modification method exhibited a very high monovalent cation perm-selectivity compared with Nafion treated by the plasma polymerization method.

Keywords: Surface modification; membrane; perm-selectivity; plasma.

1. INTRODUCTION

Because of their relatively high ionic selectivity, ion-exchange membranes have recently received considerable attention as materials for use in redox-flow batteries [1, 2], electrolyzers [3, 4], and sensors [5]. Although ion-exchange membranes

---

*The majority of this paper has been previously published in [6, 7, 8, 9, 10, 11]. Reprinted by permission of the publishers, The Electrochemical Society, Elsevier Science B.V., The Chemical Society of Japan and The Electrochemical Society of Japan.

**To whom correspondence should be addressed.

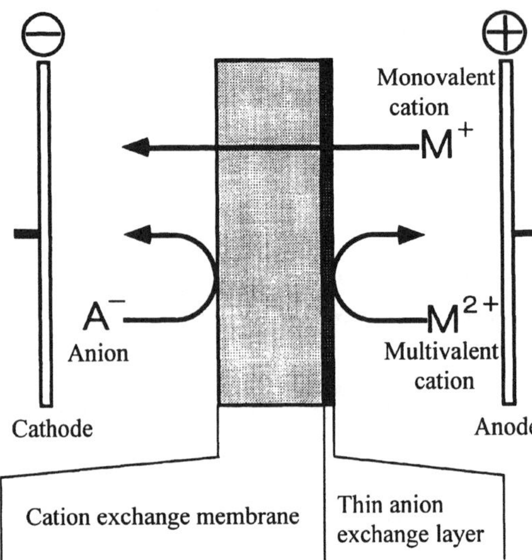

Figure 1. Principle of the enhancement of monovalent cation perm-selectivity through a cation-exchange membrane.

have a high selectivity for counter-ions over co-ions, their selectivity among different counter-ions is generally low. In some cases, such as their use in redox-flow batteries and brine electrolysis, the system performance is strongly dependent on the perm-selectivity among different counter-ions [2, 4].

We have investigated the surface modification of a cation-exchange membrane by utilizing a glow-discharge (plasma) polymerization technique in order to enhance the perm-selectivity among different counter-ions [6–11]. The principle of the enhancement of monovalent cation perm-selectivity [4, 12] is shown in Fig. 1. A thin layer of an anion exchanger is deposited on the surface of a cation-exchange membrane, e.g. Nafion. Due to the electrostatic repulsion from the fixed anion in the cation-exchange membrane, the anion cannot permeate through the membrane. Similarly, the transport of monovalent cations and multivalent cations is also suppressed by the electrostatic repulsion from fixed cations in the thin anion-exchange overlayer on the cation-exchange membrane. However, since the repulsion exerted on monovalent cations from the fixed cations is weaker than that on multivalent cations [12], monovalent cations can be preferentially transported. A high fixed cation density in the thin anion-exchange overlayer will give rise to a high perm-selectivity of monovalent cations. This enhancement will be achieved, however, at the expense of membrane conductance.

2. EXPERIMENTAL

2.1. Plasma polymerization

The apparatus used to carry out plasma polymerization consisted of a glass reactor equipped with two capacitively-coupled inner disk electrodes. A Nafion 117 sheet was utilized as the substrate. The substrate was placed in a nominal 'after-glow' region, 2 cm downstream from the edge of the electrodes. 4-Vinylpyridine vapor at a flow rate of 10 cm^3 (STP)/min was introduced into the reactor and the pressure was maintained at 67 Pa. Plasma polymerization was carried out at an RF power of 5 W.

2.2. Plasma-induced surface modification

A schematic diagram of the process for the plasma-induced surface modification is shown in Fig. 2. A Nafion 117 sheet was placed on an RF electrode and sputtered by an oxygen or argon plasma at various flow rates and RF powers in order to produce radical sites on the Nafion surface. After sputtering, 4-vinylpyridine or 3-(2-aminoethylaminopropyl)trimethoxysilane vapor was introduced into the reactor and reacted with the radical sites. Thus, a thin layer containing pyridyl or amino groups was formed on the Nafion surface.

2.3. Characterization of the plasma polymer

The rate of polymer deposition ($\mu g\, cm^{-2}\, h^{-1}$) was determined by weighing before and after polymerization. Plasma polymers deposited on the substrate were characterized by Fourier transform infrared (FT-IR) spectroscopy (Shimadzu, Model FTIR-4100), using the reflection method. To obtain IR spectra, a gold reflectance layer was deposited onto the glass substrate prior to film deposition by DC sputtering. The thickness of the gold layer was about 0.2 μm. The substrate temperature was not controlled during the plasma polymerization [9].

2.4. Quaternization of the thin layer containing pyridyl or amino groups

The thin layer containing pyridyl groups was quanternized by treatment with a 1 vol% 1-bromopropane/propylene carbonate solution at 50°C for 48 h [10, 13]. After soak-

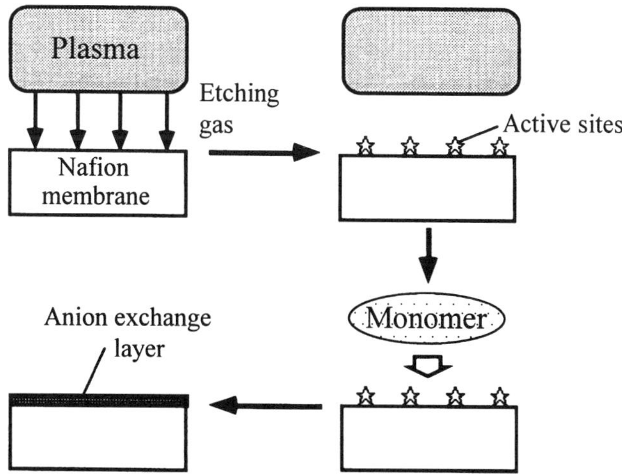

Figure 2. Schematic diagram of the process for plasma-induced surface modification of the Nafion membrane.

Table 1.
Compositions of the electrolytes used for perm-selectivity measurements

Catholyte	Anolyte
0.2 M CH_3COOH	0.2 M CH_3COOH
0.2 M CH_3COOLi	0.2 M CH_3COOLi
0.8 M LiCl	0.7 M LiCl
	0.05 M $FeCl_2$

ing, the films were washed with propylene carbonate to remove excess 1-bromopropane. The films thus treated were then dried under reduced pressure (10^{-1} Pa) to remove propylene carbonate.

The transference number of the anion of the plasma layer after qúaternization by 1-bromopropane was estimated using a concentration cell in an aqueous 0.1 M KCl/ 0.01 M KCl system. A thin plasma layer was deposited onto a microporous poly(propylene) film (average pore size 0.02×0.2 μm), Duragard® 2400 (Polyplastics Co. Inc.), and the membrane was placed as the separator of a two-compartment cell after quaternization.

2.5. *Characterization of the surface-modified Nafion film*

The ohmic resistance of the plasma-modified Nafion membrane was measured in 1.0 M HCl solution using a two-compartment cell. A movable Luggin capillary, connected to a Ag/AgCl reference electrode, was inserted into each compartment. By moving the Luggin capillary tips, the ohmic drop of the solution was measured under constant current electrolysis conditions (3.18 mA cm^{-2}). The ohmic resistance of the membrane was obtained by extrapolating the voltage profile to the position of both surfaces of the plasma-modified Nafion surface, by assuming zero membrane thickness.

Since Nafion has the least affinity for Li^+ among the alkali metal cations [14], Li^+ was selected to elucidate the enhancement of monovalent cation perm-selectivity. The perm-selectivity of the plasma-modified Nafion membrane was evaluated from the transference number of Fe^{2+} in a Li^+–Fe^{2+} system. The membrane was placed as the separator between the two compartments of a glass cell. Each compartment was filled with an electrolyte solution listed in Table 1. One compartment contains Fe^{2+} ions. A total charge of 100 C was passed at a constant current of 10 mA through the system. After the electrolysis, the total amount of Fe^{2+} in the catholyte was measured by absorption spectrometry.

3. RESULTS AND DISCUSSION

3.1. *IR spectroscopic characterization of plasma polymers*

FT-IR spectra of 4-vinylpyridine and plasma-polymerized 4-vinylpyridine are shown in Figs 3A and 3B, respectively [9]. The spectrum of 4-vinylpyridine exhibits ab-

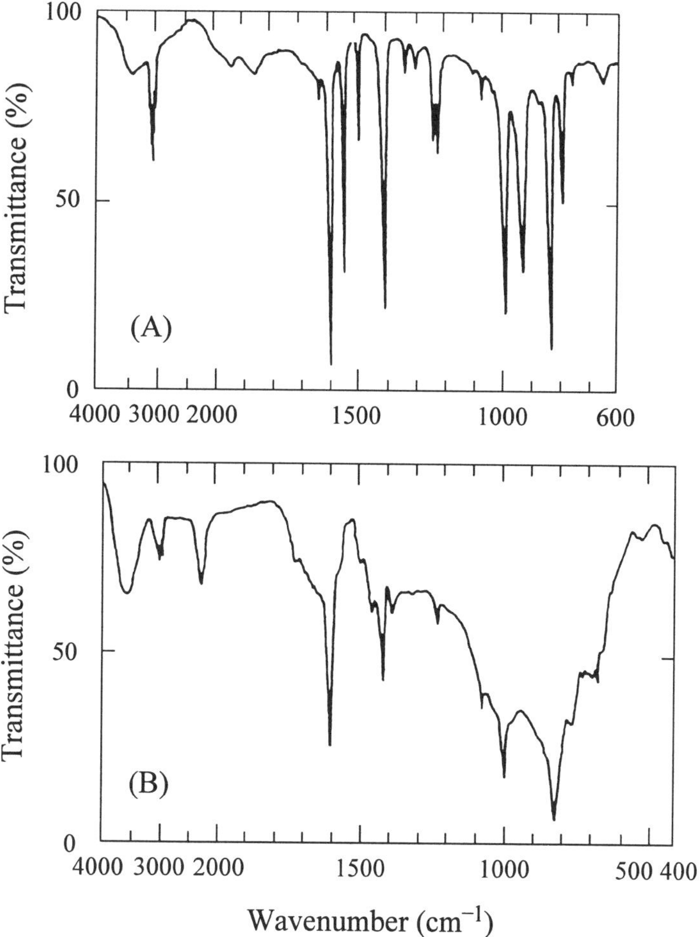

Figure 3. IR spectra of (A) 4-vinylpyridine and (B) plasma-polymerized 4-vinylpyridine. Flow rate of 4-vinylpyridine: 10 cm^3 (STP)/min; RF power: 5 W; polymerization pressure: 53 Pa.

sorption peaks at 3050 cm^{-1} (C—H vibration), 1600, 1500, and 1420 cm^{-1} (C=C ring stretching vibration), 1540 cm^{-1} (C=C stretching vibration of olefin group), and 920–720 cm^{-1} (CH out-of-plane deformation). On the other hand, the characteristic peak for the olefin group, which occurs at 1540 cm^{-1}, was completely absent in the spectrum of plasma-polymerized 4-vinylpyridine.

Plasma polymerization consists of two different processes, namely plasma-induced polymerization and plasma-state polymerization. Plasma-induced polymerization proceeds by a chain propagation mechanism (conventional polymerization mechanism). This type of polymerization can be initiated by a radical, an ion, or a highly energetic species in the glow region. However, in the case of plasma-state polymerization (atomic polymerization), the monomer molecule is first decomposed to form active species, followed by recombination of the generated reactive species to form a polymer. The disappearance of the olefin group indicates that plasma-induced polymerization has occurred.

Furthermore, the spectrum of plasma-polymerized 4-vinylpyridine exhibits new absorption peaks at 3340 cm^{-1} (N—H vibration) and 2240 cm^{-1} (C≡N vibration), which indicates that cleavage of the pyridine rings in 4-vinylpyridine has taken place during the plasma polymerization process. These results indicate also that plasma-state polymerization proceeds favorably during plasma polymerization and that the polymer formed should be crosslinked.

3.2. *Characterization of the quaternized plasma polymer layer*

Since the perm-selectivity of lithium ion in the Li^{+}–Fe^{2+} system arises from electrostatic repulsion between fixed cations in the thin anion-exchange overlayer prepared by plasma polymerization and cations in the electrolyte, the plasma layer with a high anion-exchange capacity increases the perm-selectivity of lithium ion for the Nafion. Therefore the anion transference number of the plasma layer deposited on Duragard was investigated after quaternization. The Cl^{-} transference number of Duragard used as a porous substrate was 0.51. As shown in Fig. 4, the quaternized plasma polymer layer exhibited a value beyond 0.8 at low polymerization pressure. The transference number had a tendency to increase with decreasing pressure. These results indicate that polymers formed at low pressure are highly branched and highly crosslinked. The crosslinking decreases the water content of the quaternized plasma polymer. The low water content leads to a high Cl^{-} transference number.

3.3. *Perm-selectivity and resistance of the surface-modified Nafion by the plasma polymerization process*

The transference number of Fe^{2+} through plasma-modified Nafion is shown in Fig. 5. The thicknesses of all plasma-polymerized layers were selected to be 0.24 mm by

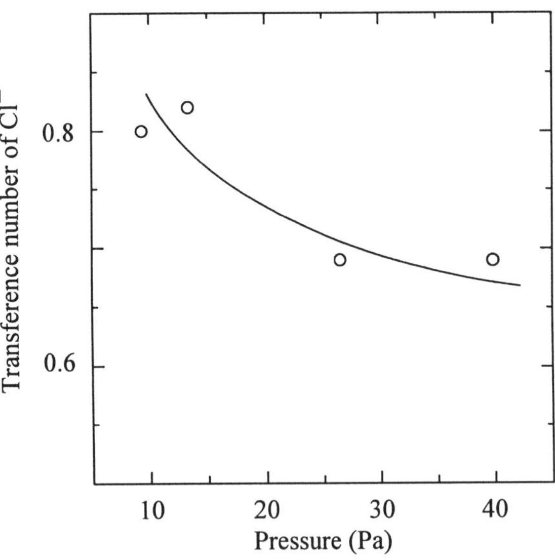

Figure 4. The transference number of Cl^{-} through the quaternized plasma polymer layer deposited on the porous poly(propylene) substrate.

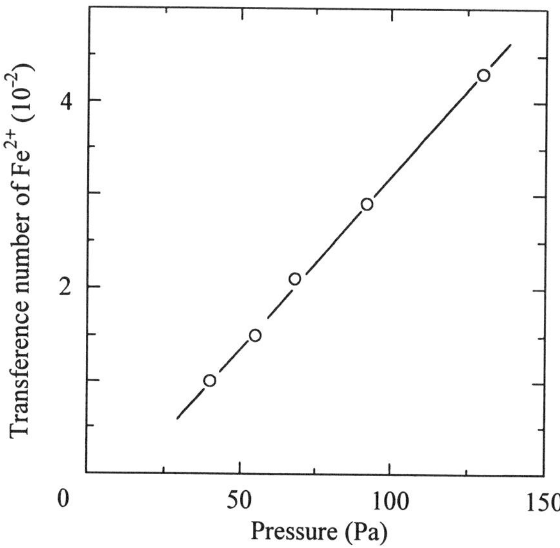

Figure 5. Dependence of the transference number of Fe^{2+} on pressure. Flow rate of 4-vinylpyridine: 10 cm^3 (STP)/min; RF power: 5 W; average film thickness: 0.24 μm.

controlling the polymerization time. The transference number of Fe^{2+} decreased with decreasing pressure. As described above, at low pressure, the crosslinking of the deposited polymer proceeds, resulting in low swelling. The low swelling exerts a strong electrostatic force on ions in the solution and gives a higher monovalent cation perm-selectivity.

Figure 6 shows the dependence of the membrane resistance on the polymerization pressure. The decrease of resistance with the increase of pressure can also be explained in terms of crosslinking and swelling; high crosslinking lowers the swelling, which leads to a high resistance [15]. The results of Figs 5 and 6 indicate that enhancement of perm-selectivity will be achieved at a high activation level in the plasma, but this will be at the expense of increased membrane resistance.

The membrane resistance of the plasma-modified Nafion is shown in Fig. 7 as a function of the thickness of the plasma polymer layer. The membrane resistance of the plasma-modified Nafion is directly proportional to the thickness of the membrane. However, the resistance obtained by extrapolating to zero membrane thickness is not zero but, rather, 7.6 Ω cm^2. The resistance of the oxygen-sputtered Nafion used as the substrate is only 1.5 Ω cm^2. These facts indicate that the membrane resistance consists of two different components; namely, the bulk resistance due to the plasma polymer layer and interfacial resistance between the plasma polymer layer and the Nafion membrane, which is 6.1 Ω cm^2.

The interfacial resistance might be ascribed to neutral layers containing no or few ion-exchange groups, or might be caused by stable ion pairing between the sulfo group of the Nafion membrane and the quaternized amino or pyridyl group. The former layer might be formed by decomposition of the sulfo group in the Nafion membrane caused by the attack of highly energetic species in the plasma glow region.

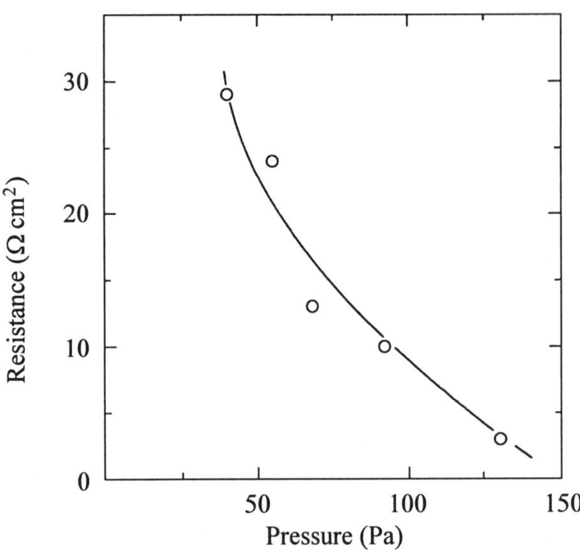

Figure 6. Dependence of the membrane resistance on pressure. Flow rate of 4-vinylpyridine: 10 cm^3 (STP)/min; RF power: 5 W; average film thickness: 0.24 μm.

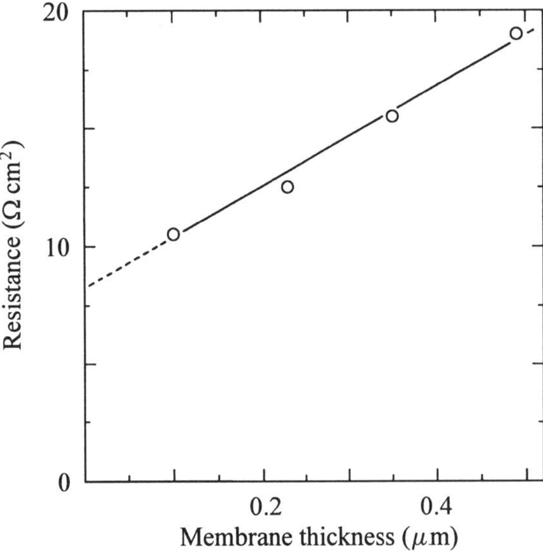

Figure 7. Variation of the membrane resistance with the membrane thickness of the plasma polymer layer. Flow rate of 4-vinylpyridine: 10 cm^3 (STP)/min; RF power: 5 W; polymerization pressure: 67 Pa.

The latter layer might be formed by the penetration of nitrogen-containing species into the Nafion membrane and then the formation of stable ion pairs between the nitrogen-containing species (amino group and/or pyridyl group) and the fixed sulfonic acid group in Nafion.

3.4. Perm-selectivity and resistance of the surface-modified Nafion by the plasma-induced surface modification process

The ESCA N 1*s* peaks of Nafion modified by reactions with 4-vinylpyridine and 3-(2-aminoethyl)aminopropyltrimethoxysilane indicate that a nitrogen atom was introduced onto the Nafion surface. However, Si was not detected on the Nafion surface modified by the reaction with 3-(2-aminoethyl)aminopropyltrimethoxysilane. Due to the weakness of the Si—C bond in 3-(2-aminoethyl)aminopropyltrimethoxysilane, the bond is cleaved and the Si-containing group evaporates.

The t_{Fe} results through Nafion modified by reactions with 4-vinylpyridine and 3-(2-aminoethyl)aminopropyltrimethoxysilane are summarized in Fig. 8. Oxygen was used as the sputtering gas. The gas flow rate and sputtering pressure were kept at 10 cm^3 (STP)/min and 67 Pa, respectively.

t_{Fe} of the untreated Nafion membrane was 0.32. These results show that t_{Fe} for Nafion treated by plasma-induced surface modification is significantly lower than t_{Fe} for untreated Nafion.

Modification by the reaction with 3-(2-aminoethyl)aminopropyltrimethoxysilane was more effective than that by 4-vinylpyridine. This may be ascribed to the fact that the Si—C bond energy is lower than the C—C bond energy. The lower bond energy facilitated the reaction of 3-(2-aminoethyl)aminopropyltrimethoxysilane with radicals on the surface, compared with the reaction of 4-vinylpyridine; the number of amino groups fixed on Nafion was higher for the former than for the latter.

As shown in Fig. 8, t_{Fe} decreased with an increase in the RF power. An increase in the RF power leads to an increase in the concentration of reactive species in the glow region. Radical sites on the Nafion surface increased with an increase in the concentration of reactive species. The large concentration of radical sites introduced on

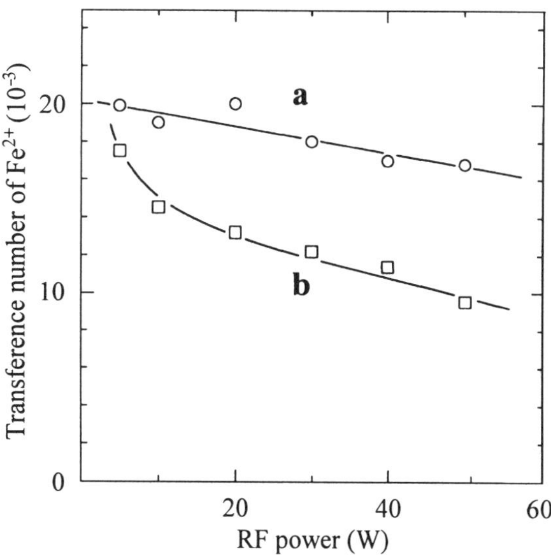

Figure 8. Dependence of the transference number of Fe^{2+} on the sputtering power. O_2: 10 cm^3 (STP)/min; pressure: 67 Pa; 1 min. (a) 4-vinylpyridine; (b) 3-(2-aminoethyl)aminopropyltrimethoxysilane.

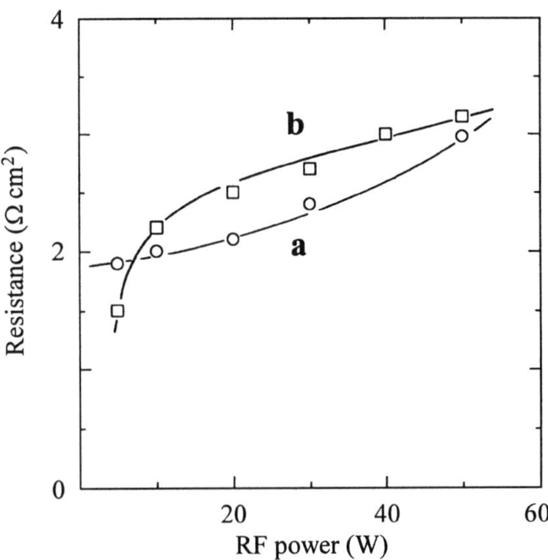

Figure 9. Dependence of the membrane resistance on the sputtering power. O_2: 10 cm^3 (STP)/min; pressure: 67 Pa; 1 min. (a) 4-vinylpyridine; (b) 3-(2-aminoethyl)aminopropyltrimethoxysilane.

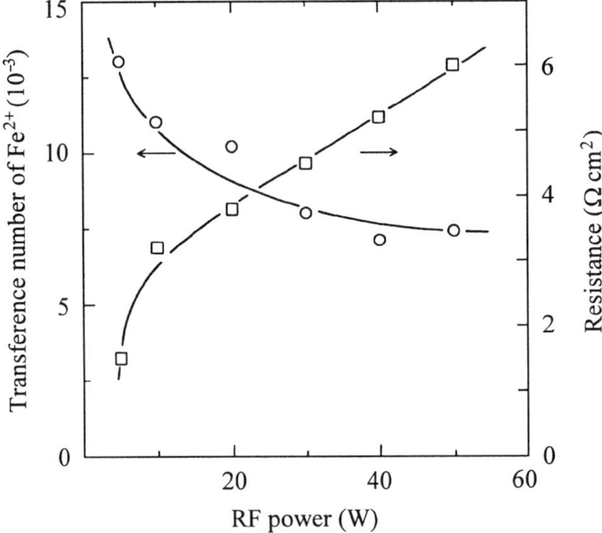

Figure 10. Dependence of the transference number of Fe^{2+} and the membrane resistance of Nafion modified by 3-(2-aminoethyl)aminopropyltrimethoxysilane on the sputtering power. Ar: 10 cm^3 (STP)/min; pressure: 67 Pa; 1 min.

the Nafion surface treated in a high RF power region leads to a remarkable lowering of t_{Fe}.

The membrane resistance of surface-modified Nafion by reactions with 4-vinylpyridine and 3-(2-aminoethyl)aminopropyltrimethoxysilane is illustrated in Fig. 9. The membrane resistance is about 2–3 Ω cm^2, which is considerably lower than the value of the Nafion surface modified by a plasma polymerization method (Fig. 7), while t_{Fe} is almost unchanged. These results reveal that the implantation of cationic species containing nitrogen was suppressed in the present plasma-induced surface modification.

The t_{Fe} results through Nafion modified by the reaction with 3-(2-aminoethyl)aminopropyltrimethoxysilane are summarized in Fig. 10. Also indicated in Fig. 10 is the membrane resistance. Argon was used as the sputtering gas. On comparing Figs 8 and 10, it can be seen that t_{Fe} for Nafion modified by sputtering using Ar is slightly lower than that when oxygen is used. Radical sites produced by ion bombardment react not only with 3-(2-aminoethyl)aminopropyltrimethoxysilane, but also with oxygen, which is a known radical scavenger. Oxygen remains in the reactor after sputtering is finished. Therefore, the surface density of amino groups sputtered by oxygen is lower than that in the case of argon sputtering, leading to a higher t_{Fe}. The membrane resistance increased with increasing RF power. The increase in the concentration of radical sites on the Nafion surface causes contradictory effects: the increase of the cationic group enhances the cation selectivity but it increases the membrane resistance. Optimization must be considered between the two effects — improvement of perm-selectivity and increase of the membrane resistance — for the better performance of perm-selectivity of the membrane as a separator.

REFERENCES

1. A. Reiner and K. Ledjeff, *J. Membr. Sci.* **36**, 535 (1988).
2. K. Nozaki, H. Kaneko, A. Negishi and T. Ozawa, *Denki Kagaku* **51**, 189 (1983).
3. Z. Ogumi, T. Mizoe, C. Zhen and Z. Takehara, *Bull. Chem. Soc. Jpn.* **64**, 1261 (1991).
4. Y. Mizutani, *J. Membr. Sci.* **54**, 233 (1990).
5. A. B. Laconti and H. J. Marget, *J. Electrochem. Soc.* **118**, 506 (1971).
6. Z. Ogumi, Y. Uchimoto, M. Tsujikawa and Z. Takehara, *J. Electrochem. Soc.* **136**, 1247 (1989).
7. Z. Ogumi, Y. Uchimoto, M. Tsujikawa, Z. Takehara and F. R. Foulkes, *J. Electrochem. Soc.* **137**, 1430 (1990).
8. Z. Ogumi, Y. Uchimoto, M. Tsujikawa and Z. Takehara, *Bull. Chem. Soc. Jpn.* **63**, 2150 (1990).
9. Z. Ogumi, Y. Uchimoto, M. Tsujikawa and Z. Takehara, *J. Membr. Sci.* **54**, 163 (1990).
10. K. Yasuda, T. Yoshida, Y. Uchimoto, Z. Ogumi and Z. Takehara, *Chem. Lett.* **1992**, 2013 (1992).
11. Z. Ogumi, Y. Uchimoto, K. Yasuda, T. Yoshida and Z. Takehara, *Denki Kagaku* **60**, 462 (1992).
12. H. Ohya, K. Emori, T. Ohto, Y. Negishi and K. Matsumoto, *Denki Kagaku* **53**, 462 (1985).
13. R. M. Fouss, M. Watanabe and B. D. Coleman, *J. Polym. Sci.* **35**, 5 (1960).
14. H. L. Yeager and A. Steck, *Anal. Chem.* **51**, 862 (1979).
15. A. Herrera and H. L. Yeager, *J. Electrochem. Soc.* **134**, 2446 (1987).

Polymer Surface Modification: Relevance to Adhesion, pp. 491–504
K. L. Mittal (Ed.)

A surface acoustic wave sensor study of polyimide thin film surface treatments — effect on water uptake

D. W. GALIPEAU,[1,*] P. R. STORY,[1] C. FEGER[2] and K.-W. LEE[2]

[1] *Department of Electrical Engineering, South Dakota State University, Brookings, SD 57007, USA*
[2] *IBM T. J. Watson Research Center, PO Box 218, Yorktown Heights, NY 10598, USA*

Revised version received 28 April 1994

Abstract—The effects of surface treatments on the water uptake in thin (1 μm) polyimide (PI) films were studied using a surface acoustic wave (SAW) sensor, X-ray photoelectron spectroscopy (XPS), external reflectance infrared (ERIR) spectroscopy, and contact angle measurements. Surface modification of PI films can affect film properties such as water uptake and adhesion. These properties, in turn, affect the performance and reliability of the devices in which these films are used. The ability to nondestructively study the results of various surface modification techniques *in situ*, prior to deposition of a metal layer for example, would be of particular benefit in the fabrication process. The results of this work indicate that the SAW sensor can measure extremely small amounts (< 0.003 μg) of water uptake in thin (1.2 μm) PI films. Also, that the water uptake of PI films, as measured by the SAW sensor, is particularly sensitive to sputter cleaning, sputtering/KOH, and Teflon AF surface treatment. The SAW, XPS, ERIR, and contact angle studies of the Teflon AF treated PI indicate that the concentration of Teflon AF is very high in the surface region of the PI and decreases into the bulk of the film. This work suggests utility of the SAW sensor as a nondestructive and *in situ* method for monitoring the surface properties of thin polymers in process control applications.

Keywords: Polyimide; surface treatment; water uptake; surface acoustic wave.

1. INTRODUCTION

Enhancing the characteristics of polyimide (PI) films is important for the improved performance and reliability of the devices in which these films are used. Currently, PI films are used in the microelectronics industry as intermetal dielectrics, multilayer thin film substrates for multichip modules (MCMs) and protective coatings [1–4]. PI films possess several attractive features for these applications including high temperature stability, a low dielectric constant, excellent planarizing characteristics and compatibility with microelectronic fabrication processes. However, film adhesion and moisture permeation continue to be important areas of concern.

*To whom correspondence should be addressed.

Investigators have reported improved adhesion of PI films resulting from surface treatments to the PI [5, 6]. It is also known that PI films can absorb significant amounts of water [7] and that water uptake in PI can affect adhesion and cause metal corrosion of devices [8–10]. Preliminary studies, using a surface acoustic wave (SAW) sensor, have shown a dramatic reduction in the water uptake of PI films, resulting from sputtering (sputter cleaning) of the PI [11]. Microbalance measurements by Clearfield *et al.* on similarly treated PI films have shown a similar effect on the water uptake [10].

Since surface treatments affect both the adhesion and water uptake of PI films, a technique that can study the effects of surface treatments quantitatively, nondestructively and *in situ* would be particularly useful during the microelectronic device fabrication process. The SAW sensor has strong potential for this application because of its small size, sensitivity, and the fact that it is fabricated using standard microelectronic fabrication techniques. The SAW sensor used in this work also offers advantages over the microbalance, the most commonly used technique for water uptake measurements in films, including: insensitivity to movement and vibration; and extreme sensitivity (0.001 μg [12] versus 0.1 μg for a microbalance) to water uptake in very thin (< 1 μm) and small surface area (< 1 cm^2) films. Since the PI films used in integrated circuits are typically in the 1 μm thickness range, this latter advantage could be extremely useful.

SAW sensors have been used extensively to study the vapor absorption and the thermomechanical properties of polymer films [12–16]. More recently, SAW sensors were used to nondestructively study the relative adhesion of PI films on quartz by measuring differences in the water uptake of the films as a function of interface characteristics [17]. The difference in the water uptake of these films was attributed to additional water present at the interfaces of the films with weaker adhesion. This work demonstrated the extreme sensitivity of the SAW sensor for measuring water uptake in thin films, especially when used in the dual delay line mode. The purpose of this work was to examine the utility of the SAW sensor as a method for measuring surface modification of polymer films and the effect of sputtering, sputtering/KOH, and Teflon AF surface treatments on the water uptake of PI films. The results of X-ray photoelectron spectroscopy (XPS), external reflectance infrared (ERIR), and contact angle measurements on the Teflon AF treated PI films are also reported.

2. EXPERIMENTAL PROCEDURE

A schematic of a dual delay line SAW sensor with two PI films is shown in Fig. 1. The dual delay line design is useful when accurate comparisons of the responses between two films are required. ST-quartz was used as the substrate because of its piezoelectric properties, temperature stability, and similarity to SiO_2 in integrated circuits (ICs). The interdigital transducers (IDTs) on the sensor are metal films (copper or aluminum) approximately 150 nm thick, that were patterned using standard microelectronic fabrication techniques. The substrate was 2.52 cm wide, 3.15 cm long and 0.9 mm thick. The operating frequency of the sensor was 80 MHz and the wavelength was 40 μm. The number of finger pairs for each IDT was 80; the aperture, 4 mm;

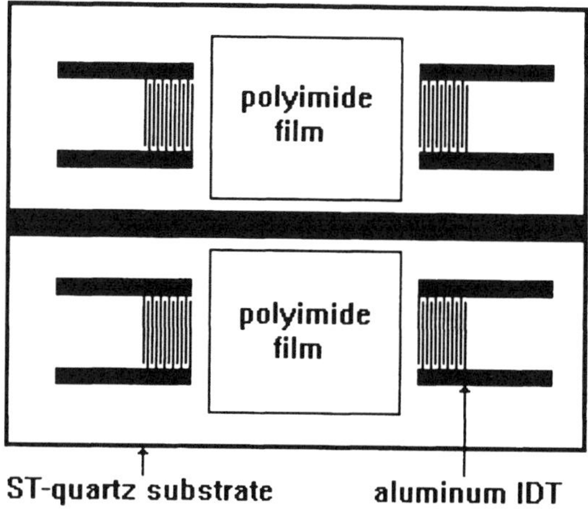

Figure 1. The dual delay line SAW sensor with two PI film samples prepared using methods 1 and 2 (see text).

and path length, 11 mm. Acoustic damping was used on the back side of the IDTs and on the bottom surface of the substrate to prevent edge reflections and bulk wave propagation, respectively. Since the sensor operated at about 30 dB of insertion loss, triple transit distortion was not a problem.

The PI used in this study was based on PMDA-ODA (DuPont 5878). Three different methods were used to deposit the PI films onto the SAW sensors. For methods 1 and 2, poly(pyromellitic dianhydride–oxydianiline)amic acid (PAA) in 1-methyl-2-pyrrolidinone (NMP) was spin-coated onto the entire SAW sensor and dried at 80°C for 30–60 minutes. The PAA was then cured using standard procedures (150°C, 30 min; 230°C, 30 min; 300°C, 30 min; and 400°C, 60 min). The final PI film pattern was obtained by covering the film to be left on the SAW delay path with a 0.1 μm copper mask. The unmasked PI was then removed by oxygen reactive ion etching (RIE). The copper mask was subsequently removed with standard photolithographic etching solution. The copper mask was deposited using one of two methods. In the first method the copper was evaporated through a Mylar contact mask onto the PI. In the second method the PI was sputter cleaned with argon for about one minute to improve the adhesion of the copper mask, and then the copper was evaporated through the contact mask onto the PI. For both methods the final size of the PI films after etching was 6.0 mm by 6.0 mm. Both processes provided the PI film pattern shown in Fig. 1. The cured PI had a final thickness of approximately 1.2 μm. These films were stored for over eighteen months and were reexamined in order to assess long term changes in water uptake.

Method 3 was designed to eliminate the need for masking and etching of the polyimide. The steps for fabricating films with this method were as follows. Those areas of the sensor not to be covered with PI were dipped into a solution of 6% Teflon AF 1600 (DuPont) in Fluorinert 48 (3M). The sensor was then dried at 80°C. The PAA was then spin-coated onto the substrate and dried at 80°C. Since the PAA does not adhere to the Teflon AF, a PAA strip 6 mm wide was left across the width of the

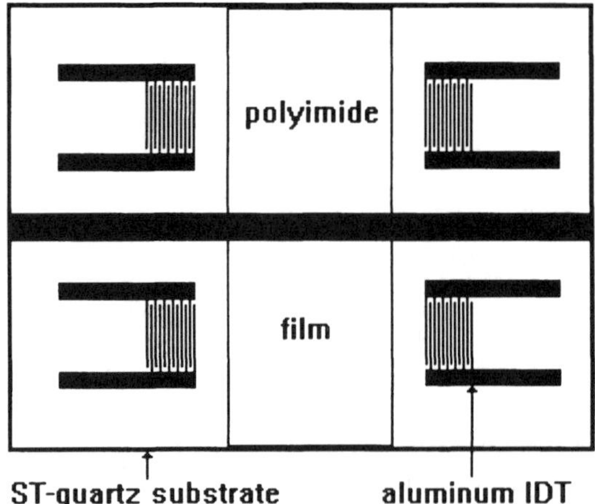

Figure 2. The dual delay line SAW sensor with a strip of PI across both delay lines prepared using method 3 (see text).

sensor and in the delay path of both delay lines as shown in Fig. 2. The remaining Teflon AF was removed by washing the entire sensor repeatedly in Fluorinert for a total exposure of about one hour. The PAA was then cured, as described in method one, to 400°C. This process resulted in 'Teflon AF treated' PI films since Teflon AF was apparently deposited on the PI during the Fluorinert washing, as suggested in the results section. The PI films were stored at room temperature and humidity (about 22°C and 30% to 75% relative humidity).

The operating principle of the SAW sensor, as used to measure thin film properties, is as follows. When an RF signal at the appropriate frequency is applied to the input transducer(s), the piezoelectric effect causes a series of surface acoustic waves or SAWs to be launched onto the propagation (delay) path(s) of the sensor. These waves then propagate along the delay path(s) to the output IDT(s) where they are converted back to an electrical signal. A film which has been deposited on the delay path can affect both the velocity and attenuation of the SAWs. Similarly, a subtle change in a film's properties can also affect the velocity and attenuation of the SAWs. These changes can be monitored in one of two ways: first, by using the input signal as a reference to which the output signal of a single channel is compared (single delay line); second, by using the output signal from one channel as a reference to which the output of a second channel is compared (dual delay line). The first method allows for small changes in film properties to be detected when the film is perturbed. It can be used to study the effect of humidity on a film. The second method allows for extremely small differences in the responses of two films to be measured. This second method can be used to compare films with small differences in water uptake.

A vector voltmeter was used to measure the phase difference change and attenuation as a function of relative humidity (RH). The vector voltmeter was used versus an oscillator method due to its stability and acceptable precision (0.1°). The phase change is directly related to changes in the SAW velocity and can be measured with a precision of under 10 parts per million. The electronic setup and humidity control system were similar to those previously described [11, 17]. The relative humidity (RH) was controlled by mixing dry compressed air and air which was passed through a bubbler (approximately 100% RH) with rotameters to provide the desired RH. The total air flow rate through the chamber was held constant at 500 cm^3/min. All experiments were done at room temperature (about 22°C). Although this humidity control system is not very precise (about 3%), it was adequate since the measurements taken were to examine either large responses of individual films or the comparative and simultaneous response of two films. The humidity in the test chamber was monitored by an Omega HX93 humidity sensor (3% accuracy, 90% response in 10 sec.).

The PI (PMDA-ODA) samples used for the XPS and contact angle measurements were prepared as follows. Polyamic acid was spin-coated onto a Si wafer and baked at 85°C for 60 min. The thickness was around 50 nm. The substrate was then dipped into a Teflon AF solution (6% in Fluorinert, FC-77 from 3M) for either 30 min or 12 h, and then rinsed with the same solvent seven times (to remove the unabsorbed Teflon AF polymer) followed by curing at 400°C for 60 min. Final thickness after curing was around 30 nm. The advancing and receding water contact angles were measured with a Rame-Hart goniometer and the XPS spectra were obtained on a Perkin-Elmer 5500 ISS/XPS system. Mg Kα (Magnesium Kα band electrons) excitation was used.

3. RESULTS AND DISCUSSION

PI films with an untreated surface (prepared using method 1) exhibited a relatively large phase change when exposed to changes in relative humidity (RH) as shown in Fig. 3 for a typical film. The RH was varied from 0% to 100% and back to 0% in steps of 25%. Steps of 25% RH were used to assess the linearity of the response. As the RH was increased the phase decreased. This negative phase change directly

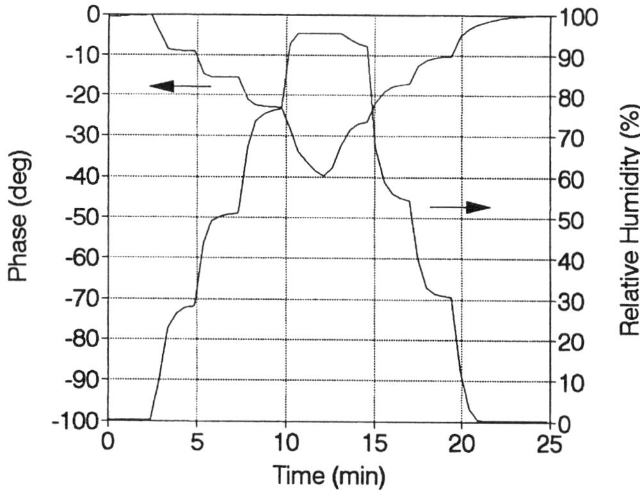

Figure 3. SAW humidity response of untreated PI film (0 to 100% RH).

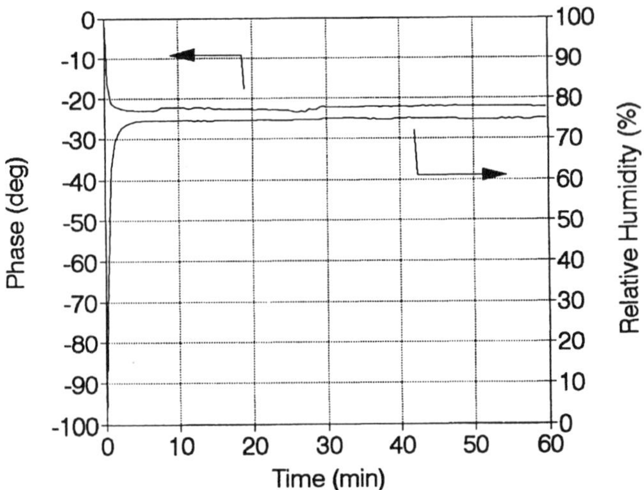

Figure 4. SAW humidity response of untreated PI film (75% RH).

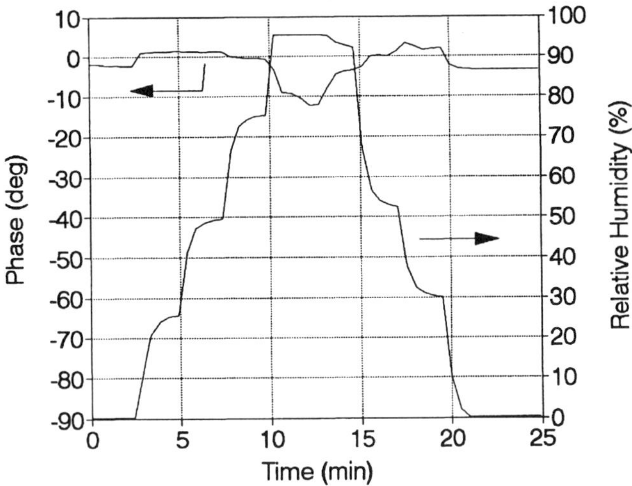

Figure 5. SAW humidity response of sputtered PI film.

corresponds to a decrease in the SAW velocity or 'loading' effect caused by water being absorbed in the film. As can be seen, there was a direct correlation between the decrease in phase and the increase in RH. The maximum phase change of about $-40°$ from Fig. 3 was equivalent to about a 2.0% increase in the mass of the PI film [17]. Exposure of the film to 75% RH for one hour indicates that there is no detectable change in film mass after about three minutes as shown in Fig. 4. The typical phase change for a sputtered PI film (prepared using method 2) is shown in Fig. 5. The phase change for the sputtered film was small, and in both the positive and negative directions. This indicates that only small amounts of water were absorbed and that it was absorbed only at humidity levels greater than 75%. This experiment suggests that sputtering changes the surface chemistry of the PI from hydrophilic to hydrophobic. If water molecules cannot be adsorbed on the film surface then they can no longer be absorbed into the bulk of the film. The water diffusion rate through sputtered PI has been shown by other investigators to be reduced for thicker films [10]. They also reported the existence of a graphite-like surface layer that inhibits water uptake in these sputtered films. The SAW sensor results suggest that for thinner films, water uptake is almost eliminated for RH below 75%. The observed results for the sputtered films were similar to the response of PI films covered with a metal mask or uncured PI films which also exhibited a small positive and negative phase change [18].

Aqueous potassium hydroxide (KOH) is known to hydrolyze PI and PAA [19]. This process also makes the film surface more hydrophilic [20, 21]. Surface treatment of the sputtered PI with KOH may activate some of the surface imide functional groups of the PI and re-expose surface oxygen. This would bring the surface chemistry of the sputtered PI film closer to that of unsputtered film. A one molar solution of KOH applied to the PI films for ten minutes at room temperature had only a small effect on the SAW humidity response. A two molar KOH solution, applied for ten minutes at room temperature, had a large effect on the SAW response as shown in Fig. 6. The maximum phase change was about −76°, a significantly larger change than the −40° seen for the untreated PI film (Fig. 3). Work done by Lee *et al.* [20] indicates that a PI surface modified with KOH yields a potassium polyamate surface which is very hydrophilic compared to an untreated PI surface [21]. Also, microscopic inspection of the sputtered/KOH treated PI films revealed a non-uniform surface. Pitting and cracking defects (probably caused by the sputtering) were present. It is reasonable to assume that these factors are responsible for the larger response of the sputtered/KOH treated PI films as compared to the untreated PI films. However, the modified PI film has become a complex system that is not well understood. Therefore the response differences at the various relative humidity levels can not be explained. Comparisons

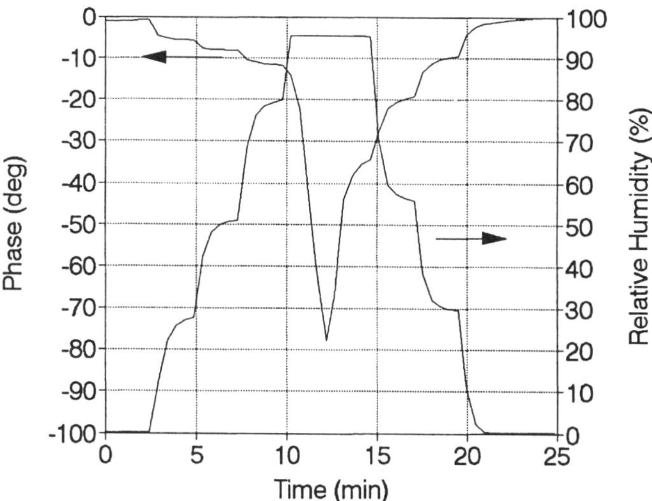

Figure 6. SAW humidity response of sputtered-KOH treated PI film.

of the water uptake response for films stored 18 months versus the response for films stored under three months [11] indicate as much as a 30% reduction in the water uptake for the older untreated and KOH treated PI films.

PI films prepared using method 3 (with Teflon AF) were examined using similar methods. The SAW sensor humidity response for a 'Teflon AF treated' film is shown in Fig. 7. The -5° maximum phase change indicates little water uptake in the film. At 75% RH, it can be seen that the water uptake rapidly reaches equilibrium at about -2° of phase change. Exposure of this film to 75% RH for periods of one hour (similar to Fig. 4) produced no further change in the water uptake than that shown in Fig. 7 for 75% RH. Surface rubbing of the PI with a cotton tipped probe, to determine if the Teflon AF was only on the surface of the PI, resulted in a very small change in the water uptake as shown in Fig. 8. This suggests that the Teflon AF treatment is resistant to physical modification since at least some of the surface layer of the PI was removed, based on observations of scratches on the PI caused by the rubbing. The effect of KOH treatment on Teflon AF treated films is shown in Fig. 9. The large change in water uptake is similar to, although somewhat smaller than that seen for the sputtered/KOH treated PI in Fig. 6. Since KOH does not normally react with Teflon AF, these results suggest that the KOH reacts with the PI as in the sputtered case, re-allowing water uptake in the PI film. It is also postulated that the Teflon AF may be washed away during the KOH treatment process.

The theoretical mechanism for the PI-SAW sensor humidity response has been examined previously [17]. The dominant parameters for the SAW humidity response are density and elastic constant changes in the film caused by water uptake in the polyimide. The theoretical predictions, which include the effect on phase caused by changes in film density, elastic constants, and stress, compare well with experimental results for newly prepared untreated PI films.

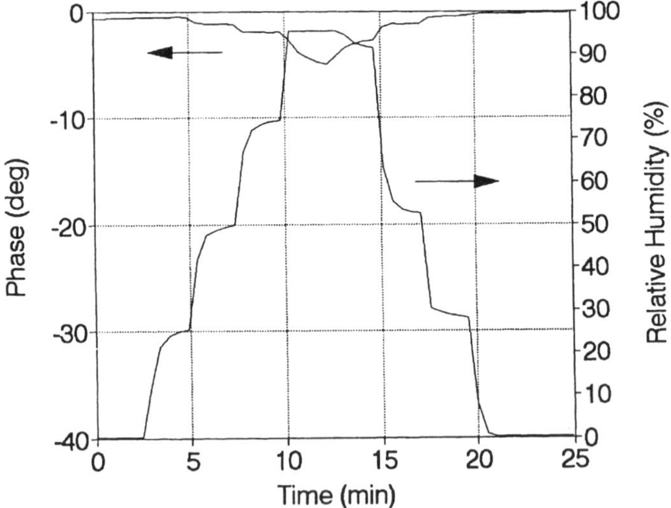

Figure 7. SAW humidity response of Teflon AF treated PI film.

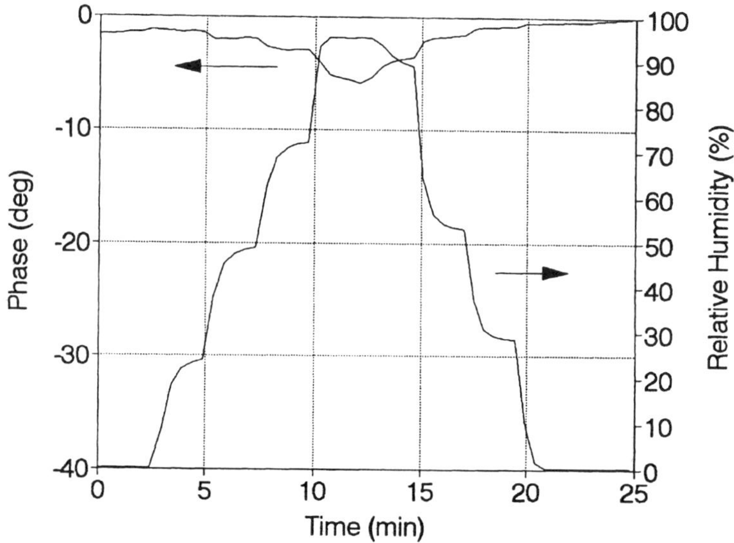

Figure 8. SAW humidity response of Teflon AF treated PI film after surface rubbing.

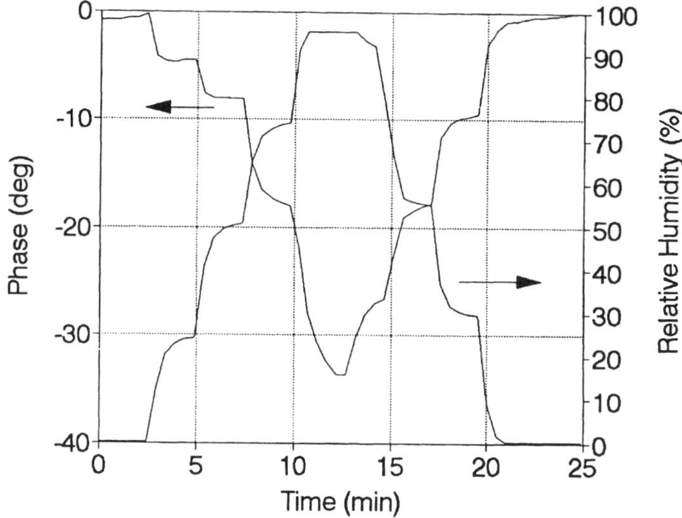

Figure 9. SAW humidity response of Teflon AF treated PI film after KOH treatment.

XPS, ERIR, and contact angle studies were also conducted in order to better understand the interaction and location of the Teflon in the PI film. Figure 10a shows the XPS spectrum of Teflon AF. The expected F_{1s} and O_{1s} peaks were observed. The advancing and receding water contact angles on a Teflon AF film are 125°/107°, indicating that it has a very low surface energy. A control experiment was carried out to test if the Fluorinert solvent absorbs into the fully cured PI (400°C for 60 min). The PI film was dipped into the solvent for 30 min, and then baked at 85°C for 60 min.

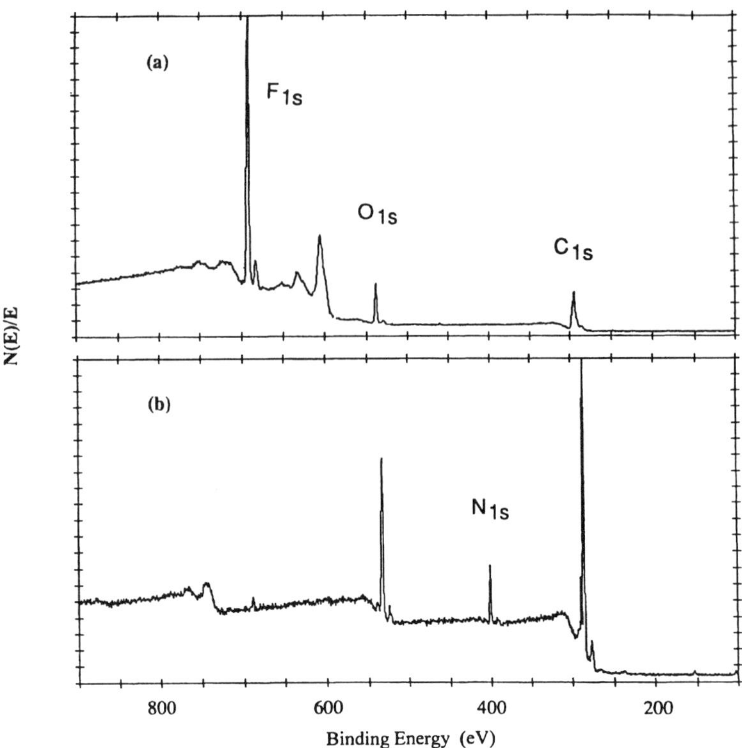

Figure 10. XPS spectra of (a) Teflon AF and (b) PMDA-ODA control. The control experiment was carried out by dipping a PMDA-ODA film into Fluorinert solvent for 30 min, and then baking the film at 85 °C for 60 min.

The XPS spectrum (Fig. 10b) of this sample (45° electron take-off angle) displays a negligible F_{1s} peak which appears at around −690 eV, indicating that the solvent absorption into the PI is negligible. Figure 10b is very similar to the spectrum of untreated PMDA-ODA.

A PI film treated with Teflon AF solution for 12 h, as described in the experimental section, was also studied. The advancing and receding water contact angles of this sample were 118°/104°, respectively, which are quite close to those of the Teflon AF film. Contact angles of the untreated PI film were 85°/38°. The XPS spectra of this sample were obtained at three different electron take-off angles: 15°, 45°, and 75° from the sample surface. As this angle increases, the XPS sampling depth increases. The approximate sampling depths are 1–1.5 nm, 3 nm, and 5 nm for 15° (Fig. 11a), 45° (Fig. 11b), and 75° (Fig. 11c), respectively. The F atom corresponds to Teflon AF, while the N atom to PMDA-ODA. The ratio of these two atomic concentrations is related to the amount of each material at the surface. The F/N ratio decreases as the sampling depth increases (38 for 15°, 10 for 45°, and 4.8 for 75°). The intensity of N_{1s} peak in Fig. 11a is negligible and the spectrum is very similar to that of Teflon AF (Fig. 10a). As shown in Fig. 11c, more PMDA-ODA repeat units are present in the outer 5 nm layer than Teflon AF units. These XPS results and the contact angle measurements indicate that the PI

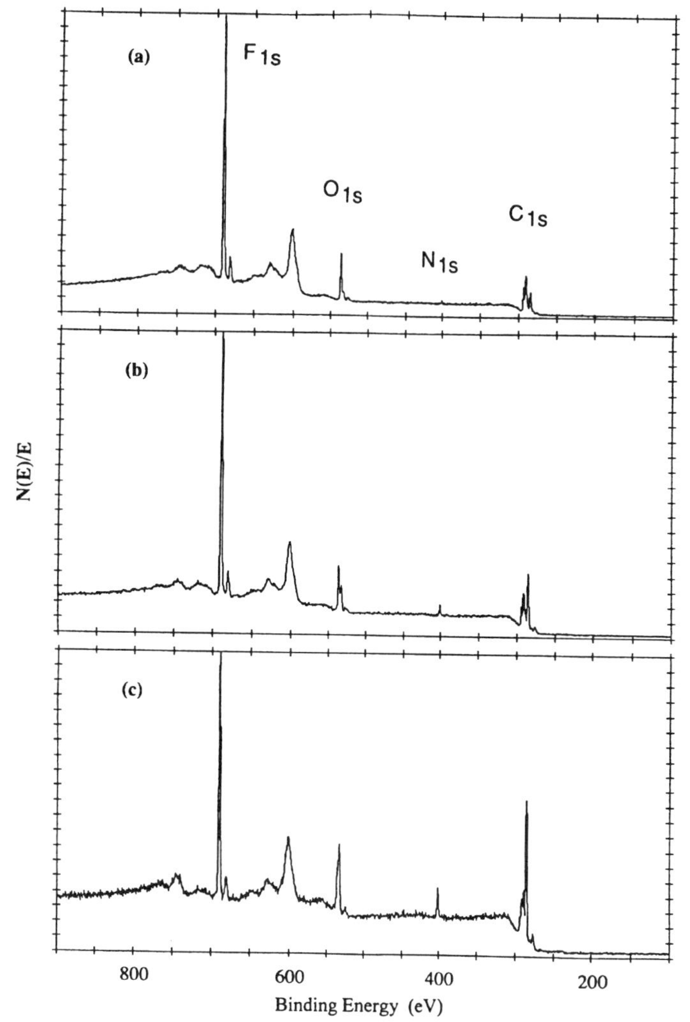

Figure 11. XPS spectra of PMDA-ODA with absorbed Teflon AF: (a) 15° electron take-off angle from the sample surface, (b) 45°, and (c) 75°.

surface contains mostly Teflon AF, and that the concentration of Teflon AF decreases into the bulk of the PI film where the PMDA-ODA molecules are the major component.

Figure 12 shows the C_{1s} XPS spectra of (a) Teflon AF, (b) PMDA-ODA with absorbed Teflon AF, and (c) PMDA-ODA. The electron take-off angle is 75° from the sample surface which represents a relatively deeper layer (5 nm). The carbons of Teflon AF appear in the range of 288–296 eV while those of PMDA-ODA appear in the range of 284–290 eV as shown in Figs 12a and 12c, respectively. The spectrum of the PMDA-ODA with absorbed Teflon AF (Fig. 12b) displays more carbons due to PMDA-ODA than Teflon AF, indicating that the inner layer contains more PMDA-ODA.

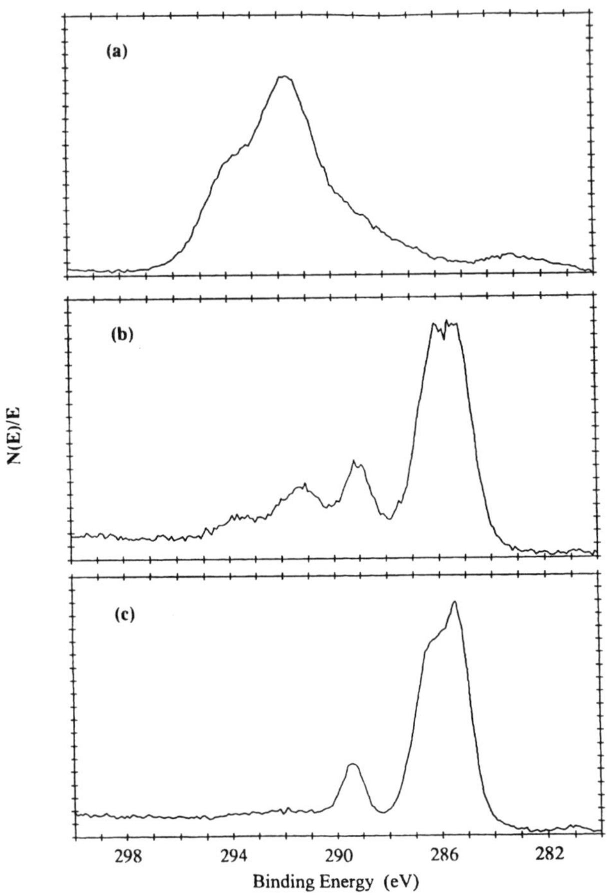

Figure 12. C_{1s} XPS spectra of (a) Teflon AF, (b) PMDA-ODA with absorbed Teflon AF, and (c) PMDA-ODA. The electron take-off angle is 75° from the sample surface.

The XPS experiments were repeated for a PI film treated for 30 min with the Teflon AF solution. The results were very similar to those for the 12 h treatment. ERIR spectra were also obtained for this film. The IR incidence angle was 15° from the sample surface. Figures 13a and 13c are for PMDA-ODA and Teflon AF, respectively. Figure 13b is for the Teflon AF treated PI. The small peaks at 1314, 1283, and 992 cm^{-1} (indicated with arrows) correspond to Teflon AF. This spectrum suggests that some Teflon AF was present below the surface of the PI film.

The SAW, XPS, ERIR, and contact angle measurements are in agreement and indicate that the concentration of Teflon AF is very high on the surface of the PI for the 'Teflon AF treated' films. The SAW and ERIR measurements further indicate that the Teflon AF may be present below the surface of the PI films but at a much lower concentration than at the surface. This could result from polymer chain mobility during curing. However, the exact mechanism for the interaction of the Teflon AF with the PI and the chemistry of the KOH treatments for both sputtered and Teflon AF treated PI films is not well understood.

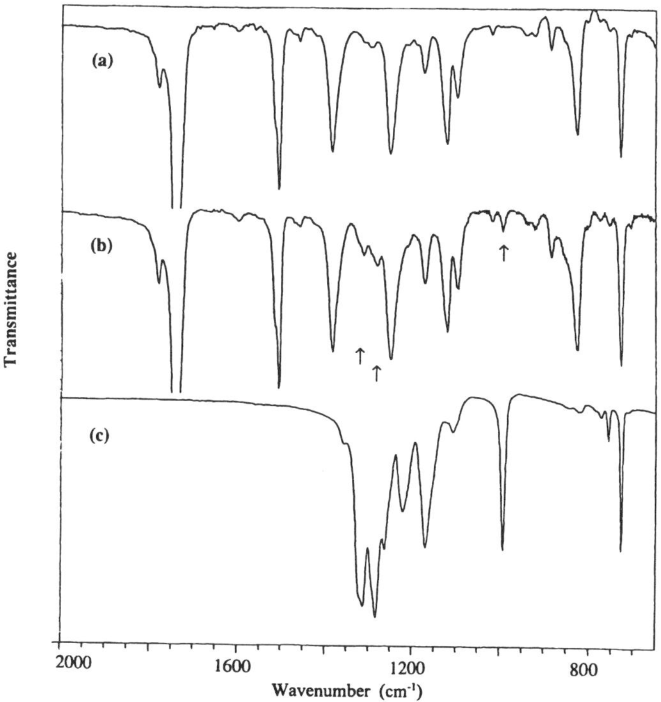

Figure 13. ERIR spectra of (a) PMDA-ODA, (b) Teflon AF-absorbed PMDA-ODA, and (c) Teflon AF. The small peaks at 1314, 1283, and 992 cm^{-1} indicated with arrows in (b) correspond to Teflon AF. The IR incidence angle was 15° from the sample surface.

4. CONCLUSIONS

SAW sensor studies of PI films have shown that this technique can detect very small changes in the water uptake characteristics of small (surface area < 1 cm^2) thin (< 1 μm) PI films resulting from sputtering, Teflon AF, and KOH surface treatments. The sensitivity (< 0.003 μg) of the SAW sensor is significantly better than a typical microbalance. While the reduced water uptake in sputtered PI films, and the increased water uptake in sputtered/KOH treated films has been known, the reduced water uptake in Teflon AF treated PI films has not been previously reported. Teflon AF treated PI films were also examined by XPS, ERIR and contact angle measurements. The results indicated that the Teflon AF concentration was very high at the surface of the PI film. This was in agreement with the results of the SAW sensor technique which showed that Teflon AF treatment reduces the total water uptake in PI films by an order of magnitude compared to the untreated PI film. The Teflon AF treatment could have applications where PI is used as a protective layer in microelectronic applications. Reduced water uptake of the PI should improve device reliability.

The SAW sensor's high sensitivity to the water uptake in thin PI films, small size and ease of fabrication suggests its use as an *in situ* monitor of PI fabrication process control in microelectronics.

Acknowledgements

This work was supported, in part, by research and equipment grants from South Dakota State University. Thanks are due to Kelly Hines for improving the data acquisition system.

REFERENCES

1. D. P. Seraphim, R. Lasky and C.-Y. Li (Eds), *Principles of Electronic Packaging*. McGraw-Hill, New York (1989).
2. R. R. Tummala and E. J. Rymaszewski (Eds), *Microelectronics Packaging Handbook*. Van Nostrand Reinhold, New York (1989).
3. D. S. Soane and Z. Martynenko, *Polymers in Microelectronics*. Elsevier, Amsterdam (1989).
4. M. L. Minges and C. A. Dostal, *Electronics Materials Handbook, Vol. 1, Packaging*. ASM International, Materials Park, Ohio (1989).
5. B. J. Bachman and M. J. Vasile, in: *Proc. IEEE 38th Electronic Components Conf.*, pp. 444–451 (1988).
6. I.-H. Loh and J. K. Hirvonen, *Mater. Res. Soc. Symp. Proc.* **108**, 241–246 (1988).
7. D. D. Denton, J. B. Camou and S. D. Senturia, in: *Moisture and Humidity*, pp. 505–513. Proc. of the Intl. Symp. on Moisture and Humidity, Instrument Society of America, Research Triangle Park (1985).
8. L. B. Rothman, *J. Electrochem. Soc.* **127**, 2216–2220 (1980).
9. D. G. Kim, T. S. Oh, S. Molis, S. Kowalczyk and J. Kim, *Mater. Res. Soc. Symp. Proc.* **203**, 65–70 (1991).
10. H. M. Clearfield, B. K. Furman, F. Bailey, N. Sheth and S. Purushothaman, *Mater. Res. Soc. Symp. Proc.* **264**, 237–242 (1992).
11. D. W. Galipeau, C. Feger and J. F. Vetelino, *Mater. Res. Soc. Symp. Proc.* **239**, 617–622 (1992).
12. D. S. Ballantine, Jr and H. Wohltjen, in: *Chemical Sensors and Microinstrumentation*, R. W. Murray (Ed.), pp. 222–236. American Chemical Society, Washington, DC (1989).
13. C. T. Chuang and R. M. White, in: *Proc. IEEE Ultrasonics Symposium*, Oct. 14–16, 1981, Chicago, IL, pp. 159–162 (1981).
14. J. A. Groetsch III and R. E. Dessy, *J. Appl. Polym. Sci.* **28**, 161–178 (1983).
15. S. J. Martin, G. C. Frye, A. J. Ricco and T. E. Zipperian, in: *Proc. IEEE Ultrasonics Symposium*, Oct. 14–16, 1987, Denver, Colorado, pp. 563–567 (1987).
16. J. G. Brace, T. S. Sanfelippo and S. G. Joshi, *Sensors and Actuators* **14**, 47–68 (1988).
17. D. W. Galipeau, J. F. Vetelino and C. Feger, *J. Adhesion Sci. Technol.* **7**, 1335–1345 (1993).
18. D. W. Galipeau, PhD Dissertation, University of Maine, Orono (1992).
19. J. G. Stephanie and P. G. Rickerl, Soc. of Plastics Eng. ANTEC, May 1991, Montreal, pp. 1696–1699.
20. K.-W. Lee, S. P. Kowalczyk and J. M. Shaw, *Macromolecules* **23**, 2097–2100 (1990).
21. K.-W. Lee, S. P. Kowalczyk and J. M. Shaw, *Langmuir* **7**, 2450–2453 (1991).

Polymer Surface Modification: Relevance to Adhesion, pp. 505–523
K. L. Mittal (Ed.)

A reactive acrylic adhesive for bonding polyolefins

JEFFREY T. FIELDS,[†] ANDREW GARTON[‡] and JAMES P. BELL*

Polymer Science Program, Institute of Materials Science, U-136, University of Connecticut, 97 North Eagleville Road, Storrs, CT 06269, USA

Revised version received 10 November 1994

Abstract—An acrylic adhesive was developed for forming strong, water resistant structural joints with polyolefins. This two-component, lightly crosslinked, methyl methacrylate (MMA) based adhesive consisted of an anaerobic curing system in one part with a copper (II) salt catalyst in the other. Bonds formed with low density polyethylene (LDPE) resulted in substrate failure upon block shear testing throughout the open time of the adhesive (45 min). The interdiffusion of the monomers into the substrates, and their subsequent polymerization was followed using several infrared spectroscopy (IR) techniques. The interphase of mixed LDPE and adhesive was determined to be as thick as 1.7 mm using IR microscopy. It was concluded that the strong adhesion in the aforementioned joints was the result of the interpenetration of the adhesive into the substrates.

Keywords: Anaerobic; interpenetration; two-part adhesive; interphase; polyolefins; acrylic adhesive; joint strength.

1. INTRODUCTION

While polyethylenes have long been desirable for their chemical resistance, it is often this property that limits their use in many applications that require adhesive bonding. Their inherent low surface tension (~35 mN/m) and limited reactivity make it difficult for adhesives to wet or react with polyethylene (PE) surfaces. Consequently, rigorous surface treatments have often been used to prepare the PE surfaces for adhesive bonding. These methods of surface modification include corona discharge [1], flame [2], and plasma [3] treatments, and are not always practicable or economical procedures.

*To whom correspondence should be addressed.

[†]Present address: Franklin International, 2020 Bruck Street, Columbus, OH 43207, USA.

[‡]Deceased.

In recent years, several authors have reported strong, durable bonding to polyethylenes and other polymers through the use of reactive adhesives [4–10]. The mechanism of adhesion presented in these papers involves the interdiffusion of the monomeric adhesive or coating into the substrate, and subsequent polymerization of the adhesive within the substrate. The resulting entangled adhesives provide strong adhesion to the substrates with significant durability and resistance to the environment. This mechanism has been referred to as interpenetration (for lack of a better term) in many of these reports, and this trend will be continued here. The same term has also been used to describe the analogous mechanism of adhesion between silane coupling agents and thermoplastic matrices [11–13]. Thus, there is precedent for the use of this nomenclature, even in the absence of the formation of a true interpenetrating network.

This paper describes the development of an anaerobic adhesive for forming strong durable structural joints with polyolefins. Anaerobic adhesives are typically single-component acrylic adhesives that polymerize at or below room temperature in the absence of oxygen, i.e. when the adhesive is placed between two substrates [14]. These adhesives have found wide usage as thread lockers for nuts and bolts, sealants, and for the *in situ* formation of gaskets. The first commercially successful anaerobic adhesives were developed by Krieble in 1959 [15] and consisted of methacrylate monomers, amines and hydroperoxides. Since this time hundreds of patients and papers have been issued on the subject.

One particular anaerobic curing system consists of cumene hydroperoxide (CHP) (structure I) as an initiator with benzoic sulfimide or saccharine (BS) (structure II) and N,N′-dimethyl-*p*-toluidine (DMPT) (structure III) as so-catalysts. The cure of this system has recently been studied by several authors [16–18]; a slight variation of this adhesive will be employed in this work with Cu(II)2-ethylhexanoate (CuEH) (structure IV) being used as an additional catalyst in the second part of the adhesive, rather than as a surface pre-treatment or primer. Methyl methacrylate (MMA) (structure V) is included as the primary monomer due to its high compatibility with the polyolefin substrates (see Table 1) [19]. A multifunctional co-monomer (triethylene glycol dimethacrylate — TRIEGMA) (structure VI) is also included in the formulation to enhance the thermal stability of the adhesive.

Table 1.
Solubility parameters of the monomers and polymers used in this project [19]

Material	δ (MPa)$^{1/2}$	δ^d (MPa)$^{1/2}$	δ^p (MPa)$^{1/2}$
MMA	18.0	—	—
PMMA	22.7	18.6	10.5
LDPE	16.2	16.2	0
PP	17.2	17.2	0

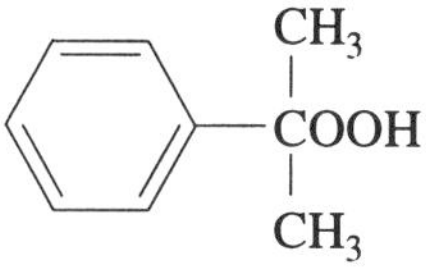

I
Cumene hydroperoxide (CHP)

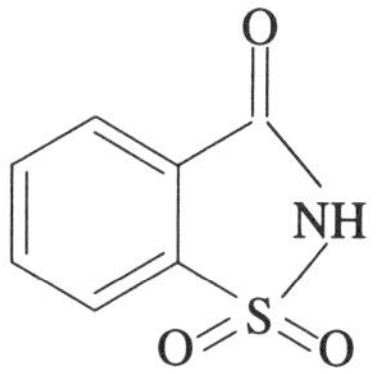

II
Benzoic sulfimide (BS)

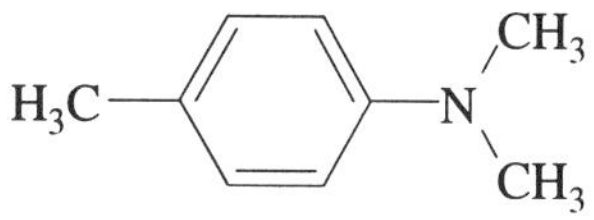

III
N,N′-dimethyl-*p*-toluidine (DMPT)

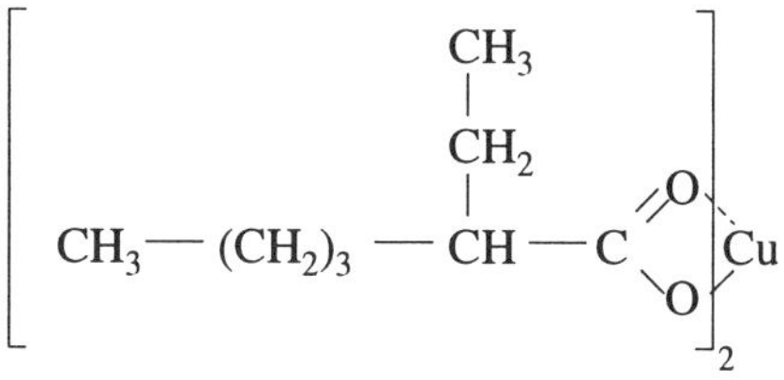

IV
Cu(II)2-ethyl-hexanoate (CuEH)

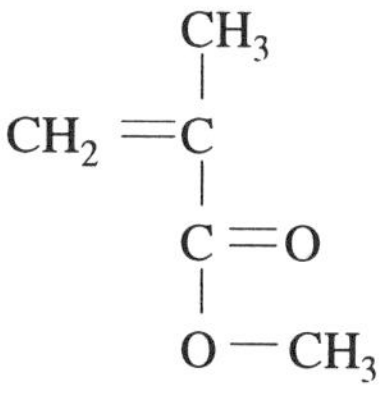

V
Methyl methacrylate (MMA)

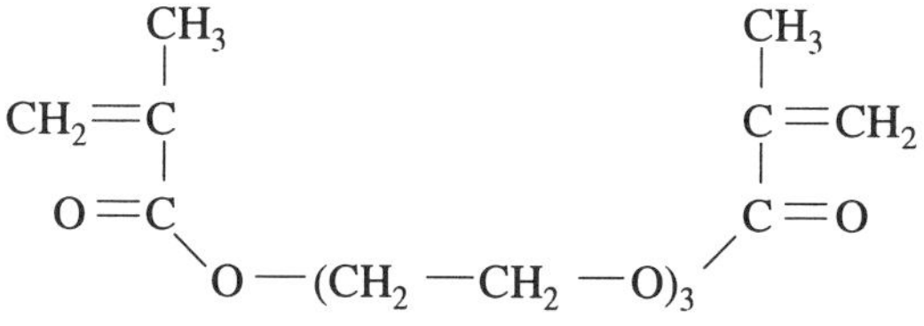

VI
Triethylene glycol dimethacrylate (TRIEGMA)

2. EXPERIMENTAL

The CHP, DMPT, BS and CuEH were all used as received from Aldrich. The MMA (Aldrich) and TRIEGMA (Scientific Polymer Products) were both passed through inhibitor removal columns, provided by their respective suppliers, prior to use. The formulation of the adhesive is presented in Table 2. The catalytic solution (part II) was measured out first and placed in a sonic bath until the CuEH went into solution. The main component (part I) of the adhesive was then measured out into a 20 ml vial. In an effort to dissolve the BS, the mixture of part I was placed in a sonic bath for

Table 2.
Adhesive formulation

Chemicals	MW (g/mol)	Amount	mmoles
Part I	—	—	—
methyl methacrylate	100.1	8.0 ml	NA
triethyleneglycol dimethacrylate	286.0	0.25 g	0.87
N,N-dimethyl-*p*-toluidine	135.0	0.25 g	1.9
cumene hydroperoxide	152.2	0.10 g	0.53
benzoic sulfimide	183.2	0.34 g	1.9
Part II	—	—	—
methyl methacrylate	100.1	5.0 ml	NA
Cu(II)2-ethylhexanoate	350.0	0.25 g	0.71

5 min. Upon removal from the bath (the BS was still not soluble), 0.10 ml of the catalytic solution were added to part I. This is referred to as time zero. At 2–3 min reaction time enough heat was produced from the polymerization reaction to dissolve the BS. Bonds were then formed for testing at 3, 10, 20, 30, and 40 min reaction time. At 45 min reaction time the adhesive began to gel and was not usable for further bond formation.

The temperature of the adhesive, resulting from the exothermic polymerization reaction, was monitored in the bulk adhesive. This was accomplished simply by measuring the temperature of the adhesive versus reaction time using a thermometer that was immersed in the adhesive.

All of the polyolefin substrates were obtained from Scientific Polymer Products. The low density polyethylene (LDPE), high density polyethylene (HDPE), and polypropylene (PP) sheets (0.64 cm thick) were cut into strips (1.4 × 2.5 cm) for lap joint formation. These strips were Soxhlet extracted with acetone for at least 24 h to remove any surface dirt or contamination. Spacers were applied to the substrates in order to obtain a uniform bondline. These shims consisted of an aluminum foil with a pressure sensitive underside and a fluorinated ethylene-propylene backing for a total thickness of 0.015 cm. The bonding areas were then wiped with acetone, followed by either a toluene or a primer (1% by weight CuEH in toluene) wipe (cotton swab) before application of the adhesive. The solvents were allowed to evaporate for about 20 min prior to joint formation. The adhesive was then applied to the prepared substrates which were clamped together with a ~1.25 cm overlap. The 11.4 cm wide joints were allowed to cure for at least 3 days and then each was cut into 3–2.5 cm wide bonds. These were then tested in block shear according to ASTM D4501-91 [20]. Unless otherwise noted the strengths presented here were the average of at least 9 samples.

The cure of the adhesive was followed using transmission FT-Near Infrared Spectroscopy (NIR) of actual bonds. This was achieved using 0.034 cm thick LDPE substrates as the windows (2.5 × 2.5 cm) for the NIR studies. These were prepared as were the substrates for block shear testing (described in the previous paragraph). Spacers were applied around 3 sides of the LDPE squares so as to provide a constant thickness and a seal to prevent leakage. The adhesive was then applied to the substrates at the proper reaction time and they were clamped together to form a bond for NIR study. A Mattson Galaxy 2050 NIR spectrometer equipped with a INSB detector was employed to follow the disappearance of the C=C−H overtone stretching band ($\sim$6165 cm^{-1}) versus time [21]. A methyl combination band at $\sim$4680 cm^{-1} was used as a reference to normalize for any differences in thickness. Each spectrum was the average of 16 scans taken at 8 cm^{-1} resolution.

The thermal properties of the adhesive were determined using Differential Scanning Calorimetry (DSC) on the cured adhesive. The adhesive samples were prepared by forming bonds between two Teflon FEP substrates and then allowing them to cure for at least 3 days. The bondlines were then easily removed from these 'non-stick' surfaces for sampling by DSC. A Perkin Elmer DSC-7 was used to determine the thermal properties. The samples were heated from 40°C to 300°C at 10°C/min, rapidly quenched back to 40°C, and then reheated using the same heating rate and temperature range.

Adhesive uptake and desorption experiments were performed on $\sim$100 μm LDPE films from Scientific Polymer Products. These films were also extracted, wiped with acetone, and wiped with toluene or a primer prior to use. The prepared films were soaked in the reacting adhesive for 10 min beginning at the indicated reaction time. Upon removal they were immersed in nitromethane (a good solvent for the adhesive, but a poor solvent for LDPE) for 10 s, physically wiped with a nitromethane cotton swab, and then reimmersed in the nitromethane for 5 s. The films were then allowed to dry for 10 min (total time from removal out of adhesive) before being sampled using Fourier Transform Infrared Spectroscopy (FTIR). A Mattson Polaris spectrometer equipped with a mercury cadmium telluride (MCT) detector was used to follow the desorption of the adhesive from the films versus time. Each transmission spectrum was the average of 32 scans taken at 4 cm^{-1} resolution.

Infrared microscopy was used to study the interphase that was formed between the adhesive and substrates. Cross sections $\sim$125 μm thick were microtomed from cured, unstressed joints for investigation using a Nicolet 60 SX FTIR spectrometer equipped with a SpectraTech IR-Plan infrared microscope. Background for this technique has been reviewed elsewhere [22]. A 25 $\times$ 400 μm sampling area was used and was situated so that the long axis was parallel with the interface. The initial spectra were taken as far away as 2 mm from the interface. The sampling spot was then moved in increments as small as 25 μm towards the bondline. The relative amount of adhesive was determined by the size of the carbonyl stretch absorption. Each spectrum was the average of 128 scans taken at a 4 cm^{-1} resolution.

Further confirmation of the presence of interpenetration was gathered using optical microscopy. Cross sections (prepared in the same manner as in the previous paragraph) of unstressed, stressed and fractured LDPE/LDPE joints were examined under tungsten light using a Nikon Labophot — Pol optical microscope. In most cases the gold color of the adhesive provided adequate contrast between the LDPE and the adhesive. Observed images were recorded using a Sony CCD-IRIS color video camera in conjunction with a Sony Mavigraph color video printer. Measurements of the interphase size were then made from the micrographs.

Durability testing was performed on selected LDPE/LDPE joints that were subjected to block shear, but not fractured. These specimens were placed in boiling water and were periodically checked for failure. After over 3 months time, the remaining unfailed joints were re-tested in block shear. These results were then compared to previously published data.

3. RESULTS

The mixing of the two components (Table 2) of the adhesive brought about a significant change in appearance. The CuEH in part II initially caused the adhesive to become green; however, this color was short lived for within ~10 s the adhesive became purple. The purple color remained for about 1 min, and then the material began to lighten to orange. This color change coincided with the maximum exotherm of the adhesive (Fig. 1) and the dissolution of the BS. The shift to orange was completed at 2–3 min reaction time. From this point on, the adhesive began to cool until about 20 min reaction time, where a plateau was reached in the temperature of the adhesive. A further darkening of the adhesive toward a reddish color was associated with this process. At ~45 min reaction time the adhesive began to gel and to lose its 'stickiness' and effectiveness as an adhesive.

The plot in Fig. 1 clearly shows the brief exotherm produced during the initial stages of the polymerization reaction. The consequences of this were that the bonds formed at 3 min reaction time and, to a lesser extent, 10 min reaction time were subjected to an adhesive at an elevated temperature. These higher temperatures should enhance

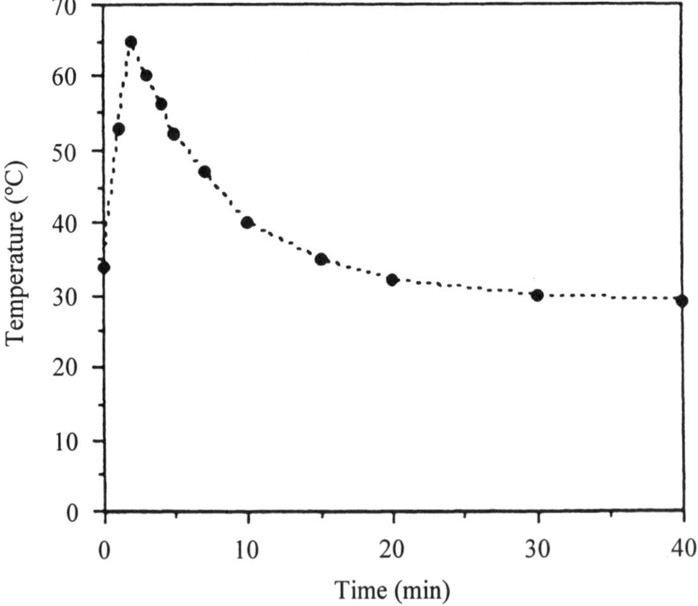

Figure 1. Plot of temperature of the reacting acrylic adhesive versus reaction time.

the interdiffusion of the adhesive and the substrates, and this will be discussed further in the following paragraphs.

LDPE/LDPE joints were formed from the reacting adhesive at 3, 10, 20, 30, and 40 min reaction time since the nature of the adhesive was constantly changing. Both primed and unprimed substrates were used to form joints for block shear testing. The results of this are included as Table 3. The joint strengths of primed HDPE/HDPE and PP/PP bonds formed at 3 min reaction time are also included in Table 3, although these values are the average of only 3 samples. HDPE and PP joints formed at other reaction times exhibited little to no strength. All of the joints that are represented in Table 3 exhibited some degree of substrate failure. This was quite drastic in some cases, as illustrated in Fig. 2, where 'clumps' of LDPE can be seen to be strongly adhered to the darker adhesive, and the permanent, macroscopic deformation of the LDPE substrates is obvious.

NIR spectroscopy was used to follow the polymerization reaction of the adhesive. This technique allowed for the *in situ* sampling of the adhesive while it was between the substrates. The overtone of the vinyl C—H stretching provided a sharp band free from interference, and this is illustrated in Fig. 3. The methyl combination band at 4680 cm^{-1} provided a good reference since it resulted only from the adhesive and did not change during the cure. This also permitted the normalization of the data for differences in thickness. The percent conversion was determined by ratioing these bands to each other and to the ratio of the same bands in an unreacted adhesive. Figure 4 illustrates the range of cures that were observed from these reactions. In general the reactions had fast initial cure rates for about the first 1 to 2 h after bond formation. These then leveled off to a much slower rate until the reaction was completed. The adhesives normally reached only about 80–90% conversion in the 24-h period, which was not surprising with the presence of the crosslinking

Table 3.
Summary of block shear strengths for polyolefin/acrylic/polyolefin joints

Reaction time	Treatment	Joint strength (MPa)	1 Standard deviation (MPa)
LDPE-3 min	primed	8.41	1.10
LDPE-3 min	unprimed	7.17	1.06
LDPE-10 min	primed	9.10	1.49
LDPE-10 min	unprimed	7.52	1.70
LDPE-20 min	primed	7.31	1.21
LDPE-20 min	unprimed	7.03	1.37
LDPE-30 min	primed	6.90	1.37
LDPE-30 min	unprimed	6.89	1.74
LDPE-40 min	primed	5.18	1.06
LDPE-40 min	unprimed	5.79	1.61
PP-3 min	primed	12.4	—
HDPE-3 min	primed	2.82	—

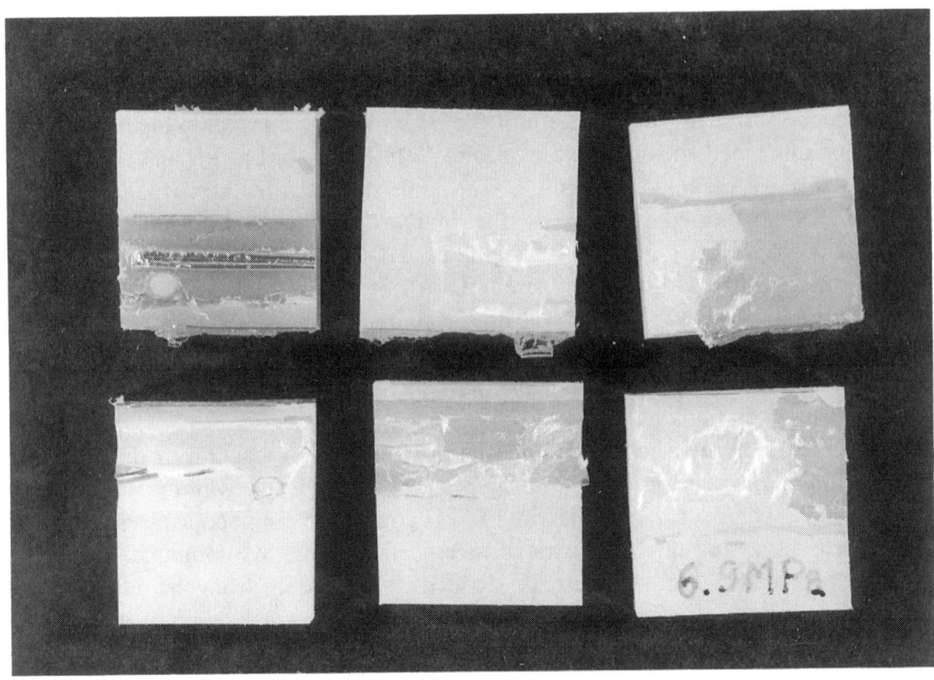

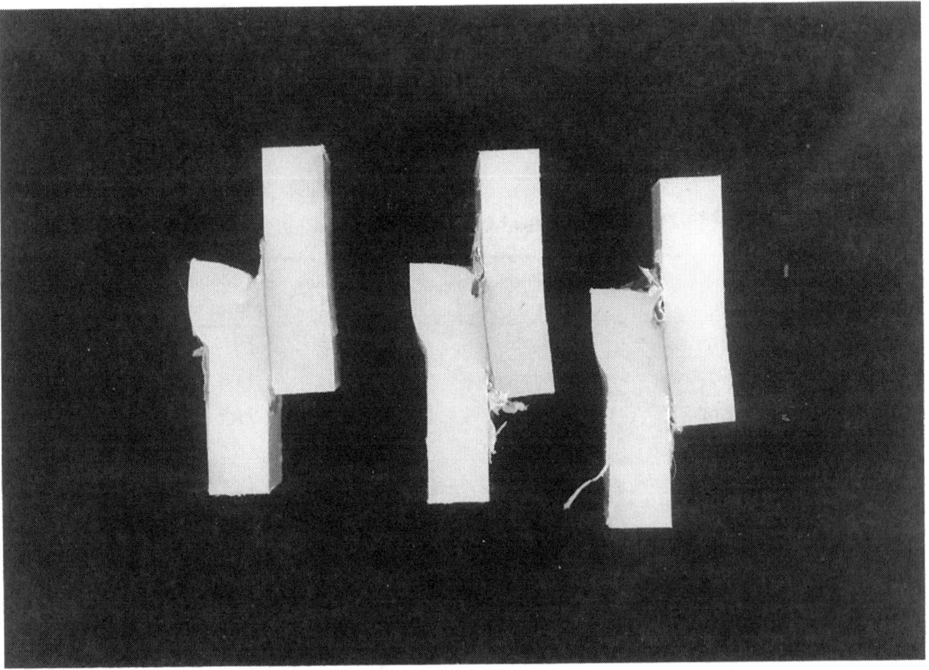

Figure 2. Representative LDPE bond failure surfaces (above) and stressed, but unfractured LDPE/LDPE joints (below) both displaying the substantial substrate failure.

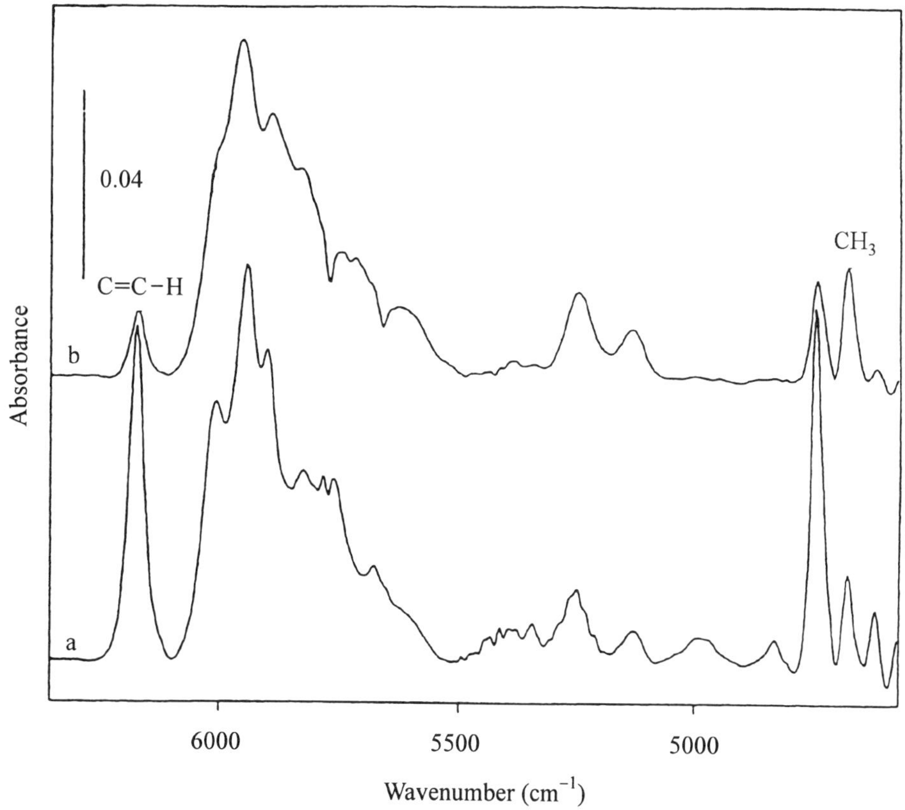

0.04
C=C−H
CH3
b
a
Absorbance
6000
5500
5000
Wavenumber (cm−1)

Figure 3. NIR spectrum of a primed bond formed at 10 min reaction time a) at the time of formation and b) 24 h after formation.

100
80
60
40
20
Conversion (%)
10 min unprimed
40 min primed
0
6
12
18
24
Time (h)

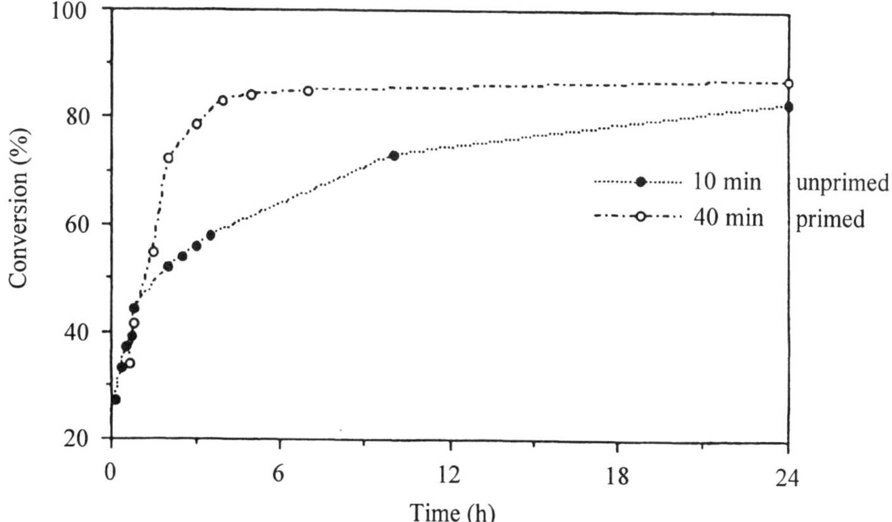

Figure 4. Plot of percent conversion versus time for an unprimed bond formed at 10 min reaction time (filled circles) and a primed bond formed at 40 min reaction time (open circles).

monomer (vitrification prior to complete reaction was likely to limit the mobility of the reactants). The overall cure rate of this adhesive was very slow for all reaction times, as it took several hours to reach the plateau region of the cure. However, no attempt was made to optimize or accelerate the cure rate. Most anaerobic adhesives cure in a matter of minutes. A summary of the initial cure rates (slope of the linear regions over the first hour) are shown in Table 4. Additionally, no significant difference was observed between the cure of the primed bonds versus the unprimed bonds.

The glass transition temperatures T_g of the cured adhesives were determined using DSC. Two distinct T_gs were detected in the initial run of each sample. The first was at ~110°C and was assigned to linear polymethylmethacrylate (PMMA). The second, much larger transition occurred at about 215°C and was assigned to the T_g of the crosslinked adhesive. After quenching, the second run had only one transition, falling between 215–220°C. In addition, an exotherm was observed in each case around 80°C. These changes were interpreted to indicate that the initial heating of the samples provided enough mobility to promote further reaction of any remaining double bonds upon subsequent heating. Table 5 summarizes these results.

Model interpenetration studies performed on LDPE films yielded surprisingly useful data. Transmission FTIR was implemented to follow the uptake and desorption

Table 4.
Summary of initial cure rates

Bond formation time	Initial cure rate (%/h)
3 min	21.1
10 min	17.3
20 min	17.0
30 min	27.8
40 min	28.5

Table 5.
Results from the DSC characterization of the adhesive after 3 days cure, starting at the indicated reaction time

Sample	Run	$1^{st} - T_g$ (°C)	$2^{nd} - T_g$ (°C)
3 min	1	108.0	216.7
3 min	2	—	217.9
10 min	1	111.4	213.5
10 min	2	—	216.4
20 min	1	111.4	215.5
20 min	2	—	217.5
30 min	1	111.8	214.0
30 min	2	—	218.6
40 min	1	110.1	216.1
40 min	2	—	218.6

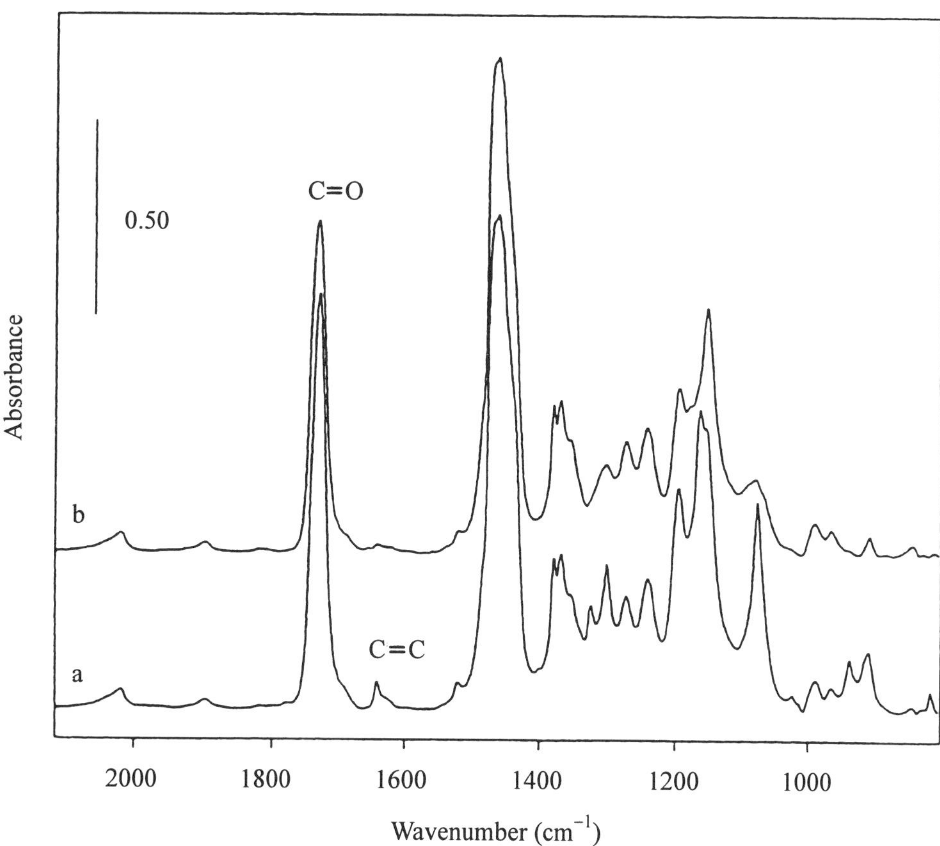

Figure 5. Transmission spectra of an LDPE film that was soaked in the reacting adhesive for 10 min, a) after removal from the adhesive and b) 24 h later.

processes of the adhesive in the LDPE films. Figure 5 shows the nature of the transmission spectrum that was obtained initially after the soak, and the spectrum of the same sample 24 h after removal from the adhesive. The carbonyl band from the adhesive was apparent at ~1728 cm^{-1} initially, as was the vinyl absorbance at 1640 cm^{-1}. Although some loss was observed after 24 h, a considerable amount of the adhesive was still present and could be clearly observed in the spectrum. This indicated that, although the cure of this particular adhesive was very slow, it was sufficient to allow polymerization within the substrate before desorption could take place. From these data alone it was difficult to determine if the penetrated adhesive was polymerized or not, since a small amount of unreacted vinyl bonds remained. A Soxhlet extraction of the film using THF (a good solvent for the PMMA) was also attempted in an effort to remove any unreacted or loosely held molecules. The subsequent spectrum was identical to the spectrum before the extraction was performed. In addition, any unreacted monomer was likely to have evaporated from the film. Representative curves illustrating the adhesive desorption over a 24 h period are also included in Fig. 6. The initial and final uptakes for both primed and unprimed films are summarized in Table 6.

Table 6.
Initial and final uptake concentrations for the adhesive in LDPE and PP soaked at the indicated reaction times

Sample	Initial uptake (mg/cm^2)	Final uptake (mg/cm^2)
3 min primed	0.439	0.424
3 min unprimed	0.407	0.281
10 min primed	0.343	0.179
10 min unprimed	0.266	0.113
20 min primed	0.248	0.103
20 min unprimed	0.225	0.0738
30 min primed	0.212	0.0902
30 min unprimed	0.226	0.0587
40 min primed	0.246	0.0948
40 min unprimed	0.205	0.0547
10 min primed PP	0.146	0.0510

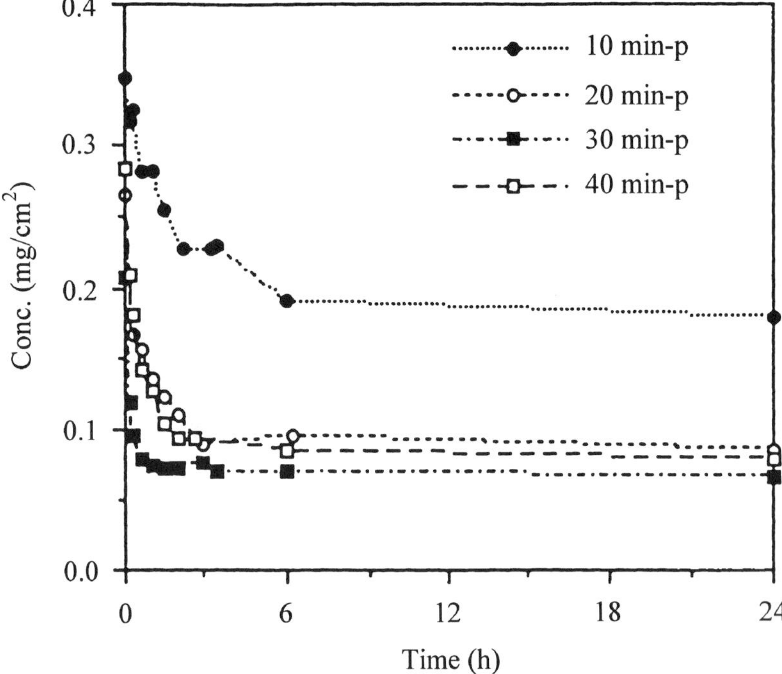

Figure 6. Plot of the desorption of the adhesive from primed LDPE films.

The interpenetration of the adhesive and substrate was confirmed using FTIR microscopy on bond cross-sections. The results presented here were somewhat unexpected; on a cross-section of a primed bond formed at 10 min reaction time the

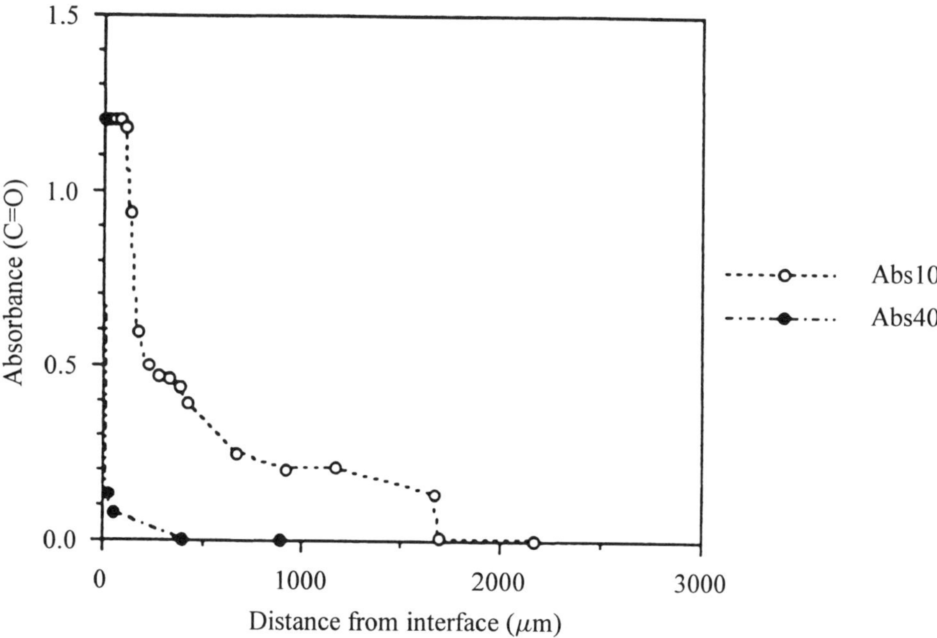

Figure 7. Plot of the carbonyl absorbance versus the distance from the interface for the cross-sections of primed bonds formed at 10 (open circles) and 40 (filled circles) minutes.

carbonyl absorption band due to the adhesive was detected almost 1.7 mm away from the adhesive/LDPE interface. Of course the concentration of the adhesive was minimal at that point, but it did show an exponential increase as the interface was approached. In fact, the last several measurements made within 100 μm of the bond-line had concentrations of adhesive so high that the carbonyl stretch was saturated and the data were beyond the region of linearity of Beer's Law. The adhesive was in a polymeric state, as the 1640 cm^{-1} vinyl band was noticeably absent. While the interphase was as thick as 1.7 mm, the majority of the interpenetrated material was localized to within the first 300–500 μm. A plot of the detected absorbance of the carbonyl stretch versus distance from the interface is included as Fig. 7. Also included in this plot are the results of the same experiment performed on a primed bond formed at 40 min reaction time. This interphase region was also significant, but was only a fraction of that observed in the 10 min case. Here the interpenetrated region was only about 50–100 μm in size with a much sharper increase in adhesive concentration. Again the absence of a vinyl band indicated that the adhesive was in a polymeric state. These data were consistent with the concentrations of the uptake experiments and explained the differences in joint strengths that were for these two reaction times.

As a proof that what was being detected was due to the adhesive and not the result of some other factor, an additional scan of the interface was performed in an area where there was no adhesive. This was because a spacer (pressure sensitive aluminum foil with an FEP coating) was adhered to the middle of the bonding surface in order to indicate the location of the interface for the optical microscopy studies. This prohib-

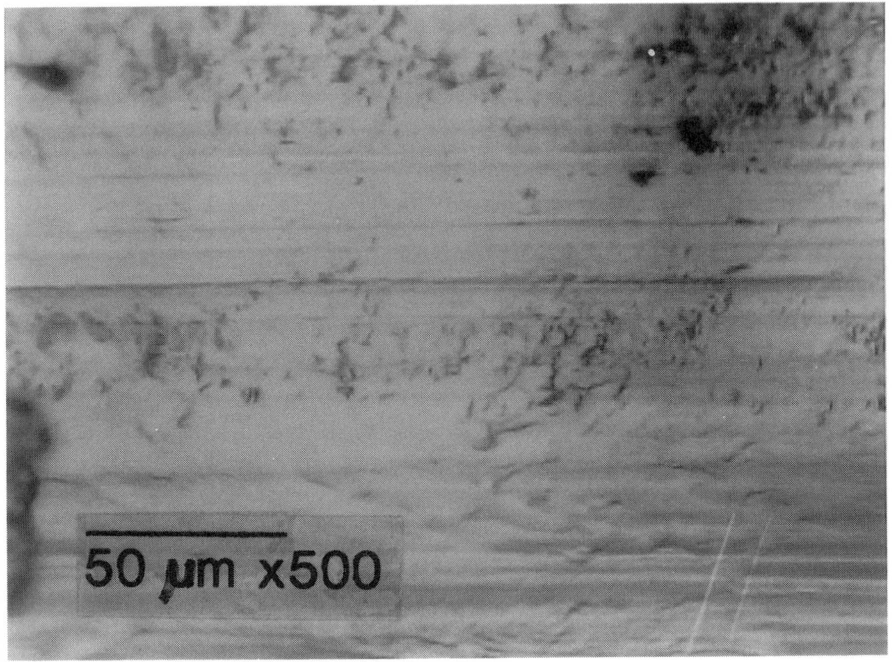

Figure 8. Optical photomicrograph of a cross-section of a stressed and fractured LDPE/acrylic/LDPE joint.

ited contact between the adhesive and substrate in this area. The scan of this region yielded no evidence of adhesive, lending further weight to the previously mentioned results.

Further evidence of interpenetration was gathered using optical microscopy. While not all of the samples that were tested had sufficient natural contrast between the two materials to give clear images, some samples yielded clear evidence of the interpenetration. Figure 8 is one such case. This was the cross-section of a fractured, primed 10 min bond magnified 500 times. In this micrograph the gold color of the adhesive can be observed beyond what appeared to be the interface between the adhesive and LDPE. This color gradually faded with distance from the interface indicating the presence of a concentration gradient. From the scale it was determined that this interphase was visibly 25–30 μm in thickness. It should be noted that the gold color of the adhesive was the result of the DMPT in the formulation. Thus, the smaller interpenetrated region detected optically represented the distance interdiffused by the DMPT molecule. The large size of this molecule, with reference to MMA, explains the smaller optically detectable region.

Further durability testing of stressed, but unfractured joints, was performed by soaking them in boiling water. These tests revealed that the high degree of interpenetration exhibited by this system led to extremely durable bonds that were not subject to moisture or high temperatures (100°C in this case). These joints re-

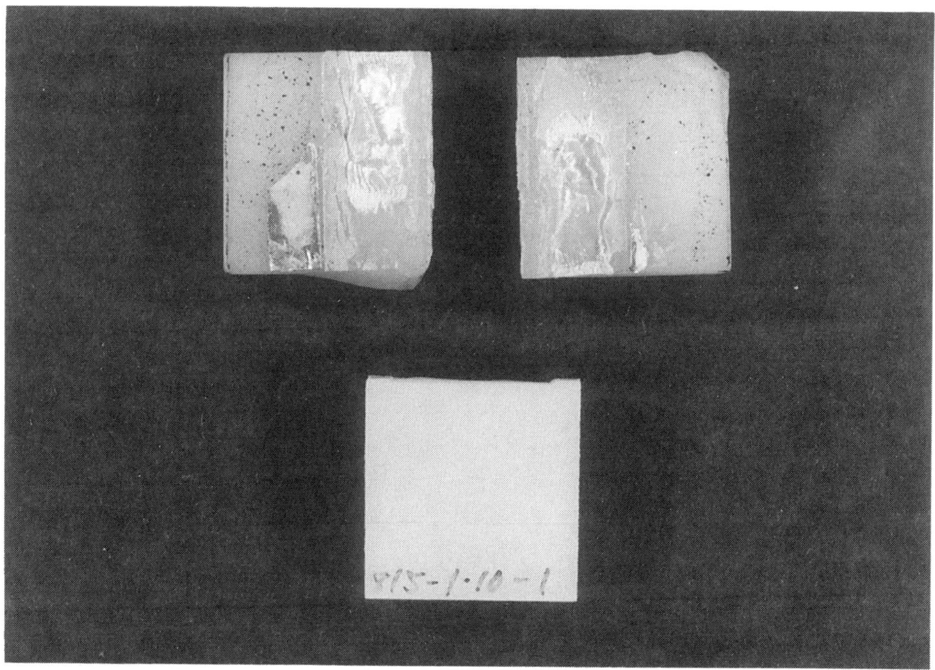

Figure 9. Representative failure surfaces from the durability studies in comparison to a virgin (not subjected to the durability test conditions) piece of LDPE labeled 815-1-10-1.

mained intact for over 3 months under these conditions. Subsequent retesting of the bonds led to considerable drops in the block shear strengths, although failure of the substrate was still evident. One example of the drop in block shear strengths (before and after comparisons could not be made in many cases because the labels on the bonds were erased due to the harsh conditions) was seen in a primed 10 min bond that had a block shear strength of 8.6 MPa prior to the soak, but only 3.0 MPa after 3 months in the boiling water. The failure surfaces from this joint are presented in Fig. 9 with a virgin piece of LDPE to illustrate the degradation of the substrate. While the loss in joint strength was significant, it was concluded that the adhesion between the two materials was not lost as significant substrate failure was observed. Rather, the lower bond strengths were the result of the severely degraded LDPE and adhesive. It is obvious in this figure that the sample is much darker than the virgin substrate and that the usual gold color of the adhesive has been replaced by a more opaque shade of brown. Further evidence of the harsh conditions experienced by this bond can also be seen in a piece of spacer remaining on the surface. This aluminum foil is coated with Teflon FEP (m.p. $\sim$260°C) which is much more thermally stable. The degradation of this material can be seen in its loss of clarity.

There appeared to be a clear trend of increasing bond strength with increased interpenetration. This is shown in Fig. 10 as a plot of the joint strengths versus the uptake concentrations found in Table 6.

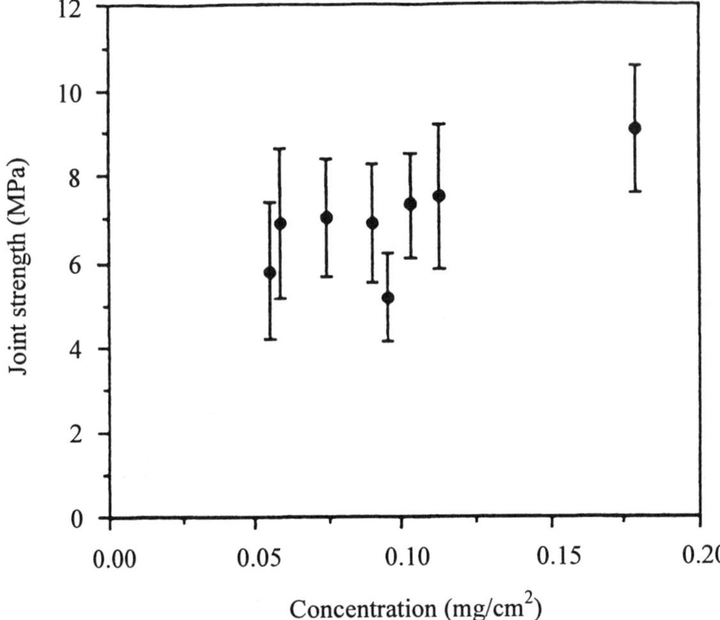

Figure 10. Plot of the average block shear strengths (LDPE joints) versus the average uptake concentrations of the adhesive for the various reaction times and treatments (error bars are one standard deviation).

4. DISCUSSION

The data clearly show that this adhesive system is capable of forming strong, durable bonds with polyolefin substrates, and that this adhesion is the result of the interpenetration (as defined earlier in this paper) of the adhesive and the substrate. However, there is definitely a difference in the relative bondability of the three polyolefins that were tested here (LDPE > HDPE > PP). This trend could not be explained by invoking the relative solubilities (Table 1) [19] or permeabilities. Literature values for ethyl acetate and oxygen are presented to show the trend for a similar molecule and that the trend is the same for a distinctly different molecule (Table 7) [23] of the adhesive in the substrates. According to the solubility data the PP should be most miscible with the MMA, followed by the LDPE and HDPE, while in the permeability case the PP was still more permeable than the HDPE. Since the permeability is the product of the solubility and the diffusivity, the relative diffusivities of these materials were investigated. Previous studies have displayed an inverse relationship between the log of the diffusivity and the loss modulus [24] or the chain mobility at the temperature of interest. Hence, loss modulus (G'') data for these polyolefins were extracted from plots of storage modulus (G') and logarithmic decrement versus temperature [25]. The G'' values of LDPE, HDPE and PP at 30°C and 70°C are shown in Table 8. These temperatures were chosen because they approximate the range of temperatures that are experienced in the bulk adhesive. The loss moduli shown in Table 8 do show the same trend that was exhibited in the joint strengths. They also give some indication as to why the HDPE and PP had significant bond strengths at 3 min reaction time.

At 70 °C (approximately the temperature of the adhesive at 3 min reaction time) the G″ values for the HDPE and PP approach that of the LDPE at 30 °C. The conclusion was then drawn that the interpenetration mechanism is inversely dependent upon the chain mobility of these substrates at the application temperature, as measured by G″, and therefore directly related to the diffusivity of the adhesive in the substrate.

It was further shown that there was a direct dependence of the joint strengths upon the extent of interpenetration (as measured in the uptake experiments). However, it should be noted that the increase in joint strength with uptake concentration of the adhesive was minimal. In fact, statistical T-tests of the different sets of data yielded no significant differences between the neighboring sets of data, i.e. the joint strengths of unprimed bonds formed at 20 min were statistically similar to the unprimed bonds formed at 10 min and 30 min. Thus, since sufficient interpenetration was achieved in the worst case (40 min reaction time), additional interpenetration only slightly increased the joint strength. This can be explained further by looking at the relative sizes of the interphases of the best (10 min reaction time) and worst cases. These were shown to differ by at least and order of magnitude (~100 μm to ~1700 μm), but the bond strengths associated with these interphases were only a few MPa apart (9.10 MPa to 5.18 MPa). It would seem logical that the joint strengths would be most dependent upon the number of adhesive polymer chains actually crossing the interface and the immediate entanglements of those chains with other adhesive or substrate molecules. Consequently the adhesive that has diffused very far into the substrate, as in the 10 min reaction time case, will have little to no effect upon the ultimate joint strength. In fact, too much interpenetration may actually have an adverse effect upon the final properties by lowering the strength of the interphase. This may explain the fact that the LDPE/LDPE bond strengths at 3 min were slightly lower

Table 7.
Permeabilities of oxygen and ethyl acetate in LDPE, HDPE, and PP [23]

Polymer	O_2 permeability (cm^2/s-Pa)	Temperature (°C)	Ethyl acetate permeability (g-mm/m^2-d)	Temperature (°C)
LDPE	2.2×10^{-13}	25	6.5	21.1
HDPE	0.30×10^{-13}	25	1.7	22.8
PP	1.7×10^{-13}	30	5.8	22.8

Table 8.
Loss moduli of LDPE, HDPE, and PP at 30 °C and 70 °C. Calculated from [25]

Polymer	G″@30 °C (MPa)	G″@70 °C (MPa)
LDPE	45	8.0
HDPE	130	89
PP	210	150

than at 10 min, although considerably higher uptake concentrations were measured (Table 6).

The T-tests also revealed that there were no significant differences between the sets of data obtained with primed substrates at one reaction time and unprimed substrates at the same reaction time, with the exceptions of 3 and 10 min reaction times. This was expected because considerably more CuEH was added in part II of the adhesive than was applied as the primer. We feel that the higher temperatures of the adhesive at 3 and 10 min may have enhanced the effectiveness of the primer at these times. This is being investigated further. In any case, it is apparent that the use of the primer is not necessary to form strong durable bonds.

Durability tests also provided some very significant and somewhat surprising results. While not conclusive evidence on its own, the resistance of these bonds to very harsh conditions lends considerable support to the mechanism of interpenetration. It is unlikely that other more traditional mechanisms of adhesion would have been to withstand the attack by the moisture or the higher temperatures, with the possible exception of mechanical interlocking which is really a similar mechanism to interpenetration, but on a macroscopic scale.

DSC data indicated that additional reactions occurred upon heating. It follows that the heat which should act to accelerate the disentanglement of the adhesive regions and the substrate, thereby decreasing the adhesion, could also be contributing to the further reaction of the adhesive, thereby increasing the adhesion. Ultimately, any loss in joint strength appeared to be the result of the degradation of both the adhesive and substrate.

5. CONCLUSIONS

LDPE/LDPE block shear strengths as high as 10 MPa, with failure occurring in the substrate, were achieved using this adhesive system. The adhesion in these bonds was not susceptible to moisture or high temperatures (100°C), and resulted from the interpenetration of the adhesive into the substrate. The extent of interpenetration had a direct relationship with the final joint strengths, and was dependent upon the percent conversion and the temperature of the adhesive at the time of application. Strong HDPE/HDPE and PP/PP joints, with failure in the substrate, were also formed at 3 min reaction time, when the adhesive was at an elevated temperature. The trend of substrate bondability, LDPE > HDPE > PP was shown to be dependent upon the relative chain mobility of these substrates.

We gratefully acknowledge the financial support of the Connecticut Department of Higher Education and the Plastics Institute of America. We also thank Loctite Corporation, especially D. Billy Yang and David Wolf, for their assistance.

REFERENCES

1. H. E. Wechsburg and J. B. Webber, *Modern Plastics* **36**, 101 (1959).
2. R. C. Snodgren, *Handbook of Surface Preparation*. Palmerton, New York (1974).

3. R. H. Hansen and H. Schonhorn, *J. Polym. Sci.* **B4**, 203 (1966).
4. A. Garton and J. Yang, *Polym. Mater. Sci. Eng.* **62**, 916 (1990).
5. L. N. Lewis and D. Katsamberis, *J. Appl. Polym. Sci.* **42**, 1551–1556 (1991).
6. J. Yang and A. Garton, *Polym. Mater. Sci. Eng.* **65**, 255 (1991).
7. J. Yang and A. Garton, in: *Proc. 15th Ann. Meeting Adhesion Soc.*, p. 43. Hilton Head, South Carolina (Feb. 1992).
8. J. Yang and A. Garton, *J. Appl. Polym. Sci.* **48**, 359–370 (1993).
9. D. Jia, Y. Pang and X. Liang, *J. Polym. Sci.: Part B: Polym. Phys.* **32**, 817–823 (1994).
10. J. Yang and A. Garton, *J. Adhesion* **46**, 67 (1994).
11. E. P. Plueddemann, *Silane Coupling Agents*. Plenum Press, New York (1982).
12. M. K. Chaudhury, T. M. Gentle and E. P. Plueddemann, *J. Adhesion Sci. Technol.* **1**, 29–38 (1987).
13. A. J. Gellman, B. M. Naasz, R. G. Schmidt, M. K. Chaudhury and T. M. Gentle, *J. Adhesion Sci. Technol.* **4**, 597–601 (1990).
14. J. M. Rooney and B. M. Malofsky, in: *Handbook of Adhesives*, I. Skeist (Ed.), p. 451. Van Nostrand, New York (1990).
15. V. Krieble, US Patent (to American Sealants Company) 2,895,950 (1959).
16. P. Beaunez, G. Helary and G. Sauvet, *J. Polym. Sci.: Part A: Polym. Chem.* **32**, 1459–1480 (1994).
17. St. Wellmann and H. Brockmann, *Intl. J. Adhesion Adhesives* **14**, 47–55 (1994).
18. Y. Okamoto, *J. Adhesion* **32**, 227–235 (1990).
19. E. A. Grulke, in: *Polymer Handbook*, 3rd edn, J. Brandrup and E. H. Immergut (Eds), pp. VII 519–559. J. Wiley & Sons, New York (1989).
20. ASTM Standard Test D 4501-91 (1991).
21. H. W. Siesler, in: *Structure-Property Relations in Polymers*, M. W. Urban and C. D. Craver (Eds), pp. 41–87. American Chemical Society, Washington, DC (1993).
22. A. Garton, *Infrared Spectroscopy of Polymer Blends, Composites, and Surfaces*. Hanser, New York (1992).
23. S. Pauly, in: *Polymer Handbook*, 3rd edn, J. Brandrup and E. H. Immergut (Eds), pp. VI/435–VI/449. J. Wiley & Sons, New York (1989).
24. J. P. Bell, *J. Appl. Polym. Sci.* **12**, 627 (1968).
25. L. E. Nielsen, in: *Mechanical Properties of Polymers*, pp. 182–185. Rheinhold Publishing Corp., New York (1962).

Polymer Surface Modification: Relevance to Adhesion, pp. 525–535
K. L. Mittal (Ed.)

Laser annealing and surface modification of plasma polymer–metal composite films

A. HEILMANN,* J. WERNER, F. HOMILIUS and F. MÜLLER

Technical University Chemnitz-Zwickau, Institute of Physics, D-09107 Chemnitz, Germany

Revised version received 3 February 1995

Abstract—Plasma polymer films with encapsulated metal particles were prepared by simultaneous plasma polymerization and metal evaporation. Laser annealing (Nd–YAG, 1064 nm) causes dramatic changes in the particle size and shape without material ablation, but with changes in the surface topography. This results in changes in the optical plasma resonance absorption, as demonstrated by UV–visible–near-infrared (NIR) spectroscopy. The invariable transmission in the UV region of the plasma polymer matrix after laser annealing confirms that there is no material ablation. Transmission electron microscopy (TEM) and scanning electron microscopy (SEM) demonstrated particle size changes and modifications of the plasma polymer surface due to laser annealing.

Keywords: Laser annealing; plasma polymerization; composite thin films; optical properties; surface modification.

1. INTRODUCTION

Plasma polymerization is a very effective method for the deposition of thin polymer films [1, 2]. Applications of plasma polymers, e.g. as protective thin film coatings [3], as electron beam resists [4] or for UV photoablation [5], have been discussed. Plasma polymer films show very good adhesion to different substrates and form continuous layers at a very low thickness. Depending on the monomers and the deposition parameters, the plasma polymer films contain a certain amount of free radicals and various functional end-groups. The so-called cauliflower morphology at the plasma polymer surface allows its use as intermediate films for adhesion promotion of overcoats.

After deposition, the plasma polymer can be modified by thermal annealing or additional polymerization stimulated by UV irradiation. Surface modification is also possible by laser irradiation. For laser ablation or laser annealing, it is useful to encapsulate metal particles in the plasma polymer to obtain an absorption of the films near the laser wavelength [6].

*To whom correspondence should be addressed.

Metal particles can be encapsulated in a plasma polymer matrix in different ways. During the plasma polymerization, metal sputtering or metal evaporation was used to encapsulate different metals, e.g. gold [7, 8] or silver [9]. Such plasma polymer–metal composite films have the advantages of the plasma polymer films, e.g. good adhesion and long-term stability, and the special optical and electrical properties of small metal particles too. Depending on the plasma polymerization conditions, monomers, and encapsulated metal particles, films with very different structural, electrical, and optical properties can be obtained.

The aim of this work was to explore the possibilities of laser-stimulated structural changes in plasma polymer films with various encapsulated metal particles, with the objective of using these films as intermediate layers for polymer metallization.

2. FILM DEPOSITION AND LASER ANNEALING

2.1. Film deposition

The plasma polymer–metal composite films were prepared by simultaneous or alternating plasma polymerization and metal evaporation [9]. Hexamethyldisiloxane (HMDSO) and hexamethyldisilazane (HMDSN) were plasma-polymerized at a pressure of 1–50 Pa with a monomer flow rate between 0.1 and 0.01 Pa l/s and with power densities between 0.1 and 5 W/cm^2. The growth rate of the plasma polymer was approximately 2 nm/s. The films were deposited on quartz substrates (for the optical measurements), as well as on silicon wafers.

The metals (Ag, Au, Cu, Sn) were evaporated thermically during the plasma polymerization or between two plasma polymerization steps. By using transmission electron microscopy (TEM) investigations, it was shown that the metals were encapsulated in the polymer matrix as small particles with sizes starting at about 3 nm [10, 11].

The thickness of the plasma polymer–metal composite films depends on the monomer flow rate, the polymerization power, and the amount of evaporated metal. The thickness was measured by both optical interferometry and scan depth profiling, and it varied between 50 nm and 1 μm. Table 1 presents the preparation parameters and thicknesses of the films investigated in this study.

The filling factor represents the metal volume part in the total film. Plasma polymer–metal composite films can be prepared with a constant or with a continuously varying filling factor. The films with a continuously varying filling factor can be used for investigations of the influence of the metal amount on the optical properties and have the advantage that the surrounding plasma polymer matrix is invariable.

Depending on the deposition parameters, films with two- or three-dimensional particle distributions in the polymer matrix were observed [12]. In films with a three-dimensional particle distribution, the particles are encapsulated on top of each other and are distributed throughout the film volume. A two-dimensional particle distribution means that the particles are mostly encapsulated in a plane parallel to the substrate. All the plasma polymer–metal composite films in Table 1 show a two-dimensional particle distribution after deposition. It is possible to describe these films

Table 1.
Preparation parameters, thicknesses, and laser irradiation parameters of plasma polymer films and plasma polymer–metal composite films

Film	Monomer	Polymerization power density (W/cm^2)	Encapsulated metal	Thickness (nm)	Laser power (W)	Figure
1	HMDSN	0.4	Silver	400	8.6	1a
2	HMDSN	0.4	Copper	250	4.4	1b
3	HMDSN	0.4	—	400		2a
4	HMDSN	0.4	—	550		2a
5	HMDSN	0.1	—	750		2b
6	HMDSN	1.0	—	400		2b
7	HMDSN	0.4	Silver	400	8.6	3a
8	HMDSN	0.4	Gold	400	8.6	3b
9	HMDSO	0.4	Copper	250	1.1	3c
10	HMDSO	1.0	Tin	550	1.1	3d
11	HMDSN	0.4	Silver	120	8.6	4a,b
12	HMDSN	0.4	Silver	120	8.6	5a,b

as a multilayer structure consisting of a plasma polymer film, a plasma polymer–metal composite film, and a second plasma polymer film. The film thickness in Table 1 is the total film thickness; the composite film thickness is about 10–20% of the total thickness.

Film adhesion on quartz substrates was measured by a simple Scotch tape test. Most films showed best adhesion equivalent to the value 5 of the Scotch tape test. Film adhesion was seen to decrease with the increase of the filling factor. Especially in the case of films with a two-dimensional particle distribution, the films delaminated at the interface between the composite film and the plasma polymer film. For films with a three-dimensional particle distribution, the adhesion was good, even with high filling factors.

2.2. Laser annealing

For laser annealing, the films were deposited on glass or quartz substrates. The samples were irradiated using a Nd–YAG laser (1064 nm). The continuous wave mode (CW mode) was used for thermally induced effects. In order to avoid material ablation, the laser power was varied between 8.6 W (e.g. film 7) and 1.1 W (film 9) using a constant beam diameter of 100 μm. For the optical investigations, areas of at least 1 cm^2 were irradiated with equidistant laser lines.

In the case of films with a high metal filling factor, the glass substrates can shatter because of the high absorption of the films at the laser wavelength. Therefore, the scanning speed and scan distance have to be optimized for the laser annealing. After irradiation, two different effects were observed using the light microscope. Figure 1a is a micrograph of film 1 showing the laser irradiation starting from the right side to

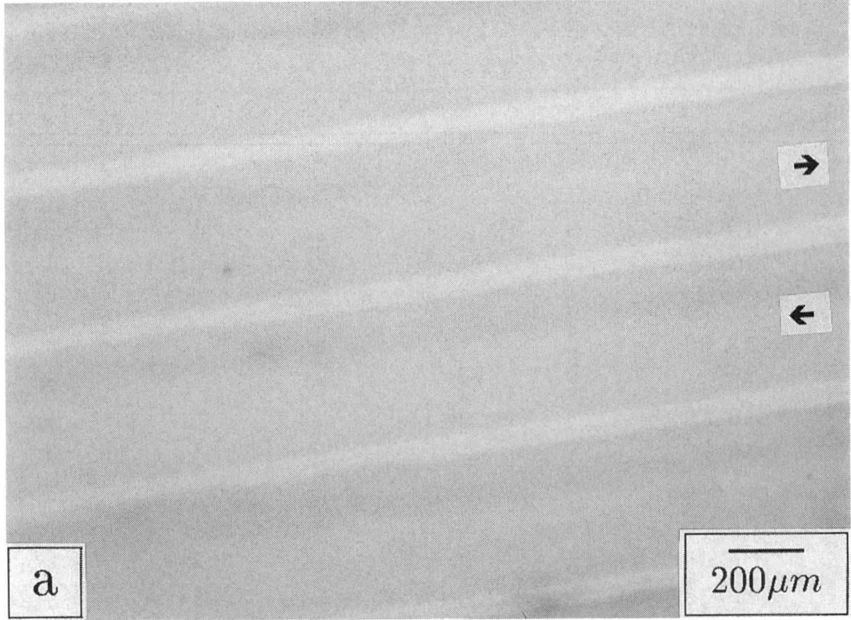

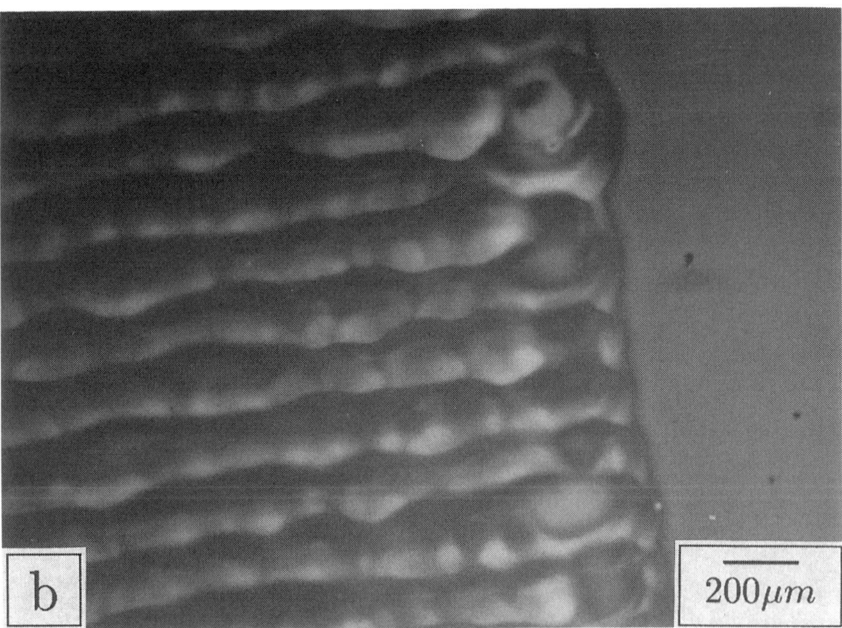

Figure 1. Optical microscope pictures of the laser-annealed plasma polymer–silver composite film 1 (a) and plasma polymer–copper composite film 2 (b).

the left side (arrow) with a scanning distance of 300 μm. The yellow regions represent the film as deposited, and the orange regions the laser-annealed film. To anneal the whole film, the distance between two lines of the laser beam was too long.

After laser irradiation with high power densities, the films have a very rough surface. The left part of Fig. 1b shows the surface modification of a plasma polymer–copper composite film (film 2) by laser annealing; the right part shows non-irradiated areas. For the optical measurements, the scanning distance between two laser lines was decreased to 100 μm to anneal the whole films. Furthermore, optical measurements were carried out only on films without the surface modification effect shown in Fig. 1b.

3. RESULTS

3.1. Optical properties

The optical properties of pure plasma polymer films were characterized by a steadily increasing optical transmission from the UV region to the NIR region [13, 14]. Especially in the UV region between 50 000 and 35 000 cm^{-1}, the transmission depends on the film thickness, the monomer used, and the preparation parameters.

The UV–visible–NIR spectra were recorded between 50 000 and 40 000 cm^{-1} with a conventional double-beam spectrometer and by using an uncoated quartz substrate in the reference beam. Figure 2 shows the optical transmission of four plasma polymer films made from hexamethyldisilazane (HMDSN). The increase of the transmission from the NIR to the UV spectral region depends on the chemical composition of the monomers, and different UV transmission behaviours for films deposited from the same monomer were observed depending on the power density and monomer flow rate [15].

The film structures were investigated by IR spectroscopy [15, 16]. Especially in the case of films deposited at high power densities, the molecule oscillations were very broad, had low intensities, and gave no information about differences in the chemical structure before and after laser annealing.

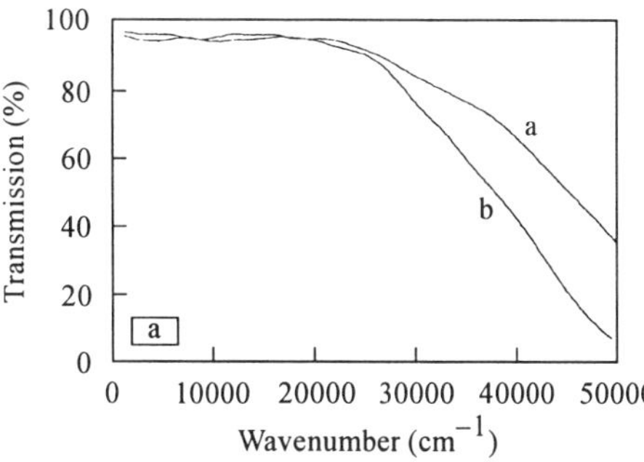

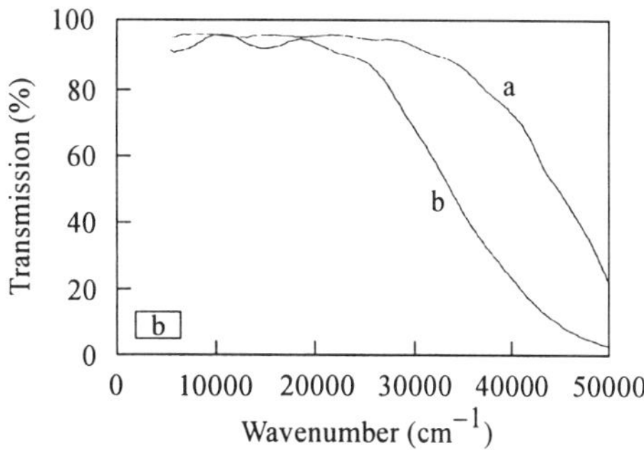

Figure 2. Optical transmission spectra of plasma polymer films from HMDSN. Figure 2a shows spectra from films with thicknesses of 400 nm [film 3, (a)] and 550 nm [film 4, (b)]. Figure 2b shows spectra from films deposited at polymerization power densities of 0.1 W/cm^2 [film 5, (a)] and 1.0 W/cm^2 [film 6, (b)].

For various plasma polymer thicknesses (Fig. 2a) as well as various plasma polymerization power densities (Fig. 2b), the optical transmission at 9400 cm^{-1} (near the wavelength of the Nd–YAG laser) was higher than 90%. Considering the interference conditions, this is close to the possible maximum value of 96%. At higher thickness, interferences were found in the transmission spectrum (Fig. 2b, curve b). These interferences could be used for calculations of the film thickness and refractive index. Both figures demonstrate that there is only a very small absorption in the plasma polymer films at 9400 cm^{-1} and that the laser irradiation does not have any effect.

Due to the plasma resonance absorption of small metal particles [17], the encapsulation of metal particles in the polymer matrix causes transmission minima in the visible and NIR spectral regions. The spectral range, the intensity, and the half-peak width of the transmission minima depend on the nature of the metal as well as on the size and shape of the encapsulated particles [18]. In the following, the optical transmission spectra of plasma polymer films with various encapsulated particles before and after laser annealing are described.

Figure 3 shows the transmission spectra of samples from four plasma polymer films with various encapsulated metal particles before (a) and after (b) laser irradiation. For the sample from a plasma polymer–silver composite film (film 7, Fig. 3a), the transmission minimum was found at 18 000 cm^{-1}. After laser annealing, it shifted

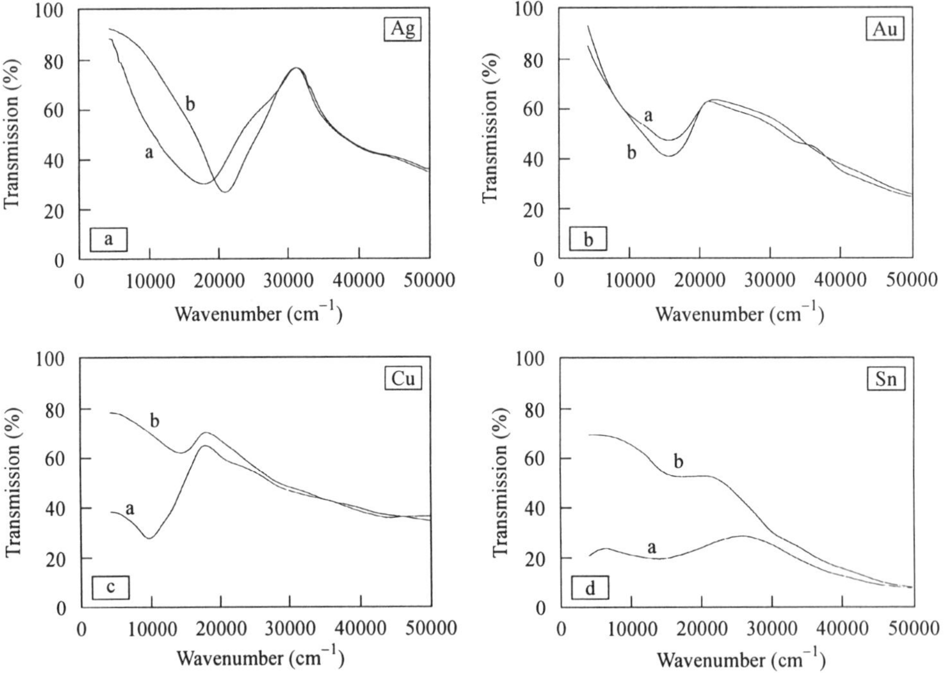

Figure 3. Optical transmission spectra of plasma polymer–metal composite films before (a) and after (b) laser annealing. (3a) Plasma polymer–silver composite film (film 7); (3b) plasma polymer–gold composite film (film 8); (3c) plasma polymer–copper composite film (film 9); (3d) plasma polymer–tin composite film (film 10).

to higher wavenumbers (20 500 cm^{-1}) and its half-width decreased from 15 000 to 11 000 cm^{-1}, and the relative intensity increased. The changes in the transmission spectra are the result of the silver particle size and shape changes discussed later based on the TEM micrographs in Section 3.2.

There are no changes in the transmission spectra at wavenumbers higher than 30 000 cm^{-1}. In this spectral region, the transmission was only determined by the optical behaviour of the plasma polymer film. As there are no changes in the spectra in this region, then changes in the polymer film and polymer film ablation can be excluded as there is a strong dependence of the optical transmission on the polymer structure and thickness as shown in Fig. 2. Figure 3a shows that changes in the silver particles' size and shape are the only reason for the changes in the transmission spectra due to laser annealing.

The laser annealing of a plasma polymer–gold composite film (film 8, Fig. 3b) shows results similar to those of the silver composite film. After deposition, the plasma resonance absorption was found at 16 000 cm^{-1}. The laser annealing causes an increase in the intensity of the transmission minimum but no spectral shift is observed. There are no changes in the plasma polymer transmission before or after laser annealing at wavenumbers higher than 22 000 cm^{-1}. This indicates that only changes in the gold particles' size and shape have occurred.

For the plasma polymer–copper composite film (film 9, Fig. 3c) the situation is different because of the encapsulation of copper particles and copper particles with copper oxide shells in the polymer matrix due to the presence of oxygen during the deposition of plasma polymer films from HMDSO. In the spectrum after deposition, the transmission minimum was found at 9500 cm^{-1}. After laser annealing, the transmission minimum shifted to higher wavenumbers (14 000 cm^{-1}) and its intensity decreased substantially. The laser irradiation of the plasma polymer–copper composite film resulted in a dramatic change in the visible part of the spectrum, and in addition to the changes in the particle size, changes in the copper/copper oxide ratio, because of diffusion of oxygen from the film surface, also have to be considered. The spectrum in the UV region demonstrates again that there are no changes in the plasma polymer matrix.

Finally, for plasma polymer films with encapsulated tin (film 10, Fig. 3d), after deposition only a weak transmission minimum between 10 000 and 20 000 cm^{-1} was observed. The plasma resonance absorption of the tin particles is rather small. After laser irradiation, the transmission increased, especially in the visible region, e.g. at 9400 cm^{-1} from 21% up to 70%. These are the largest changes, but for this sample, material ablation cannot be excluded. The spectrum in the UV region shows a partial ablation of the plasma polymer or changes in the plasma polymer film structure.

The transmission spectra in Fig. 3 show that there is a great variety of optical behaviours of the films, depending on the nature of the encapsulated particles before and after laser annealing. For plasma polymer composite films containing silver, gold, or copper, it is possible to avoid material ablation and only changes in the particle size and shape occur.

3.2. Structural changes

After the optical measurements, changes in the size and shape of the encapsulated metal particles were investigated. Transmission electron microscopy (TEM) allows investigation of the particles' size and shape changes. With scanning electron microscopy (SEM), one should be able to investigate the sample surface before and after laser irradiation.

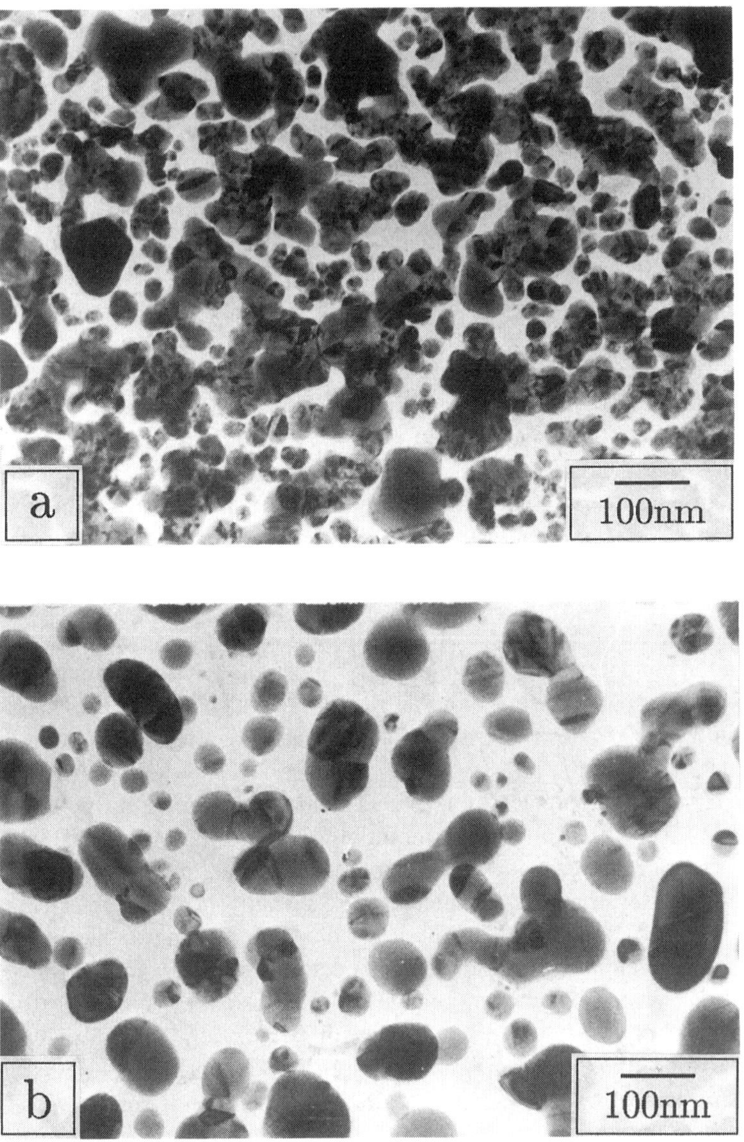

Figure 4. TEM pictures of a plasma polymer–silver composite film (film 11). The left-hand picture (a) shows a region of the sample after deposition and the right-hand picture (b) a region after laser annealing.

For the TEM investigations, the plasma polymer–metal composite thin films were deposited on silicon wafers with a predeposited thin film of potassium bromide. Parts of the substrates were laser-irradiated. After that, the KBr support was dissolved in distilled water and the films were picked up with the usual copper grids. So, both an irradiated and a non-irradiated area could be found on a single sample. The TEM investigations were carried out using a JEM 100 CX electron microscope at 100 kV.

Figure 4 shows two TEM micrographs of the plasma polymer–silver composite film (film 11). The picture shows non-irradiated (Fig. 4a) and irradiated (Fig. 4b) areas.

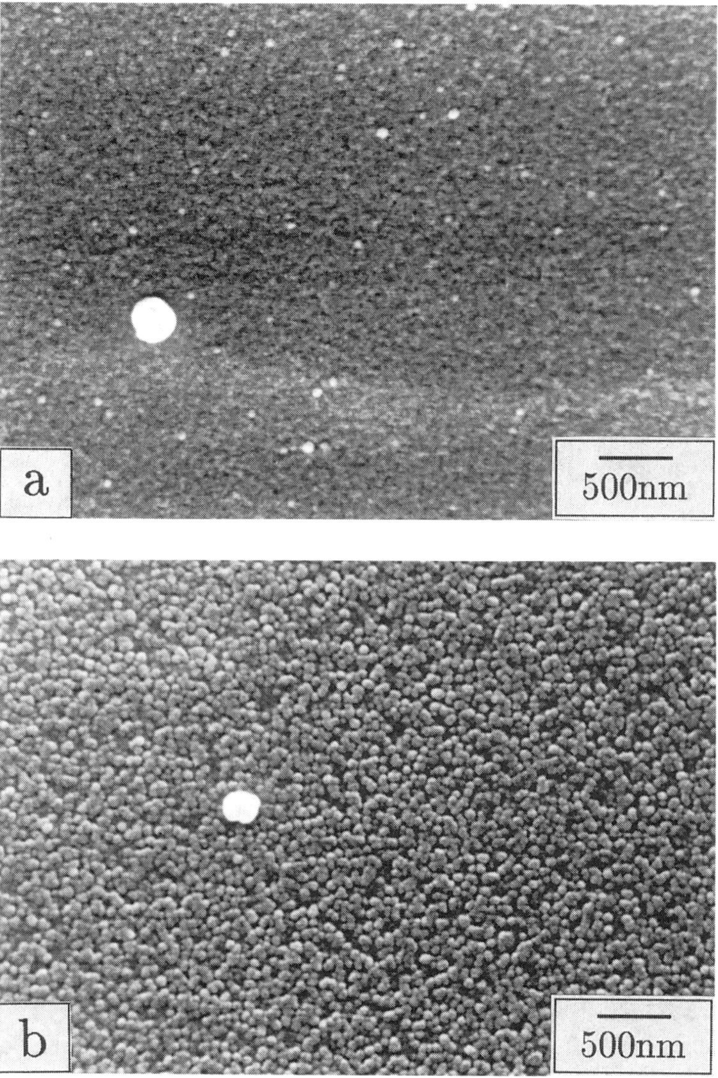

Figure 5. SEM pictures of a plasma polymer–silver composite film (film 12). Figure 5a shows the surface of a non-irradiated area and Fig. 5b that of a laser-irradiated area.

Substantial changes in the size and shape of the silver particles are visible. After deposition, most of the particles are irregular in shape. After laser annealing, the particles are more spherical and more separated from each other. Furthermore, their number has decreased. These changes were designated as coalescence and were also observed by thermal annealing to 500 K [19]. The similarity of the laser annealing to thermal annealing makes it obvious that laser annealing stimulates the same kind of coalescence process as thermal annealing. The coalescence due to thermal annealing is influenced by the surface diffusion and recrystallization of the metal particles and is the subject of additional work.

With the presented TEM micrographs, the changes in the optical spectra of the plasma polymer–silver composite film in Fig. 3a can be explained. The filling factor of the sample used for electron microscopy is comparable to the sample used for the optical measurements in Fig. 3a. The shift of the transmission minimum to higher wavenumbers results from the decrease in particle size. The more uniform and spherical shape of the silver particles leads to a decrease in the half-width of the transmission minimum.

Furthermore, SEM was used to observe the changes in the film surface due to laser annealing. The micrographs were recorded at 30 kV with a Philips SEM 515 scanning electron microscope. Before investigation, the films were coated with a platinum layer to prevent sample charging. Figure 5a shows a non-irradiated area and Fig. 5b an irradiated region. In Fig. 5b a rough surface is observed. On laser annealing, the silver particles change their shape from flat objects parallel to the substrate to larger encapsulated spherical silver particles with diameters of about 100 nm. The result is that the vertical diameter of the silver spheres reaches a value close to the film thickness. Because of this, the laser-irradiated surface is much rougher than the non-irradiated surface.

4. CONCLUSIONS

The results presented on laser irradiation with a Nd–YAG laser at 1064 nm show that it is possible to realize changes in the optical properties and microstructure of plasma polymer composite films with different encapsulated metal particles. Material ablation was excluded because the same optical properties were observed for the plasma polymer matrix before and after laser irradiation. Similar to thermal annealing, laser annealing causes coalescence of the metal particles within the films.

Laser irradiation of plasma polymer–metal composite films offers the possibility of manipulating the microstructure and surface topography of already deposited films. Laser irradiation of electrical conducting films above the percolation threshold could induce a substantial decrease in the electrical DC conductivity, as is known for thermal annealing [7], and could allow direct writing of a non-conducting structure in a conducting region. The laser-induced changes in the particle size could provide a tool for manipulating films in order to modify the metal particle–insulator interface for better metal–polymer adhesion.

REFERENCES

1. H. Yasuda, *J. Polym. Sci., Macromol. Rev.* **16**, 199 (1981).
2. R. d'Agostino (Ed.), *Plasma Deposition, Treatment, and Etching of Polymers*. Academic Press, San Diego, CA (1990).
3. T. J. Lin, B. H. Chun, H. K. Yasuda, D. J. Yang and J. A. Antonelli, *J. Adhesion Sci. Technol.* **5**, 893 (1991).
4. S. A. Gangal, M. Hori, S. Morita and S. Hattori, *Thin Solid Films* **149**, 341 (1987).
5. J. H. Brannon, D. Scholl and E. Kay, *Appl. Phys. A* **52**, 160 (1991).
6. W. R. Creasy, J. A. Zimmerman, W. Jacob and E. Kay, *J. Appl. Phys.* **72**, 2462 (1992).
7. E. Kay, *Z. Phys. D* **3**, 251 (1986).
8. L. Martinu, H. Biederman and J. Zemek, *Vacuum* **35**, 171 (1985).
9. A. Heilmann and C. Hamann, *Prog. Colloid Polym. Sci.* **85**, 102 (1991).
10. A. Heilmann, C. Hamann, G. Kampfrath and Do Ngoc Uan, *Phys. status solidi A* **114**, 551 (1989).
11. C. Reinhardt, A. Heilmann, W. Grünewald and C. Hamann, *Thin Solid Films* **235**, 57 (1993).
12. W. Grünewald, A. Heilmann and J. Werner, *Fresenius J. Anal. Chem.* **349**, 238 (1994).
13. A. Heilmann, G. Kampfrath and V. Hopfe, *J. Phys. D, Appl. Phys.* **21**, 986 (1988).
14. T. Nakano, S. Koike and Y. Ohki, *J. Phys. D, Appl. Phys.* **23**, 711 (1990).
15. F. Homilius, A. Heilmann, J. Werner and W. Grünewald, *Phys. status solidi A* **137**, 145 (1993).
16. S. Y. Park and N. Kim, *J. Appl. Polym. Sci., Appl. Polym. Symp.* **46**, 91 (1990).
17. U. Kreibig and L. Genzel, *Surface Sci.* **156**, 678 (1985).
18. J. Perrin, B. Despax and E. Kay, *Phys. Rev. B* **32**, 719 (1985).
19. A. Heilmann, J. Werner, O. Stenzel and F. Homilius, *Thin Solid Films* **246**, 77 (1994).

Polymer Surface Modification: Relevance to Adhesion, pp. 537–547
K. L. Mittal (Ed.)

Surface and interface charge effects in metal–polymer adhesion

A. F. ADADUROV

Radiation Dynamic Group, Physical–Technical Department, Kharkov State University, 310052, PO Box 60, Kharkov, Ukraine

Revised version received 8 April 1995

Abstract—The objective of the present work was to study the electrostatic component of adhesion in metal film–polymer systems under low energy (some keV) photon irradiation. In this energy region electrostatic forces can be influenced directly by fast secondary electrons. Quantum yields of such electrons for Al, Ag, Au, polyethylene, Mylar, PVC, and Teflon were calculated with the special Monte Carlo code and used as input for an analytical model of the charge profile formation near the metal film–polymer interface. The dependence of adhesion characteristics on photon energy and atomic number of materials (metal films, polymer substrates) is discussed.

Keywords: Kilovoltage photons; secondary electrons; interface charge; Monte Carlo method; analitical model.

1. INTRODUCTION

It is well known that adhesion of metal films to polymers is determined by chemical and/or van der Waals forces. Their relative contribution markedly depends on the characteristics of the metal films and polymer surfaces involved. Chemical bonding is usually accompanied by a significant charge displacement (electrical double layer formation). Therefore, electrostatic forces are of great importance in adhesion. 'Even if the electrostatic component is small compared with the strength of van der Waals or homopolar bonds, its contribution in adhesion can be significant' [1]. It is obvious that any way to influence the interface charge state is important in adhesion and must be thoroughly investigated.

Previous experiments indicated that the electrostatic component of adhesion in metal film–dielectric (polymer) systems could be increased during and after X-ray irradiation. This effect was accounted for by the bulk ionization of materials involved. As a result, the charge carrier concentration was increased affecting the double electric layer formation [1]. X-ray (or γ-ray) irradiation is not a unique way to increase the charge carrier concentration in the material considered. The same result could be obtained with electrons as primary particles. However, in that case a decrease in adhesion force was observed [2]. A simple bulk ionization is unlikely to account for such different results (adhesion improvement in one case and a decrease in another). More precise models are required providing more detailed accounting of radiation transport and charge buildup. Unfortunately, it is difficult to study experimentally the influence of ionizing radiation on the electrostatic component of adhesion. Radiation interaction with matter is usually accompanied by various additional factors (for example, material destruction) which can also change the adhesion force.

In the present work we will consider the electrostatic component of adhesion due to the charge buildup at or near the metal film–polymer interfaces.[1] This charge results from direct displacement of fast secondary electrons caused by X-ray irradiation. Such a process, in addition to simple bulk ionization, could be the cause of various radiation effects [3, 4]. We shall restrict ourselves to the case of low energy (less than 10 keV) photons and secondary electrons. The reason for investigating this energy region is that the range of low energy electron is of the order of several micrometers, i.e. approximately, the same as electrical double layer width.

We wish to state here that one important purpose of the present work is to bring electron transport calculations to the attention of the specialists in adhesion science and technology. For this purpose we will first briefly review the physical basis of irradiation-induced charge carrier generation in solids and the mathematical model used to describe the electron flux near the interface. Then we will discuss the Monte Carlo code used in our investigation and numerical results concerning some characteristics of secondary electrons in commonly used polymers. Finally, we will describe some cases where low energy photon irradiation is shown to have a marked influence on metal film–polymer bond strength.

2. CHARGE CARRIER GENERATION AND TRANSPORT

The effects of γ-rays on irradiated media occur via fast secondary electrons. Within the photon energy range up to about 10 MeV, virtually all of the photon energy-loss interactions take place by means of three competing interactions:

1. The photoelectric effect, in which most of the incident photon's energy is transferred to a single electron.
2. The Compton effect, in which only part of the primary photon's energy is transferred to a single atomic electron and a 'Compton scattered' photon is emitted, whose direction and energy are determined from relativistic momentum-energy conservation.
3. Pair production, in which an electron–positron pair is produced.

For relatively light materials (such as polymers), the photoelectric effect is dominant for photon energy E_γ up to 40 keV and the Compton effect predominates from $E_\gamma = 40$ keV to 10 MeV. Most photon radiations of practical interest fall within this range and, therefore, pair production is not an important process in our approach.

[1]The analysis has been made without reference to radiation conductivity and potential change. One can take into account those effects using results presented below.

Also, coherent or Raleigh scattering which contributes up to about 10% to the total photon cross section (attenuation coefficient), but does not contribute to the energy and charge deposition, will not be considered here. Relative probabilities of the above interactions are determined by corresponding cross sections and are usually expressed in terms of linear (or mass) attenuation coefficient μ. Tables of photon mass-attenuation and mass-energy absorption coefficients for some polymers are given in [5]. For metals we have used tables [6].

The photoelectric effect is an interaction of the photon with the atom as a whole. The angular emission probability of the photoelectrons is fairly well described by Fisher's formula [7]:

$$\Phi(\theta) \approx (1 - B)^2 (1 - B\cos\theta)^{-4} \sin^3\theta,$$

where parameter B depends only on E_γ.

Photoelectron energy is equal to the initial photon energy less the energy with which the electron was bound to the atom. The binding energy is released subsequently by the emission of Auger electrons or fluorescent radiation. We have used the simple equation for probability of this process: $P(Z) = X^4/(Z^4 + X^4)$ where Z is the atomic number of the material and parameter $X = 33$, 89, and 113 for K, L, and M shells, respectively [11]. Angular distribution of both Auger electrons and fluorescent quanta was supposed to be isotropic.

The Compton effect, described by the well-known Klein–Nishina formula [7], gives rise to a broad distribution of electron energies, since only part of the photon energy is transmitted to the recoil electron. The energy of the Compton scattered photon depends on its emission angle measured with respect to the incident photon direction. The maximum energy of the recoil electron is $T_e = 2\alpha E_\gamma/(1 + 2\alpha)$, where $\alpha = E_\gamma/m_0c^2$.

Thus, in γ- or X-irradiated materials a flux of photons passing through a medium drives with it a flux of fast secondary electrons continually generated by the photo-effect or the Compton effect. These 'photo–Compton' currents are present in every material exposed to space or nuclear radiation. In the presence of external or internal electric and magnetic fields, motions of secondary electrons in irradiated material can be described by Newton's law with the Lorentz force in conventional form and with drag force F_D to account for energy loss in the medium:

$$\frac{d\vec{p}}{dt} = -e\left[\vec{E} + \frac{\vec{v}}{c} \times \vec{B}\right] - \vec{F}_D, \tag{1}$$

where

e — electron charge,
$\vec{p}$ — electron momentum,
$\vec{v}, c$ — the velocities of the electron and light, respectively,
$\vec{E}, \vec{B}$ — electric and magnetic fields, respectively.

Since the influence of magnetic field is not considered here, the corresponding term in brackets is neglected. As for the remaining term, one should note the following:

In the general case, it may be assumed that a quiescent space charge exists in an electrical double layer formed when the metal/polymer couple is built. The amount of charge and its spatial extent are unique to the particular metal and polymer. The internal electric field $\vec{E}$ of the double layer can be rather intense and can be expected to influence the distribution of irradiation-induced charge carriers within a few micrometers of the metal film–polymer interface, which is a region of concern in this paper. Inclusion of pre-existing space charge would require a model for the interaction of deeply trapped quiescent charge with the irradiation-induced secondary electrons. The latter will determine the manner in which the system approaches equilibrium. It is to be noted that the stopping power $\vec{F}_D$ of light materials within the energy region considered is of order of 10^8 V/cm [16, 17] corresponding to the electrical double layer field $\vec{E}$ for the case of a metal–metal interface. Corresponding electric fields in polymers are much less, i.e. of the order of 10^5–10^6 V/cm [1]. Hence in the following we expect quiescent and secondary electronic charges to be simply additive in the interface region and assume secondary electron transport not to be affected by the electrical double layer field $\vec{E}$. So the first term in (1) is also neglected.

For F_D we have used the Bethe–Bloch formula [20] with the modification proposed by Sugijama [19] with formal neglecting of the exchange effect. Such an approach provides a good agreement with the values of stopping power obtained more precisely in [16] and [17].

3. MODEL DESCRIPTION

It is desirable first to review the qualitative features of electron transport in the single plane parallel plate.

Assume that X-rays with flux density Φ_0 impinge normally on the back surface of the plane parallel plate. Let $\Phi(x)$, μ and R be the photon flux density at given point, X-ray absorption coefficient and electron range, respectively. In the bulk of the irradiated target (plate), i.e. at a distance more than R from the surface, $\phi_e(x) \approx \Phi(x)\mu R$. If X-ray flux is spatially uniform over the plate thickness, then $\Phi(x) = \Phi_0$ and therefore $\phi_e(x) =$ constant. Since the charge generation rate at a given point is proportional to the derivative of $\phi_e(x)$ with respect to x, we conclude that in the bulk region, the total charge deposited is nearly zero and the absorbed energy (dose) is a constant E_{eq} determined only by photon attenuation cross sections. (Absorbed energy could be estimated as the electron flux multiplied by the stopping power of the material.) This is well known in dosimetry, an electron equilibrium phenomenon, where the charge entering a small volume of the material is approximately equal to that leaving that volume. However, near the surfaces (at a distance less than R) the equilibrium is disturbed resulting in a local volume charge deposition [8].

The total electron flux near a surface can be split into three components:

Φ_1 — the flux of secondary electrons crossing the surface and therefore leaving the target,

Φ_2 — the flux of secondary electrons generated in the surface region and absorbed in the bulk,

Φ_3 — the flux of secondary electrons generated in the bulk and absorbed in the surface region.

When electron scattering is small, as in light materials, such as polymers, then $\Phi_1 \approx \Phi_3 \approx \eta^{\mathrm{F}}$ — the quantum yield of secondary electrons in the forward direction (i.e. the direction of primary X-rays). The flux Φ_2 is proportional to the backward yield η^{B}. Therefore in that case (i.e. in the absence of scattering) the charge generation rate Q near the forward surface is equal to the backward yield of secondary electrons. Analogously, the charge generation rate near the back surface is equal to the forward yield.

Turning to a planar interface of bilayer construction (a metal film and a polymer) let us note that an additional component Φ_4 of the total electron flux arises in the case. It is the flux of secondary electrons generated in one material (layer) and absorbed in another. The direction of this flux is opposite to that of Φ_1.

Let us assume that photons always impinge normally on the surface of the first layer (it may be either metal or polymer). Then, taking into account the multiple reflection at the interface, the sign and magnitude of charge generation rate Q_i ($i = 1$ or 2 — the layer number) can be estimated as:

$$
\begin{aligned}
Q_1 &= \Phi_0 e\left[\left(\eta_1^{\mathrm{B}} - \eta_1^{\mathrm{F}}\right)\frac{1}{1+\beta_1} - \eta_2^{\mathrm{B}}\frac{1-\beta_1}{1-\beta_1\beta_2} + \eta_1^{\mathrm{F}}\frac{1-\beta_2}{1-\beta_1\beta_2}\right],\\
Q_2 &= \Phi_0 e\left[\left(\eta_2^{\mathrm{F}} - \eta_2^{\mathrm{B}}\right)\frac{1}{1+\beta_2} - \eta_1^{\mathrm{F}}\frac{1-\beta_2}{1-\beta_1\beta_2} + \eta_2^{\mathrm{B}}\frac{1-\beta_1}{1-\beta_1\beta_2}\right],
\end{aligned} \tag{2}
$$

where β is the diffusion reflection coefficient. (For $4 \leqslant Z \leqslant 92$, $\beta \approx 0.475Z^{0.177} - 0.40$ [15].)

The first term in (2) equals zero is an isotropic case, where $\eta^{\mathrm{F}} \approx \eta^{\mathrm{B}}$ (or $\Phi_2 = \Phi_3$). The second term is the charge caused by electrons generated in the other layer (Φ_4). The third term determines the electron leakage from a given layer through the interface.

An equation for the mean energy absorbed near the interface region can likewise be expressed in the form:

$$
\begin{aligned}
E_1 &= \Phi_0\left[\left(\overline{E}_1^{\mathrm{F}} - \overline{E}_1^{\mathrm{B}}\right)\frac{1}{1+\beta_2} - \overline{E}_1^{\mathrm{F}}\frac{1-\beta_2}{1-\beta_1\beta_2} + \overline{E}_2^{\mathrm{B}}\frac{1-\beta_1}{1-\beta_1\beta_2}\right],\\
E_2 &= \Phi_0\left[\left(\overline{E}_2^{\mathrm{B}} - \overline{E}_2^{\mathrm{F}}\right)\frac{1}{1+\beta_2} + \overline{E}_1^{\mathrm{F}}\frac{1-\beta_2}{1-\beta_1\beta_2} - \overline{E}_2^{\mathrm{B}}\frac{1-\beta_1}{1-\beta_1\beta_2}\right].
\end{aligned} \tag{3}
$$

Numerous experimental and theoretical data showed that $Q(x)$ and $E(x)$ approximately fitted curves of the exponential form:

$$
\begin{aligned}
Q_i(x) &= Q_i \exp\left(-x/\delta_{\mathrm{e}}\right)/\delta_{\mathrm{e}},\\
E_i(x) &= E_i \exp\left(-x/\delta_{\mathrm{e}}\right)/\delta_{\mathrm{e}} + E_{\mathrm{eq}},
\end{aligned} \tag{4}
$$

where

x is the distance from the interface,
Q_i and E_i are obtained from (2) and (3),
$\delta_e^{F,B} = \eta^{F,B}/\mu$.
The quantity $\delta_e^{F,B}$ could be defined as a mean projection of electron range in the primary photon direction. One can obtain it using results presented below for a simple estimation for the width of the interface region, where dose and charge perturbations occur.

Many equations formally similar to (2) and especially (3) are described in the literature [9, 11, 14, 15]. Ultimately the only significant difference between them is the η^F and η^B estimation. In the present paper we have used a somewhat unusual approach by replacing the analytical representation of these quantities by the previously calculated values. Such an approach allows us to combine the simplicity of analytical estimations with the accuracy of numerical ones. So far, detailed data on low energy secondary electron yields are not presented in literature,[2] so we have obtained them using the special code TOKLOW described below.

4. MONTE CARLO CODE

Because of the stochastic nature of most radiation processes, the Monte Carlo method is the powerful tool for various problems of radiation transport, in particular, the material charging problem. In general, this method makes it possible to solve the problems with any initial and/or boundary conditions. Unfortunately, a Monte Carlo simulation requires the generation of a large number of particles to obtain the uncertainty desired of the deposited charge value. Consequently, a long computer time is needed to follow all these particles.

As far as secondary electrons are concerned, the problem consists in the fact that part of them can be absorbed before leaving the target. As we are only interested in these leaving electrons, it is reasonable to trace only those particles that have a non-zero probability of crossing the boundary, i.e. only those originating at a distance less then δ_e from it. As the matter of fact only about one third of these particles leave the target. However, information on whether a particular particle should be traced or not can be obtained only at the end of the simulation of a given 'history' which requires considerable computing time.

The program TOKLOW is based on, as proposed by the author, the modification of the Monte Carlo method — 'Method of translations and rotations'. This method takes into account the material isotropy and uniformity which considerably increases the performance of a Monte Carlo code at a given level of statistical uncertainty. The only input needed is: initial energy E_0, atomic weight A and atomic number Z, as well as the photon's attenuation coefficients for the initial energy and for absorption edges. Here we will only briefly review the main characteristics of the Monte Carlo code TOKLOW. The physical basis and the formula used are described above. A more complete discussion can be found in [13].

[2]For the high energy case (0.02–10 MeV) see tables in [10].

In the TOKLOW the electron histories are simulated in a conventional manner using the logarithmic energy step [21]: $E_{i+1} = kE_i$. The trajectory obtained is in the form of a broken line with the nodal points x_i for which energies E_i and scattering angles θ_i are determined. These points in the program are interpreted as the points of intersection with the surface of the semi-infinite target. To put it another way, the method consists in 'translation' of the trajectory in semi-infinite media in such a manner that each node consequently fits the boundary. The main advantage of this approach is that it enables us to trace only 'valuable' trajectories. Moreover, each such trajectory, called 'total', while tracing from initial point with the E_0 to the rest point with the cutoff energy E_c gives N contributions to the value of electron flux at the surface:

$$N = 1 + \ln(E_c/E_0)/\ln k.$$

The electron cutoff energy in all the calculations was taken to be 0.1 keV, so at $E_0 = 10$ keV and $k = 2^{-1/m}$ $(m = 36)$ we obtain $N = 240$.

The results presented in this paper are based on 1000 'total' trajectories providing a statistical error of less than 5%. In the worst case (Ag, 10 keV), the computing time was less than 10 min on an IBM AT-286 personal computer.

5. RESULTS AND DISCUSSION

The values of secondary electron yields for often used metals and polymers are presented in Table 1. These results have been obtained by means of Monte Carlo program TOKLOW and are based on 1000 total trajectories providing statistical uncertainty not exceeding 10% at a confidence level 0.95.

We see that in the energy region considered the backward yield values are nearly the same as forward ones. It means that the secondary electron flux is nearly isotropic, so the initial photon direction is not important. For convenience let us assume that photons penetrate the metal first.

Substituting the numbers from Table 1 into (2) we can obtain the interface charge rate at various energies. This dependence for all Al and Ag next to Mylar and Teflon are shown in Figs 1–4. (For energies greater than 10 keV we have used yield values tabulated in [10].)

But before analyzing the dependencies it is useful to remember the rules of charge formation near the interface of materials with different Z, given earlier by Kooi and Kuznezov [11]. These are:

1. When photons penetrate a high Z material (metal) first, the charge in a low Z material (polymer) is negative for the photoeffect and positive for the Compton effect.
2. When photons penetrate a low Z material (polymer) first, the charge in a high Z material (metal) is negative for the photoeffect and negative for the Compton effect.
3. The signs of the charge on opposite sides of the interface are the same for the photoeffect and are different for the Compton effect.

Table 1.
Secondary electron yield, 10^{-3} electron/photon, from specific polymers and metals vs. photon energy[a]

Material	Energy (keV)					
	1	2	4	6	8	10
Polyethylene	1.82	0.75	0.313	0.17	0.12	0.009
	1.47	0.54	0.199	0.11	0.06	0.05
Mylar	3.67	1.33	0.61	0.36	0.26	0.19
	2.67	1.11	0.44	0.25	0.16	0.11
Teflon	5.73	2.52	1.08	0.71	0.52	0.39
	5.21	2.02	0.84	0.49	0.31	0.25
Polyvinyl chloride	3.11	1.25	3.57	2.12	1.64	1.45
	2.58	0.975	3.13	1.53	1.12	0.91
Hostaflon[b]	4.95	2.08	2.36	1.34	1.07	0.94
	4.27	1.78	2.16	1.16	0.81	0.66
Al	1.59	14.53	6.33	2.02	1.65	1.37
	1.45	13.62	5.69	1.65	1.21	0.95
Ag	14.7	6.38	16.4	7.97	6.05	4.65
	13.8	6.58	15.5	8.35	5.34	4.42
Au	12.2	7.20	19.3	13.0	9.74	7.54
	11.4	7.40	20.4	12.2	8.93	7.01

[a]Top value — forward yield η^F, bottom value — backward yield η^B.
[b]This represents $(C_2F_3Cl)_n$ polymer.

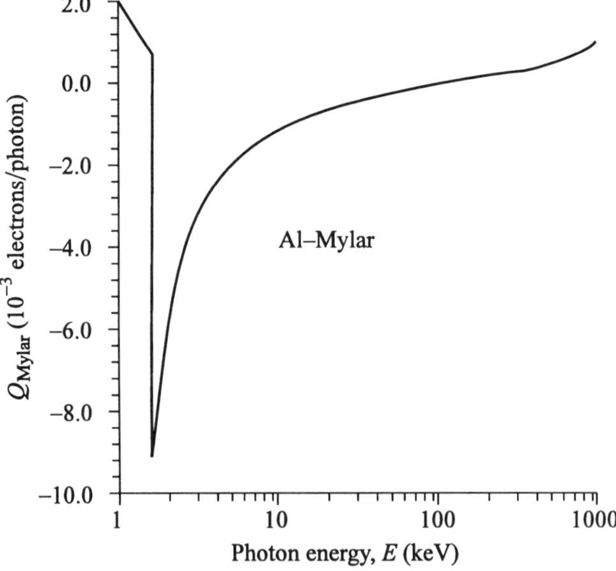

Figure 1. Monte Carlo calculations of the charge generation in Mylar vs. photon energy for Al–Mylar interface.

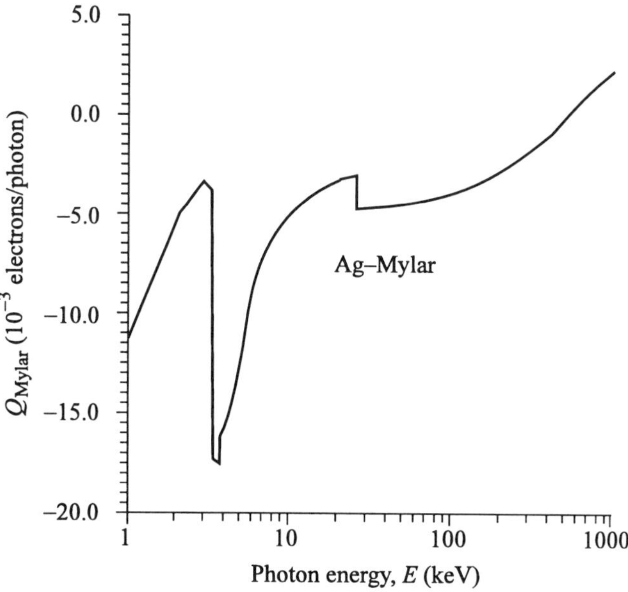

Figure 2. Monte Carlo calculations of the charge generation in Mylar vs. photon energy for Ag–Mylar interface.

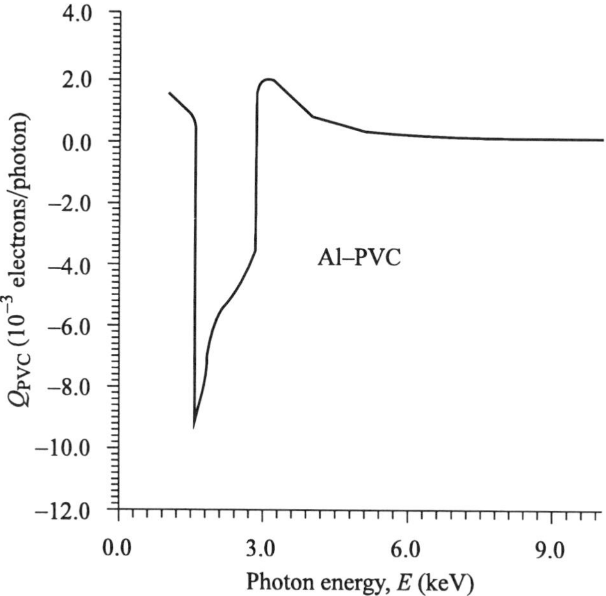

Figure 3. Monte Carlo calculations of the charge generation in PVC vs. photon energy for Al–PVC interface.

Now let us look at Fig. 1. First of all it must be noted that the above rules are held for Mylar near the interface with Al for all energies greater than approximately 2 keV. In the high energy region (approximately above 0.4 MeV) where the Compton effect

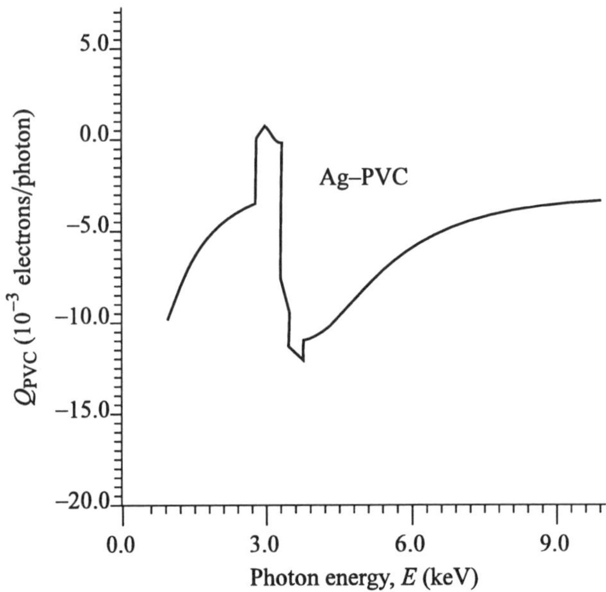

Figure 4. Monte Carlo calculations of the charge generation in PVC vs. photon energy for Ag–PVC interface.

dominates there is a positive charge generation in the polymer. When secondary electrons are generated by the photoeffect whose cross section varies as Z^4 (10 keV $\leqslant E_\gamma \leqslant$ 0.3 MeV) there is a negative charge generation in the polymer.

However, this trend may not hold at certain energies in the low-energy region (below 10 keV). Thus in Fig. 1 we see an abrupt drop in the interface charge value at 1.56 keV — K-absorption edge for Al. A more rapid charge change occurs when several absorption edges are in the energy region considered. In Fig. 2 one can see the step charge changing in Mylar near the interface with Ag, when we pass through K and L levels energies of silver. (To analyze the data in these figures we must bear in mind that energies of $L_1 - L_3$ edges for Ag are, respectively, 3.806, 3.524, and 3.351 keV, and K-absorption edge energies for Al, Cl, and Ag are 1.56, 2.824, and 25.54 keV.)

In general, as the photon energy increases and as we pass through an edge, the electron flux in the material increases. If that flux is contributing to the charge injected in the adjacent layer, the total charge in the interface region of the layer decreases. The same effect in the target material (in our case, polymer) leads to abrupt increase in interface charge. Both cases we can observe in Fig. 3 (Aluminum–PVC) where step decreasing of charge in the polymer at 1.56 keV is due to K-edge of Al and charge increasing at 2.824 keV is due to K-edge of Cl. According to Fig. 4 an analogous trend is held in the case of Ag next to PVC.

It is interesting to note that a drop in energy of slightly more than 1 keV alters the charge profiles considerably even for materials which are close in atomic numbers, e.g. Teflon (CF_2) and C_2F_3Cl.

6. SUMMARY

In general, we find that in contrast to the results reported earlier at high energies, the region below 10 keV is dominated by X-ray absorption edges. To summarize, the interface charge in the interface region of metal film–polymer structure strongly depends on photon energy, materials involved and the direction of irradiation. High atomic number elements do not always cause a negative charge in the adjacent materials of low atomic number; a positive charge is found at such interfaces at energies near X-ray absorption edges. Charge value can vary by more than ten times for a slight variation in photon energy.

Thus, the initial electrical double layer is considerably affected under low energy photon irradiation and the resulting induced charge can have marked influence on metal–polymer adhesion (or adhesion failure).

REFERENCES

1. B. V. Derjaguin, N. A. Krotova and V. P. Smilga, *Adhesion of Solids*. Consultans Bureau, New York (1978).
2. A. F. Adadurov, Ju. N. Borysenko, V. T. Grytsyna and V. T. Lazurik, *Electronnaja Obrabotka Metallov* **4**, 21–23 (1984).
3. C. L. Longmire, *IEEE Trans. Nucl. Sci.* **NS-22**, 2340–2344 (1975).
4. Y. S. Horowitz, M. Moscovitch, J. M. Mack, H. Hsu and E. Kearsley, *Nucl. Sci. Eng.* **94**, 233–240 (1986).
5. J. H. Hubbell, *Int. J. Appl. Radiat. Isot.* **33**, 1269–1290 (1982).
6. E. Storm and H. Israel, *Photon Cross Sections from 0.001 to 100 MeV for Elements 1 Through 100*, Los Alamos Scientific Laboratory, New Mexico (1967).
7. K. Siegbahn (Ed.), *Alpha-, Beta- and Gamma-ray Spectroscopy*, Vol. 1. North Holland, Amsterdam (1965).
8. A. F. Adadurov, V. T. Lazurik, B. A. Shilobreev and M. V. Jakovlev, *Voprosy Atomnoj Nauki i Techniki. Ser. Obschaja i Jadernaja Fizika* **3** (43), 3–7 (1988).
9. W. L. Chadsey, C. W. Wilson and V. W. Pine, *IEEE Trans. Nucl. Sci.* **NS-22**, 2345–2350 (1975).
10. A. F. Akkerman, M. Ya. Grudskii and V. V. Smirnov, Vtorichnoje electronnoje izluchenije iz tverdych tel pod dejstviem gamma kvantov (Secondary Electrons Radiation from Solids Under Gamma–Quanta Irradiation), Energoatomizdat, Moscow (1986).
11. C. F. Kooi and N. Kuznezov, *IEEE Trans. Nucl. Sci.* **NS-20**, 97–104 (1973).
12. I. A. Wall and E. A. Burke, *IEEE Trans. Nucl. Sci.* **NS-17**, 305–309 (1970).
13. M. Ya. Grudskii, N. M. Roldugin, V. V. Smirnov, A. F. Adadurow and V. T. Lazurik, *Nucl. Instrum. Meth. Phys. Res. B* **227**, 126–134 (1984).
14. J. C. Garth, *IEEE Trans. Nucl. Sci.* **NS-25**, 1598–1606 (1978).
15. E. A. Burke and J. C. Garth, *IEEE Trans. Nucl. Sci.* **NS-23**, 1838–1845 (1976).
16. A. F. Akkerman, V. A. Botvin, M. Ya. Grudskii and V. V. Smirnov, *Phys. Stat. Solidi* **110** (b), 285–297 (1982).
17. J. C. Ashley, C. J. Tung and R. H. Ritchie, *IEEE Trans. Nucl. Sci.* **NS-25**, 1598–1606 (1978).
18. A. F. Akkerman and G. Ya. Chernov, *Phys. Stat. Solidi* **89** (b), 329–341 (1978).
19. H. Sugijama, *Jpn J. Appl. Phys.* **15**, 1779–1790 (1976).
20. *ICRU Report 37* (International Commission on Radiation Units and Measurement), Bethesda, Maryland (1984).
21. M. J. Berger, in: *Methods in Computational Physics*, Vol. 1, B. Alder, S. Fernbach and M. Rotenberg (Eds), p. 135. Academic Press, New York (1963).